Cracking the AP* PHYSICS C Exam

Cracking the AP* PHYSICS C Exam

2008 Edition

Paul Waechtler and Steven A. Leduc

PrincetonReview.com

Random House, Inc. New York

The Princeton Review, Inc.
2315 Broadway
New York, NY 10024
E-mail: editorialsupport@review.com

ISBN: 978-0375-42854-8
ISSN: 1937-6391

Editor: Mariwyn Curtin
Production Editor: Heather Brady
Production Coordinator: Effie Hadjiioannou

Printed in the United States of America.

10 9 8 7 6 5 4 3 2 1

2008 Edition

ACKNOWLEDGEMENTS

I am grateful for the overall support of the physics education community. Specifically, Mary Beth Barrett and Alan Brix have motivated me with their creativity and enthusiasm. I want to thank my department chair, Gerry Munley, for having confidence in me throughout my teaching career and offering me new opportunities every year. I also want to thank the New Trier community for providing an outstanding teaching experience each and every year.

I especially want to thank my editors, Mariwyn Curtin and Briana Gordon, for answering my many questions patiently and thoroughly. I also want to thank Suzanne Barker, Heather Brady, Effie Hadjiioannou, Kim Howie, Mary Kinzel, and Stephen White for their outstanding production work.

I am grateful to my wife, Lara, for her patience, encouragement and love. I am also grateful for my two daughters, Emily and Abigail, as they motivate me with their boundless curiosity.

—Paul Waechtler

My thanks and appreciation to John Katzman, Steve Quattrociocchi, Paul Maniscalco, Kris Gamache, Tricia McCloskey, Andy Lutz, and Suellen Glasser for making me feel at home. Thanks also to Art Brown for his thoughtful and valuable input.

Special thanks to Paul Kanarek for his friendship, counsel, and encouragement.

—Steven A. Leduc

DEDICATION

To the memory of my mother,
Nancy Elizabeth Waechtler.
—Paul Waechtler

To the memory of my great aunt,
Norma Perron Lamb Piette.
—Steven A. Leduc

CONTENTS

INTRODUCTION

WHAT IS THE PRINCETON REVIEW?

The Princeton Review is an international test preparation company with branches in all major U.S. cities and several abroad. In 1981, John Katzman started teaching an SAT prep course in his parents' living room. Within five years, The Princeton Review had become the largest SAT prep program in the country.

Our phenomenal success in improving students' scores on standardized tests is due to a simple, innovative, and radically effective philosophy: Study the test, not just what the test claims to test. This approach has led to the development of techniques for taking standardized tests based on the principles the test writers themselves use to write the tests.

The Princeton Review has found that its methods work not just for cracking the SAT, but for any standardized test. We've already successfully applied our system to the GMAT, LSAT, MCAT, and GRE, to name just a few. Obviously, you need to be well versed in physics to do well on the AP Physics C Exam, but you should remember that any standardized test is partly a measure of your ability to think like the people who write standardized tests. This book will help you brush up on your AP Physics and prepare for the exam using our time-tested principle: Crack the system based on how the test is created.

We also offer books and online services that cover an enormous variety of education and career-related topics. If you're interested, check out our website at PrincetonReview.com.

The AP (Advanced Placement) Examinations are given in May of each year and offer students the opportunity to earn advanced placement or college credit for work they've done in high school or independently.

There are two versions of the AP Physics Exam; they're called Physics B and Physics C. In case you're curious, the Physics B Exam covers a wide range of topics, but the questions are not too mathematically sophisticated. The Physics C Exam covers a smaller range of topics, but the questions are more challenging and calculus is used widely. The Physics C Exam is actually composed of two separate exams: one in Mechanics and one in Electricity and Magnetism (E & M). You can take just the Mechanics, just the E & M, or both. Separate scores are reported for the Mechanics and E & M sections.

You have probably decided to take the Physics C exam, which contains two sections: a multiple-choice section and a free-response section. Questions in the multiple-choice section are each followed by five possible responses (only one of which is correct), and your job, of course, is to choose the right answer. Each right answer is worth one point, but one-quarter of a point is subtracted for each wrong answer. Therefore, there's little to be gained by randomly guessing on a question you know nothing about; but if you're able to eliminate one or more answer choices, then it's a good strategy to guess among the remaining choices. There are 70 total multiple-choice questions on the full Physics C Exam, and the time limit is 90 minutes. You may *not* use a calculator on the multiple-choice section.

The free-response section consists of six multi-part questions, which require you to actually write out your solutions, showing your work. The total amount of time for this section is 90 minutes, so you have an average of 15 minutes per question. Unlike the multiple-choice section, which is scored by computer, the free-response section is graded by high school and college teachers. They have guidelines for awarding partial credit, so you don't need to correctly answer every part to get points. You are allowed to use a calculator (programmable or graphing calculators are okay, but ones with a typewriter-style keyboard are not) on the free-response section, and a table of equations is provided for your use on this section. The two sections—multiple-choice and free-response—are weighed equally, so each is worth 50 percent of your grade. See page 423 for the format of the AP Physics C Exam.

Grades on the AP Physics Exam are reported as a number: either 1, 2, 3, 4, or 5. The descriptions for each of these five numerical scores are as follows:

AP Exam Grade	Description
5	Extremely well qualified
4	Well qualified
3	Qualified
2	Possibly qualified
1	No recommendation

Colleges are generally looking for a 4 or 5, but some may grant credit for a 3. How well do you have to do to earn such a grade? Each test is curved, and specific cut-offs for each grade vary a little from year to year, but here's a rough idea of how many points you must earn—as a percentage of the maximum possible raw score—to achieve each of the grades 2 through 5:

AP Exam Grade	Percentage Needed
5	≥ 75%
4	≥ 60%
3	≥ 45%
2	≥ 35%

The percentages needed are usually a little lower for the E&M section of Physics C.

So, what's on the exams and how do you prepare for them? Here's a listing of the major topics covered on the AP Physics B and C Exams, along with an approximate percentage of the questions in each topic (what the College Board calls the *percentage goal*):

Topic	Physics B	Physics C
Newtonian Mechanics	**35%**	**50%**
Kinematics	7%	9%
Newton's Laws	9%	10%
Work, Energy, and Power	5%	7%
Linear Momentum	4%	6%
Circular Motion and Rotation	4%	9%
Oscillations and Gravitation	6%	9%
Fluid Mechanics and Thermal Physics	**15%**	
Fluid Mechanics	6%	
Temperature and Heat	2%	
Kinetic Theory and Thermodynamics	7%	
Electricity and Magnetism	**25%**	**50%**
Electrostatics	5%	15%
Conductors and Capacitors	4%	7%
Electric Circuits	7%	10%
Magnetic Fields	4%	10%
Electromagnetism	5%	8%
Waves	**5%**	
Optics	**10%**	
Atomic and Nuclear Physics	**10%**	

Naturally, it's important to be familiar with the topics—to understand the basics of the theory, to know the definitions of the fundamental quantities, and to recognize and be able to use the equations. Then, you must acquire practice at applying what you've learned to answering questions like you'll see on the Exam. This book is designed to review all of the content areas covered on the Exam, illustrated by hundreds of examples. Also, each chapter (except the first) is followed by practice multiple-choice and free-response questions, and perhaps even more important, *answers and explanations are provided for every example and question in this book*. You'll learn as much—if not more—from actively reading the solutions as you will from reading the text and examples. Also, two full-length practice tests (with solutions) are provided at the end of this book. The difficulty level of the examples and questions in this book is at or slightly above AP level, so if you have the time and motivation to attack these questions and learn from the solutions, you should feel confident that you can do your very best on the real thing. Though Physics B has been mentioned in this chapter as a point of comparison, the rest of this book applies directly to the AP Physics C Exam.

SOME FRIENDLY ADVICE

Here are a few simple strategies: On the multiple-choice section, do not linger over any one question. Go through the exam and answer the questions on the topics you know well, leaving the tough ones for later when you make another pass through the section. All the questions are worth the same amount, so you don't want to run out of time and not get to questions you could have answered because you spent too much time agonizing over a few complex questions. No one is expected to answer all of them, so maximize the number you get right.

Make a copy of the table on page 424 and use it when you work on the problems in this book. A similar table will be provided when you take the exam, so you do not need to memorize everything on the page. However, it will help to be familiar with the table when you take the exam. Also, you should work the multiple-choice questions at the end of the chapters and the practice tests without a calculator. Students are not permitted calculators for the multiple-choice portion of the AP Physics test, so it is important that you practice mental arithmetic.

On the free-response section, be sure to show the graders what you're thinking. Write clearly—that is *so* important—and show your steps. If you make a mistake in one part and carry an incorrect result to a later part of the question, you can still earn valuable points if your method is correct. But the graders cannot give you credit for work they can't follow or can't read. And, where appropriate, be sure to include units on your final answers.

The most important advice we can give you for the free-response section of the AP Physics Exam is to read the questions carefully and answer according to exactly what the questions are asking you to do. Credit for the answers depends not only on the quality of the solutions but also on how they are explained. On the AP Physics Exam, the words "justify," "explain," "calculate," "what is," "determine," and "derive" have specific meanings, and the graders are looking for very precise approaches in your explanations in order to assign maximum credit.

Questions that ask you to "justify" are looking for you to both show an understanding in words of the principles underlying physical phenomena and to perform the mathematical operations needed to arrive at the correct answer. The word "justify" as well as the word "explain" requires that you support your answers with text, equations, calculations, diagrams, or graphs. In some cases, the text or equations must elucidate physics fundamentals or laws, while in other cases they will serve to analyze the behavior of different values or different types of variables in the equation.

The word "calculate" requires you to show numerical or algebraic work to arrive at the final answer. In contrast, "what is" and "determine"questions signify that full credit may be given without showing mathematical work. Just remember, showing work that leads to the correct answer is always a good idea when possible, especially since showing work may still earn you partial credit even if the answer is not correct.

"Derive" questions are looking for a more specific approach, which entails beginning the solution with one or more fundamental equations and then arriving at the final answer through the proper use of mathematics, usually involving some algebra.

Your answers should be concise and focused and should not contain irrelevant or off-the-point information. If you make a mistake, you may either cross it out or erase it. The graders will not score crossed-out work. And, as we mentioned before, partial solutions may receive partial credit, so you should definitely show all your work, especially since correct answers without supporting work may lose credit. This is particularly true when you are asked to "justify" your answer, as graders are looking for some evidence of how you arrived at your solution. Finally, make sure that all of your numerical answers are in the appropriate units.

Sample test questions are also available directly from the College Board. You can request information about ordering such materials by contacting them by mail, phone, or on the web:

Advanced Placement Program
PO Box 6671
Princeton, NJ 08541-6671
(609) 771-7300/(888) 225-5427
http://www.collegeboard.com

We wish you all the best as you study for the AP Physics C Exam. Good luck!

Vectors

INTRODUCTION

Vectors will show up all over the place in our study of physics. Some physical quantities that are represented as vectors are: displacement, velocity, acceleration, force, momentum, and electric and magnetic fields. Since vectors play such a recurring role, it's important to become comfortable working with them; the purpose of this chapter is to provide you with a mastery of the fundamental vector algebra we'll use in subsequent chapters. For now, we'll restrict our study to two-dimensional vectors (that is, ones that lie flat in a plane).

DEFINITION

A **vector** is a quantity that involves both magnitude and direction and obeys the **commutative law for addition,** which we'll explain in a moment. A quantity that does not involve direction is a **scalar**. For example, the quantity *55 miles per hour* is a scalar, while the quantity *55 miles per hour, to the north* is a vector. Other examples of scalars include: mass, work, energy, power, temperature, and electric charge.

Vectors can be denoted in several ways, including:

$$\mathbf{A}, A, \vec{A}, \vec{\mathrm{A}}$$

In textbooks, you'll usually see one of the first two, but when it's handwritten, you'll see one of the last two.

Displacement (which is net distance traveled plus direction) is the prototypical example of a vector:

$$\underbrace{\mathbf{A}}_{\text{displacement}} = \underbrace{4 \text{ miles}}_{\text{magnitude}} \underbrace{\text{to the north}}_{\text{direction}}$$

When we say that vectors obey the commutative law for addition, we mean that if we have two vectors of the same type, for example another displacement,

$$\mathbf{B} = \underbrace{3 \text{ miles}}_{\text{magnitude}} \underbrace{\text{to the east}}_{\text{direction}}$$

then **A** + **B** must equal **B** + **A**. The vector sum **A** + **B** means *the vector* **A** *followed by* **B**, while the vector sum **B** + **A** means *the vector* **B** *followed by* **A**. That these two sums are indeed identical is shown in the following figure:

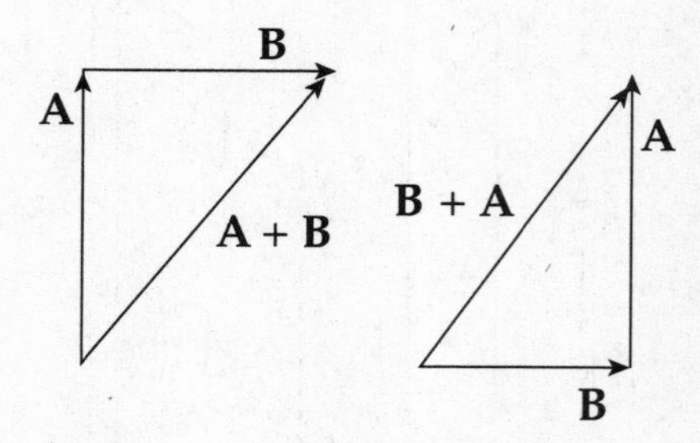

Two vectors are equal if they have the same magnitude and the same direction.

VECTOR ADDITION (GEOMETRIC)

The figure above illustrates how vectors are added to each other geometrically. Place the tail (the initial point) of one vector at the tip of the other vector, then connect the exposed tail to the exposed tip. The vector formed is the sum of the first two. This is called the "tip-to-tail" method of vector addition.

Example 1.1 Add the following two vectors:

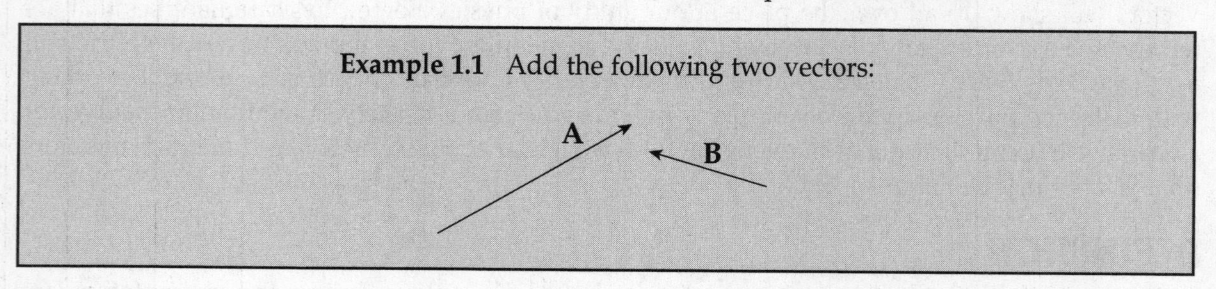

Solution. Place the tail of **B** at the tip of **A** and connect them:

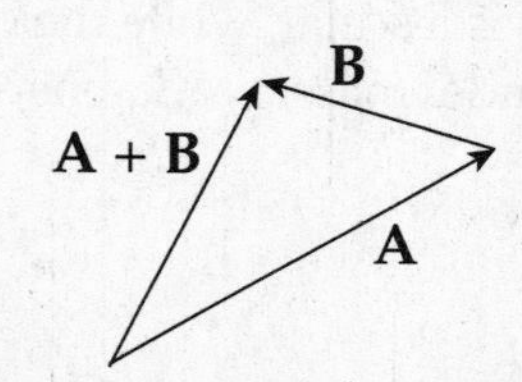

SCALAR MULTIPLICATION

A vector can be multiplied by a scalar (that is, by a number), and the result is a vector. If the original vector is **A** and the scalar is k, then the scalar multiple $k\mathbf{A}$ is as follows:

$$\text{magnitude of } k\mathbf{A} = |k| \times (\text{magnitude of } \mathbf{A})$$

$$\text{direction of } k\mathbf{A} = \begin{cases} \text{the same as } \mathbf{A} \text{ if } k \text{ is positive} \\ \text{the opposite of } \mathbf{A} \text{ if } k \text{ is negative} \end{cases}$$

Example 1.2 Sketch the scalar multiple 2**A**, $\frac{1}{2}$**A**, –**A**, and –3**A** of the vector **A**:

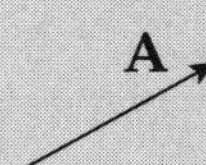

Solution.

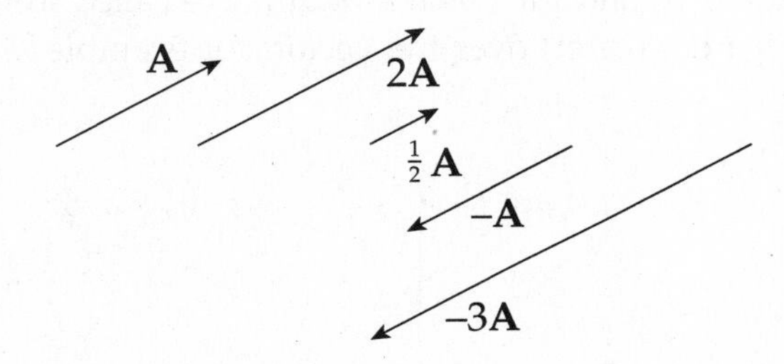

VECTOR SUBTRACTION (GEOMETRIC)

To subtract one vector from another, for example, to get **A** – **B**, simply form the vector –**B**, which is the scalar multiple (–1)**B**, and add it to **A**:

$$\mathbf{A} - \mathbf{B} = \mathbf{A} + (-\mathbf{B})$$

Example 1.3 For the two vectors **A** and **B**, find the vector **A** – **B**.

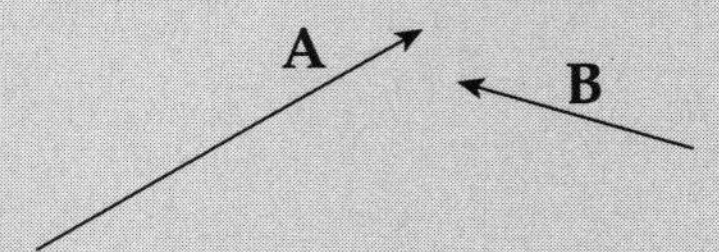

Solution. Flip **B** around—thereby forming –**B**—and add that vector to **A**:

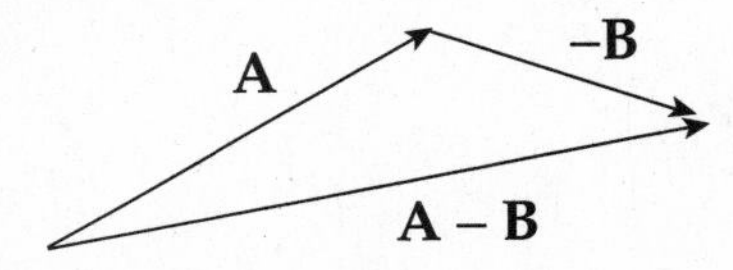

STANDARD BASIS VECTORS

Two-dimensional vectors, that is, vectors that lie flat in a plane, can be written as the sum of a horizontal vector and a vertical vector. For example, in the following diagram, the vector **A** is equal to the horizontal vector **B** plus the vertical vector **C**:

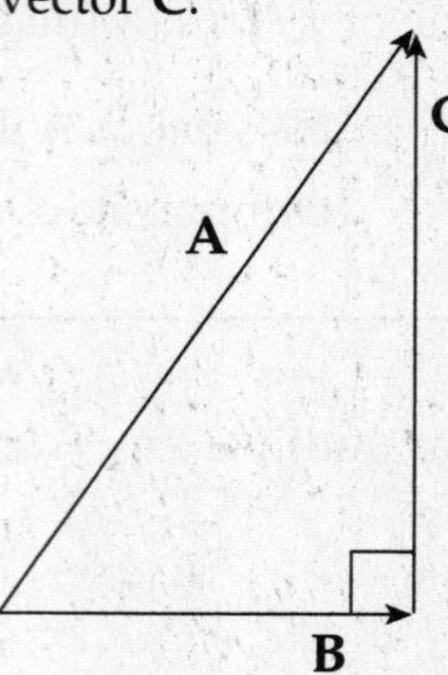

The horizontal vector is always considered a scalar multiple of what's called the **horizontal basis vector, i**, and the vertical vector is a scalar multiple of the **vertical basis vector, j**. Both of these special vectors have a magnitude of 1, and for this reason, they're called **unit vectors.** Unit vectors are often represented by placing a hat (caret) over the vector; for example, the unit **vectors i** and **j** are sometimes denoted $\hat{\mathbf{i}}$ and $\hat{\mathbf{j}}$.

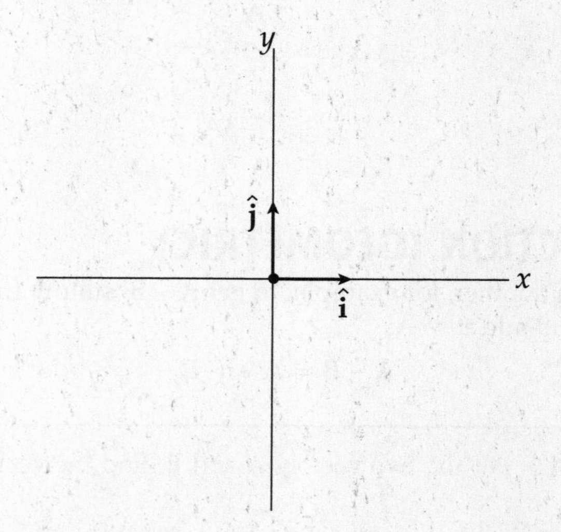

For instance, the vector **A** in the figure below is the sum of the horizontal vector $\mathbf{B} = 3\hat{\mathbf{i}}$ and the vertical vector $\mathbf{C} = 4\hat{\mathbf{j}}$.

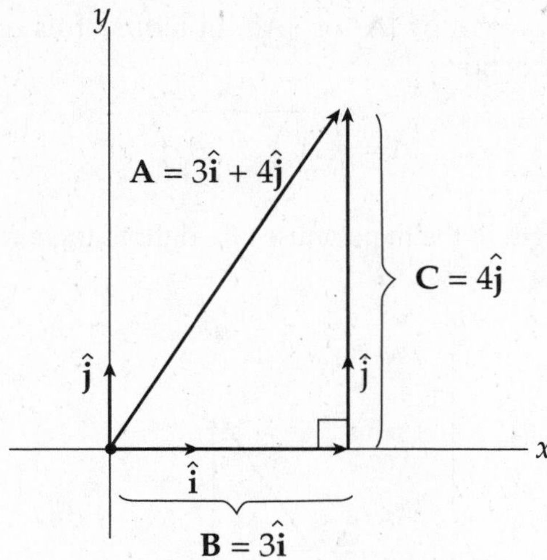

The vectors **B** and **C** are called the **vector components** of **A**, and the scalar multiples of $\hat{\mathbf{i}}$ and $\hat{\mathbf{j}}$ which give **A**—in this case, 3 and 4—are called the **scalar components** of **A**. So vector **A** can be written as the sum $A_x\hat{\mathbf{i}} + A_y\hat{\mathbf{j}}$, where A_x and A_y are the scalar components of **A**. The component A_x is called the **horizontal** scalar component of **A**, and A_y is called the **vertical** scalar component of **A**.

VECTOR OPERATIONS USING COMPONENTS

The use of components makes the vector operations of addition, subtraction, and scalar multiplication pretty straightforward:

Vector addition: *Add the respective components*

$$\mathbf{A} + \mathbf{B} = (A_x + B_x)\hat{\mathbf{i}} + (A_y + B_y)\hat{\mathbf{j}}$$

Vector subtraction: *Subtract the respective components*

$$\mathbf{A} - \mathbf{B} = (A_x - B_x)\hat{\mathbf{i}} + (A_y - B_y)\hat{\mathbf{j}}$$

Scalar multiplication: *Multiply each component by k*

$$k\mathbf{A} = (kA_x)\hat{\mathbf{i}} + (kA_y)\hat{\mathbf{j}}$$

Example 1.4 If $\mathbf{A} = 2\hat{\mathbf{i}} - 3\hat{\mathbf{j}}$ and $\mathbf{B} = -4\hat{\mathbf{i}} + 2\hat{\mathbf{j}}$, compute each of the following vectors: **A** + **B**, **A** – **B**, 2**A**, and **A** + 3**B**.

Solution. It's very helpful that the given vectors **A** and **B** are written explicitly in terms of the standard basis vectors $\hat{\mathbf{i}}$ and $\hat{\mathbf{j}}$:

$$\mathbf{A} + \mathbf{B} = (2 - 4)\hat{\mathbf{i}} + (-3 + 2)\hat{\mathbf{j}} = -2\hat{\mathbf{i}} - \hat{\mathbf{j}}$$

$$\mathbf{A} - \mathbf{B} = [2 - (-4)]\hat{\mathbf{i}} + (-3 - 2)\hat{\mathbf{j}} = 6\hat{\mathbf{i}} - 5\hat{\mathbf{j}}$$

$$2\mathbf{A} = 2(2)\hat{\mathbf{i}} + 2(-3)\hat{\mathbf{j}} = 4\hat{\mathbf{i}} - 6\hat{\mathbf{j}}$$

$$\mathbf{A} + 3\mathbf{B} = [2 + 3(-4)]\hat{\mathbf{i}} + [-3 + 3(2)]\hat{\mathbf{j}} = -10\hat{\mathbf{i}} + 3\hat{\mathbf{j}}$$

MAGNITUDE OF A VECTOR

The magnitude of a vector can be computed with the Pythagorean theorem. The magnitude of vector **A** can be denoted in several ways: A or $|\mathbf{A}|$ or $\|\mathbf{A}\|$. In terms of its components, the magnitude of $\mathbf{A} = A_x\hat{\mathbf{i}} + A_y\hat{\mathbf{j}}$ is given by the equation

$$A = \sqrt{(A_x)^2 + (A_y)^2}$$

which is the formula for the length of the hypotenuse of a right triangle with sides of lengths A_x and A_y.

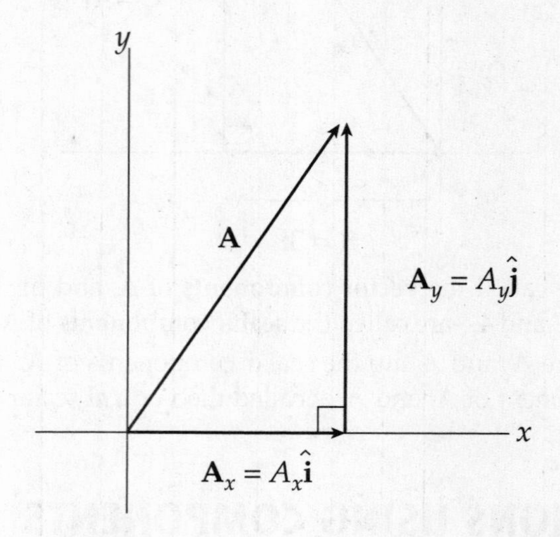

DIRECTION OF A VECTOR

The direction of a vector can be specified by the angle it makes with the positive x axis. You can sketch the vector and use its components (and an inverse trig function) to determine the angle. For example, if θ denotes the angle that the vector $\mathbf{A} = 3\hat{\mathbf{i}} + 4\hat{\mathbf{j}}$ makes with the $+x$ axis, then $\tan\theta = 4/3$, so $\theta = \tan^{-1}(4/3) = 53.1°$. Make sure your calculator is in the correct mode, either radian or degree, when using trig functions.

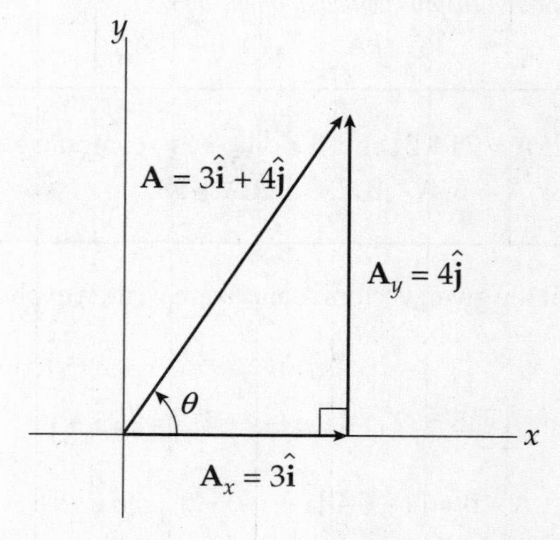

If **A** makes the angle θ with the $+x$ axis, then its x- and y-components are $A \cos \theta$ and $A \sin \theta$, respectively (where A is the magnitude of **A**).

$$\mathbf{A} = \underbrace{(A \cos\theta)}_{A_x}\hat{\mathbf{i}} + \underbrace{(A \sin\theta)}_{A_y}\hat{\mathbf{j}}$$

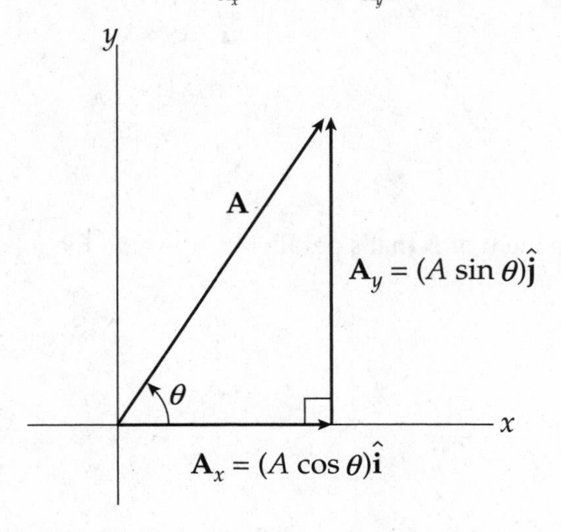

In general, any vector in the plane can be written in terms of two perpendicular component vectors. For example, vector **W** (shown below) is the sum of two component vectors whose magnitudes are $W \cos \theta$ and $W \sin \theta$:

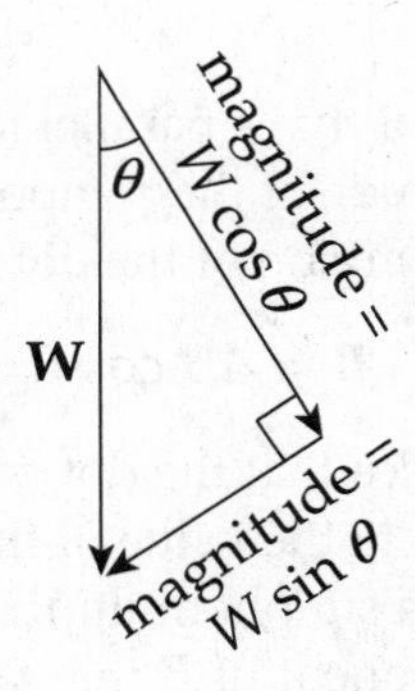

THE DOT PRODUCT

A vector can be multiplied by a scalar to yield another vector, as you saw in Example 1.2. For instance, we can multiply the vector **A** by the scalar 2 to get the scalar multiple 2**A**. The product is a vector that has twice the magnitude and, since the scalar is positive, the same direction as **A**.

Additionally, we can form a scalar by multiplying two vectors. The product of the vectors in this case is called the **dot product** or the **scalar product**, since the result is a scalar. Several physical concepts (work, electric and magnetic flux) require that we multiply the magnitude of one vector by the magnitude of the component of the other vector that's parallel to the first. The dot product was invented specifically for this purpose.

Consider these two vectors, **A** and **B**:

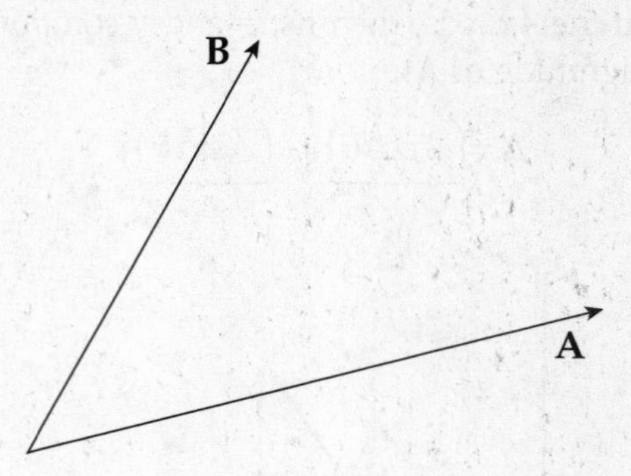

In order to find the component of **B** that's parallel to **A**, we do the following:

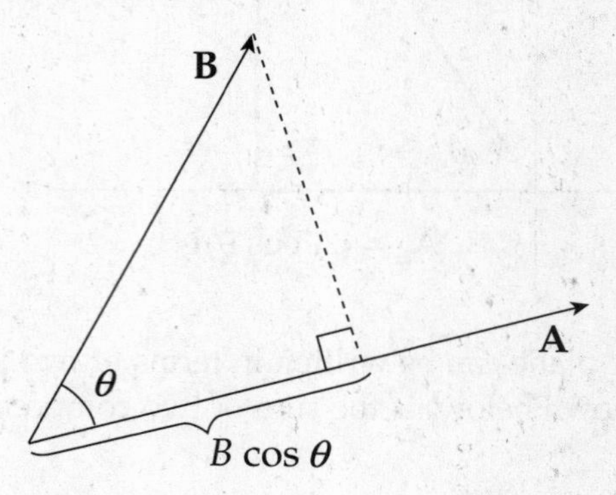

As this figure shows, the component of **B** that's parallel to **A** has the magnitude $B \cos \theta$. So if we multiply the magnitude of **A** by the magnitude of the component of **B** that's parallel to **A**, we would form the product $A(B \cos \theta)$. This is the definition of the dot product of the vectors **A** and **B**:

$$\mathbf{A} \cdot \mathbf{B} = AB \cos \theta$$

where θ is the angle between **A** and **B**. Notice that the dot product of two vectors is a scalar.

The angle between the vectors is crucial to the value of the dot product. If $\theta = 0$, then the vectors are already parallel to each other, so we can simply multiply their magnitudes: $\mathbf{A} \cdot \mathbf{B} = AB$. If **A** and **B** are perpendicular, then there is no component of **B** that's parallel to **A** (or vice versa), so the dot product should be zero. And if θ is greater than 90°,

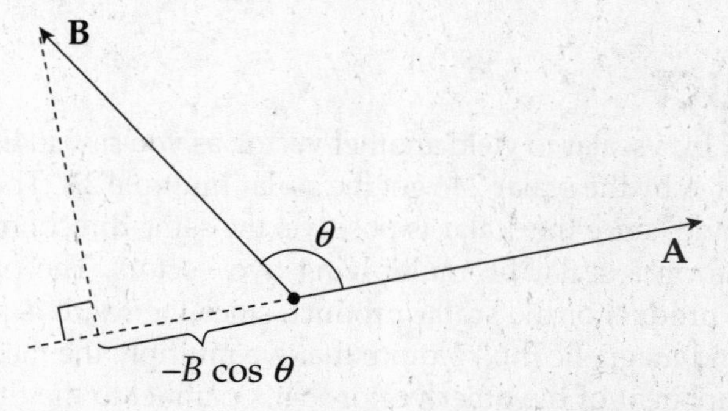

then the component of one that's *parallel* to the other is actually *antiparallel* (backwards), and this will give the dot product a negative value (because $\cos \theta < 0$ if $90° < \theta < 180°$).

The value of the dot product can, of course, be figured out using the definition above ($AB \cos \theta$) if θ is known. If θ is not known, the dot product can be calculated from the components of **A** and **B** in this way:

$$\mathbf{A} \cdot \mathbf{B} = (A_x \hat{\mathbf{i}} + A_y \hat{\mathbf{j}}) \cdot (B_x \hat{\mathbf{i}} + B_y \hat{\mathbf{j}}) = A_x B_x + A_y B_y$$

This means that, to form the dot product, simply add the product of the scalar x components and the product of the scalar y components.

Example 1.5

(a) What is the dot product of the vectors $\hat{\mathbf{i}}$ and $\hat{\mathbf{j}}$?

(b) of $\hat{\mathbf{i}}$ and $\hat{\mathbf{i}}$?

Solution.

(a) Since $\hat{\mathbf{i}}$ and $\hat{\mathbf{j}}$ are perpendicular to each other ($\theta = 90°$), their dot product must be zero, because $\cos 90° = 0$. (In fact, unless **A** or **B** already has magnitude zero, it's also true that two vectors are perpendicular to each other when their dot product is zero.)

(b) Because $\hat{\mathbf{i}}$ and $\hat{\mathbf{i}}$ are parallel to each other, their dot product is just the product of the magnitudes, which is $1 \cdot 1 = 1$.

Example 1.6 If $\mathbf{A} = -2\hat{\mathbf{i}} + 4\hat{\mathbf{j}}$ and $\mathbf{B} = 6\hat{\mathbf{i}} + B_y\hat{\mathbf{j}}$, find the value of B_y such that the vectors **A** and **B** will be perpendicular to each other.

Solution. Two vectors are perpendicular to each other if their dot product is zero. The dot product of **A** and **B** can be determined as follows: Multiply the scalar x components, $(-2)(6) = -12$, multiply the scalar y components, $(4)(B_y)$, and add them: $-12 + 4B_y$. Setting this equal to 0, we see that B_y must equal 3.

THE CROSS PRODUCT

Some physical concepts (torque, angular momentum, magnetic force) require that we multiply the magnitude of one vector by the magnitude of the component of the other vector that's *perpendicular* to the first. The cross product was invented for this specific purpose.

Consider these two vectors **A** and **B**:

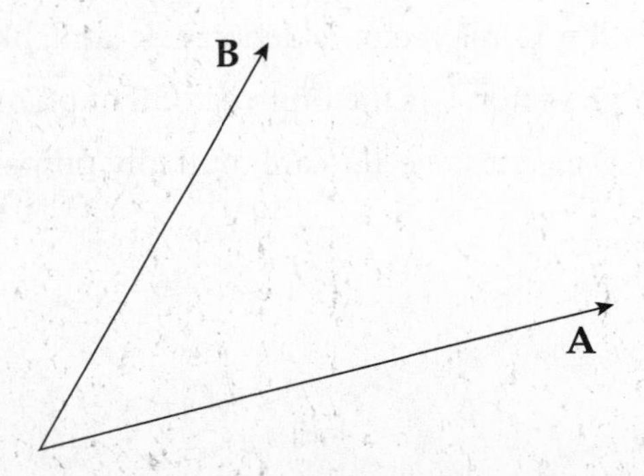

In order to find the component of one of these vectors that's perpendicular to the other one, we do the following:

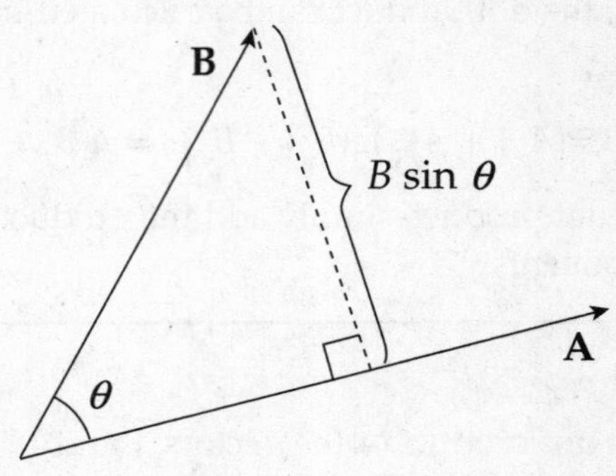

As this figure shows, the component of **B** that's perpendicular to **A** has magnitude $B \sin \theta$. Therefore, if we multiply the magnitude of **A** by the magnitude of the component of **B** that's perpendicular to **A**, we form the product $A(B \sin \theta)$. This is the magnitude of what's called the **cross product** of the vectors **A** and **B**, denoted $\mathbf{A} \times \mathbf{B}$:

$$|\mathbf{A} \times \mathbf{B}| = AB \sin \theta$$

where θ is the angle between **A** and **B** (such that $0° \leq \theta \leq 180°$).

The equation above gives the magnitude of the cross product. The cross product of two vectors is another vector that's always perpendicular to both **A** and **B**, with its direction determined by a procedure known as the *right-hand rule*. The direction of $\mathbf{A} \times \mathbf{B}$ is perpendicular to the plane that contains **A** and **B**, but this leads to an ambiguity, since there are two directions perpendicular to a plane (one on either side; they point in opposite directions). The following description resolves this ambiguity.

Make sure you are using your right hand. Point your index finger in the direction of the first vector, **A**, then point your middle finger in the direction of the second vector, **B**, and your thumb now points in the direction of the cross product, **A × B**.

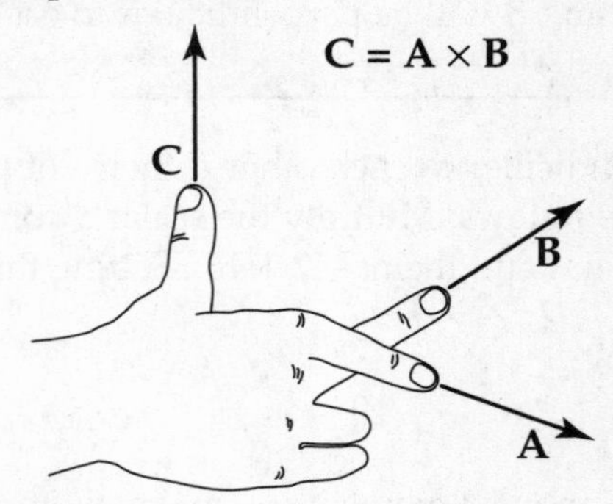

This is called the **right-hand rule**.

You know that the standard basis vectors $\hat{\mathbf{i}}$ and $\hat{\mathbf{j}}$ can be used to write any two-dimensional vector, but now that we have introduced the cross product we need a third basis unit vector to write three-dimensional vectors. This third unit vector is denoted $\hat{\mathbf{k}}$ and, like $\hat{\mathbf{i}}$ and $\hat{\mathbf{j}}$, it points along the direction of a coordinate axis. The vector $\hat{\mathbf{i}}$ is the unit vector that points along the x axis, $\hat{\mathbf{j}}$ along the y, and now $\hat{\mathbf{k}}$ along the z. The three basis vectors are mutually perpendicular:

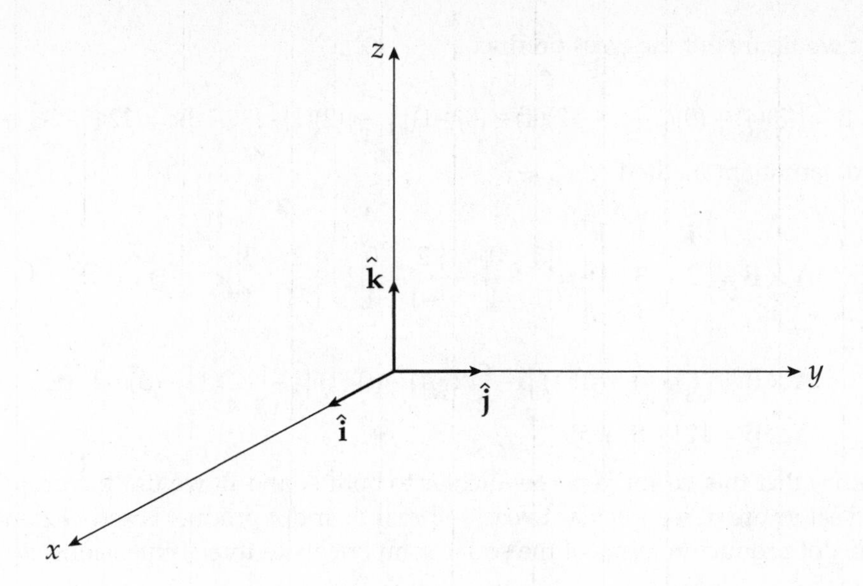

The coordinate axes shown define a *right-handed* coordinate system, because the directions of the x, y, and z axes obey the right-hand rule. That is, $\hat{\mathbf{i}} \times \hat{\mathbf{j}}$ points in the direction of $\hat{\mathbf{k}}$. In fact, $\hat{\mathbf{i}} \times \hat{\mathbf{j}}$ actually equals $\hat{\mathbf{k}}$, since the magnitude of $\hat{\mathbf{i}} \times \hat{\mathbf{j}}$ is 1. As a *right-handed* coordinate system, $\hat{\mathbf{j}} \times \hat{\mathbf{k}} = \hat{\mathbf{i}}$ and $\hat{\mathbf{k}} \times \hat{\mathbf{i}} = \hat{\mathbf{j}}$.

Unlike the dot product, the cross product is not commutative; $\mathbf{A} \times \mathbf{B}$ is not equal to $\mathbf{B} \times \mathbf{A}$. This is because applying the right-hand rule to determine the direction of $\mathbf{B} \times \mathbf{A}$ would give a vector that points in the direction opposite to that of $\mathbf{A} \times \mathbf{B}$. Therefore, $\mathbf{B} \times \mathbf{A} = -(\mathbf{A} \times \mathbf{B})$.

The cross product can be computed directly from the scalar components of **A** and **B** without first determining the angle θ, as follows: If $\mathbf{A} = A_x\hat{\mathbf{i}} + A_y\hat{\mathbf{j}} + A_z\hat{\mathbf{k}}$ and $\mathbf{B} = B_x\hat{\mathbf{i}} + B_y\hat{\mathbf{j}} + B_z\hat{\mathbf{k}}$, then the cross product (magnitude and direction) of **A** and **B** is

$$\mathbf{A} \times \mathbf{B} = (A_yB_z - A_zB_y)\hat{\mathbf{i}} + (A_zB_x - A_xB_z)\hat{\mathbf{j}} + (A_xB_y - A_yB_x)\hat{\mathbf{k}}$$

This formula requires a lot of memorization. Another method for determining the cross product is to realize that the cross product is the determinant of the following 3 × 3 matrix.

$$\mathbf{A} \times \mathbf{B} = \begin{vmatrix} \hat{\mathbf{i}} & \hat{\mathbf{j}} & \hat{\mathbf{k}} \\ A_x & A_y & A_z \\ B_x & B_y & B_z \end{vmatrix} = \begin{vmatrix} A_y & A_z \\ B_y & B_z \end{vmatrix}\hat{\mathbf{i}} - \begin{vmatrix} A_x & A_z \\ B_x & B_z \end{vmatrix}\hat{\mathbf{j}} + \begin{vmatrix} A_x & A_y \\ B_x & B_y \end{vmatrix}\hat{\mathbf{k}}$$

By taking the determinant of each 2 × 2 matrix you will realize that you get the same formula as shown above. This may appear to be equally difficult to memorize, however you can just memorize two facts. Each 2 × 2 matrix is missing the column associated with the basis vector that multiples it and the terms alternate in sign.

Example 1.7 Calculate the cross product of the vectors $\mathbf{A} = 2\hat{\mathbf{i}} + 3\hat{\mathbf{j}}$ and $\mathbf{B} = -\hat{\mathbf{i}} + \hat{\mathbf{j}} + 4\hat{\mathbf{k}}$, and verify that it's perpendicular to both **A** and **B**.

Solution. First we figure out the cross product:

$$\mathbf{A} \times \mathbf{B} = [(3)(4) - (0)(1)]\,\hat{\mathbf{i}} - [(2)(4) - (0)(-1)]\,\hat{\mathbf{j}} + [(2)(1) - (3(-1)]\hat{\mathbf{k}} = 12\,\hat{\mathbf{i}} - 8\,\hat{\mathbf{j}} + 5\hat{\mathbf{k}}$$

Using the determinant method

$$\mathbf{A} \times \mathbf{B} = \begin{vmatrix} \hat{\mathbf{i}} & \hat{\mathbf{j}} & \hat{\mathbf{k}} \\ 2 & 3 & 0 \\ -1 & 1 & 4 \end{vmatrix} = \begin{vmatrix} 3 & 0 \\ 1 & 4 \end{vmatrix} \hat{\mathbf{i}} - \begin{vmatrix} 2 & 0 \\ -1 & 4 \end{vmatrix} \hat{\mathbf{j}} + \begin{vmatrix} 2 & 3 \\ -1 & 1 \end{vmatrix} \hat{\mathbf{k}}$$

$$\mathbf{A} \times \mathbf{B} = \left[(3)(4) - (0)(1)\right]\hat{\mathbf{i}} - \left[(2)(4) - (0)(1)\right]\hat{\mathbf{j}} + \left[(2)(1) - (3)(-1)\right]\hat{\mathbf{k}}$$

$$\mathbf{A} \times \mathbf{B} = 12\hat{\mathbf{i}} - 8\hat{\mathbf{j}} + 5\hat{\mathbf{k}}$$

Now, to verify that this vector is perpendicular to both **A** and **B**, we use the property of the dot product: Two vectors are perpendicular to each other if their dot product is zero. Extending the computation of the dot product in terms of the scalar components to three dimensions, we get:

$$(\mathbf{A} \times \mathbf{B}) \cdot \mathbf{A} = (12\,\hat{\mathbf{i}} - 8\,\hat{\mathbf{j}} + 5\hat{\mathbf{k}}) \cdot (2\,\hat{\mathbf{i}} + 3\,\hat{\mathbf{j}}) = (12)(2) + (-8)(3) + (5)(0) = 0$$

$$(\mathbf{A} \times \mathbf{B}) \cdot \mathbf{B} = (12\,\hat{\mathbf{i}} - 8\,\hat{\mathbf{j}} + 5\hat{\mathbf{k}}) \cdot (-\hat{\mathbf{i}} + \hat{\mathbf{j}} + 4\hat{\mathbf{k}}) = (12)(-1) + (-8)(1) + (5)(4) = 0$$

Kinematics

INTRODUCTION

Kinematics is the study of an object's motion in terms of its displacement, velocity, and acceleration. Questions such as *How far does this object travel?* or *How fast and in what direction does it move?* or *At what rate does its speed change?* all properly belong to kinematics. In the next chapter, we will study **dynamics**, which delves more deeply into *why* objects move the way they do.

POSITION, DISTANCE, AND DISPLACEMENT

Position is an object's relation to a coordinate axis system. **Distance** is a scalar that represents the total amount traveled by an object. **Displacement** is an object's change in position. It's the vector that points from the object's initial position to its final position, regardless of the path actually taken. Since displacement means *change in position*, it is generically denoted $\Delta\mathbf{s}$, where Δ denotes *change in* and **s** means spatial location. (The letter **p** is not used for position because it's reserved for another quantity: **momentum.**) If it's known that the displacement is horizontal, then it can be called $\Delta\mathbf{x}$; if the displacement is vertical, then it's $\Delta\mathbf{y}$. The magnitude of this vector is the *net* distance traveled and, sometimes, the word *displacement* refers just to this scalar quantity. Since a distance is being measured, the SI unit for displacement is the meter $[\Delta s] = \text{m}$.

Example 2.1 Traveling along a single axis, a car starts 10 m from the origin. The car then moves 8 m directly away from the origin and then turns around and moves 12 m back toward the origin. Determine the final position of the car, the distance the car traveled and the displacement of the car.

Solution. The car ends at the 6 m mark, so the final position is 6 m. The car moves a total of 20 m, so the distance traveled is 20 m. The displacement of the car only refers to the car's final position minus its initial position. Because the car ended 4 m behind where it started its displacement would be –4 m.

Example 2.2 An infant crawls 5 m east, then 3 m north, then 1 m east. Find the magnitude of the infant's displacement.

Solution. Although the infant crawled a *total* distance of 5 + 3 + 1 = 9 m, this is not the displacement, which is merely the *net* distance traveled.

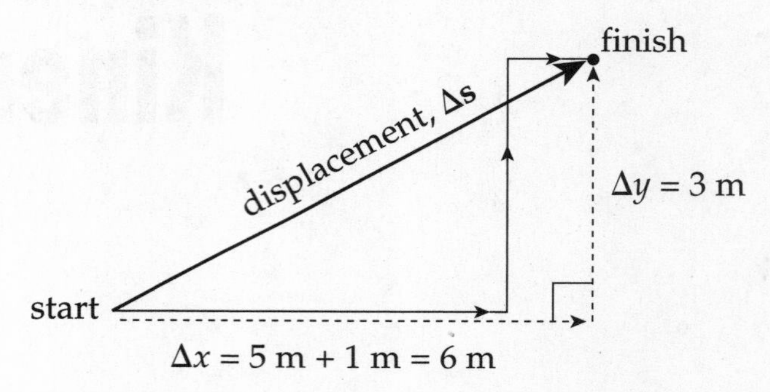

Using the Pythagorean theorem, we can calculate that the magnitude of the displacement is

$$\Delta s = \sqrt{(\Delta x)^2 + (\Delta(y))^2} = \sqrt{(6\text{ m})^2 + (3\text{ m})^2} = \sqrt{45\text{ m}^2} = 6.7\text{ m}$$

Example 2.3 In a track-and-field event, an athlete runs exactly once around an oval track, a total distance of 500 m. Find the runner's displacement for the race.

Solution. If the runner returns to the same position from which she left, then her displacement is zero.

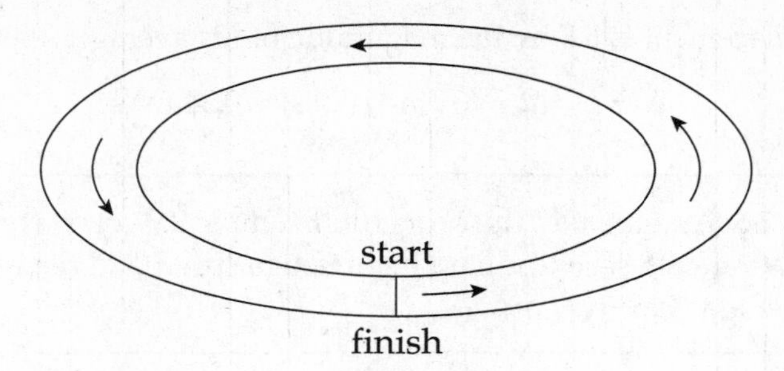

The *total* distance covered is 500 m, but the net distance—the displacement—is 0.

SPEED AND VELOCITY

When we're in a moving car, the speedometer tells us how fast we're going; it gives us our speed. But what does it mean to have a speed of say, 10 m/s? It means that we're covering a distance of 10 meters every second. By definition, **average speed** is the ratio of the total distance traveled to the time required to cover that distance:

$$\text{average speed} = \frac{\text{total distance}}{\text{time}}$$

The car's speedometer doesn't care in what direction the car is moving (as long as the wheels are moving forward). You could be driving north, south, east, west, whatever; the speedometer would make no distinction. *55 miles per hour, north* and *55 miles per hour, east* register the same on the speedometer: 55 miles per hour. Speed is a scalar.

However, we will also need to include *direction* in our descriptions of motion. We just learned about displacement, which takes both distance (net distance) and direction into account. The single concept that embodies both speed and direction is called **velocity**, and the definition of average velocity is:

$$\text{average velocity} = \frac{\text{displacement}}{\text{time}}$$

$$\bar{\mathbf{v}} = \frac{\Delta \mathbf{s}}{\Delta t}$$

(The bar over the **v** means *average*.) Because $\Delta\mathbf{s}$ is a vector, $\bar{\mathbf{v}}$ is also a vector, and because Δt is a *positive* scalar, the direction of $\bar{\mathbf{v}}$ is the same as the direction of $\Delta\mathbf{s}$. The magnitude of the velocity vector is called the object's **speed**, and is expressed in units of meters per second (m/s).

Note the distinction between speed and velocity. In everyday language, they're often used interchangeably. However, in physics, *speed* and *velocity* are technical terms whose definitions are not the same. *Velocity is speed plus direction.*[1]

Example 2.4 If the infant in Example 2.2 completes his journey in 20 seconds, find the magnitude of his average velocity.

[1] Technical note: The magnitude of the velocity is the speed. However (and this is perhaps a bit unfortunate and can be confusing), the magnitude of the average velocity is *not* called the average speed. Average speed is defined as the *total* distance traveled divided by the elapsed time. On the other hand, the magnitude of the average velocity is the *net* distance traveled divided by the elapsed time.

Solution. Since the displacement is 6.7 m, the magnitude of his average velocity is

$$\overline{v} = \Delta s/\Delta t = (6.7 \text{ m})/(20 \text{ s}) = 0.34 \text{ m/s}$$

Example 2.5 Assume that the runner in Example 2.3 completes the race in 1 minute and 18 seconds. Find her average speed and the magnitude of her average velocity.

Solution. *Average speed is total distance divided by elapsed time.* Since the length of the track is 500 m, the runner's average speed was (500 m)/(78 s) = 6.4 m/s. However, since her displacement was zero, her average velocity was zero also: $\overline{v} = \Delta s/\Delta t = (0 \text{ m})/(78 \text{ s}) = 0 \text{ m/s}$.

Example 2.6 Is it possible to move with constant speed but not constant velocity? Is it possible to move with constant velocity but not constant speed?

Solution. The answer to the first question is *yes*. For example, if you set your car's cruise control at 55 miles per hour but turn the steering wheel to follow a curved section of road, then the direction of your velocity changes (which means your velocity is not constant), even though your speed doesn't change.

The answer to the second question is *no*. Velocity means speed and direction; if the velocity is constant, then that means both speed and direction are constant. If speed were to change, then the velocity vector's magnitude would change (by definition), which immediately implies that the vector changes.

ACCELERATION

When you step on the gas pedal in your car, the car's speed increases; step on the brake and the car's speed decreases. Turn the wheel, and the car's direction of motion changes. In all of these cases, the velocity changes. To describe this change in velocity, we need a new term: **acceleration**. In the same way that velocity measures the rate-of-change of an object's position, acceleration measures the rate-of-change of an object's velocity. An object's average acceleration is defined as follows:

$$\text{average acceleration} = \frac{\text{change in velocity}}{\text{time}}$$

$$\overline{\mathbf{a}} = \frac{\Delta \mathbf{v}}{\Delta t}$$

The units of acceleration are meters per second, per second: $[a] = \text{m/s}^2$. Because $\Delta\mathbf{v}$ is a vector, $\overline{\mathbf{a}}$ is also a vector; and because Δt is a *positive* scalar, the direction of $\overline{\mathbf{a}}$ is the same as the direction of $\Delta\mathbf{v}$.

Furthermore, if we take an object's original direction of motion to be positive, then an increase in speed corresponds to a positive acceleration, while a decrease in speed corresponds to a negative acceleration (deceleration).

Note that an object can accelerate even if its speed doesn't change. (Again, it's a matter of not allowing the everyday usage of the word *accelerate* to interfere with its technical, physics usage.) This is because acceleration depends on $\Delta\mathbf{v}$, and the velocity vector $\mathbf{v}$ changes if (1) speed changes, or (2) direction changes, or (3) both speed and direction change. For instance, a car traveling around a circular racetrack is constantly accelerating even if the car's *speed* is constant, because the direction of the car's velocity vector is constantly changng.

> **Example 2.7** A car is traveling in a straight line along a highway at a constant speed of 80 miles per hour for 10 seconds. Find its acceleration.

Solution. Since the car is traveling at a constant velocity, its acceleration is zero. If there's no change in velocity, then there's no acceleration.

> **Example 2.8** A car is traveling along a straight highway at a speed of 20 m/s. The driver steps on the gas pedal and, 3 seconds later, the car's speed is 32 m/s. Find its average acceleration.

Solution. Assuming that the direction of the velocity doesn't change, it's simply a matter of dividing the change in velocity, 32 m/s – 20 m/s = 12 m/s, by the time interval during which the change occurred: $\bar{a} = \Delta v/\Delta t = (12 \text{ m/s}) / (3 \text{ s}) = 4 \text{ m/s}^2$.

> **Example 2.9** Spotting a police car ahead, the driver of the car in the previous example slows from 32 m/s to 20 m/s in 2 seconds. Find the car's average acceleration.

Solution. Dividing the change in velocity, 20 m/s – 32 m/s = –12 m/s, by the time interval during which the change occurred, 2 s, give us $\bar{\mathbf{a}} = \Delta\mathbf{v}/\Delta t = (-12 \text{ m/s}) / (2 \text{ s}) = -6 \text{ m/s}^2$. The negative sign here means that the direction of the acceleration is opposite the direction of the velocity, which describes slowing down.

UNIFORMLY-ACCELERATED MOTION AND THE BIG FIVE

The simplest type of motion to analyze is motion in which the acceleration is *constant* (possibly equal to zero). Although true uniform acceleration is rarely achieved in the real world, many common motions are governed by approximately constant acceleration and, in these cases, the kinematics of uniformly accelerated motion provide a pretty good description of what's happening. Notice that if the acceleration is constant, then taking an average yields nothing new, so $\bar{a} = a$.

Another restriction that will make our analysis easier is to consider only motion that takes place along a straight line. In these cases, there are only two possible directions of motion. One is positive, and the opposite direction is negative. Most of the quantities we've been dealing with—displacement, velocity, and acceleration—are vectors, which means that they include both a magnitude and a direction. With straight-line motion, direction can be specified simply by attaching a + or – sign to the magnitude of the quantity. Also, if we agree to start our clocks at $t_0 = 0$ then $\Delta t = t - 0 = t$. We will use $\Delta t = t$ to keep the notation cleaner and to match the AP equation sheet. Therefore, although we will often abandon the use of bold letters to denote the vector quantities of displacement, velocity, and acceleration, the fact that these quantities include direction will still be indicated by a positive or negative sign.

Let's review the quantities we've seen so far. The fundamental quantities are displacement (Δx), velocity (v), and acceleration (a). Acceleration is a change in velocity, from an initial velocity (v_i or v_0) to a final velocity (v_f or simply v—with no subscript). And, finally, the motion takes place during some elapsed time interval, Δt. Also, if we agree to start our clocks at $t_0 = 0$ then $\Delta t = t - 0 = t$. We will use $\Delta t = t$ to keep the notation cleaner and to match the AP equation sheet. Therefore, we have five kinematics quantities: Δx, v_0, v, a, and t.

These five quantities are related by a group of five equations that we call the *Big Five*. They work in cases where acceleration is uniform, which are the cases we're considering.

		Variable that is missing
Big Five #1	$\Delta x = x - x_0 = \overline{v}t$	a
Big Five #2	$v = v_0 + at$	Δx
Big Five #3	$x = x_0 + v_0 t + \frac{1}{2}at^2$	v
Big Five #4	$x = x_0 + vt - \frac{1}{2}at^2$	v_0
Big Five #5	$v^2 = v_0^2 + 2a(x - x_0)$	t

The change in the position of an object is the displacement: $\Delta x = x - x_0$. Equations #2, #3, and #5 are given on the Advanced Placement Physics C Equations Sheet. They are written here exactly as they will appear on the sheet, which is why we are using Δx rather than Δs for the displacement.

It is best to memorize these equations, as they are needed for the multiple-choice section of the exam, but you are only allowed to use the AP equations sheet for the free-response section. Because the acceleration is constant, the average velocity is: $\overline{v} = \frac{1}{2}(v_o + v)$.

Each of the Big Five equations is missing one of the five kinematic quantities. The way you decide which equation to use when solving a problem is to determine which of the kinematic quantities is missing from the problem—that is, which quantity is neither given nor asked for—and then use the equation that doesn't contain that variable. For example, if the problem never mentions the final velocity—v is neither given nor asked for—the equation that will work is the one that's missing v. That's Big Five #3.

Big Five #1 and #2 are simply the definitions of $\overline{v}$ and $\overline{a}$ written in forms that don't involve fractions. The other Big Five equations can be derived from these two definitions and the equation $\overline{v} = \frac{1}{2}(v_0 + v)$, using a bit of algebra.

Example 2.10 An object with an initial velocity of 4 m/s moves along a straight axis under constant acceleration. Three seconds later, its velocity is 14 m/s. How far did it travel during this time?

Solution. We're given v_0, t, and v, and we're asked for Δx. So a is missing; it isn't given and it isn't asked for, and we use Big Five #1:

$$\Delta x = \bar{v}t$$

$$\Delta x = \frac{1}{2}(14+4)(3) = 27 \text{ m}$$

That is, it's okay to leave off the units in the middle of the calculation *as long as you remember to include them in your final answer*. Leaving units off of your final answer will cost you points on the AP exam.

Example 2.11 A car that's initially traveling at 10 m/s accelerates uniformly for 4 seconds at a rate of 2 m/s², in a straight line. How far does the car travel during this time?

Solution. We're given v_0, t, and a, and we're asked for Δx. So, v is missing; it isn't given and it isn't asked for, and we use Big Five #3:

$$\Delta x = v_0 t + \frac{1}{2}a(t)^2 = (10 \text{ m/s})(4 \text{ s}) + \frac{1}{2}(2 \text{ m/s}^2)(4 \text{ s})^2 = 56 \text{ m}$$

Example 2.12 A rock is dropped off a cliff that's 80 m high. If it strikes the ground with an impact velocity of 40 m/s, what acceleration did it experience during its descent?

Solution. If something is *dropped*, then that means it has no initial velocity: $v_0 = 0$. So, we're given v_0, Δx, and v, and we're asked for a. Since t is missing, we use Big Five #5:

$$v^2 = v_0^2 + 2a\Delta x \Rightarrow v^2 = 2a\Delta x \quad (\text{since } v_0 = 0)$$

$$a = \frac{v^2}{2\Delta x} = \frac{(40 \text{ m/s})^2}{2(80 \text{ m})} = 10 \text{ m/s}^2 \text{ downward}$$

Note that since a has the same sign as Δx, the acceleration vector points in the same direction as the displacement vector. This makes sense here, since the object moves downward and the acceleration it experiences is due to gravity, which also points downward.

KINEMATICS WITH GRAPHS

So far, we have dealt with kinematics problems algebraically, but you should also be able to handle kinematics questions in which information is given graphically. The two most popular graphs in kinematics are position-vs.-time graphs and velocity-vs.-time graphs. For example, consider an object that's moving along an axis in such a way that its position x as a function of time t is given by the following position-vs.-time graph:

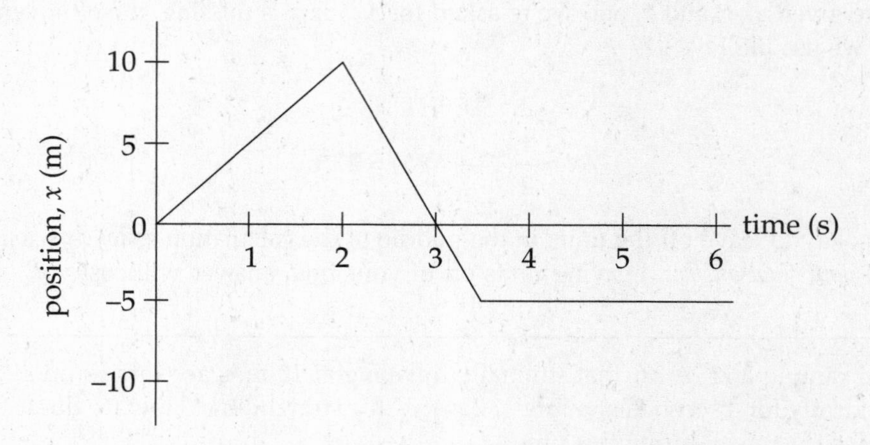

What does this graph tell us? It says that at time $t = 0$, the object was at position $x = 0$. Then, in the next two seconds, its position changed from $x = 0$ to $x = 10$ m.

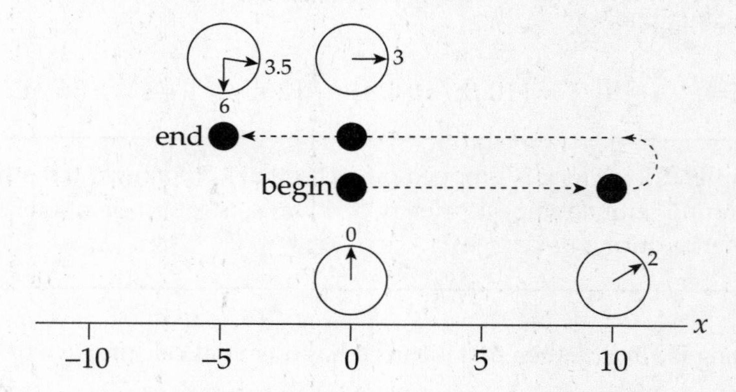

Then, at time $t = 2$ s, it reversed direction and headed back toward its starting point, reaching $x = 0$ at time $t = 3$ s, and continued, reaching position $x = -5$ m at time $t = 3.5$ s. Then the object remained at this position, $x = -5$ m, at least through time $t = 6$ s. Notice how economically the graph embodies all this information!

We can also determine the object's average velocity (and average speed) during particular time intervals. For example, its average velocity from time $t = 0$ to time $t = 2$ s is equal to the object's displacement, $10 - 0 = 10$ m, divided by the elapsed time, 2 s:

$$\bar{v} = \frac{\Delta x}{\Delta t} = \frac{(10-0)\text{ m}}{(2-0)\text{ s}} = 5\text{ m/s}$$

Note, however, that the ratio that defines the average velocity, $\Delta x/\Delta t$, also defines the slope of the x vs. t graph. Therefore, we know the following important fact:

The slope of a position-vs.-time graph gives the velocity.

What was the average velocity from time t = 2 s to time t = 3.5 s? The slope of the line segment joining the point (t, x) = (2 s, 10 m) to the point (t, x) = (3.5 s, – 5m) is

$$\overline{v} = \frac{\Delta x}{\Delta t} = \frac{(-5-10)\text{ m}}{(3.5-2)\text{ s}} = -10 \text{ m/s}$$

The fact that $\overline{v}$ is negative tells us that the object's displacement was negative during this time interval; that is, it moved in the negative x direction. The fact that $\overline{v}$ is negative agrees with the observation that the slope of a line that falls to the right is negative. What is the object's average velocity from time t = 3.5 s to time t = 6 s? Since the line segment from t = 3.5 s to t = 6 s is horizontal, its slope is zero, which implies that the average velocity is zero, but we can also figure this out from looking at the graph, since the object's position did not change during that time.

Finally, let's figure out the object's average velocity and average speed for its entire journey (from t = 0 to t = 6 s). The average velocity is

$$\overline{v} = \frac{\Delta x}{\Delta t} = \frac{(-5-0)\text{ m}}{(6-0)\text{ s}} = -0.83 \text{ m/s}$$

This is the slope of the imagined line segment that joins the point (t, x) = (0 s, 0 m) to the point (t, x) = (6 s, –5 m). The average speed is the total distance traveled by the object divided by the elapsed time. In this case, notice that the object traveled 10 m in the first 2 s, then 15 m (albeit backward) in the next 1.5 s; it covered no additional distance from t = 3.5 s to t = 6 s. Therefore, the total distance traveled by the object is d = 10 + 15 = 25 m, which took 6 s, so

$$\text{average speed} = \frac{d}{\Delta t} = \frac{25 \text{ m}}{6 \text{ s}} = 4.2 \text{ m/s}$$

Let's next consider an object moving along a straight axis in such a way that its velocity, v, as a function of time, t, is given by the following velocity-vs.-time graph:

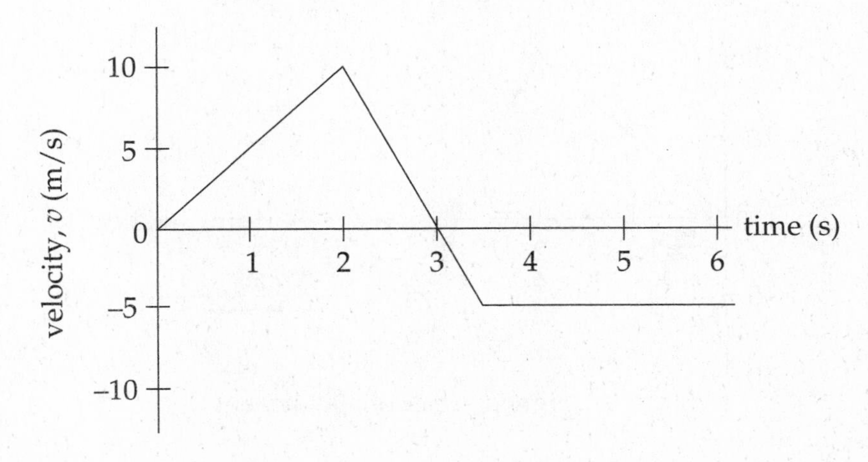

What does this graph tell us? It says that, at time $t = 0$, the object's velocity was $v = 0$. Over the first two seconds, its velocity increased steadily to 10 m/s. At time $t = 2$ s, the velocity then began to decrease (eventually becoming $v = 0$, at time $t = 3$ s). The velocity then became negative after $t = 3$ s, reaching $v = -5$ m/s at time $t = 3.5$ s. From $t = 3.5$ s on, the velocity remained a steady -5 m/s.

What can we ask about this motion? First, the fact that the velocity changed from $t = 0$ to $t = 2$ s tells us that the object accelerated. The acceleration during this tme was

$$a = \frac{\Delta v}{\Delta t} = \frac{(10-0)\text{ m/s}}{(2-0)\text{ s}} = 5\text{ m/s}^2$$

Note, however, that the ratio that defines the acceleration, $\Delta v/\Delta t$, also defines the slope of the v vs. t graph. Therefore,

The slope of a velocity-vs.-time graph gives the acceleration.

What was the acceleration from time $t = 2$ s to time $t = 3.5$ s? The slope of the line segment joining the point $(t, v) = (2\text{ s}, 10\text{ m/s})$ to the point $(t, v) = (3.5\text{ s}, -5\text{ m/s})$ is

$$a = \frac{\Delta v}{\Delta t} = \frac{(-5-10)\text{ m/s}}{(3.5-2)\text{ s}} = -10\text{ m/s}^2$$

The fact that a is negative tells us that the object's velocity change was negative during this time interval; that is, the object accelerated in the negative direction. In fact, after time $t = 3$ s, the velocity became more negative, indicating that the direction of motion was negative at increasing speed. What is the object's acceleration from time $t = 3.5$ s to time $t = 6$ s? Since the line segment from $t = 3.5$ s to $t = 6$ s is horizontal, its slope is zero, which implies that the acceleration is zero, but you can also see this from looking at the graph; the object's velocity did not change during this time interval.

Another question can be asked when a velocity-vs.-time graph is given: How far did the object travel during a particular time interval? For example, let's figure out the displacement of the object from time $t = 4$ s to time $t = 6$ s. During this time interval, the velocity was a constant -5 m/s, so the displacement was $\Delta x = v\Delta t = (-5\text{ m/s})(2\text{ s}) = -10$ m.

Geometrically, we've determined the area between the graph and the horizontal axis. After all, the area of a rectangle is *base* × *height* and, for the shaded rectangle shown below, the *base* is Δt, and the *height* is v. So, *base* × *height* equals $\Delta t \times v$, which is displacement.

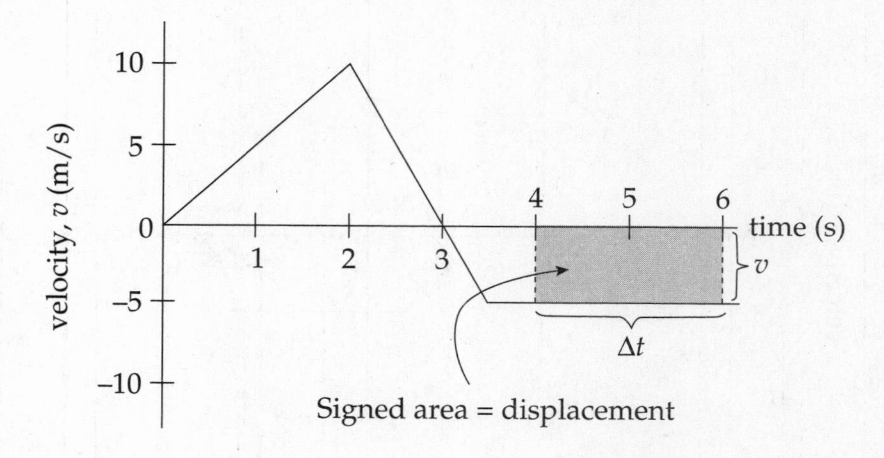

We say *signed area* because regions below the horizontal axis are negative quantities (since the object's velocity is negative, its displacement is negative). Therefore, counting areas above the horizontal axis as positive and areas below the horizontal axis as negative, we can make the following claim:

Given a velocity-vs.-time graph, the area between the graph and the t axis equals the object's displacement.

What is the object's displacement from time $t = 0$ to $t = 3$ s? Using the fact that displacement is the area bounded by the velocity graph, we figure out the area of the triangle shown below:

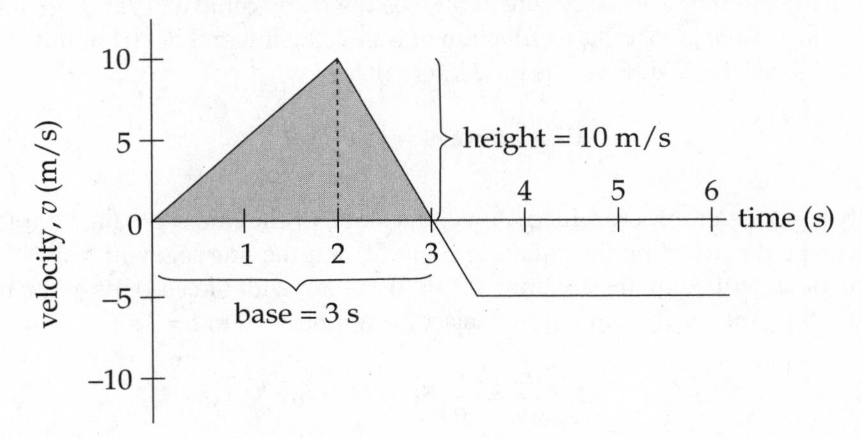

Since the area of a triangle is $\left(\frac{1}{2}\right) \times$ base $\times$ height, we find that $\Delta x = \frac{1}{2}(3 \text{ s})(10 \text{ m/s}) = 15 \text{ m}$.

A Note About Calculus

The use of graphs for solving kinematics questions provides a link to some basic definitions in calculus. The slope of a curve is the geometric definition of the **derivative** of a function. Since the slope of a position-vs.-time graph gives the velocity, the derivative of a position function gives the velocity. That is, given an equation of the form $x = x(t)$ for the position x of an object as a function of time t, the derivative of $x(t)$—with respect to t—gives the object's velocity $v(t)$:

$$v(t) = \frac{dx}{dt}$$

Often, a derivative with respect to time is denoted by a dot above the quantity, so an alternative notation for the equation above is $v(t) = \dot{x}(t)$.

Next, since the slope of a velocity-vs.-time graph gives the acceleration, the derivative of a velocity function gives the acceleration. That is, given an equation of the form $v = v(t)$ for the velocity v of an object as a function of time t, the derivative of $v(t)$—with respect to t—gives the object's acceleration $a(t)$:

$$a(t) = \frac{dv}{dt}$$

Combining this equation with the previous one, we see that the acceleration is the second derivative of the position:

$$a(t) = \frac{dv}{dt} = \frac{d}{dt}\left(\frac{dx}{dt}\right) = \frac{d^2x}{dt^2}$$

Using the dot notation, these last two equations may be written as $a(t) = \dot{v}(t)$ and $a(t) = \ddot{x}(t)$, respectively.

The signed area bounded by a curve and the horizontal axis is the geometric definition of the **definite integral** of a function. Since the area bounded by a velocity-vs.-time graph gives the displacement, the definite integral of a velocity function gives the displacement. That is, given the equation $v = v(t)$ for the velocity v of an object as a function of time t, the integral of $v(t)$ from time $t = t_1$ to time $t = t_2$ equals the displacement during this time interval:

$$\text{displacement} = \int_{t_1}^{t_2} v(t)\,dt$$

For the velocity-vs.-time graph examined above, the slope of the line segment from $(t, v) = (0, 0)$ to $(t, v) = (2, 10)$ can be described by the equation $v(t) = 5t$, and the line segment from $(t, v) = (2, 10)$ to $(t, v) = (3, 0)$ can be described by the equation $v(t) = -10t + 30$ (with t in seconds and v in m/s in both equations). Therefore, the acceleration of the object from time $t = 0$ to $t = 2$ s is

$$a = \frac{dv}{dt} = \frac{d}{dt}(5t) = 5 \text{ m/s}^2$$

as we found above.

Also, the displacement of the object from time $t = 0$ to $t = 3$ s is the value of the integral from $t = 0$ to $t = 3$ s, which we can get by adding the integral of $v(t) = 5t$ from $t = 0$ to $t = 2$ s to the integral of $v(t) = -10t + 30$ from $t = 2$ s to $t = 3$ s:

$$\begin{aligned}
\text{displacement} &= \int_0^2 5t\,dt + \int_2^3 (-10t + 30)\,dt \\
&= \left[\frac{5}{2}t^2\right]_0^2 + \left[-5t^2 + 30t\right]_2^3 \\
&= \frac{5}{2}(2^2 - 0^2) + \left[(-5 \cdot 3^2 + 30 \cdot 3) - (-5 \cdot 2^2 + 30 \cdot 2)\right] \\
&= 10 + [45 - 40] \\
&= 15 \text{ m}
\end{aligned}$$

which also agrees with the value computed above.

Example 2.13 The velocity of an object as a function of time is given by the following graph:

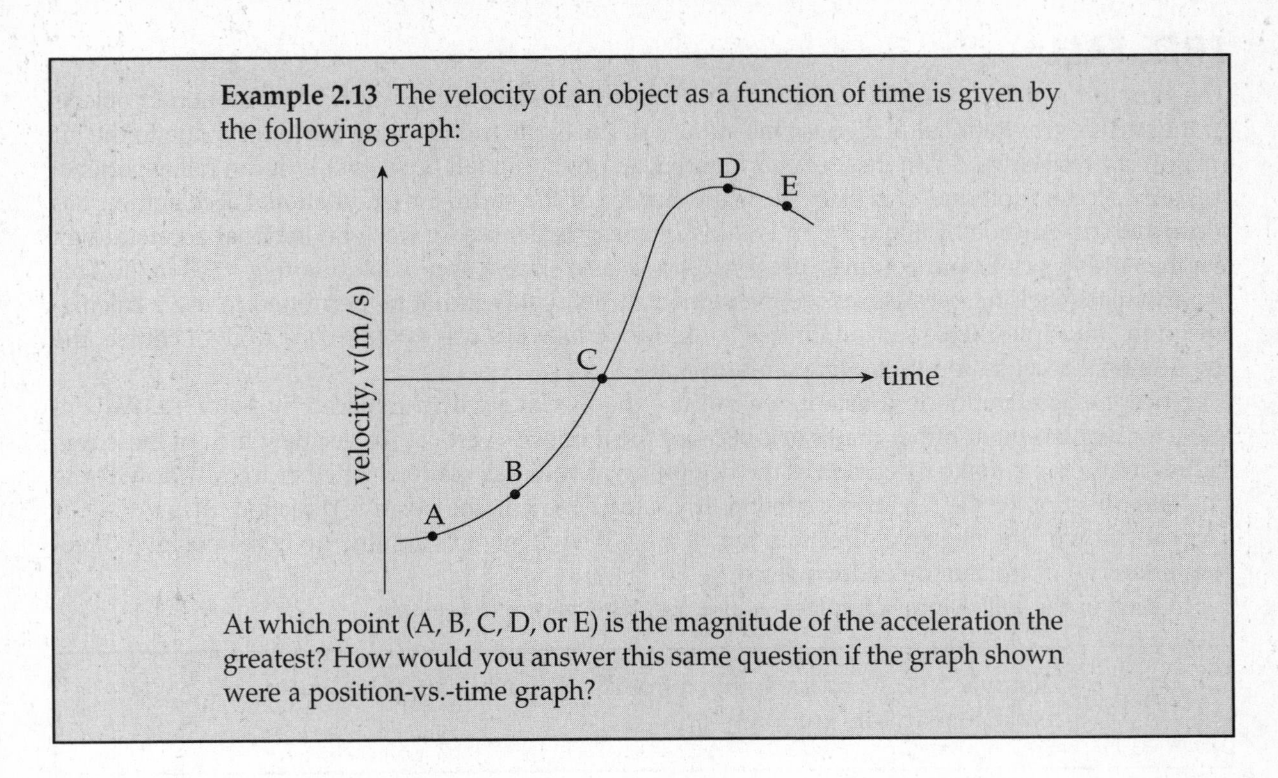

At which point (A, B, C, D, or E) is the magnitude of the acceleration the greatest? How would you answer this same question if the graph shown were a position-vs.-time graph?

Solution. The acceleration is the slope of the velocity-vs.-time graph. Although this graph is not composed of straight lines, the concept of slope still applies; at each point, the slope of the curve is the slope of the tangent line to the curve. The slope is essentially zero at Points A and D (where the curve is flat), small and positive at B, and small and negative at E. The slope at Point C is large and positive, so this is where the object's acceleration is the greatest.

If the graph shown were a position-vs.-time graph, then the slope would be the velocity. The slope of the given graph starts at zero (around Point A), slowly increases to a small positive value at B, continues to slowly increase to a large positive value at C, then, at around Point D, this large positive slope decreases quickly to zero. Of the points designated on the graph, Point D is the location of the greatest slope change, which means that this is the point of the greatest velocity change. Therefore, this is the point at which the magnitude of the acceleration is greatest.

Average vs. Instantaneous Quantities

We have a variety of methods to determine the velocity and the acceleration of an object. We also need to distinguish between average quantities and instantaneous quantities. Average velocity is the displacement over some time interval, while instantaneous velocity is how fast the object is traveling at a specific instant of time. The speedometer on a car gives us the instantaneous *speed* (it does not indicate the direction, so it does not indicate the instantaneous velocity). Typically, it is easiest to solve for average quantities using the definitions mentioned before. It is easiest to solve for instantaneous quantities by using one of the Big Five equations, or taking the slope of a given graph or derivative of a given function. Make sure you carefully read the question and understand whether they are asking for an average quantity or an instantaneous quantity.

FREE FALL

The simplest real-life example of motion under pretty constant acceleration is the motion of objects in the earth's gravitational field, near the surface of the earth and ignoring any effects due to the air (mainly air resistance). With these effects ignored, an object can fall *freely*, that is, it can fall experiencing only acceleration due to gravity. Near the surface of the earth, the gravitational acceleration has a constant magnitude of about 9.8 m/s^2; this quantity is denoted g (for gravitational acceleration). On the AP Physics Exam, you may use $g = 10$ m/s^2 as a simple approximation to $g = 9.8$ m/s^2. This is particularly helpful because, as we mentioned earlier, you will not be permitted to use a calculator on the multiple-choice section. In this book, we will always use $g = 10$ m/s^2. And, of course, the gravitational acceleration vector, **g**, points *downward*.

Since the acceleration is constant, we can use the Big Five with a replaced by $+g$ or $-g$. We will use y for displacement rather than s or x because the motion is vertical. To decide which of these two values to use for a, make a decision at the beginning of your calculations whether to call "down" the positive direction or the negative direction. If you call "down" the positive direction, then $a = +g$. If you call "down" the negative direction, then $a = -g$. We will always assume up is the positive direction unless all of the motion is downward.

In each of the following examples, we'll ignore effects due to the air.

Example 2.14 A rock is dropped from an 80-meter cliff. How long does it take to reach the ground?

Solution. Since all of the rock's motion is *down*, we call *down* the positive direction, so $a = +g$. We're given v_0, Δy, and a, and asked for t. So v is missing; it isn't given and it isn't asked for, and we use Big Five #3:

$$y = y_0 + v_0 t + \frac{1}{2}gt^2 \Rightarrow y = \frac{1}{2}gt^2 \quad (\text{since } y_0 = 0, \text{ and } v_0 = 0)$$

$$t = \sqrt{\frac{2y}{g}} = \sqrt{\frac{2(80)}{10}} = 4.0 \text{ s}$$

Example 2.15 A baseball is thrown straight upward with an initial speed of 20 m/s. How high will it go?

Solution. Since the ball travels upward, call *up* the positive direction. Therefore, *down* is the negative direction, so $a = -g$. The ball's velocity drops to zero at the instant the ball reaches its highest point, so we're given a, v_0, and v, and asked for Δy. Since t is missing, we use Big Five #5:

$$v^2 = v_0^2 + 2a\Delta y \Rightarrow 2a\Delta y = -v_0^2 \ (\text{because } v = 0)$$

$$\Delta y = -\frac{v_0^2}{2a}$$

$$= -\frac{v_0^2}{2(-g)} = -\frac{(+20 \text{ m/s})^2}{2(-10 \text{ m/s}^2)} = 20 \text{ m}$$

> **Example 2.16** One second after being thrown straight down, an object is falling with a speed of 20 m/s. How fast will it be falling 2 seconds after it was traveling 20 m/s?

Solution. Because all the motion is downward, call *down* the positive direction, so $a = +g$ and $v_0 = +20$ m/s. We're given v_0, a, and t, and asked for v. Since Δy is missing, we use Big Five #2:

$$v = v_0 + at = (+20 \text{ m/s}) + (+10 \text{ m/s}^2)(2 \text{ s}) = 40 \text{ m/s}$$

> **Example 2.17** If an object is thrown straight upward with an initial speed of 8 m/s and takes 3 seconds to strike the ground, from what height was the object thrown?

Solution. The figure below shows the path of the ball. We will use *up* as positive because not all the motion is downward. Therefore $a = -g$. We're given a, v_0, and t, and we need to find Δy. Since v is missing, we use Big Five #3:

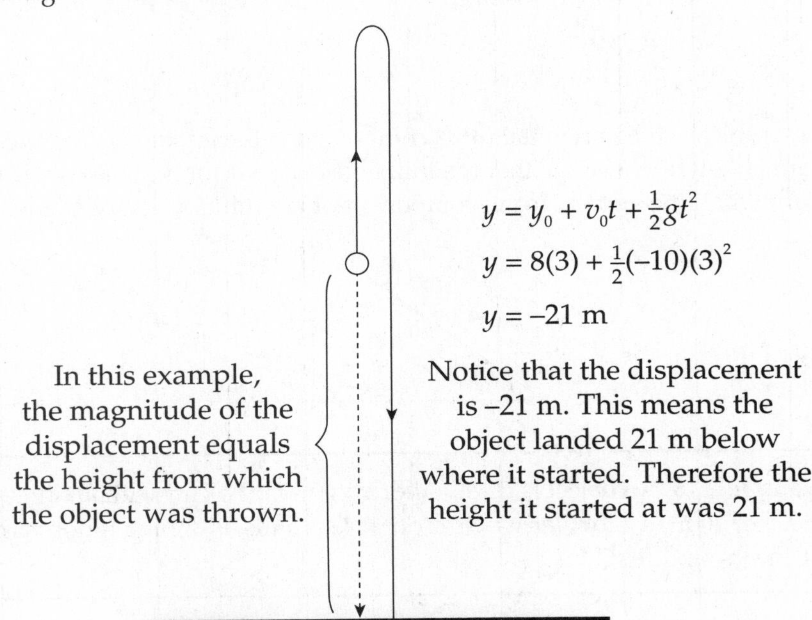

PROJECTILE MOTION

In general, an object that moves near the surface of the earth will not follow a straight-line path (for example, a baseball hit by a bat, a golf ball struck by a club, or a tennis ball hit from the baseline). If we launch an object at an angle other than straight upward and consider only the effect of acceleration due to gravity, then the object will travel along a parabolic trajectory.

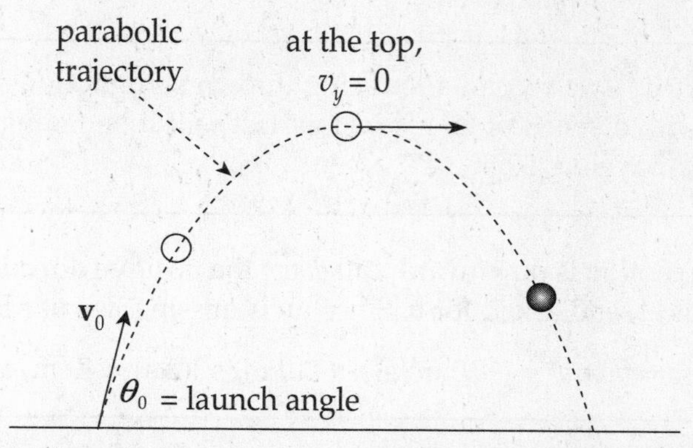

To simplify the analysis of parabolic motion, *we analyze the horizontal and vertical motions* ***separately***, using the Big Five. This is the key to doing projectile motion problems. Calling *down* the negative direction, we have

Horizontal motion:	Vertical motion:
$\Delta x = v_{0x}t$	$\Delta y = v_{0y}t + \frac{1}{2}(-g)t^2$
$v_x = v_{0x}$ (constant!)	$v_y = v_{0y} + (-g)t$
$a_x = 0$	$a_y = -g$
	$v_y^2 = v_{0y}^2 + 2(-g)\Delta y$

The quantity v_{0x}, which is the horizontal (or x) component of the initial velocity, is equal to $v_0 \cos \theta_0$, where θ_0 is the **launch angle**, the angle that the initial velocity vector, $\mathbf{v}_0$, makes with the horizontal. Similarly, the quantity v_{0y}, the vertical (or y) component of the initial velocity, is equal to $v_0 \sin \theta_0$.

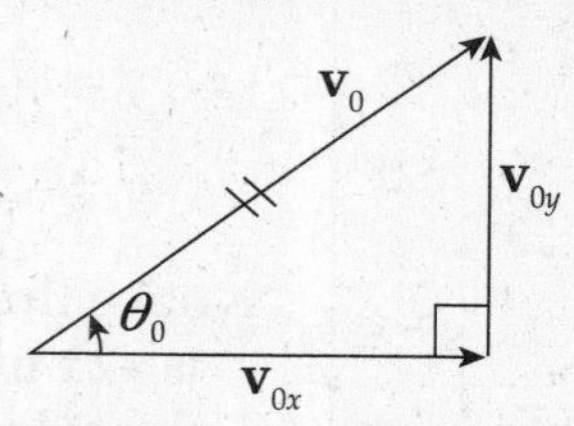

Example 2.18 An object is thrown horizontally off a cliff with an initial speed of 10 m/s. How far will it drop in 4 seconds assuming it does not hit the ground first?

Solution. The first step is to decide whether this is a *horizontal* question or a *vertical* question, since you must consider these motions separately. The question, *How far will it drop?* is a *vertical* question, so the set of equations we will consider are those listed above under *vertical motion*. Next, *How far...?* implies that we will use the first of the vertical-motion equations, the one that gives vertical displacement, Δy.

Now, since the object is thrown horizontally, there is no vertical component to its initial velocity vector $\mathbf{v}_0$; that is, $v_{0y} = 0$. Therefore,

$\mathbf{v}_0 = \mathbf{v}_{0x}$

Δy

$$\Delta y = v_{0y}t + \frac{1}{2}(-g)t^2 \rightarrow \Delta y = \frac{1}{2}(-g)t^2 \text{ (because } v_{0y} = 0\text{)}$$
$$= \frac{1}{2}(-10)(4^2)$$
$$= -80\,\text{m}$$

The fact that Δy is negative means that the displacement is *down*. Also, notice that the information given about v_{0x} is irrelevant to the qustion.

> **Example 2.19** From a height of 100 m, a ball is thrown horizontally with an initial speed of 15 m/s. How far does it travel horizontally in the first 2 seconds?

Solution. The question, *How far does it travel horizontally...?* immediately tells us that we should use the first of the horizontal-motion equations listed above:

$$\Delta x = v_{0x}t = (15\text{ m/s})(2\text{ s}) = 30\text{ m}$$

The information that the initial vertical position is 100 m above the ground is irrelevant (except for the fact that it's high enough that the ball doesn't strike the ground before the two seconds have elapsed).

> **Example 2.20** A projectile is traveling in a parabolic path for a total of 6 seconds. How does its horizontal velocity 1 s after launch compare to its horizontal velocity 4 s after launch?

Solution. The only acceleration experienced by the projectile is due to gravity, which is purely vertical, so that there is no horizontal acceleration. If there's no horizontal acceleration, then the horizontal velocity cannot change during flight, and the projectile's horizontal velocity 1 s after it's launched is the same as its horizontal velocity 3 s later.

> **Example 2.21** An object is projected upward with a 30° launch angle and an initial speed of 40 m/s. How long will it take for the object to reach the top of its trajectory? How high is this?

Solution. When the projectile reaches the top of its trajectory, its velocity vector is momentarily horizontal; that is, $v_y = 0$. Using the vertical-motion equation for v_y, we can set it equal to 0 and solve for t:

$$v_y \overset{\text{set}}{=} 0 \Rightarrow v_{0y} + (-g)t = 0$$

$$t = \frac{v_{0y}}{g} = \frac{v_0 \sin\theta_0}{g} = \frac{(40 \text{ m/s})\sin 30^\circ}{10 \text{ m/s}^2} = 2 \text{ s}$$

At this time, the projectile's vertical displacment is

$$\Delta y = v_{0y}t + \frac{1}{2}(-g)t^2 = (v_0 \sin\theta_0)t + \frac{1}{2}(-g)t^2$$

$$= [(40 \text{ m/s}) \sin 30^\circ](2 \text{ s}) + \frac{1}{2}(-10 \text{ m/s}^2)(2 \text{ s})^2$$

$$= 20 \text{ m}$$

Example 2.22 An object is projected upward with a 30° launch angle and an initial speed of 60 m/s. For how many seconds will it be in the air? How far will it travel horizontally?

Solution. The total time the object spends in the air is equal to twice the time required to reach the top of the trajectory (because the parabola is symmetrical). So, as we did in the previous example, we find the time required to reach the top by setting v_y equal to 0, and now double that amount to time:

$$v_y \overset{\text{set}}{=} 0 \Rightarrow v_{0y} + (-g)t = 0$$

$$t = \frac{v_{0y}}{g} = \frac{v_0 \sin\theta_0}{g} = \frac{(60 \text{ m/s})\sin 30^\circ}{10 \text{ m/s}^2} = 3 \text{ s}$$

Therefore, the *total* flight time (that is, up and down) is $T = 2t = 2 \times (3 \text{ s}) = 6 \text{ s}$.

Now, using the first horizontal-motion equation, we can calculate the horizontal displacement after 6 seconds:

$$\Delta x = v_{0x}T = (v_0 \cos\theta_0)T = [(60 \text{ m/s})\cos 30^\circ](6 \text{ s}) = 310 \text{ m}$$

By the way, the full horizontal displacement of a projectile is called the projectile's **range**.

A Note About Notation

We are trying to be as consistent as possible with the test by using equations identical to those on the AP Equations Sheet that you will be given to use during the free-response section of the exam. When the motion is vertical we use y instead of x, and occasionally use s to indicate when the motion is in two dimensions. It is important that you judge which variable best represents the values you are given and those you need to solve for.

KINEMATICS WITH CALCULUS

As we mentioned earlier, velocity is the time derivative of position, and acceleration is the time derivative of velocity. Equivalently, the integral of acceleration is the change in velocity, and the integral of velocity is displacement. These relationships can be summarized in the following diagram:

$$\text{position} \xrightarrow{\text{differentiate}} \text{velocity} \xrightarrow{\text{differentiate}} \text{acceleration}$$
$$\Delta x(t) \xleftarrow{\text{integrate}} \Delta v\ (t) \xleftarrow{\text{integrate}} a(t)$$

Example 2.23 The position of an object (measured in meters from the origin, where $x = 0$) moving along a straight line is given as a function of time t (measured in seconds) by the equation $x(t) = 4t^2 - 6t - 40$. Find

(a) its velocity at time t,

(b) its acceleration at time t,

(c) the time at which the object is at the origin, and

(d) the object's velocity and acceleration at the time calculated in (c).

Solution.

(a) The velocity as a function of t is the derivative of the position function:

$$v(t) = \dot{x}(t) = \frac{d}{dt}\left(4t^2 - 6t - 40\right) = 8t - 6 \text{ (in m/s)}$$

(b) The acceleration as a function of t is the derivative of the velocity function:

$$a(t) = \dot{v}(t) = \frac{d}{dt}(8t - 6) = 8 \text{ m/s}^2$$

Notice that the acceleration is constant (because it doesn't depend on t).

(c) The object is at the origin when $x(t)$ is equal to 0; that is, when

$$4t^2 - 6t - 40 = 0$$
$$2(2t^2 - 3t - 20) = 0$$
$$2(2t + 5)(t - 4) = 0$$
$$t = -\frac{5}{2} \text{ or } 4 \text{ seconds}$$

(d) Disregarding the negative value for t, we can say that the object passes through the origin at $t = 4$ s. At this time, the object's velocity is

$$v(4) \equiv v(t)\big|_{t=4} = (8t - 6)\big|_{t=4} = 8 \cdot 4 - 6 = 26 \text{ m/s}$$

The object's acceleration is a constant 8 m/s^2 throughout its motion, so, in particular, at $t = 4$ s, the acceleration is 8 m/s^2.

Example 2.24 An object is moving along the x-axis with an acceleration given by the function, $a(t) = (4t + 7)\ \text{m/s}^2$. At time $t_0 = 0$, the object is at $x = 6$ m, and it is moving at 2 m/s. How fast will the object be traveling at time $t = 4$ s? Where will the object be at time $t = 4$ s?

Solution. By integrating the acceleration with respect to time, we find the velocity as a function of time.

$$v(t) = \int a(t)dt = \int (4t + 7)dt = 2t^2 + 7t + c$$

We determine the value of c, the constant of integration, by using the given initial velocity, $v(0) = 2$ m/s.

$$v(0) = 2 \Rightarrow (2t^2 + 7t + c)\big|_{t=0} = 2 \Rightarrow c = 2$$

(c is just the initial velocity, v_0). So the velocity function is given by: $v(t) = 2t^2+7t+2$ (in m/s). Therefore, we can evaluate the function at $t = 4$ s and determine the velocity at that moment of time.

$$v(4) = (2t^2 + 7t + 2)\big|_{t=4} (2(4)^2 + 7(4) + 2) = 62\ \text{m/s}$$

Integrating this velocity function with respect to time will give us the position function.

$$x(t) = \int v(t)dt = \int (2t^2 + 7t + 2)dt = \frac{2}{3}t^3 + \frac{7}{2}t^2 + 2t + c_1$$

Again we determine the constant of integration, c_1, by using the given initial position, $x(0) = 6$ m.

$$x(0) = 6 \Rightarrow \left(\frac{2}{3}t^3 + \frac{7}{2}t^2 + 2t + c_1\right)\Big|_{t=0} = 6 \Rightarrow c_1 = 6$$

(c_1, is just the initial position, x_0). So the position function is given by:

$$x(t) = \left(\frac{2}{3}t^3 + \frac{7}{2}t^2 + 2t + 6\right) \text{in meters}$$

Therefore, we can evaluate the function at $t = 4$ s and determine the position at that moment of time.

$$x(4) = \left(\frac{2}{3}t^3 + \frac{7}{2}t^2 + 2t + 6\right)\Big|_{t=4} = \left(\frac{2}{3}(4)^3 + \frac{7}{2}(4)^2 + 2(4) + 6\right) = 112.\overline{66}\ \text{m}$$

CHAPTER 2 REVIEW QUESTIONS

SECTION I: MULTIPLE CHOICE

1. An object that's moving with constant speed travels once around a circular path. Which of the following is/are true concerning this motion?

    I. The displacement is zero.
    II. The average speed is zero.
    III. The acceleration is zero.

    (A) I only
    (B) I and II only
    (C) I and III only
    (D) III only
    (E) II and III only

2. At time $t = t_1$, an object's velocity is given by the vector $\mathbf{v}_1$ shown below:

    A short time later, at $t = t_2$, the object's velocity is the vector $\mathbf{v}_2$:

    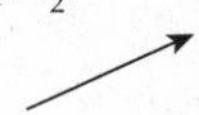

    If $v_2 = v_1$, which one of the following vectors best illustrates the object's average acceleration between $t = t_1$ and $t = t_2$?

    (A) 

    (B) 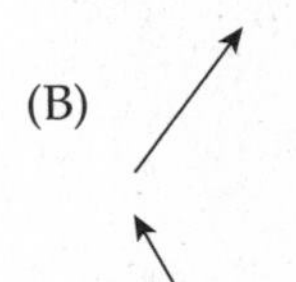

    (C)

    (D) 

    (E) 

3. Which of the following is/are true?

    I. If an object's acceleration is constant, then it must move in a straight line.
    II. If an object's acceleration is zero, then its speed must remain constant.
    III. If an object's speed remains constant, then its acceleration must be zero.

    (A) I and II only
    (B) I and III only
    (C) II only
    (D) III only
    (E) II and III only

4. A baseball is thrown straight upward. What is the ball's acceleration at its highest point?

    (A) 0
    (B) $\frac{1}{2}g$, downward
    (C) g, downward
    (D) $\frac{1}{2}g$, upward
    (E) g, upward

5. How long would it take a car, starting from rest and accelerating uniformly in a straight line at 5 m/s^2, to cover a distance of 200 m?

    (A) 9.0 s
    (B) 10.5 s
    (C) 12.0 s
    (D) 15.5 s
    (E) 20.0 s

6. A rock is dropped off a cliff and strikes the ground with an impact velocity of 30 m/s. How high was the cliff?

    (A) 15 m
    (B) 20 m
    (C) 30 m
    (D) 45 m
    (E) 60 m

7. A stone is thrown horizontally with an initial speed of 10 m/s from a bridge. If air resistance could be ignored, how long would it take the stone to strike the water 80 m below the bridge?

(A) 1 s
(B) 2 s
(C) 4 s
(D) 6 s
(E) 8 s

8. A soccer ball, at rest on the ground, is kicked with an initial velocity of 10 m/s at a launch angle of 30°. Calculate its total flight time, assuming that air resistance is negligible.

(A) 0.5 s
(B) 1 s
(C) 1.7 s
(D) 2 s
(E) 4 s

9. A stone is thrown horizontally with an initial speed of 30 m/s from a bridge. Find the stone's total speed when it enters the water 4 seconds later. (Ignore air resistance.)

(A) 30 m/s
(B) 40 m/s
(C) 50 m/s
(D) 60 m/s
(E) 70 m/s

10. Which one of the following statements is true concerning the motion of an ideal projectile launched at an angle of 45° to the horizontal?

(A) The acceleration vector points opposite to the velocity vector on the way up and in the same direction as the velocity vector on the way down.
(B) The speed at the top of the trajectory is zero.
(C) The object's total speed remains constant during the entire flight.
(D) The horizontal speed decreases on the way up and increases on the way down.
(E) The vertical speed decreases on the way up and increases on the way down.

11. The position of an object moving in a straight line is given by $x = (7 + 10t - 6t^3)$ m, where t is in seconds. What is the object's velocity at 4 seconds?

(A) −8 m/s
(B) −62 m/s
(C) 298 m/s
(D) −278 m/s
(E) none of the above

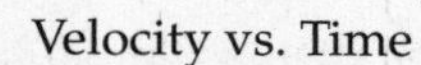

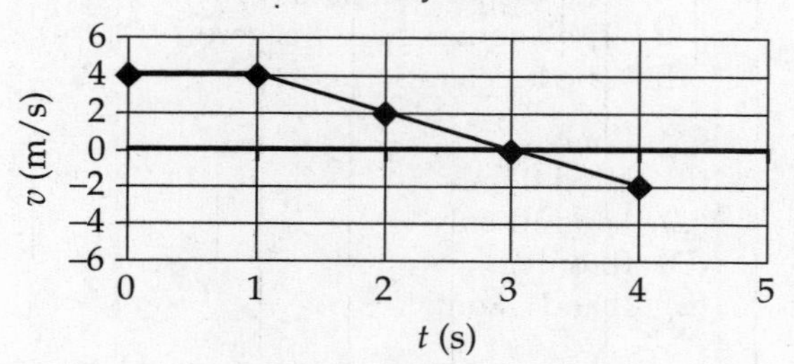

12. The velocity of an object moving is a straight line is graphed above. If $x = 3.0$ m at $t = 0$ s, what is the position of the particle at $t = 3.0$ s?

(A) 6 m
(B) 10 m
(C) 8 m
(D) 11 m
(E) −5 m

SECTION II: FREE RESPONSE

1. This question concerns the motion of a car on a straight track; the car's velocity as a function of time is plotted below.

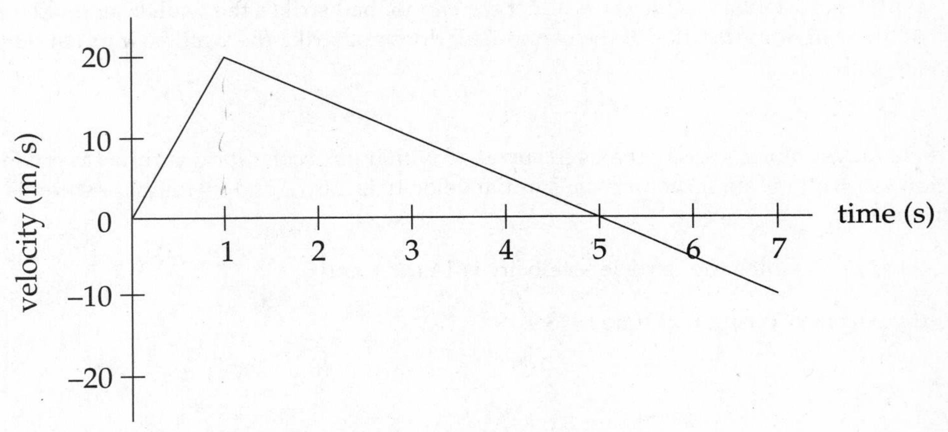

(a) Describe what happened to the car at time $t = 1$ s.

(b) How does the car's average velocity between time $t = 0$ and $t = 1$ s compare to its average velocity between times $t = 1$ s and $t = 5$ s?

(c) What is the displacement of the car from time $t = 0$ to time $t = 7$ s?

(d) Plot the car's acceleration during this interval as a function of time.

(e) Plot the object's position during this interval as a function of time. Assume that the car begins at $s = 0$.

2. Consider a projectile moving in a parabolic trajectory under constant gravitational acceleration. Its initial velocity has magnitude v_0, and its launch angle (with the horizontal) is θ_0. Solve the following in terms of given quantities and the acceleration of gravity, g.

(a) Calculate the maximum height, H, of the projectile.

(b) Calculate the (horizontal) range, R, of the projectile.

(c) For what value of θ_0 will the range be maximized?

(d) If $0 < h < H$, compute the time that elapses between passing through the horizontal line of height h in both directions (ascending and descending); that is, compute the time required for the projectile to pass through the two points shown in this figure:

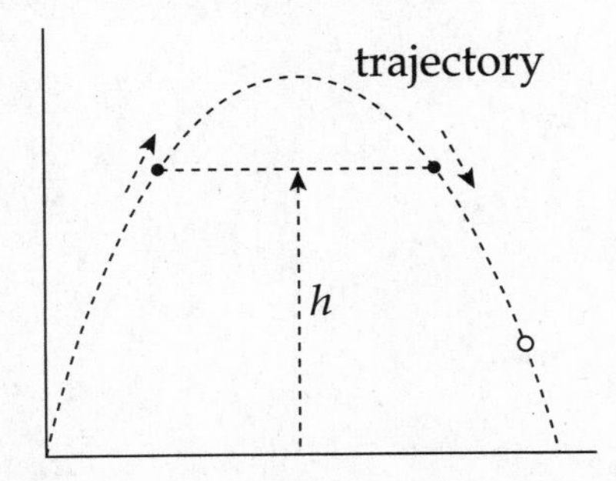

3. A cannonball is shot with an initial speed of 50 m/s at a launch angle of 40° toward a castle wall 220 m away. The height of the wall is 30 m. Assume that effects due to the air are negligible. (For this problem, use $g = 9.8 \text{ m/s}^2$.)

   (a) How long will it take the cannonball to reach the vertical plane of the wall?

   (b) Will the cannonball strike the wall? If the cannonball strikes the wall, how far below the top of the wall does it strike? If the cannonball does not strike the wall, how much does it clear the wall by?

4. A particle moves along a straight axis in such a way that its acceleration at time t is given by the equation $a(t) = 6t$ (m/s²). If the particle's initial velocity is 2 m/s and its initial position is $x = 4$ m, determine

   (a) the time at which the particle's velocity is 14 m/s, and

   (b) the particle's position at time $t = 3$ s.

KINEMATICS SUMMARY

DEFINITIONS

- **Position** refers to where an object is relative to a coordinate axes system.
- **Distance** refers to the total measure of the ground traveled by an object.
- **Displacement** is how far an object is from where it started, final position – initial position.

- $\text{average speed} = \dfrac{\text{distance}}{\text{time}}$
- $\bar{v} = \dfrac{\text{displacement}}{\text{time}}$
- $\bar{a} = \dfrac{\text{change in velocity}}{\text{time}}$

CONSTANT ACCELERATION EQUATIONS:

$$\Delta x = x - x_0 = \bar{v}t$$

$$v = v_0 + at$$

$$x = x_0 + v_0 t + \frac{1}{2}at^2$$

$$x = x_0 + vt - \frac{1}{2}at^2$$

$$v^2 = v_0^2 + 2a(x - x_0)$$

GRAPHS

- The slope of a position vs. time graph is the velocity.
- The area between the t-axis and the velocity function is the displacement.
- The slope of a velocity vs. time graph is the acceleration.

MOTION FUNCTIONS

- The derivate of $x(t)$ is the velocity.
- The integral of $v(t)$ is the displacement.
- The derivate of $v(t)$ is the acceleration.

FREE FALL AND PROJECTILES

- The acceleration due to gravity, g, is a constant 9.8 m/s^2 downward, for all objects close to the surface of the earth. Note that the test directions state that to simplify calculations you may use 10m/s^2 in all problems.
- It is important that you analyze the vertical (constant acceleration) and horizontal (constant velocity) motions separately. You can use the constant acceleration equations with a replaced by g and x replaced by y to indicate the acceleration is in the vertical direction. The motion in the x direction has a constant velocity, so the only equation you need for that is $x = v_x t$.

3

Newton's Laws

INTRODUCTION

In the previous chapter we studied the vocabulary and equations that describe motion. Now we will learn why things move the way they do; this is the subject of **dynamics**.

An interaction between two bodies—a push or a pull—is called a **force**. If you lift a book, you exert an upward force (created by your muscles) on it. If you pull on a rope that's attached to a crate, you create a *tension* in the rope that pulls the crate. When a skydiver is falling through the air, the earth is exerting a downward pull called *gravitational force*, and the air exerts an upward force called *air resistance*. When you stand on the floor, the floor provides an upward, supporting force called the *normal force*. If you slide a book across a table, the table exerts a *frictional force* against the book, so the book slows down and then stops. Static cling provides a directly observable example of the *electrostatic force*. Protons and neutrons are held together in the nuclei of atoms by the *strong nuclear force* and radioactive nuclei decay through the action of the *weak nuclear force*.

The Englishman Sir Isaac Newton published a book in 1687 called *The Mathematical Principles of Natural Philosophy*—referred to nowadays as simply *The Principia*—which began the modern study of physics as a scientific discipline. Three of the laws that Newton stated in *The Principia* form the basis for dynamics and are known simply as *Newton's Laws of Motion*.

THE FIRST LAW

Newton's First Law says that *an object will continue in its state of motion unless compelled to change by a force impressed upon it*. That is, unless an unbalanced force acts on an object, the object's velocity will not change: If the object is at rest, then it will stay at rest; and if it is moving, then it will continue to move at a constant speed in a straight line.

Basically, no force means no change in velocity. This property of objects, their natural resistance to changes in their state of motion, is called **inertia**. In fact, the First Law is often referred to as the **Law of Inertia**.

THE SECOND LAW

Newton's Second Law predicts what will happen when a force *does* act on an object: The object's velocity will change; the object will accelerate. More precisely, it says that its acceleration, **a**, will be directly proportional to the strength of the total—or *net*—force ($\mathbf{F}_{net}$) and inversely proportional to the object's mass, m:

$$\mathbf{F}_{net} = m\mathbf{a}$$

This is the most important equation in mechanics!

The **mass** of an object is the quantitative measure of its inertia; intuitively, it measures how much matter is contained in an object. Two identical boxes, one empty and one full, have different masses. The box that's full has the greater mass, because it contains more stuff; more stuff, more mass. Mass is measured in *kilograms*, abbreviated kg. (Note: An object whose mass is 1 kg weighs about 2.2 pounds.) It takes twice as much force to produce the same change in velocity of a 2 kg object than of a 1 kg object. Mass is a measure of an object's inertia, its resistance to acceleration.

Forces are represented by vectors; they have magnitude and direction. If several different forces act on an object simultaneously, then the net force, $\mathbf{F}_{net}$, is the vector sum of all these forces. (The phrase *resultant force* is also used to mean *net force*.)

Since $\mathbf{F}_{net} = m\mathbf{a}$, and m is a *positive* scalar, the direction of **a** always matches the direction of $\mathbf{F}_{net}$. Finally, since $F = ma$, the units for F equal the units of m times the units of a:

$$\begin{aligned}[F] &= [m][a] \\ &= \text{kg}\cdot\text{m/s}^2\end{aligned}$$

A force of 1 $\text{kg}\cdot\text{m/s}^2$ is renamed 1 **newton** (abbreviated N). A medium-sized apple weighs about 1 N.

THE THIRD LAW

This is the law that's commonly remembered as, *to every action, there is an equal, but opposite, reaction*. More precisely, if Object 1 exerts a force on Object 2, then Object 2 exerts a force back on Object 1, equal in strength but opposite in direction. These two forces, $\mathbf{F}_{1\text{-on-}2}$ and $\mathbf{F}_{2\text{-on-}1}$, are called an **action/reaction pair**.

Example 3.1 What net force is required to maintain a 5000 kg object moving at a constant velocity of magnitude 7500 m/s?

Solution. The First Law says that any object will *continue* in its state of motion unless a force acts on it. Therefore, *no* net force is required to maintain a 5000 kg object moving at a constant velocity of magnitude 7500 m/s. Here's another way to look at it: Constant velocity means $\mathbf{a} = \mathbf{0}$, so the equation $\mathbf{F}_{net} = m\mathbf{a}$ immediately gives $\mathbf{F}_{net} = \mathbf{0}$.

Example 3.2 How much force is required to cause an object of mass 2 kg to have an acceleration of 4 m/s²?

Solution. According to the Second Law, $F_{net} = ma = (2 \text{ kg})(4 \text{ m/s}^2) = 8 \text{ N}$.

Example 3.3 An object feels two forces; one of strength 8 N pulling to the left and one of strength 20 N pulling to the right. If the object's mass is 4 kg, what is its acceleration?

Solution. Forces are represented by vectors and can be added and subtracted. Therefore, an 8 N force to the left added to a 20 N force to the right yields a net force of 20 – 8 = 12 N to the right. Then Newton's Second Law gives $\mathbf{a} = \mathbf{F}_{net}/m = (12 \text{ N to the right})/(4 \text{ kg}) = 3 \text{ m/s}^2$ to the right.

WEIGHT

Mass and weight are not the same thing—there is a clear distinction between them in physics—but they are often used interchangeably in everyday life. The **weight** of an object is the gravitational force exerted on it by the earth (or by whatever planet it happens to be on). Mass, by contrast, is an intrinsic property of an object that measures its inertia. An object's mass does not change with location. Put a baseball in a rocket and send it to the Moon. The baseball's *weight* on the Moon is less than its weight here on Earth (because the Moon's gravitational pull is weaker than the earth's due to its much smaller mass), but the baseball's mass would be the same.

Since weight is a force, we can use $\mathbf{F} = m\mathbf{a}$ to compute it. What acceleration would the gravitational force impose on an object? The gravitational acceleration, of course! Therefore, setting $\mathbf{a} = \mathbf{g}$, the equation $\mathbf{F} = m\mathbf{a}$ becomes

$$\mathbf{F}_w = m\mathbf{g}$$

This is the equation for the weight of an object of mass m. ($\mathbf{F}_g$ and $\mathbf{F}_w$, are both commonly used to represent the force of gravity. Weight is often symbolized merely by $\mathbf{w}$, rather than $\mathbf{F}_w$.) Notice that mass and weight are proportional but not identical. Furthermore, mass is measured in kilograms, while weight is measured in newtons.

Example 3.4 What is the mass of an object that weighs 500 N?

Solution. Since weight is m multiplied by g, mass is F_w (weight) divided by g. Therefore,

$$m = F_w/g = (500 \text{ N})/(10 \text{ m/s}^2) = 50 \text{ kg}$$

Example 3.5 A person weighs 150 pounds. Given that a pound is a unit of weight equal to 4.45 N, what is this person's mass?

Solution. This person's weight in newtons is (150 lb)(4.45 N/lb) = 667.5 N, so his mass is

$$m = F_w/g = (667.5 \text{ N})/(10 \text{ m/s}^2) = 66.75 \text{ kg}$$

Example 3.6 A book whose mass is 2 kg rests on a table. Find the magnitude of the force exerted by the table on the book.

Solution. The book experiences two forces: The downward pull of the earth's gravity and the upward, supporting force exerted by the table. Since the book is at rest on the table, its acceleration is zero, so the net force on the book must be zero. Therefore the magnitude of the support force must equal the magnitude of the book's weight, which is $F_w = mg = (2 \text{ kg})(10 \text{ m/s}^2) = 20$ N.

Example 3.7 A can of paint with a mass of 6 kg hangs from a rope. If the can is to be pulled up to a rooftop with an acceleration of 1 m/s², what must the tension in the rope be?

Solution. First draw a picture. Represent the object of interest (the can of paint) as a heavy dot, and draw the forces that act on the object as arrows connected to the dot. This is called a **free-body** (or **force**) **diagram**.

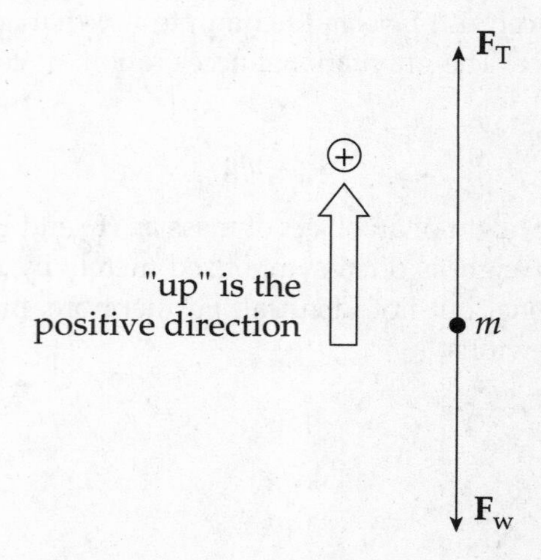

We have the tension force in the rope, $\mathbf{F}_T$ (also symbolized merely by **T**), which is upward, and the weight, $\mathbf{F}_w$, which is downward. Calling *up* the positive direction, the net force is $F_T - F_w$. The Second Law, $F_{net} = ma$, becomes $F_T - F_w = ma$, so

$$F_T = F_w + ma = mg + ma = m(g + a) = 6(10 + 1) = 66 \text{ N}$$

> **Example 3.8** A can of paint with a mass of 6 kg hangs from a rope. If the can is to be pulled up to a rooftop with a constant velocity of 1 m/s, what must the tension in the rope be?

Solution. The phrase "constant velocity" automatically means $a = 0$ and, therefore, $F_{net} = 0$. In the diagram above, $\mathbf{F}_T$ would need to have the *same* magnitude as $\mathbf{F}_w$ in order to keep the can moving at a constant velocity. Thus, in this case, $F_T = F_w = mg = (6)(10) = 60$ N.

> **Example 3.9** What force must be exerted to lift a 50 N object with an acceleration of 10 m/s²?

Solution. First draw a free-body diagram:

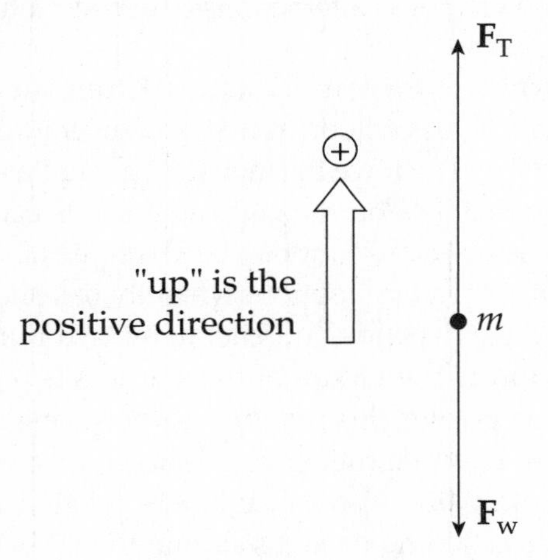

We have the tension force, $\mathbf{F}_T$, which is upward, and the weight, $\mathbf{F}_w$, which is downward. Calling *up* the positive direction, the net force is $F_T - F_w$. The Second Law, $F_{net} = ma$, becomes $F_T - F_w = ma$, so $F_T = F_w + ma$. Remembering that $m = F_w/g$, we find that

$$F_T = F_w + ma = F_w + \frac{F_w}{g}a = 50 \text{ N} + \frac{50 \text{ N}}{10 \text{ m/s}^2}(10 \text{ m/s}^2) = 100 \text{ N}$$

THE NORMAL FORCE

When an object is in contact with a surface, the surface exerts a contact force on the object. The component of the contact force that's *perpendicular* to the surface is called the **normal force** on the object. (In physics, the word *normal* means *perpendicular*.) The normal force is what prevents objects from falling through tabletops or you from falling through the floor. The normal force is denoted by $\mathbf{F}_N$, or simply by **N**. (If you use the latter notation, be careful not to confuse it with N, the abbreviation for the newton.)

> **Example 3.10** A book whose mass is 2 kg rests on a table. Find the magnitude of the normal force exerted by the table on the book.

Solution. The book experiences two forces: The downward pull of Earth's gravity and the upward, supporting force exerted by the table. Since the book is at rest on the table, its acceleration is zero, so the net force on the book must be zero. Therefore, the magnitude of the support force must equal the magnitude of the book's weight, which is $F_w = mg = (2)(10) = 20$ N. This means the normal force must be 20 N as well: $F_N = 20$ N. (Note that this is a repeat of Example 3.6, except now we have a name for the "upward, supporting force exerted by the table"; it's called the **normal force**.)

FRICTION

When an object is in contact with a surface, the surface exerts a contact force on the object. The component of the contact force that's *parallel* to the surface is called the **friction force** on the object. Friction, like the normal force, arises from electrical interactions between atoms that comprise the object and those that comprise the surface.

We'll look at two main categories of friction: (1) **static friction** and (2) **kinetic (sliding) friction**. If you attempt to push a heavy crate across a floor, at first you meet with resistance, but then you push hard enough to get the crate moving. The force that acted on the crate to cancel out your initial pushes was static friction, and the force that acts on the crate as it slides across the floor is kinetic friction. Static friction occurs when there is no relative motion between the object and the surface (no sliding); kinetic friction occurs when there *is* relative motion (when there's sliding).

The strength of the friction force depends, in general, on two things: The nature of the surfaces and the strength of the normal force. The nature of the surfaces is represented by the **coefficient of friction,** denoted by μ (*mu*). The greater this number is, the stronger the friction force will be. For example, the coefficient of friction between rubber-soled shoes and a wooden floor is 0.7, but between rubber-soled shoes and ice, it's only 0.1. Also, since kinetic friction is generally weaker than static friction (it's easier to keep an object sliding once it's sliding than it is to start the object sliding in the first place), there are two coefficients of friction; one for static friction (μ_s) and one for kinetic friction (μ_k). For a given pair of surfaces, it's virtually always true that $\mu_k < \mu_s$. The strengths of these two types of friction forces are given by the following equations:

$$F_{\text{static friction, max}} = \mu_s F_N$$

$$F_{\text{kinetic friction}} = \mu_k F_N$$

Note that the equation for the strength of the static friction force is for the *maximum* value only. This is because static friction can vary, precisely counteracting weaker forces that attempt to move an object. For example, suppose an object feels a normal force of $F_N = 100$ N and the coefficient of static friction between it and the surface it's on is 0.5. Then, the *maximum* force that static friction can exert is (0.5)(100 N) = 50 N. However, if you push on the object with a force of, say, 20 N, then the static friction force will be 20 N (in the opposite direction), *not* 50 N; the object won't move. The net force on a stationary object must be zero. Static friction can take on all values, up to a certain maximum, and you must overcome the maximum static friction force to get the object to slide. The direction of $\mathbf{F}_{\text{kinetic friction}} = \mathbf{F}_{\text{f (kinetic)}}$ is opposite to that of motion (sliding), and the direction of $\mathbf{F}_{\text{static friction}} = \mathbf{F}_{\text{f (static)}}$ is opposite to that of the intended motion.

> **Example 3.11** A crate of mass 20 kg is sliding across a wooden floor. The coefficient of kinetic friction between the crate and the floor is 0.3.
>
> (a) Determine the strength of the friction force acting on the crate.
>
> (b) If the crate is being pulled by a force of 90 N (parallel to the floor), find the acceleration of the crate.

Solution. First draw a free-body diagram:

In part (a) $F = 0$, and in part (b) $F = 90$N for our free-body diagram.
Reminder: separate the horizontal and vertical forces and use $\Sigma F_x = ma_x$ and $\Sigma F_y = ma_y$.

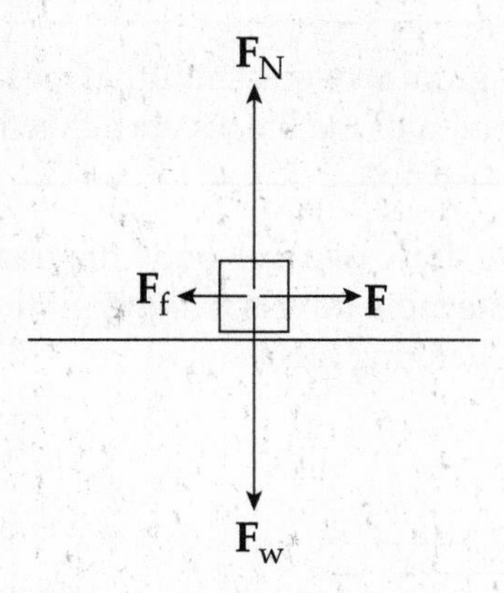

(a) The normal force on the object balances the object's weight, so $F_N = mg = (20\text{ kg})(10\text{ m/s}^2) = 200\text{ N}$. Therefore, $F_{\text{(kinetic)}} = \mu_k F_N = (0.3)(200\text{ N}) = 60\text{ N}$.

(b) The net horizontal force that acts on the crate is $F - F_f = 90\text{ N} - 60\text{ N} = 30\text{ N}$, so the acceleration of the crate is $a = F_{\text{net}}/m = (30\text{ N})/(20\text{ kg}) = 1.5\text{ m/s}^2$.

> **Example 3.12** A crate of mass 100 kg rests on the floor. The coefficient of static friction is 0.4. If a force of 250 N (parallel to the floor) is applied to the crate, what's the magnitude of the force of static friction on the crate?

Solution. The normal force on the object balances its weight, so $F_N = mg = (100\text{ kg})(10\text{ m/s}^2) = 1{,}000\text{ N}$. Therefore, $F_{\text{static friction, max}} = F_{\text{f (static), max}} = \mu_s F_N = (0.4)(1{,}000\text{ N}) = 400\text{ N}$. This is the *maximum* force that static friction can exert, but in this case it's *not* the actual value of the static friction force. Since the applied force on the crate is only 250 N, which is less than the $F_{\text{f (static), max}}$, the force of static friction will be less also: $F_{\text{f (static)}} = 250\text{ N}$, and the crate will not slide.

PULLEYS

Pulleys are devices that change the direction of the tension force in the cords that slide over them. Here we'll consider each pulley to be frictionless and massless, which means that their masses are so much smaller than the objects of interest in the problem that they can be ignored.

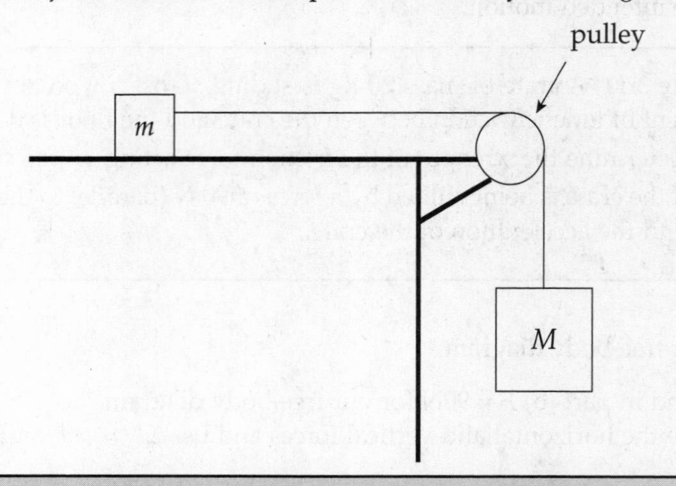

Example 3.13 In the diagram above, assume that the tabletop is frictionless. Determine the acceleration of the blocks once they're released from rest.

Solution. There are two blocks, so we draw two free-body diagrams: The positive directions for each block must coincide. If the block on the table travels to the right then the hanging block travels down. This is why down is positive for the hanging block.

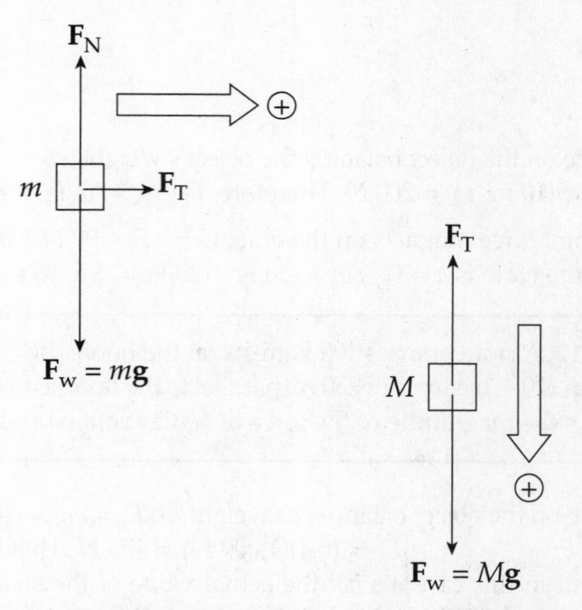

To get the acceleration of each one, we use Newton's Second Law, $\mathbf{F}_{net} = m\mathbf{a}$.

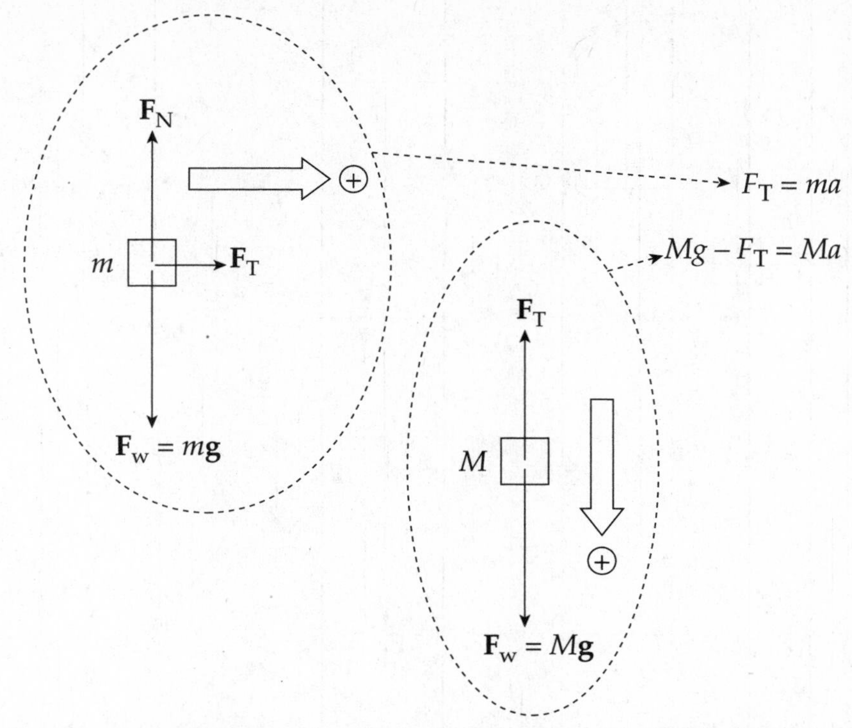

Note that there are two unknowns, F_T and a, but we can eliminate F_T by adding the two equations, and then we can solve for a.

$$\begin{aligned} F_T &= ma \\ Mg - F_T &= Ma \\ \hline Mg &= ma + Ma \\ &= a(m + M) \end{aligned}$$

Add the equations to eliminate F_T.

$$\frac{Mg}{m + M} = a$$

> **Example 3.14** Using the same diagram as in the previous example, assume that $m = 2$ kg, $M = 10$ kg, and the coefficient of kinetic friction between the small block and the tabletop is 0.5. Compute the acceleration of the blocks.

Solution. Once again, draw a free-body diagram for each object. Note that the only difference between these diagrams and the ones in the previous example is the inclusion of the force of (kinetic) friction, $\mathbf{F}_f$, that acts on the block on the table.

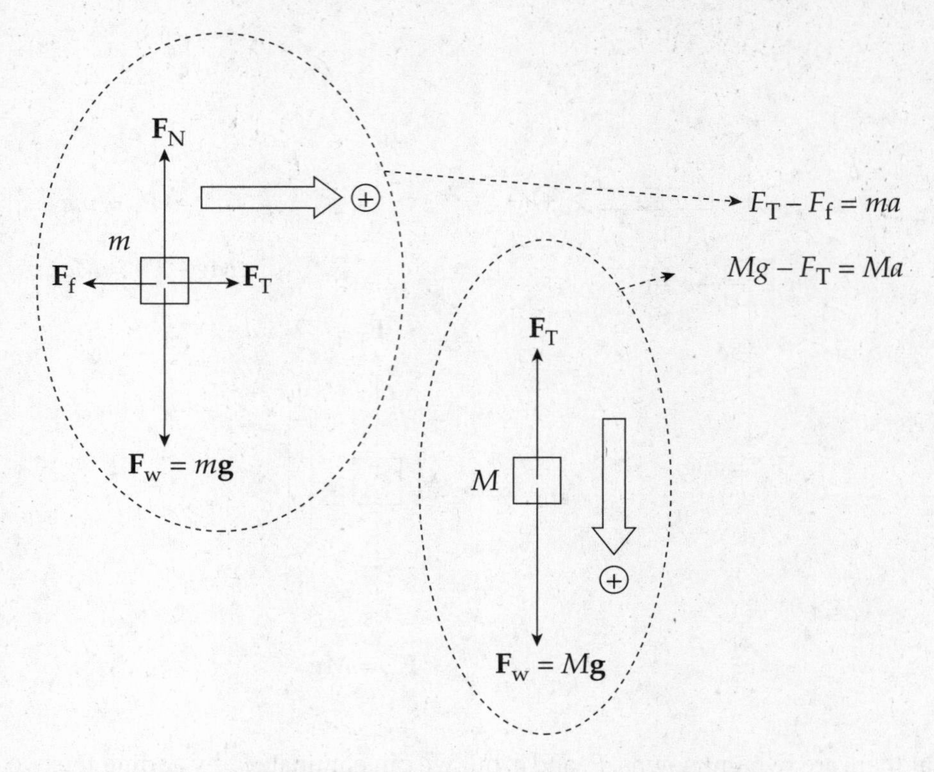

As before, we have two equations that contain two unknowns (a and F_T):

$$F_T - F_f = ma \quad (1)$$

$$Mg - F_T = Ma \quad (2)$$

Add the equations (thereby eliminating F_T) and solve for a. Note that, by definition, $F_f = \mu F_N$, and from the free-body diagram for m, we see that $F_N = mg$, so $F_f = \mu mg$:

$$Mg - F_f = ma + Ma$$

$$Mg - \mu mg = a(m + M)$$

$$\frac{M - \mu m}{m + M} g = a$$

Substituting in the numerical values given for m, M, and μ, we find that $a = \frac{3}{4}g$ (or 7.5 m/s²).

> **Example 3.15** In the previous example, calculate the strength of the tension in the cord.

Solution. Since the value of a has been determined, we can use either of the two original equations to calculate F_T. Using Equation (2), $Mg - F_T = Ma$ (because it's simpler), we find

$$F_T = Mg - Ma = Mg - M \cdot \frac{3}{4}g = \frac{1}{4}Mg = \frac{1}{4}(10)(10) = 25 \text{ N}$$

As you can see, we would have found the same answer if Equation (1) had been used:

$$F_T - F_f = ma \Rightarrow F_T = F_f + ma = \mu mg + ma = \mu mg + m \cdot \frac{3}{4}g = mg\left(\mu + \frac{3}{4}\right)$$
$$= (2)(10)(0.5 + 0.75)$$
$$= 25 \text{ N}$$

INCLINED PLANES

An **inclined plane** is basically a ramp. If an object of mass m is on the ramp, then the force of gravity on the object, $\mathbf{F}_w = m\mathbf{g}$, has two components: One that's parallel to the ramp ($mg \sin \theta$) and one that's normal to the ramp ($mg \cos \theta$), where θ is the incline angle. The force driving the block down the inclined plane is the component of the block's weight that's parallel to the ramp: $mg \sin \theta$.

When analyzing objects moving up or down inclined planes it is almost always easiest to rotate the coordinate axes such that the x-axis is parallel to the incline and the y-axis is perpendicular to the incline, as shown in the diagram. The object would accelerate in both the x and y directions as it moved down along the incline if you did not rotate the axis. However, with the rotated axes the acceleration in the y-direction is zero. Now we only have to worry about the acceleration in the x-direction.

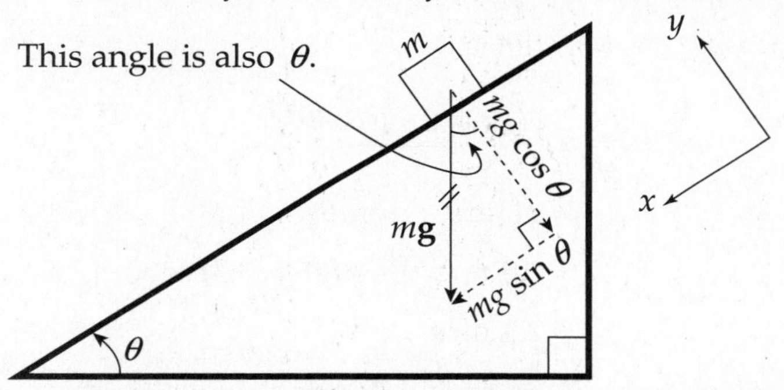

> **Example 3.16** A block slides down a frictionless, inclined plane that makes a 30° angle with the horizontal. Find the acceleration of this block.

Solution. Let m denote the mass of the block, so the force that pulls the block down the incline is $mg \sin \theta$, and the block's acceleration down the plane is

$$a = \frac{F}{m} = \frac{mg \sin \theta}{m} = g \sin \theta = g \sin 30^\circ = \frac{1}{2}g = 5 \text{ m/s}^2$$

> **Example 3.17** A block slides down an inclined plane that makes a 30° angle with the horizontal. If the coefficient of kinetic friction is 0.3, find the acceleration of the block.

Solution. First draw a free-body diagram. Notice that, in the diagram shown below, the weight of the block, $\mathbf{F}_w = m\mathbf{g}$, has been written in terms of its scalar components: $F_w \sin\theta$ parallel to the ramp and $F_w \cos\theta$ normal to the ramp:

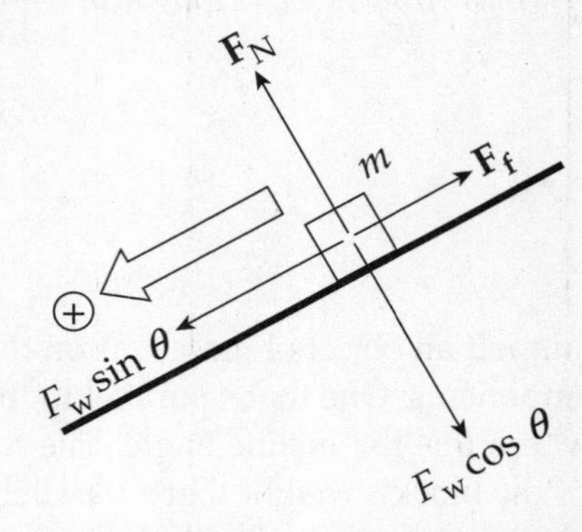

The force of friction, $\mathbf{F}_f$, that acts up the ramp (opposite to the direction in which the block slides) has magnitude $F_f = \mu F_N$. But the diagram shows that $F_N = F_w \cos\theta$, so $F_f = \mu(mg\cos\theta)$. Therefore the net force down the ramp is

$$F_w \sin\theta - F_f = mg\sin\theta - \mu mg\cos\theta = mg(\sin\theta - \mu\cos\theta)$$

Then, setting F_{net} equal to ma, we solve for a:

$$\begin{aligned} a = \frac{F_{net}}{m} &= \frac{mg(\sin\theta - \mu\cos\theta)}{m} \\ &= g(\sin\theta - \mu\cos\theta) \\ &= (10\text{ m/s}^2)(\sin 30^\circ - 0.3\cos 30^\circ) \\ &= 2.4\text{ m/s}^2 \end{aligned}$$

UNIFORM CIRCULAR MOTION

In Chapter 2, we considered two types of motion; straight-line motion and parabolic motion. We will now look at motion that follows a circular path, such as a rock on the end of a string, a horse on a merry-go-round, and (to a good approximation) the Moon around Earth and Earth around the Sun.

Let's simplify matters and consider the object's speed around its path to be constant. This is called **uniform circular motion**. You should remember that although the speed may be constant, the velocity is not, because the direction of the velocity is always changing. Since the velocity is changing, there must be acceleration. This acceleration does not change the speed of the object; it only changes the direction of the velocity to keep the object on its circular path. Also, in order to produce an acceleration, there must be a force; otherwise, the object would move off in a straight line (Newton's First Law).

The figure on the left on the next page shows an object moving along a circular trajectory, along with its velocity vectors at two nearby points. The vector $\mathbf{v}_1$ is the object's velocity at time $t = t_1$, and $\mathbf{v}_2$ is the object's velocity vector a short time later (at time $t = t_2$). The velocity vector is always tangential to the object's path (whatever the shape of the trajectory). Notice that since we are assuming constant speed, the lengths of $\mathbf{v}_1$ and $\mathbf{v}_2$ (their magnitudes) are the same.

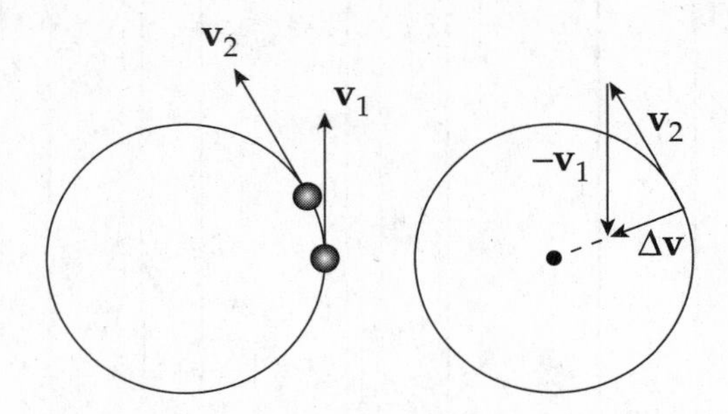

Since $\Delta\mathbf{v} = \mathbf{v}_2 - \mathbf{v}_1$ points toward the center of the circle (see the figure on the right), so does the acceleration, since $\mathbf{a} = \Delta\mathbf{v}/\Delta t$. Because the acceleration vector points toward the center of the circle, it's called **centripetal acceleration**, or $\mathbf{a}_c$. The centripetal acceleration is what turns the velocity vector to keep the object traveling in a circle. The magnitude of the centripetal acceleration depends on the object's speed, v, and the radius, r, of the circular path according to the equation

$$a_c = \frac{v^2}{r}$$

> **Example 3.18** An object of mass 5 kg moves at a constant speed of 6 m/s in a circular path of radius 2 m. Find the object's acceleration and the net force responsible for its motion.

Solution. By definition, an object moving at constant speed in a circular path is undergoing uniform circular motion. Therefore, it experiences a centripetal acceleration of magnitude v^2/r, always directed toward the center of the circle:

$$a_c = \frac{v^2}{r} = \frac{(6\text{ m/s})^2}{2\text{ m}} = 18\text{ m/s}^2$$

The force that produces the centripetal acceleration is given by Newton's Second Law, coupled with the equation for centripetal acceleration:

$$F_c = ma_c = m\frac{v^2}{r}$$

This equation gives the magnitude of the force. As for the direction, recall that because $\mathbf{F} = m\mathbf{a}$, the directions of **F** and **a** are always the same. Since centripetal acceleration points toward the center of the circular path, so does the force that produces it. Therefore, it's called **centripetal force**. The centripetal force acting on this object has a magnitude of $F_c = ma_c = (5\text{ kg})(18\text{ m/s}^2) = 90\text{ N}$.

Two common types of circular motion are offered here. The diagrams below show examples of a ball on a string traveling in a *horizontal circle* and a *vertical circle*.

Horizontal Circle

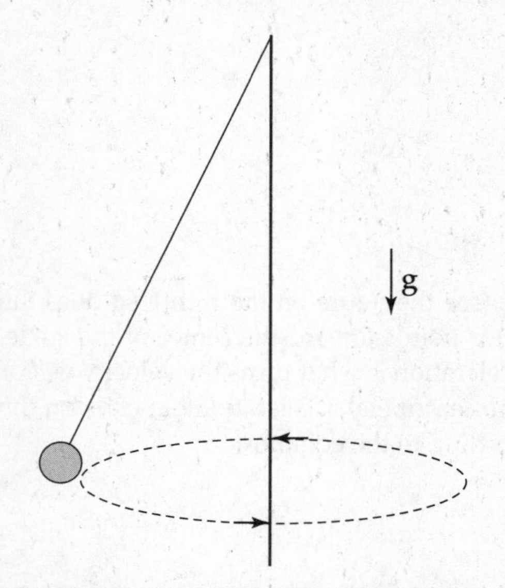

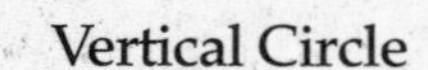

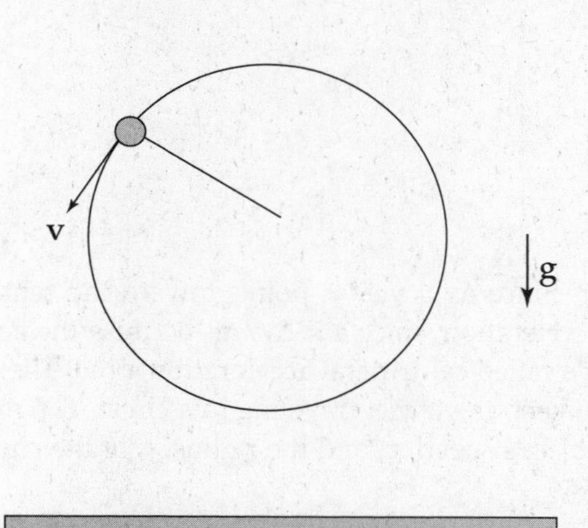

Example 3.19 A 10.0 kg mass is attached to a string that has a breaking strength of 200 N. If the mass is whirled in a horizontal circle of radius 80 cm, what maximum speed can it have?

Solution. The first thing to do in problems like this is to identify what force(s) provide the centripetal force. In this example, the tension in the string provides the centripetal force:

$$\mathbf{F}_T \text{ provides } \mathbf{F}_c \Rightarrow F_T = \frac{mv^2}{r} \Rightarrow v = \sqrt{\frac{rF_T}{m}} \Rightarrow v_{max} = \sqrt{\frac{rF_{T,\,max}}{m}}$$

$$= \sqrt{\frac{(0.80 \text{ m})(200 \text{ N})}{10 \text{ kg}}}$$

$$= 4 \text{ m/s}$$

Example 3.20 An athlete who weighs 800 N is running around a curve at a speed of 5.0 m/s in an arc whose radius of curvature, r, is 5.0 m. Find the centripetal force acting on him. What provides the centripetal force? What could happen to him if r were smaller?

Solution. Using the equation for the strength of the centripetal force, we find that

$$F_c = m\frac{v^2}{r} = \frac{F_w}{g} \cdot \frac{v^2}{r} = \frac{800 \text{ N}}{10 \text{ N/kg}} \cdot \frac{(5.0 \text{ m/s})^2}{5.0 \text{ m}} = 400 \text{ N}$$

In this case, static friction provides the centripetal force. Since the coefficient of static friction between his shoes and the ground is most likely around 1, the maximum force that static friction can exert is $\mu_s F_N \approx F_N = F_w = 800$ N. Fortunately, 800 N is greater than 400 N. But notice that if the radius of curvature of the arc were much smaller, then F_c would become greater than what static friction could handle, and he would slip.

> **Example 3.21** A roller-coaster car enters the circular-loop portion of the ride. At the very top of the circle (where the people in the car are upside down), the speed of the car is 25 m/s, and the acceleration points straight down. If the diameter of the loop is 50 m and the total mass of the car (plus passengers) is 1200 kg, find the magnitude of the normal force exerted by the track on the car at this point. Also find the normal force exerted by the track on the car when it is at the bottom of the loop. Assume it is still traveling 25 m/s at that location too.

Solution. When analyzing circular motion, consider all forces pointing toward the center to be positive and all forces pointing away to be negative. There are two forces acting on the car at its topmost point: the normal force exerted by the track and the gravitational force, both of which point downward.

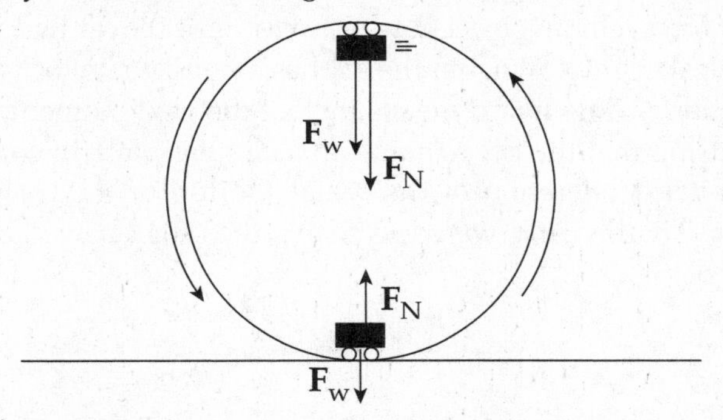

The combination of these two forces, $\mathbf{F}_N + \mathbf{F}_w$, provides the centripetal force:

$$F_N + F_w = \frac{mv^2}{r} \Rightarrow F_N = \frac{mv^2}{r} - F_w$$

$$= \frac{mv^2}{r} - mg$$

$$= m\left(\frac{v^2}{r} - g\right)$$

$$= (1200 \text{ kg})\left[\frac{(25 \text{ m/s})^2}{\frac{1}{2}(50 \text{ m})} - 10 \text{ m/s}^2\right]$$

$$= 1.8 \times 10^4 \text{ N}$$

Now we will determine the normal force exerted by the track on the car at the bottom of the loop.

$$F_c = \frac{mv^2}{r}$$

$$F_N - F_w = \frac{mv^2}{r} \Rightarrow F_N = \frac{mv^2}{r} + F_w$$

$$F_N = \frac{mv^2}{r} + mg$$

$$F_N = \frac{1200(25)^2}{25} + 1200(10)$$

$$F_N = 4.2 \times 10^4 \text{ N}$$

Notice that the normal force is much greater when the car is at the bottom of the track than when it is at the top. When it is at the top of the track gravity is helping the car travel in a circle.

Example 3.22 In the previous example, if the net force on the car at its topmost point is straight down, why doesn't the car fall straight down?

Solution. Remember that force tells an object how to *accelerate*. If the car had zero velocity at this point, then it would certainly fall straight down, but the car has a non-zero velocity (to the left) at this point. The fact that the acceleration is downward means that, at the next moment, **v** will point down to the left at a slight angle, ensuring that the car remains on a circular path, in contact with the track.

The minimum centripital acceleration of the car at the top of the track would be equal to the acceleration of gravity, $g = 9.8 \text{ m/s}^2$. If a_c were less than g then the car would fall off its circular path.

CHAPTER 3 REVIEW QUESTIONS

SECTION I: MULTIPLE CHOICE

1. A person standing on a horizontal floor feels two forces: the downward pull of gravity and the upward supporting force from the floor. These two forces

   (A) have equal magnitudes and form an action/reaction pair
   (B) have equal magnitudes but do not form an action/reaction pair
   (C) have unequal magnitudes and form an action/reaction pair
   (D) have unequal magnitudes and do not form an action/reaction pair
   (E) None of the above

2. A person who weighs 800 N steps onto a scale that is on the floor of an elevator car. If the elevator accelerates upward at a rate of 5 m/s^2, what will the scale read?

   (A) 400 N
   (B) 800 N
   (C) 1000 N
   (D) 1200 N
   (E) 1600 N

3. A frictionless inclined plane of length 20 m has a maximum vertical height of 5 m. If an object of mass 2 kg is placed on the plane, which of the following best approximates the net force it feels?

   (A) 5 N
   (B) 10 N
   (C) 15 N
   (D) 20 N
   (E) 30 N

4. A 20 N block is being pushed across a horizontal table by an 18 N force. If the coefficient of kinetic friction between the block and the table is 0.4, find the acceleration of the block.

   (A) 0.5 m/s^2
   (B) 1 m/s^2
   (C) 5 m/s^2
   (D) 7.5 m/s^2
   (E) 9 m/s^2

5. The coefficient of static friction between a box and a ramp is 0.5. The ramp's incline angle is 30°. If the box is placed at rest on the ramp, the box will

   (A) accelerate down the ramp
   (B) accelerate briefly down the ramp but then slow down and stop
   (C) move with constant velocity down the ramp
   (D) not move
   (E) Cannot be determined from the information given

6. 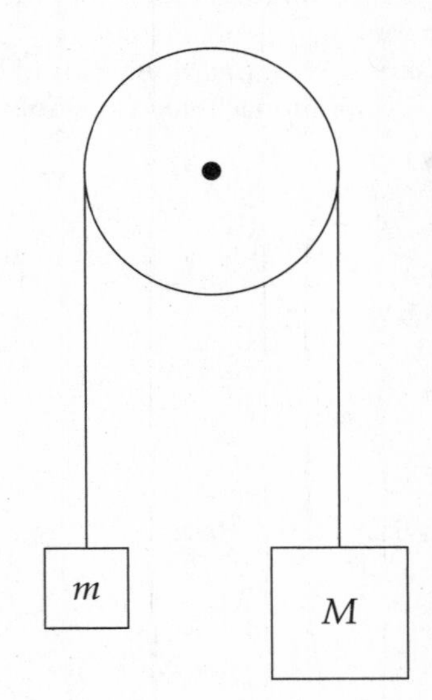

   Assuming a frictionless, massless pulley, determine the acceleration of the blocks once they are released from rest.

   (A) $\frac{m}{M+m}g$
   (B) $\frac{M}{M+m}g$
   (C) $\frac{M}{m}g$
   (D) $\frac{M+m}{M-m}g$
   (E) $\frac{M-m}{M+m}g$

7. If all of the forces acting on an object balance so that the net force is zero, then

(A) the object must be at rest
(B) the object's speed will decrease
(C) the object will follow a parabolic trajectory
(D) the object's direction of motion can change, but not its speed
(E) None of the above

8. A block of mass m is at rest on a frictionless, horizontal table placed in a laboratory on the surface of the earth. An identical block is at rest on a frictionless, horizontal table placed on the surface of the Moon. Let **F** be the net force necessary to give the Earth-bound block an acceleration of **a** across the table. Given that g_{Moon} is one-sixth of g_{Earth}, the force necessary to give the Moon-bound block the same acceleration **a** across the table is

(A) **F**/12
(B) **F**/6
(C) **F**/3
(D) **F**
(E) 6**F**

9. A crate of mass 100 kg is at rest on a horizontal floor. The coefficient of static friction between the crate and the floor is 0.4, and the coefficient of kinetic friction is 0.3. A force **F** of magnitude 344 N is then applied to the crate, parallel to the floor. Which of the following is true?

(A) The crate will accelerate across the floor at 0.5 m/s^2.
(B) The static friction force, which is the reaction force to **F** as guaranteed by Newton's Third Law, will also have a magnitude of 344 N.
(C) The crate will slide across the floor at a constant speed of 0.5 m/s.
(D) The crate will not move.
(E) None of the above

10. An object moves at constant speed in a circular path. Which of the following statements is/are true?

I. The velocity is constant.
II. The acceleration is constant.
III. The net force on the object is zero since its speed is constant.

(A) II only
(B) I and III only
(C) II and III only
(D) I and II only
(E) None of the above

Questions 11–12:

A 60 cm rope is tied to the handle of a bucket which is then whirled in a vertical circle. The mass of the bucket is 3 kg.

11. At the lowest point in its path, the tension in the rope is 50 N. What is the speed of the bucket?

(A) 1 m/s
(B) 2 m/s
(C) 3 m/s
(D) 4 m/s
(E) 5 m/s

12. What is the critical speed below which the rope would become slack when the bucket reaches the highest point in the circle?

(A) 0.6 m/s
(B) 1.8 m/s
(C) 2.4 m/s
(D) 3.2 m/s
(E) 4.8 m/s

13. An object moves at a constant speed in a circular path of radius r at a rate of 1 revolution per second. What is its acceleration?

(A) 0
(B) $2\pi^2 r$
(C) $2\pi^2 r^2$
(D) $4\pi^2 r$
(E) $4\pi^2 r^2$

SECTION II: FREE RESPONSE

1. This question concerns the motion of a crate being pulled across a horizontal floor by a rope. In the diagram below, the mass of the crate is m, the coefficient of kinetic friction between the crate and the floor is μ, and the tension in the rope is $\mathbf{F}_T$.

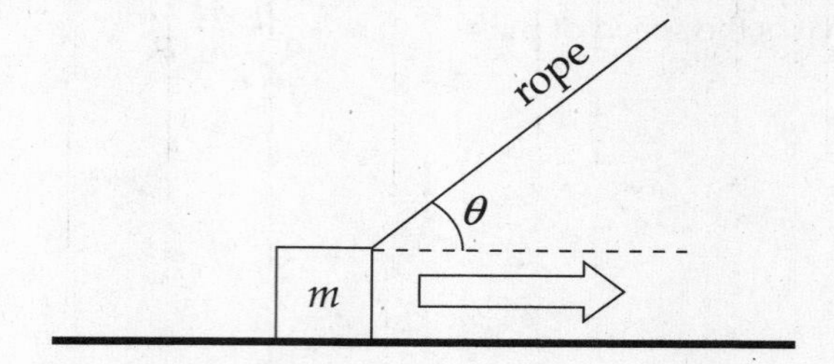

(a) Draw and label all of the forces acting on the crate.

(b) Compute the normal force acting on the crate in terms of m, F_T, θ, and g.

(c) Compute the acceleration of the crate in terms of m, F_T, θ, μ, and g.

(d) Assume that the magnitude of the tension in the rope is fixed but that the angle may be varied. For what value of θ would the resulting horizontal acceleration of the crate be maximized?

2. In the diagram below, a massless string connects two blocks—of masses m_1 and m_2, respectively—on a flat, frictionless tabletop. A force **F** pulls on Block #2, as shown:

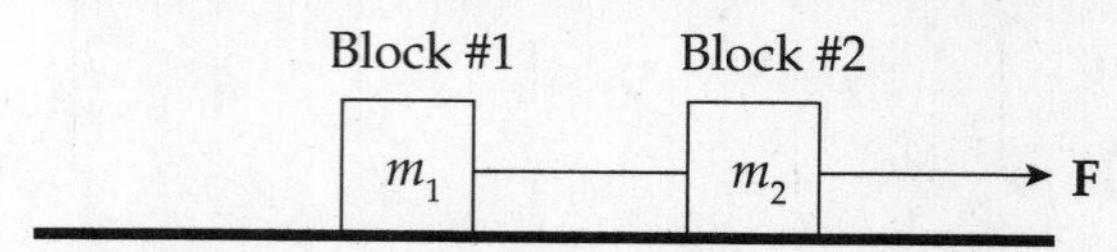

Solve for the following in terms of given quantities.

(a) Draw and label all of the forces acting on Block #1.

(b) Draw and label all of the forces acting on Block #2.

(c) What is the acceleration of Block #1?

(d) What is the tension in the string connecting the two blocks?

(e) If the string connecting the blocks were not massless, but instead had a mass of m, figure out

(i) the acceleration of Block #1, and

(ii) the difference between the strength of the force that the connecting string exerts on Block #2 and the strength of the force that the connecting string exerts on Block #1.

3. In the figure shown, assume that the pulley is frictionless and massless.

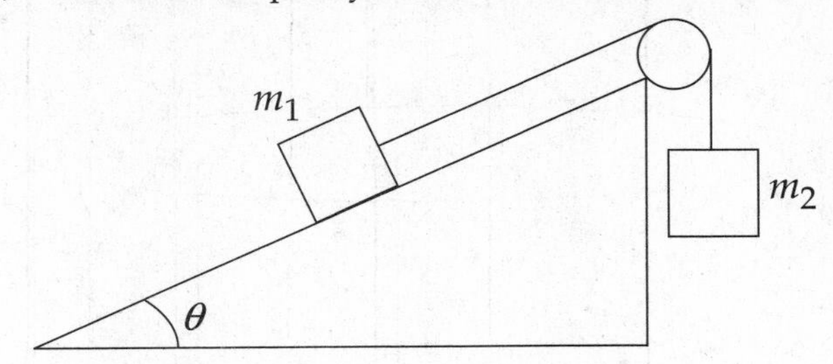

Solve for the following in terms of given quantities and the acceleration of gravity, g.

(a) If the surface of the inclined plane is frictionless, determine what value(s) of θ will cause the box of mass m_1 to

(i) accelerate up the ramp;

(ii) slide up the ramp at constant speed.

(b) If the coefficient of kinetic friction between the surface of the inclined plane and the box of mass m_1 is μ_k, derive (but do not solve) an equation satisfied by the value of θ which will cause the box of mass m_1 to slide up the ramp at constant speed.

4. A sky diver is falling with speed v_0 through the air. At that moment (time $t = 0$), she opens her parachute and experiences the force of air resistance whose strength is given by the equation $F = kv$, where k is a proportionality constant and v is her descent speed. The total mass of the sky diver and equipment is m. Assume that g is constant throughout her descent.

(a) Draw and label all the forces acting on the sky diver after her parachute opens.

(b) Determine the sky diver's acceleration in terms of m, v, k, and g.

(c) Determine the sky diver's terminal speed (that is, the eventual constant speed of descent).

(d) Sketch a graph of v as a function of time, starting at $t = 0$ and going until she lands, being sure to label important values on the vertical axis.

(e) Derive an expression for her descent speed, v, as a function of time t since opening her parachute in terms of m, k, and g.

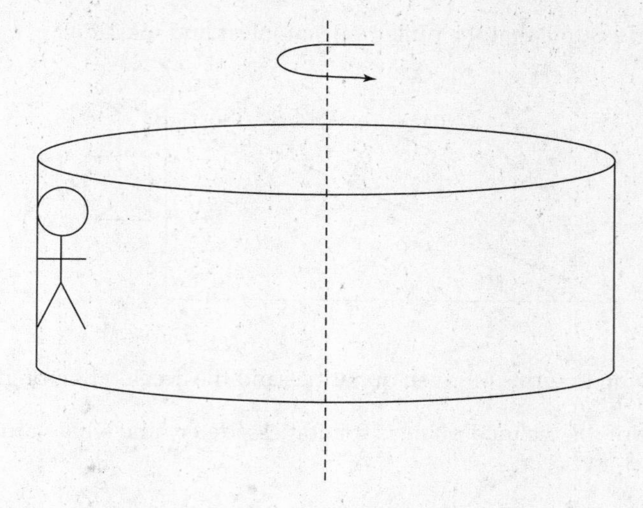

5. An amusement park ride consists of a large cylinder that rotates around its central axis as the passengers stand against the inner wall of the cylinder. Once the passengers are moving at a certain speed v, the floor on which they were standing is lowered. Each passenger feels pinned against the wall of the cylinder as it rotates. Let r be the inner radius of the cylinder.

   Solve for the following in terms of given quantities and the acceleration of gravity, g.

   (a) Draw and label all the forces acting on a passenger of mass m as the cylinder rotates with the floor lowered.

   (b) Describe what conditions must hold to keep the passengers from sliding down the wall of the cylinder.

   (c) Compare the conditions discussed in part (b) for an adult passenger of mass m and a child passenger of mass $m/2$.

6. A curved section of a highway has a radius of curvature of r. The coefficient of friction between standard automobile tires and the surface of the highway is μ_s.

   (a) Draw and label all the forces acting on a car of mass m traveling along this curved part of the highway.

   (b) Compute the maximum speed with which a car of mass m could make it around the turn without skidding in terms of μ_s, r, g, and m.

   City engineers are planning on banking this curved section of highway at an angle of θ to the horizontal.

   (c) Draw and label all of the forces acting on a car of mass m traveling along this banked turn. Do not include friction.

   (d) The engineers want to be sure that a car of mass m traveling at a constant speed v (the posted speed limit) could make it safely around the banked turn even if the road were covered with ice (that is, essentially frictionless). Compute this banking angle θ in terms of r, v, g, and m.

NEWTON'S LAWS OF MOTION SUMMARY

NEWTON'S LAWS

- **Newton's First Law (Law of Inertia)** states that objects will continue in their state of motion unless acted upon by an unbalanced force.

- **Newton's Second Law**: $F = ma$

- **Newton's Third Law** states that whenever two objects interact the force the first object exerts on the second object is equal to, but in opposite direction of, the force the second object exerts on the first object.

WEIGHT

- The weight of an object is given by: $\mathbf{F}_w = m\mathbf{g}$

THE NORMAL FORCE

- The normal force is the component of the contact force exerted on an object in contact with a surface and is *perpendicular* to the surface.

- The normal force can be denoted by $\mathbf{F}_N$ or $\mathbf{N}$.

FRICTION

- Friction is the component of the contact force exerted on an object in contact with a surface and is *parallel* to the surface.

- Static friction occurs when there is no relative motion between the object and the surface. Its strength is given by the equation $F_{\text{static friction, max}} = \mu_s F_N$

- Kinetic friction occurs when there is relative motion between the two surfaces. Its strength is given by the equation $F_{\text{kinetic friction}} = \mu_k F_N$

INCLINED PLANES

- There are two components to the force of gravity on an object on an inclined plane: the force parallel to the ramp ($mg \sin \theta$) and the force normal to the ramp ($mg \cos \theta$).

- To simplify analysis of an object moving up or down a ramp, rotate the coordinate axes so that the x-axis is parallel to the incline and the y-axis is perpendicular to the incline.

UNIFORM CIRCULAR MOTION

- The velocity is tangent to the circle.

- The centripetal acceleration points toward the center of the circle, and therefore the centripetal force must also point to the center of the circle.

  $$a_c = \frac{v^2}{r}$$

- Any force, or component of a force, that points toward the center of the circle is positive and any force, or component of a force, that points away from the center is negative when using the equation, $F_c = \frac{mv^2}{r}$.

4

Work, Energy, and Power

INTRODUCTION

It wasn't until more than one hundred years after Newton that the idea of energy became incorporated into physics, but today it permeates every branch of the subject.

It's difficult to give a precise definition of energy; there are different forms of energy because there are different kinds of forces. There's gravitational energy (a meteor crashing into the earth), elastic energy (a stretched rubber band), thermal energy (an oven), radiant energy (sunlight), electrical energy (a lamp plugged into a wall socket), nuclear energy (nuclear power plants), and mass energy (the heart of Einstein's equation $E = mc^2$). Energy can come into a system or leave it via various interactions that produce changes. One of the best definitions we know reads as follows: **Force** is the agent of change, **energy** is the measure of change, and **work** is the way of transferring energy from one system to another. And one of the most important laws in physics (the **Law of Conservation of Energy**, also known as the **First Law of Thermodynamics**) says that if you account for all its various forms, the total amount of energy in a given process will stay constant; that is, it will be *conserved*. For example, electrical energy can be converted into light and heat (this is how a light bulb works), but the amount of electrical energy *coming in* to the light bulb equals the total amount of light and heat *given off*. Energy cannot be created or destroyed; it can only be transferred (from one system to another) or transformed (from one form to another).

WORK

When you lift a book from the floor, you exert a force on it, over a distance, and when you push a crate across a floor, you also exert a force on it, over a distance. The application of force over a distance, and the resulting change in energy of the system that the force acted on, give rise to the concept of **work**. When you hold a book in your hand, you exert a force on the book (normal force) but, since the book is at rest, the force does not act through a distance, so you do no work on the book. Although you did work on the book as you lifted it from the floor, once it's at rest in your hand, you are no longer doing work on it.

Definition. If a force **F** acts over a distance d, and **F** is parallel to **d**, then the work done by **F** is the product of force and distance: $W = Fd$.

Notice that, although work depends on two vectors (**F** and **d**), work itself is *not* a vector. *Work is a scalar quanity.*

> **Example 4.1** You slowly lift a book of mass 2 kg at constant velocity a distance of 3 m. How much work did you do on the book?

Solution. In this case, the force you exert must balance the weight of the book (otherwise the velocity of the book wouldn't be constant), so $F = mg = (2 \text{ kg})(10 \text{ m/s}^2) = 20$ N. Since this force is straight upward and the displacement of the book is also straight upward, **F** and **d** are parallel, so the work done by your lifting force is $W = Fd = (20 \text{ N})(3 \text{ m}) = 60$ N·m. The unit for work, the newton-meter (N·m) is renamed a **joule**, and abbreviated J. So the work done here is 60 J.

The definition above takes care of cases in which **F** is parallel to the motion. If **F** is not parallel to the motion, then the definition needs to be generalized.

Definition. If a force **F** acts over a distance d, and θ is the angle between **F** and **d**, then the work done by **F** is the product of the component of force in the direction of the motion and the distance: $W = (\mathbf{F} \cos \theta)d$. [Notice that you can write this definition using the *dot product*: $W = \mathbf{F} \cdot \mathbf{d}$.]

> **Example 4.2** A 15 kg crate is moved along a horizontal floor by a warehouse worker who's pulling on it with a rope that makes a 30° angle with the horizontal. The tension in the rope is 200 N and the crate slides a distance of 10 m. How much work is done on the crate by the worker?

Solution. The figure below shows that $\mathbf{F}_T$ and $\mathbf{d}$ are not parallel. It's only the component of the force acting along the direction of motion, $\mathbf{F}_T \cos\theta$, that does work.

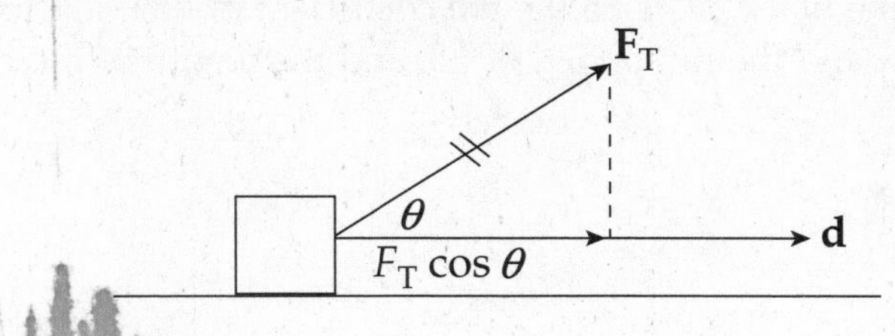

Therefore,

$$W = (F_T \cos\theta)d = (200\text{ N} \cdot \cos 30°)(10\text{ m}) = 1730\text{ J}$$

> **Example 4.3** In the previous example, assume that the coefficient of kinetic friction between the crate and the floor is 0.4.
> (a) How much work is done by the normal force?
> (b) How much work is done by the friction force?

Solution.

(a) Clearly, the normal force is not parallel to the motion, so we use the general definition of work. Since the angle between $\mathbf{F}_N$ and $\mathbf{d}$ is 90° (by definition of *normal*) and $\cos 90° = 0$, the normal force does zero work.

(b) The friction force, $\mathbf{F}_f$, is also not parallel to the motion; it's *antiparallel*. That is, the angle between $\mathbf{F}_f$ and $\mathbf{d}$ is 180°. Since $\cos 180° = -1$, and since the strength of the normal force is $F_N = F_w = mg = (15\text{ kg})(10\text{ m/s}^2) = 150\text{ N}$, the work done by the friction force is:

$$W = -F_f d = -\mu_k F_N d = -(0.4)(150\text{ N})(10\text{ m}) = -600\text{ J}$$

The two previous examples show that work, which is a scalar quantity, may be positive, negative, or zero. If the angle between $\mathbf{F}$ and $\mathbf{d}$ (θ) is less than 90°, then the work is positive (because $\cos\theta$ is positive in this case); if $\theta = 90°$, the work is zero (because $\cos 90° = 0$); and if $\theta > 90°$, then the work is negative (because $\cos\theta$ is negative). Intuitively, if a force helps the motion, the work done by the force is positive, but if the force opposes the motion, then the work done by the force is negative.

Example 4.4 A box slides down an inclined plane (incline angle = 37°). The mass of the block, m, is 35 kg, the coefficient of kinetic friction between the box and the ramp, μ_k, is 0.3, and the length of the ramp, d, is 8 m.

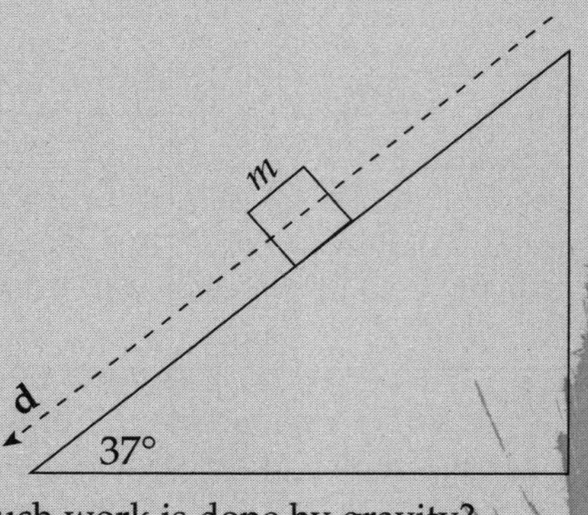

(a) How much work is done by gravity?
(b) How much work is done by the normal force?
(c) How much work is done by friction?
(d) What is the total work done?

Solution.

(a) Recall that the force that's directly responsible for pulling the box down the plane is the component of the gravitational force that's parallel to the ramp: $F_w \sin\theta = mg \sin\theta$ (where θ is the incline angle). This component is parallel to the motion, so the work done by gravity is

$$W_{\text{by gravity}} = (mg \sin\theta)d = (35 \text{ kg})(10 \text{ N/kg})(\sin 37°)(8 \text{ m}) = 1680 \text{ J}$$

Note that the work done by gravity is positive, as we would expect it to be, since gravity is helping the motion. Also, be careful with the angle θ. The general definition of work reads $W = (F \cos\theta)d$, where θ is the angle between **F** and **d**. However, the angle between $\mathbf{F}_w$ and **d** is *not* 37° here, so the work done by gravity is not $(mg \cos 37°)d$. The angle θ used in the calculation above is the incline angle.

(b) Since the normal force is perpendicular to the motion, the work done by this force is zero.

(c) The strength of the normal force is $F_w \cos\theta$ (where θ is the incline angle), so the strength of the friction force is $F_f = \mu_k F_N = \mu_k F_w \cos\theta = \mu_k mg \cos\theta$. Since $\mathbf{F}_f$ is antiparallel to **d**, the cosine of the angle between these vectors (180°) is –1, so the work done by friction is

$$W_{\text{by friction}} = -F_f d = -(\mu_k mg \cos\theta)(d) = -(0.3)(35 \text{ kg})(10 \text{ N/kg})(\cos 37°)(8 \text{ m}) = -672 \text{ J}$$

Note that the work done by friction is negative, as we expect it to be, since friction is opposing the motion.

(d) The total work done is found simply by adding the values of the work done by each of the forces acting on the box:

$$W_{\text{total}} = \Sigma W = W_{\text{by gravity}} + W_{\text{by normal force}} + W_{\text{by friction}} = 1680 + 0 + (-672) = 1008 \text{ J}$$

WORK DONE BY A VARIABLE FORCE

If a force remains constant over the distance through which it acts, then the work done by the force is simply the product of force and distance. However, if the force does not remain constant, then the work done by the force is given by a definite integral. Focusing only on displacements that are along a straight line (call it the x axis), let **F** be a force whose component in the x direction varies with position according to the equation $F = F(x)$. Then the work done by this force as it acts from position $x = x_1$ to position $x = x_2$ is equal to

$$W = \int_{x_1}^{x_2} F(x)\, dx$$

If a graph of F vs. x is given, then the work done by **F** as it acts from $x = x_1$ to $x = x_2$ is equal to the area bounded by the graph of F, the x axis, and the vertical lines $x = x_1$ and $x = x_2$:

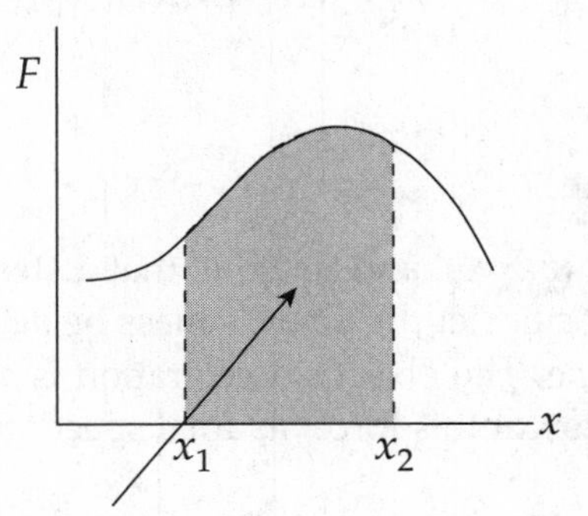

area = work done by **F**

> **Example 4.5** The force exerted by a spring when it's displaced by x from its natural length is given by the equation $F(x) = -kx$, where k is a positive constant. This equation is known as Hooke's Law. What is the work done by a spring as it pushes out from $x = -x_2$ to $x = -x_1$ (where $x_2 > x_1$)?

Solution. Since the force is variable, we calculate the following definite integral:

$$\begin{aligned} W = \int_{-x_2}^{-x_1} F(x)\, dx = \int_{-x_2}^{-x_1} (-kx)\, dx &= -\tfrac{1}{2}kx^2\Big]_{-x_2}^{-x_1} \\ &= \tfrac{1}{2}kx^2\Big]_{-x_1}^{-x_2} \\ &= \tfrac{1}{2}k\left(x_2^2 - x_1^2\right) \end{aligned}$$

Another solution would involve sketching a graph of $F(x) = -kx$ and calculating the area under the graph from $x = -x_2$ to $x = -x_1$.

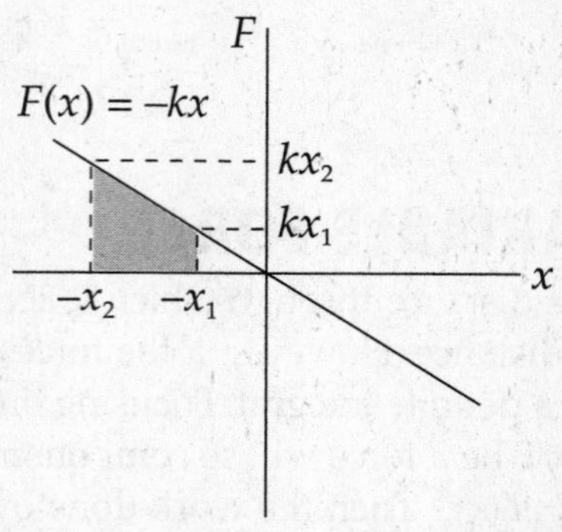

Here, the region is a trapezoid with area $A = \frac{1}{2}(\text{base}_1 + \text{base}_2) \times \text{height}$, so

$$
\begin{aligned}
W = A &= \tfrac{1}{2}(kx_2 + kx_1)(x_2 - x_1) \\
&= \tfrac{1}{2}k(x_2 + x_1)(x_2 - x_1) \\
&= \tfrac{1}{2}k\left(x_2^2 - x_1^2\right)
\end{aligned}
$$

KINETIC ENERGY

Consider an object at rest ($v_0 = 0$), and imagine that a steady force is exerted on it, causing it to accelerate. Let's be more specific; let the object's mass be m, and let **F** be the force acting on the object, pushing it in a straight line. The object's acceleration is $a = F/m$, so after the object has traveled a distance Δx under the action of this force, its final speed, v, is given by Big Five #5:

$$v^2 = v_0^2 + 2a(x - x_0) = 2a\Delta x = 2\frac{F}{m}\Delta x \quad \Rightarrow \quad F\Delta x = \tfrac{1}{2}mv^2$$

But the quantity $F\Delta x$ is the work done by the force, so $W = \frac{1}{2}mv^2$. The work done on the object has transferred energy to it, in the amount $\frac{1}{2}mv^2$. The energy an object possesses by virtue of its motion is therefore defined as $\frac{1}{2}mv^2$ and is called **kinetic energy**:

$$K = \tfrac{1}{2}mv^2$$

THE WORK–ENERGY THEOREM

Kinetic energy is expressed in joules, just like work, since in the case we just looked at, $W = K$. In fact, the derivation above can be extended to an object with a non-zero initial speed, and the same analysis will show that the total work done on an object—or, equivalently, the work done by the net force—will equal its change in kinetic energy; this is known as the **work–energy theorem**:

$$W_{\text{total}} = \Delta K$$

Note that kinetic energy, like work, is a scalar quantity.

Example 4.6 What is the kinetic energy of a ball (mass = 0.10 kg) moving with a speed of 30 m/s?

Solution. From the definition,

$$K = \tfrac{1}{2}mv^2 = \tfrac{1}{2}(0.10\text{ kg})\,(30\text{ m/s})^2 = 45\text{ J}$$

Example 4.7 A tennis ball (mass = 0.06 kg) is hit straight upward with an initial speed of 50 m/s. How high would it go if air resistance were negligible?

Solution. This could be done using the Big Five, but let's try to solve it using the concepts of work and energy. As the ball travels upward, gravity acts on it by doing negative work. [The work is negative because gravity is opposing the upward motion. $\mathbf{F}_w$ and $\mathbf{d}$ are in opposite directions, so $\theta = 180°$, which tells us that $W = (F_w \cos\theta)d = -F_w d$.] At the moment the ball reaches its highest point, its speed is 0, so its kinetic energy is also 0. The work–energy theorem says

$$W = \Delta K \;\Rightarrow\; -F_w d = 0 - \tfrac{1}{2}mv_0^2 \;\Rightarrow\; d = \frac{\tfrac{1}{2}mv_0^2}{F_w} = \frac{\tfrac{1}{2}mv_0^2}{mg} = \frac{\tfrac{1}{2}v_0^2}{g} = \frac{\tfrac{1}{2}(50\text{ m/s})^2}{10\text{ m/s}^2} = 125\text{ m}$$

Example 4.8 Consider the box sliding down the inclined plane in Example 4.4. If it starts from rest at the top of the ramp, with what speed does it reach the bottom?

Solution. It was calculated in Example 4.4 that W_{total} = 1008 J. According to the work–energy theorem,

$$W_{\text{total}} = \Delta K \;\Rightarrow\; W_{\text{total}} = K_f - K_i = K_f = \tfrac{1}{2}mv^2 \;\Rightarrow\; v = \sqrt{\frac{2W_{\text{total}}}{m}} = \sqrt{\frac{2(1008\text{ J})}{35\text{ kg}}} = 7.6\text{ m/s}$$

Example 4.9 A pool cue striking a stationary billiard ball (mass = 0.25 kg) gives the ball a speed of 2 m/s. If the average force of the cue on the ball was 200 N, over what distance did this force act?

Solution. The kinetic energy of the ball as it leaves the cue is

$$K = \tfrac{1}{2}mv^2 = \tfrac{1}{2}(0.25\text{ kg})(2\text{ m/s})^2 = 0.50\text{ J}$$

The work W done by the cue gave the ball this kinetic energy, so

$$W = \Delta K \;\Rightarrow\; W = K_f \;\Rightarrow\; Fd = K \;\Rightarrow\; d = \frac{K}{F} = \frac{0.50\text{ J}}{200\text{ N}} = 0.0025\text{ m} = 0.25\text{ cm}$$

POTENTIAL ENERGY

Kinetic energy is the energy an object has by virtue of its motion. Potential energy is independent of motion; it arises from the object's position (or the system's configuration). For example, a ball at the edge of a tabletop has energy that could be transformed into kinetic energy if it falls off. An arrow in an archer's pulled-back bow has energy that could be transformed into kinetic energy if the archer releases the arrow. Both of these examples illustrate the concept of **potential energy**, the energy an object or system has by virtue of its position or configuration. In each case, work was done on the object to put it in the given configuration (the ball was lifted to the tabletop, the bowstring was pulled back), and since work is the means of transferring energy, these things have *stored energy that can be retrieved*, as kinetic energy. This is **potential energy**, denoted by U.

Because there are different types of forces, there are different types of potential energy. The ball at the edge of the tabletop provides an example of **gravitational potential energy**, U_{grav}, which is the energy stored by virtue of an object's position in a gravitational field. This energy would be converted to kinetic energy as gravity pulled the ball down to the floor. For now, let's concentrate on gravitational potential energy.

Assume the ball has a mass m of 2 kg, and that the tabletop is $h = 1.5$ m above the floor. How much work did gravity do as the ball was lifted from the floor to the table? The strength of the gravitational force on the ball is $F_w = mg = (2 \text{ kg})(10 \text{ N/kg}) = 20$ N. The force $\mathbf{F}_w$ points downward, and the ball's motion was upward, so the work done by gravity during the ball's ascent was

$$W_{\text{by gravity}} = -F_w h = -mgh = -(20 \text{ N})(1.5 \text{ m}) = -30 \text{ J}$$

So someone performed +30 J of work to raise the ball from the floor to the tabletop. That energy is now stored and, if the ball was given a push to send it over the edge, by the time the ball reached the floor it would acquire a kinetic energy of 30 J. We therefore say that the change in the ball's gravitational potential energy in moving from the floor to the table was +30 J. That is,

$$\Delta U_{grav} = -W_{\text{by gravity}}$$

Note that potential energy, like work (and kinetic energy), is expressed in joules.

In general, if an object of mass m is raised a height h (which is small enough that g stays essentially constant over this altitude change), then the increase in the object's gravitational potential energy is

$$\Delta U_{grav} = mgh$$

An important fact that makes the above equation possible is that the work done by gravity as the object is raised does not depend on the path taken by the object. The ball could be lifted straight upward, or in some curvy path; it would make no difference. Gravity is said to be a **conservative** force because of this property.

If we decide on a reference level to call $h = 0$, then we can say that the gravitational potential energy of an object of mass m at a height h is $U_{grav} = mgh$. In order to use this last equation, it's essential that we choose a reference level for height. For example, consider a passenger in an airplane reading a book. If the book is 1 m above the floor of the plane then, to the passenger, the gravitational potential energy of the book is mgh, where $h = 1$ m. However, to someone on the ground looking up, the floor of the plane may be, say, 9000 m above the ground. So, to this person, the gravitational potential energy of the book is mgH, where $H = 9001$ m. What both would agree on, though, is that the difference in potential energy between the floor of the plane and the position of the book is $mg \times (1 \text{ m})$, since the airplane passenger would calculate the difference as $mg \times (1 \text{ m} - 0 \text{ m})$, while the person on the ground would calculate it as $mg \times (9001 \text{ m} - 9000 \text{ m})$. Differences, or changes, in potential energy are unambiguous, but values of potential energy are relative.

> **Example 4.10** A stuntwoman (mass = 60 kg) scales a 40-meter-tall rock face. What is her gravitational potential energy (relative to the ground)?

Solution. Calling the ground $h = 0$, we find

$$U_{\text{grav}} = mgh = (60 \text{ kg})(10 \text{ m/s}^2)(40 \text{ m}) = 24{,}000 \text{ J}$$

> **Example 4.11** If the stuntwoman in the previous example were to jump off the cliff, what would be her final speed as she landed on a large, air-filled cushion lying on the ground?

Solution. The gravitational potential energy would be transformed into kinetic energy. So

$$U \rightarrow K \quad \Rightarrow \quad U \rightarrow \tfrac{1}{2}mv^2 \quad \Rightarrow \quad v = \sqrt{\frac{2 \cdot U}{m}} = \sqrt{\frac{2(24{,}000 \text{ J})}{60 \text{ kg}}} = 28 \text{ m/s}$$

CONSERVATION OF MECHANICAL ENERGY

We have seen energy in its two basic forms: Kinetic energy (K) and potential energy (U). The sum of an object's kinetic and potential energies is called its **mechanical energy**, E:

$$E = K + U$$

(Note that because U is relative, so is E.) Assuming that no nonconservative forces (friction, for example) act on an object or system while it undergoes some change, then mechanical energy is conserved. That is, the initial mechanical energy, E_i, is equal to the final mechanical energy, E_f, or

$$K_i + U_i = K_f + U_f$$

This is the simplest form of the Law of Conservation of Total Energy, which we mentioned at the beginning of this section.

> **Example 4.12** A ball of mass 2 kg is gently pushed off the edge of a tabletop that is 5.0 m above the floor. Find the speed of the ball as it strikes the floor.

Solution. Ignoring the friction due to the air, we can apply Conservation of Mechanical Energy. Calling the floor our $h = 0$ reference level, we write

$$\begin{aligned} K_i + U_i &= K_f + U_f \\ 0 + mgh &= \tfrac{1}{2}mv^2 + 0 \\ v &= \sqrt{2gh} \\ &= \sqrt{2(10 \text{ m/s}^2)(5.0 \text{ m})} \\ &= 10 \text{ m/s} \end{aligned}$$

Note that the ball's potential energy decreased, while its kinetic energy increased. This is the basic idea behind conservation of mechanical energy: One form of energy decreases while the other increases.

Example 4.13 A box is projected up a long ramp (incline angle with the horizontal = 37°) with an initial speed of 10 m/s. If the surface of the ramp is very smooth (essentially frictionless), how high up the ramp will the box go? What distance along the ramp will it slide?

Solution. Because friction is negligible, we can apply Conservation of Mechanical Energy. Calling the bottom of the ramp our $h = 0$ reference level, we write

$$\begin{aligned} K_i + U_i &= K_f + U_f \\ \tfrac{1}{2}mv_0^2 + 0 &= 0 + mgh \\ h &= \frac{\tfrac{1}{2}v_0^2}{g} \\ &= \frac{\tfrac{1}{2}(10\text{ m/s})^2}{10\text{ m/s}^2} \\ &= 5\text{ m} \end{aligned}$$

Since the incline angle is $\theta = 37°$, the distance d it slides up the ramp is found in this way:

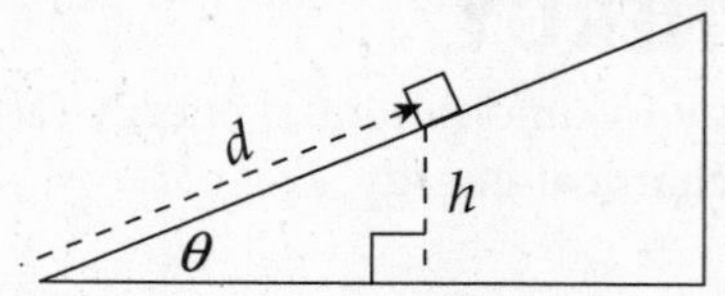

$$h = d\sin\theta$$

$$d = \frac{h}{\sin\theta} = \frac{5\text{ m}}{\sin 37°} = \frac{25}{3}\text{ m} = 8.3\text{ m}$$

Example 4.14 A skydiver jumps from a hovering helicopter that's 3000 m above the ground. If air resistance can be ignored, how fast will he be falling when his altitude is 2000 m?

Solution. Ignoring air resistance, we can apply Conservation of Mechanical Energy. Calling the ground our $h = 0$ reference level, we write

$$\begin{aligned} K_i + U_i &= K_f + U_f \\ 0 + mgH &= \tfrac{1}{2}mv^2 + mgh \\ v &= \sqrt{2g(H-h)} \\ &= \sqrt{2(10\text{ m/s}^2)(3000\text{ m} - 2000\text{ m})} \\ &= 140\text{ m/s} \end{aligned}$$

(That's over 300 mph! The terminal velocity of a human falling is about 100 mph, which shows that air resistance *does* play a role, even before the parachute is opened.)

The equation $K_i + U_i = K_f + U_f$ holds if no nonconservative forces are doing work. However, if work is done by such forces during the process under investigation, then the equation needs to be modified to account for this work as follows:

$$K_i + U_i + W_{other} = K_f + U_f$$

Example 4.15 Wile E. Coyote (mass = 40 kg) falls off a 50-meter-high cliff. On the way down, the force of air resistance has an average strength of 100 N. Find the speed with which he crashes into the ground.

Solution. The force of air resistance opposes the downward motion, so it does negative work on the coyote as he falls: $W_r = -F_r h$. Calling the ground $h = 0$, we find that

$$\begin{aligned} K_i + U_i + W_r &= K_f + U_f \\ 0 + mgh + (-F_r h) &= \tfrac{1}{2}mv^2 + 0 \\ v &= \sqrt{2h(g - F_r / m)} = \sqrt{2(50)(10 - 100/40)} = 27 \text{ m/s} \end{aligned}$$

Example 4.16 A skier starts from rest at the top of a 20° incline and skis in a straight line to the bottom of the slope, a distance d (measured along the slope) of 400 m. If the coefficient of kinetic friction between the skis and the snow is 0.2, calculate the skier's speed at the bottom of the run.

Solution. The strength of the friction force on the skier is $F_f = \mu_k F_N = \mu_k(mg\cos\theta)$, so the work done by friction is $-F_f d - \mu_k(mg\cos\theta)\cdot d$. The vertical height of the slope above the bottom of the run (which we designate the $h = 0$ level) is $h = d\sin\theta$. Therefore, Conservation of Mechanical Energy (including the negative work done by friction) gives

$$\begin{aligned} K_i + U_i + W_{friction} &= K_f + U_f \\ 0 + mgh + (-\mu_k mg\cos\theta \cdot d) &= \tfrac{1}{2}mv^2 + 0 \\ mg(d\sin\theta) + (-\mu_k mg\cos\theta \cdot d) &= \tfrac{1}{2}mv^2 \\ gd(\sin\theta - \mu_k\cos\theta) &= \tfrac{1}{2}v^2 \\ v &= \sqrt{2gd(\sin\theta - \mu_k\cos\theta)} \\ &= \sqrt{2(10)(400)[\sin 20° - (0.2)\cos 20°]} \\ &= 35 \text{ m/s} \end{aligned}$$

POTENTIAL ENERGY CURVES

The behavior of a system can be analyzed if we are given a graph of its potential energy, $U(x)$ as a function of position and its mechanical energy, E. Since $K+U=E$ we have $\frac{1}{2}mv^2+U(x)=E$, which can be solved for v, the velocity at position x:

$$v=\pm\sqrt{\frac{2}{m}[E-U(x)]}$$

For example, consider the following potential energy curve:

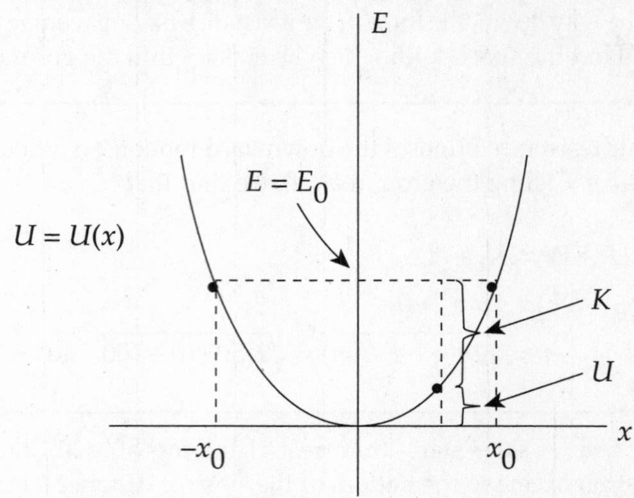

The graph shows how the potential energy, U, varies with position, x. A particular value of the total energy, $E = E_0$, is also shown. Motion of an object whose potential energy is given by $U(x)$ and which has a mechanical energy of E_0 is confined to the region $-x_0 \le x \le x_0$, because only in this range is $E_0 \ge U(x)$. At each position x in this range, the kinetic energy, $K = E_0 - U(x)$, is positive. However, if $x > x_0$ (or if $x < -x_0$), then $U(x) > E_0$, which is physically impossible because the difference $E_0 - U(x)$, which should give K, is negative.

This particular energy curve with $U(x) = \frac{1}{2}kx^2$, describes one of the most important physical systems: a simple harmonic oscillator. The force felt by the oscillator can be recovered from the potential energy curve. Recall that, in the case of gravitational potential energy we defined $\Delta U_{grav} = -W_{by\ grav}$. In general, $\Delta U = -W$. If we account for a variable force of the form $F = F(x)$, which does the work W, then over a small displacement Δx, we have $\Delta U(x) = -W = -F(x)\,\Delta x$, so $F(x) = -\Delta U(x)/\Delta x$. In the limit as $\Delta x \to 0$, this last equation becomes

$$F(x)=-\frac{dU}{dx}$$

Therefore, in this case, we find $F(x) = -(d/dx)(\frac{1}{2}kx^2) = -kx$, which specifies a linear restoring force, a prerequisite for simple harmonic motion. This equation, $F(x) = -kx$, is called **Hooke's Law** and is obeyed by ideal springs (see Example 4.5).

With this result, we can appreciate the oscillatory nature of the system whose energy curve is sketched above. If x is positive (and not greater than x_0), then $U(x)$ is increasing, so dU/dx is positive, which tells us that F is negative. So the oscillator feels a force—and an acceleration—in the negative direction, which pulls it back through the origin ($x = 0$). If x is negative (and not less than $-x_0$), then $U(x)$

is decreasing, so dU/dx is negative, which tells us that F is positive. So the oscillator feels a force—and an acceleration—in the positive direction, which pushes it back through the origin ($x = 0$).

Furthermore, the difference between E_0 and U, which is K, decreases as x approaches x_0 (or as x approaches $-x_0$), dropping to zero at these points. The fact that K decreases to zero at $\pm x_0$ tells us that the oscillator's speed decreases to zero as it approaches these endpoints, before changing direction and heading back toward the origin—where its kinetic energy and speed are maximized—for another oscillation. By looking at the energy curve with these observations in mind, you can almost see the oscillator moving back and forth between the barriers at $x = \pm x_0$.

The origin is a point at which $U(x)$ has a minimum, so the tangent line to the curve at this point is horizontal; the slope is zero. Since $F = -dU/dx$, the force F is 0 at this point [which we also know from the equation $F(x) = -kx$]; this means that this is a point of **equilibrium**. If the oscillator is pushed from this equilibrium point in either direction, the force $F(x)$ will attempt to restore it to $x = 0$, so this is a point of **stable** equilibrium. However, a point where the $U(x)$ curve has a maximum is also a point of equilibrium, but it's an **unstable** one, because if the system were moved from this point in either direction, the force would accelerate it *away* from the equilibrium position.

Consider the following potential energy curve:

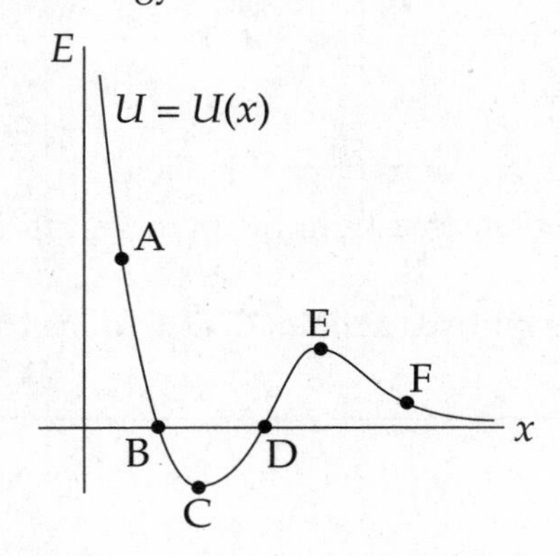

Point C is a position of stable equilibrium and E is a point of unstable equilibrium. Since $U(x)$ is decreasing at Points A and F, $F(x)$ is positive, accelerating the system in the positive x direction. Points B and D mark the barriers of oscillation if the system has a mechanical energy E_0 of 0.

Example 4.17 An object of mass $m = 4$ kg has a potential energy function

$$U(x) = (x - 2) - (2x - 3)^3$$

Where x is measured in meters and U in joules. The following graph is a sketch of the potential energy function.

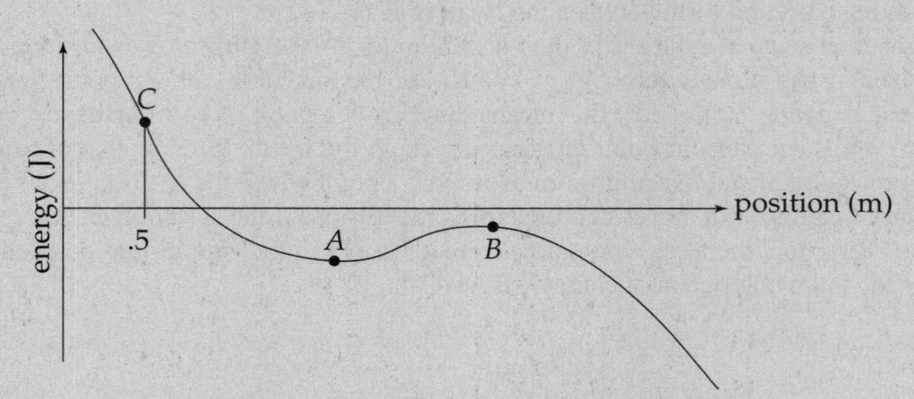

(a) Determine the positions of points A and B.

(b) If the object is released from the point B, can it reach point A or C? Explain.

(c) The particle is released at point C. Determine its speed as it passes point A.

Solution.

(a) The points A and B are a local minimum and maximum, respectively, so the derivative of $U(x)$ will be zero at these locations.

$$\frac{dU}{dx} = 1 - 3(2x - 3)^2(2) = 0$$
$$-24x^2 + 72x - 53 = 0$$
$$x = 1.30 \text{ m and } 1.70 \text{ m}$$

(b) The object has a negative total amount of mechanical energy at point B because all of its energy is potential energy. It will not be able to reach point C, because that position has a potential energy well above zero. However the object would be able to reach point A because the potential energy at A is less than the energy the object started with at point B. As the object moved from B to A its potential energy would decrease (become more negative) and its kinetic energy would increase.

(c) First we need to find how much potential energy the object has at point C, and this will define the total mechanical energy of the object. Then determine the potential energy at point A, $x = 1.3$ m, and use conservation of energy to determine the speed at point A.

$$U(0.5) = (0.5 - 2) - (2 \cdot (0.5) - 3)^3 = 6.5\text{J}$$
$$U(1.3) = -0.636\text{J}$$
$$E_C = E_A$$
$$6.5 = -0.636 + K_A$$
$$7.136 = \frac{1}{2}(4)v_A{}^2$$
$$v_A = 1.89 \text{ m/s}$$

POWER

Simply put, **power** is the rate at which work gets done (or energy gets transferred, which is the same thing). Suppose you and I each do 1000 J of work, but I do the work in 2 minutes while you do it in 1 minute. We both did the same amount of work, but you did it more quickly; you were more powerful. Here's the definition of power:

$$\text{Power} = \frac{\text{Work}}{\text{time}} \quad \text{— in symbols} \rightarrow \quad P = \frac{W}{t}$$

The unit of power is the joule per second (J/s), which is renamed the **watt**, and symbolized W (not to be confused with the symbol for work, W). One watt is 1 joule per second: 1 W = 1 J/s. Here in the United States, which still uses older units like inches, feet, yards, miles, ounces, pounds, and so forth, you still hear of power ratings expressed in horsepower (particularly of engines). One horsepower is defined as, well, the power output of a large horse. Horses can pull a 150-pound weight at a speed of $2\frac{1}{2}$ mph for quite a while. Let's assume that **F** and **d** are parallel, so that $W = Fd$; then the definition $P = W/t$ becomes $P = Fd/t$, which is Fv. Therefore,

$$P = Fv \quad \Rightarrow \quad 1 \text{ horsepower (hp)} = (150 \text{ lb})(2\tfrac{1}{2} \text{ mph})$$

Now for some unit conversions:

$$150 \text{ lb} = 150 \text{ lb} \times \frac{4.45 \text{ N}}{\text{lb}} = 667.5 \text{ N}$$
$$2\tfrac{1}{2} \text{ mph} = \frac{2\frac{1}{2} \text{ mi}}{\text{hr}} \times \frac{1609 \text{ m}}{1 \text{ mi}} \times \frac{1 \text{ hr}}{3600 \text{ s}} = 1.117 \text{ m/s}$$

Therefore,

$$1 \text{ hp} = (667.5 \text{ N})(1.117 \text{ m/s}) = 746 \text{ W}$$

By contrast, a human in good physical condition can do work at a steady rate of about 75 W (about 1/10 that of a horse!) but can attain power levels as much as twice this for short periods of time.

Example 4.18 A mover pushes a large crate (mass m = 75 kg) from the inside of the truck to the back end (a distance of 6 m), exerting a steady push of 300 N. If he moves the crate this distance in 20 s, what is his power output during this time?

Solution. The work done on the crate by the mover is $W = Fd$ = (300 N)(6 m) = 1800 J. If this much work is done in 20 s, then the power delivered is $P = W/t$ = (1800 J)(20 s) = 90 W.

Example 4.19 What must be the power output of a car engine, which moves a 1000 kg car at a constant speed of 8.0 m/s?

Solution. The equation $P = Fv$, with $F = mg$, yields

$$P = mgv = (1000 \text{ kg})(10 \text{ N/kg})(8.0 \text{ m/s}) = 80{,}000 \text{ W} = 80 \text{ kW}$$

CHAPTER 4 REVIEW QUESTIONS

SECTION I: MULTIPLE CHOICE

1. A force **F** of strength 20 N acts on an object of mass 3 kg as it moves a distance of 4 m. If **F** is perpendicular to the 4 m displacement, the work it does is equal to

   (A) 0 J
   (B) 60 J
   (C) 80 J
   (D) 600 J
   (E) 2400 J

2. Under the influence of a force, an object of mass 4 kg accelerates from 3 m/s to 6 m/s in 8 s. How much work was done on the object during this time?

   (A) 27 J
   (B) 54 J
   (C) 72 J
   (D) 96 J
   (E) Cannot be determined from the information given

3. A box of mass m slides down a frictionless inclined plane of length L and vertical height h. What is the change in its gravitational potential energy?

   (A) $-mgL$
   (B) $-mgh$
   (C) $-mgL/h$
   (D) $-mgh/L$
   (E) $-mghL$

4. An object of mass m is traveling at constant speed v in a circular path of radius r. How much work is done by the centripetal force during one-half of a revolution?

   (A) πmv^2
   (B) $2\pi mv^2$
   (C) 0
   (D) $\pi mv^2 r$
   (E) $2\pi mv^2 r$

5. While a person lifts a book of mass 2 kg from the floor to a tabletop, 1.5 m above the floor, how much work does the gravitational force do on the book?

   (A) −30 J
   (B) −15 J
   (C) 0 J
   (D) 15 J
   (E) 30 J

6. A block of mass 3.5 kg slides down a frictionless inclined plane of length 6.4 m that makes an angle of 30° with the horizontal. If the block is released from rest at the top of the incline, what is its speed at the bottom?

   (A) 5.0 m/s
   (B) 5.7 m/s
   (C) 6.4 m/s
   (D) 8.0 m/s
   (E) 10 m/s

7. A block of mass m slides from rest down an inclined plane of length s and height h. If F is the magnitude of the force of kinetic friction acting on the block as it slides, then the kinetic energy of the block when it reaches the bottom of the incline will be equal to

   (A) mgh
   (B) $mgh - Fh$
   (C) $mgs - Fh$
   (D) $mgh - Fs$
   (E) $mgs - Fs$

8. As a rock of mass 4 kg drops from the edge of a 40-meter-high cliff, it experiences air resistance, whose average strength during the descent is 20 N. At what speed will the rock hit the ground?

   (A) 8 m/s
   (B) 10 m/s
   (C) 12 m/s
   (D) 16 m/s
   (E) 20 m/s

9. An astronaut drops a rock from the top of a crater on the Moon. When the rock is halfway down to the bottom of the crater, its speed is what fraction of its final impact speed?

(A) $\frac{1}{4\sqrt{2}}$

(B) $\frac{1}{4}$

(C) $\frac{1}{2\sqrt{2}}$

(D) $\frac{1}{2}$

(E) $\frac{1}{\sqrt{2}}$

10. A force of 200 N is required to keep an object sliding at a constant speed of 2 m/s across a rough floor. How much power is being expended to maintain this motion?

(A) 50 W
(B) 100 W
(C) 200 W
(D) 400 W
(E) Cannot be determined from the information given

SECTION II: FREE RESPONSE

1. A box of mass m is released from rest at Point A, the top of a long, frictionless slide. Point A is at height H above the level of Points B and C. Although the slide is frictionless, the horizontal surface from Point B to C is not. The coefficient of kinetic friction between the box and this surface is μ_k, and the horizontal distance between Point B and C is x.

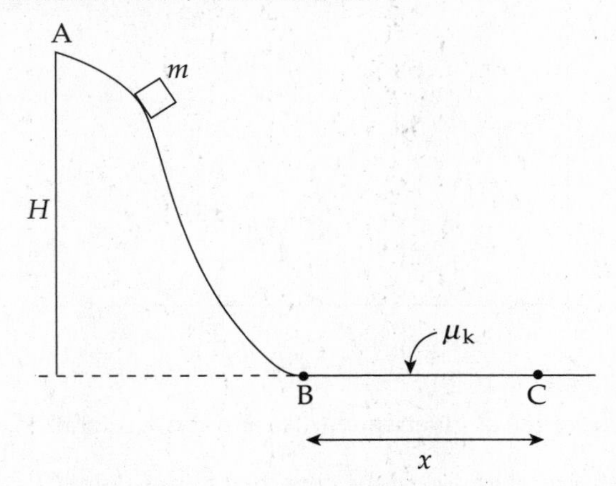

Solve for the following in terms of given quantities and the acceleration of gravity, g.

(a) Find the speed of the box when its height above Point B is $\frac{1}{2}H$.

(b) Find the speed of the box when it reaches Point B.

(c) Determine the value of μ_k so that the box comes to rest at Point C.

(d) Now assume that Points B and C were not on the same horizontal level. In particular, assume that the surface from B to C had a uniform upward slope so that Point C were still at a horizontal distance of x from B but now at a vertical height of y above B. Answer the question posed in part (c).

(e) If the slide were not frictionless, determine the work done by friction as the box moved from Point A to Point B if the speed of the box as it reached Point B were half the speed calculated in part (b).

2. The diagram below shows a roller-coaster ride which contains a circular loop of radius r. A car (mass m) begins at rest from Point A and moves down the frictionless track from A to B where it then enters the vertical loop (also frictionless), traveling once around the circle from B to C to D to E and back to B, after which it travels along the flat portion of the track from B to F (which is not frictionless).

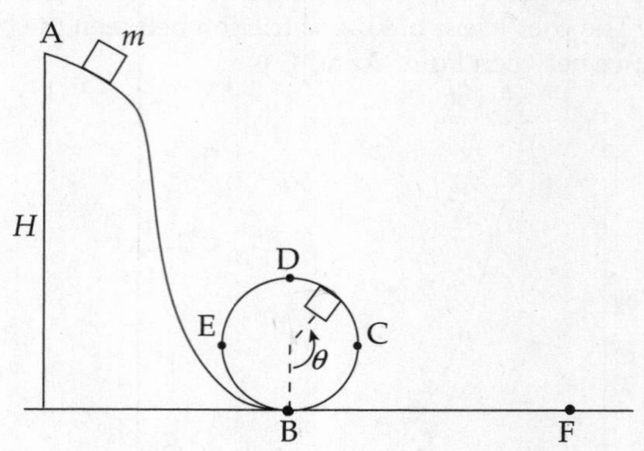

Solve for the following in terms of given quantities and the acceleration of gravity, g.

(a) Find the centripetal acceleration of the car when it is at Point C.

(b) Determine the speed of the car when its position relative to Point B is specified by the angle θ shown in the diagram.

(c) What is the minimum cut-off speed v_c that the car must have at Point D to make it around the loop?

(d) What is the minimum height H necessary to ensure that the car makes it around the loop?

(e) If $H = 6r$ and the coefficient of friction between the car and the flat portion of the track from B to F is 0.5, how far along this flat portion of the track will the car travel before coming to rest at Point F?

3. A particle of mass $m = 3$ kg has the potential energy function

$$U(x) = 3(x - 1) - (x - 3)^3$$

where x is measured in meters and U in joules. The following graph is a sketch of this potential energy function.

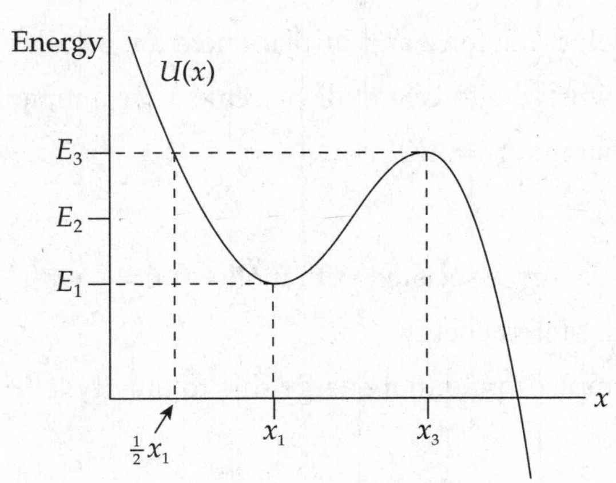

The energies indicated on the vertical axis are evenly spaced; that is, $E_3 - E_2 = E_2 - E_1$. The energy E_1 is equal to $U(x_1)$, and the energy E_3 is equal to $U(x_3)$.

(a) Determine the numerical values of x_1 and x_3.

(b) Describe the motion of the particle if its total energy is E_2.

(c) What is the particle's speed at $x = x_1$ if its total energy, E, equals 58 J?

(d) Sketch the graph of the particle's acceleration as a function of x. Be sure to indicate x_1 and x_3 on your graph.

(e) The particle is released from rest at $x = \frac{1}{2}x_1$. Find its speed as it passes through $x = x_1$.

4. The force on a 6 kg object is given by the equation: $F(x) = 3x + 5$, in newtons. The object is moving 2 m/s at the origin.

(a) Determine the work done on the object by the force when it is moved 4 m from the origin in the x direction.

(b) Determine the speed of the object when it has moved 4 meters.

WORK, ENERGY, AND POWER SUMMARY

Work

- Work is the dot product of force and displacement. $W = F \cdot x$
- When force varies with x, the work is given by the equation: $W = \int F(x)dx$
- Work is positive when the force and displacement are parallel.
- Work is negative when the force and displacement are antiparallel.
- **Work/Energy Theorem**: $W = \Delta KE$

Energy

- **Kinetic energy** is energy associated with motion. $K = \frac{1}{2}mv^2$
- **Potential energy** is stored energy.
- One common example is **potential energy due to gravity**: $U_g = mgh$
- Another common example is **potential energy due to springs**: $U_s = \frac{1}{2}kx^2$
- Work done by a **conservative force** only depends on the initial and final positions, and not on the path taken. Gravity and springs are examples of conservative forces.
- Work done by a **non-conservative force** depends on the path taken and mechanical energy is lost by heat, sound, and so on, when these forces act on a system. Friction and air resistance are examples of non-conservative forces.
- **Conservation of Mechanical Energy** states that the total mechanical energy of a system is constant when there are no non-conservative forces acting on the system. It is usually written as

  $E_i = E_f$

  $K_i + U_i = K_f + U_f$

Potential Energy Diagrams

- The potential energy can be given as $U(x)$. Then $F = -\frac{dU}{dx}$.
- If $\frac{dU}{dx} = 0$, then $F = 0$, and it is an equilibrium point.
- **Stable equilibrium** occurs when the force restores the object back toward the equilibrium point after it is disturbed.
- **Unstable equilibrium** occurs when the force moves the object further away from the equilibrium point after it is disturbed.

Power

- Power is the rate at which work is done.

  $$P = \frac{W}{t} = \frac{dW}{dt}$$

  $$P = Fv$$

5

Linear Momentum

INTRODUCTION

When Newton first expressed his Second Law, he didn't write $\mathbf{F}_{net} = m\mathbf{a}$. Instead, he expressed the law in the words, *The alteration of motion is . . . proportional to the . . . force impressed. . . .* By "motion," he meant the product of mass and velocity, a vector quantity known as **linear momentum** and denoted by $\mathbf{p}$:

$$\mathbf{p} = m\mathbf{v}$$

So Newton's original formulation of the Second Law read $\Delta\mathbf{p} \propto \mathbf{F}$, or, equivalently, $\mathbf{F} \propto \Delta\mathbf{p}$. But a large force that acts for a short period of time can produce the same change in linear momentum as a small force acting for a greater period of time. Knowing this, we can turn the proportion above into an equation, if we take the average force that acts over the time interval Δt:

$$\bar{\mathbf{F}} = \frac{\Delta\mathbf{p}}{\Delta t}$$

This equation becomes $\mathbf{F} = m\mathbf{a}$, since $\Delta\mathbf{p}/\Delta t = \Delta(m\mathbf{v})/\Delta t = m\,(\Delta\mathbf{v}/\Delta t) = m\mathbf{a}$ (assuming that m remains constant). If we take the limit as $\Delta t \to 0$, then the equation above takes the form:

$$\mathbf{F} = \frac{d\mathbf{p}}{dt}$$

Example 5.1 A golfer strikes a golf ball of mass 0.05 kg, and the time of impact between the golf club and the ball is 1 ms. If the ball acquires a velocity of magnitude 70 m/s, calculate the average force exerted on the ball.

Solution. Using Newton's Second Law, we find

$$\bar{F} = \frac{\Delta p}{\Delta t} = \frac{\Delta(mv)}{\Delta t} = m\frac{v-0}{\Delta t} = (0.05\text{ kg})\frac{70\text{ m/s}}{10^{-3}\text{ s}} = 3500\text{ N} \quad [\approx 790\text{ lb (!)}]$$

IMPULSE

The product of force and the time during which it acts is known as **impulse**; it's a vector quantity that's denoted by **J**:

$$\mathbf{J} = \bar{\mathbf{F}}\Delta t$$

In terms of impulse, Newton's Second Law can be written in yet another form:

$$\mathbf{J} = \Delta\mathbf{p}$$

Sometimes this is referred to as the **impulse–momentum theorem**, but it's just another way of writing Newton's Second Law. If **F** varies with time over the interval during which it acts, then the impulse delivered by the force $\mathbf{F} = \mathbf{F}(t)$ from time $t = t_1$ to $t = t_2$ is given by the following definite integral:

$$\mathbf{J} = \int_{t_1}^{t_2} \mathbf{F}(t)\,dt$$

If a graph of force versus time is given, then the impulse of force F as it acts from t_1 to t_2 is equal to the area bounded by the graph of F, the t-axis and the vertical lines associated with t_1 and t_2 as shown below.

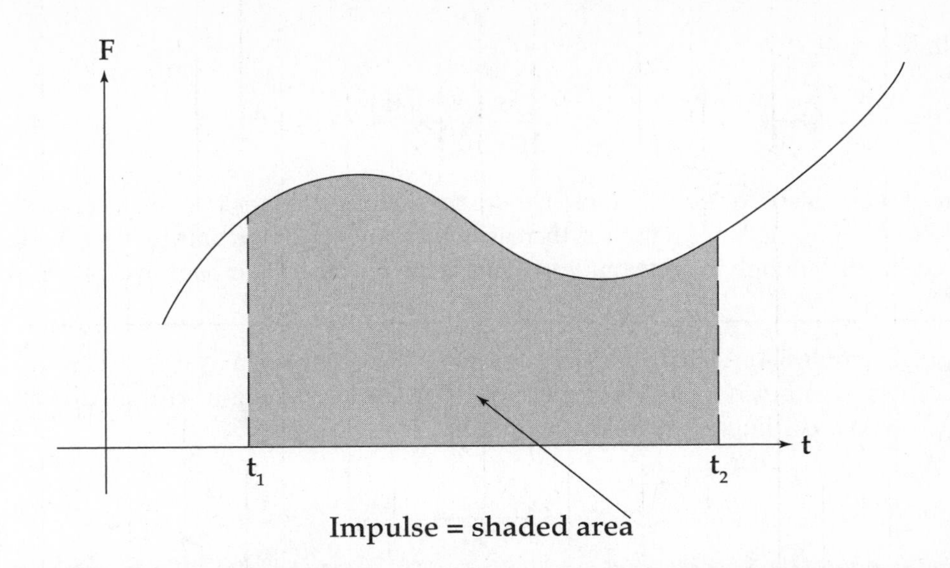

Example 5.2 A football team's kicker punts the ball (mass = 0.4 kg) and gives it a launch speed of 30 m/s. Find the impulse delivered to the football by the kicker's foot and the average force exerted by the kicker on the ball, given that the impact time is 8 ms.

Solution. Impulse is equal to change in linear momentum, so

$$J = \Delta p = p_f - p_i = p_f = mv = (0.4 \text{ kg})(30 \text{ m/s}) = 12 \text{ kg·m/s}$$

Using the equation $\bar{F} = J / \Delta t$, we find that the average force exerted by the kicker is

$$\bar{F} = J / \Delta t = (12 \text{ kg} \cdot \text{m/s}) / (8 \times 10^{-3} \text{ s}) = 1500 \text{ N} \quad [\approx 340 \text{ lb}]$$

Example 5.3 An 80 kg stuntman jumps out of a window that's 45 m above the ground.

(a) How fast is he falling when he reaches ground level?

(b) He lands on a large, air-filled target, coming to rest in 1.5 s. What average force does he feel while coming to rest?

(c) What if he had instead landed on the ground (impact time = 10 ms)?

Solution.

(a) His gravitational potential energy turns into kinetic energy: $mgh = \frac{1}{2}mv^2$, so

$$v = \sqrt{2gh} = \sqrt{2(10)(45)} = 30 \text{ m/s} \quad [\approx 70 \text{ mph}]$$

(You could also have answered this question using Big Five #5.)

(b) Using $\bar{\mathbf{F}} = \Delta \mathbf{p} / \Delta t$, we find that

$$\bar{\mathbf{F}} = \frac{\Delta \mathbf{p}}{\Delta t} = \frac{\mathbf{p}_f - \mathbf{p}_i}{\Delta t} = \frac{\mathbf{0} - m\mathbf{v}_i}{\Delta t} = \frac{-(80 \text{ kg})(30 \text{ m/s})}{1.5 \text{ s}} = -1600 \text{ N} \Rightarrow \bar{F} = 1600 \text{ N}$$

(c) In this case,

$$\bar{\mathbf{F}} = \frac{\Delta \mathbf{p}}{\Delta t} = \frac{\mathbf{p}_\mathrm{f} - \mathbf{p}_\mathrm{i}}{\Delta t} = \frac{0 - m\mathbf{v}_\mathrm{i}}{\Delta t} = \frac{-(80 \text{ kg})(30 \text{ m/s})}{10 \times 10^{-3} \text{ s}} = -240{,}000 \text{ N} \quad \Rightarrow \quad \bar{F} = 240{,}000 \text{ N}$$

This force is equivalent to about 27 tons(!), more than enough to break bones and cause fatal brain damage. Notice how crucial impact time is: Increasing the slowing-down time reduces the acceleration and the force, ideally enough to prevent injury. This is the purpose of air bags in cars, for instance.

Example 5.4 A small block of mass m = 0.07 kg, initially at rest, is struck by an impulsive force **F** of duration 10 ms whose strength varies with time according to the following graph:

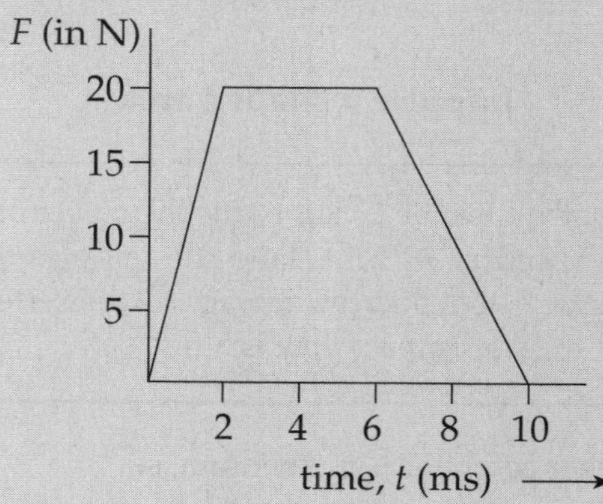

What is the resulting speed of the block?

Solution. The impulse delivered to the block is equal to the area under the F vs. t graph. The region is a trapezoid, so its area, $\frac{1}{2}(\text{base}_1 + \text{base}_2) \times \text{height}$, can be calculated as follows:

$$J = \int_0^{10} F(t)\,dt = \tfrac{1}{2}[(10 \text{ ms} - 0) + (6 \text{ ms} - 2 \text{ ms})] \times (20 \text{ N}) = 0.14 \text{ N} \cdot \text{s}$$

Now, by the impulse–momentum theorem,

$$J = \Delta p = p_\mathrm{f} - p_\mathrm{i} = p_\mathrm{f} = mv_\mathrm{f} \quad \Rightarrow \quad v_\mathrm{f} = \frac{J}{m} = \frac{0.14 \text{ N} \cdot \text{s}}{0.07 \text{ kg}} = 2 \text{ m/s}$$

CONSERVATION OF LINEAR MOMENTUM

Newton's Third Law says that when one object exerts a force on a second object, the second object exerts an equal but opposite force on the first. Since Newton's Second Law says that the impulse delivered to an object is equal to the resulting change in its linear momentum, **J** = Δ**p**, the two interacting objects experience equal but opposite momentum changes (assuming that there are no external forces), which implies that the total linear momentum of the system remains constant. In fact, given any number of interacting objects, each pair that comes in contact will undergo equal but opposite momentum changes, so the result described for two interacting objects will actually hold for any number of objects, given that the only forces they feel are from each other. This means that, in an isolated system, *the total linear momentum will remain constant*. This is the **Law of Conservation of Linear Momentum**.

Example 5.5 An astronaut is floating in space near her shuttle when she realizes that the cord that's supposed to attach her to the ship has become disconnected. Her total mass (body + suit + equipment) is 89 kg. She reaches into her pocket, finds a 1 kg metal tool, and throws it out into space with a velocity of 9 m/s directly away from the ship. If the ship is 10 m away, how long will it take her to reach it?

Solution. Here, the astronaut + tool are the system. Because of Conservation of Linear Momentum,

$$m_{\text{astronaut}}\mathbf{v}_{\text{astronaut}} + m_{\text{tool}}\mathbf{v}_{\text{tool}} = \mathbf{0}$$

$$m_{\text{astronaut}}\mathbf{v}_{\text{astronaut}} = -m_{\text{tool}}\mathbf{v}_{\text{tool}}$$

$$\mathbf{v}_{\text{astronaut}} = -\frac{m_{\text{tool}}}{m_{\text{astronaut}}}\mathbf{v}_{\text{tool}}$$

$$= -\frac{1\text{ kg}}{89\text{ kg}}(-9\text{ m/s}) = +0.101\text{ m/s}$$

Using *distance* = *rate* × *time*, we find

$$t = \frac{d}{v} = \frac{10\text{ m}}{0.101\text{ m/s}} = 98.9\text{ s}$$

COLLISIONS

Conservation of Linear Momentum is routinely used to analyze **collisions**. The objects whose collision we will analyze form the *system*, and although the objects exert forces on each other during the impact, these forces are only *internal* (they occur within the system), and the system's total linear momentum is conserved.

Collisions are classified into two major categories: (1) **elastic** and (2) **inelastic**. A collision is said to be *elastic* if kinetic energy is conserved. Ordinary macroscopic collisions are never truly elastic, because there is always a change in energy due to energy transferred as heat, deformation of the objects, and the sound of the impact. However, if the objects do not deform very much (for example, two billiard balls or a hard glass marble bouncing off a steel plate), then the loss of initial kinetic energy is small enough to be ignored, and the collision can be treated as virtually elastic. *Inelastic* collisions, then, are ones in which the total kinetic energy is different after the collision. An extreme example of inelasticism is **completely** (or **perfectly** or **totally**) **inelastic**. In this case, the objects stick together after the collision and move as one afterward. In all cases of isolated collisions (elastic or not), Conservation of Linear Momentum states that

$$\text{total } \mathbf{p}_{\text{before collision}} = \text{total } \mathbf{p}_{\text{after colision}}$$

Example 5.6 Two balls roll toward each other. The red ball has a mass of 0.5 kg and a speed of 4 m/s just before impact. The green ball has a mass of 0.3 kg and a speed of 2 m/s. After the head-on collision, the red ball continues forward with a speed of 1.7 m/s. Find the speed of the green ball after the collision. Was the collision elastic?

Solution. First remember that momentum is a vector quantity, so the direction of the velocity is crucial. Since the balls roll toward each other, one ball has a positive velocity while the other has a negative velocity. Let's call the red ball's velocity before the collision positive; then $\mathbf{v}_{red} = +4$ m/s, and $\mathbf{v}_{green} = -2$ m/s. Using a prime to denote *after the collision*, Conservation of Linear Momentum gives us the following:

$$\begin{aligned} \text{total } \mathbf{p}_{before} &= \text{total } \mathbf{p}_{after} \\ m_{red}\mathbf{v}_{red} + m_{green}\mathbf{v}_{green} &= m_{red}\mathbf{v}'_{red} + m_{green}\mathbf{v}'_{green} \\ (0.5)(+4) + (0.3)(-2) &= (0.5)(+1.7) + (0.3)\mathbf{v}'_{green} \\ \mathbf{v}'_{green} &= +1.83 \text{ m/s} \end{aligned}$$

Notice that the green ball's velocity was reversed as a result of the collision; this typically happens when a lighter object collides with a heavier object. To see whether the collision was elastic, we need to compare the total kinetic energies before and after the collision. In this case, however, an explicit calculation is not needed since both objects experienced a decrease in speed as a result of the collision. Kinetic energy was lost (so the collision was inelastic); this is usually the case with macroscopic collisions. Most of the lost energy was transferred as heat; the two objects are both slightly warmer as a result of the collision.

Example 5.7 Two balls roll toward each other. The red ball has a mass of 0.5 kg and a speed of 4 m/s just before impact. The green ball has a mass of 0.3 kg and a speed of 2 m/s. If the collision is completely inelastic, determine the velocity of the composite object after the collision.

Solution. If the collision is completely inelastic, then, by definition, the masses stick together after impact, moving with a velocity, $\mathbf{v}'$. Applying Conservation of Linear Momentum, we find

$$\begin{aligned} \text{total } \mathbf{p}_{before} &= \text{total } \mathbf{p}_{after} \\ m_{red}\mathbf{v}_{red} + m_{green}\mathbf{v}_{green} &= (m_{red} + m_{green})\mathbf{v}' \\ (0.5)(+4) + (0.3)(-2) &= (0.5 + 0.3)\mathbf{v}' \\ \mathbf{v}' &= +1.8 \text{ m/s} \end{aligned}$$

Example 5.8 An object of mass m_1 is moving with velocity v_1 toward a target object of mass m_2 which is stationary ($v_2 = 0$). The objects collide head-on, and the collision is elastic. Show that the relative velocity before the collision, $v_2 - v_1$, has the same magnitude as $v_2' - v_1'$, the relative velocity after the collision.

Solution. Since the collision is elastic, both total linear momentum and kinetic energy are conserved. Therefore,

$$m_1 v_1 = m_1 v_1' + m_2 v_2' \quad (1)$$
$$\tfrac{1}{2} m_1 v_1^2 = \tfrac{1}{2} m_1 v_1'^2 + \tfrac{1}{2} m_2 v_2'^2 \quad (2)$$

Now for some algebra. Cancel the $\frac{1}{2}$'s in the second equation and factor to get the following pair of equations:

$$m_1(v_1 - v_1') = m_2 v_2' \quad (1')$$
$$m_1(v_1 - v_1')(v_1 + v_1') = m_2 v_2'^2 \quad (2')$$

Next, dividing the second equation by the first gives $v_1 + v_1' = v_2'$, so we can write

$$v_1 - v_1' = (m_2 / m_1) v_2' \quad (1'')$$
$$v_1 + v_1' = v_2' \quad (2'')$$

Adding Equations (1″) and (2″) gives

$$2v_1 = \left(\tfrac{m_2}{m_1} + 1\right) v_2' \quad \Rightarrow \quad v_2' = \frac{2v_1}{\frac{m_2}{m_1} + 1} = \frac{2m_1}{m_1 + m_2} v_1$$

Substituting this result into Equation (2″) gives

$$v_1 + v_1' = \frac{2m_1}{m_1 + m_2} v_1 \quad \Rightarrow \quad v_1' = \left(\frac{2m_1}{m_1 + m_2} - 1\right) v_1 = \frac{m_1 - m_2}{m_1 + m_2} v_1$$

Now we have calculated the final velocities, v_1' and v_2'. To verify the claim made in the statement of the question, we notice that

$$v_2' - v_1' = \frac{2m_1}{m_1 + m_2} v_1 - \frac{m_1 - m_2}{m_1 + m_2} v_1 = \frac{m_1 + m_2}{m_1 + m_2} v_1 = v_1 - 0 = v_1 - v_2 = -(v_2 - v_1)$$

so the relative velocity after the collision, $v_2' - v_1'$, is equal (in magnitude) but opposite (in direction) to $v_2 - v_1$, the relative velocity before the collision. *This is a general property that characterizes elastic collisions.*

Example 5.9 An object of mass m moves with velocity $\mathbf{v}$ toward a stationary object of mass $2m$. After impact, the objects move off in the directions shown in the following diagram:

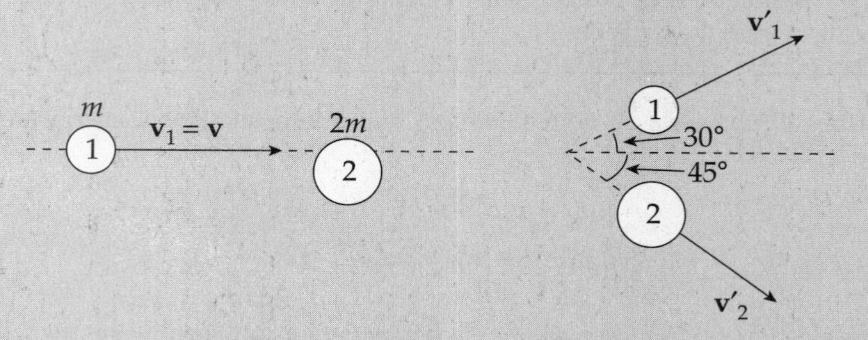

Before the collision After the collision

(a) Determine the magnitudes of the velocities after the collision (in terms of v).

(b) Is the collision elastic? Explain your answer.

Solution.

(a) Conservation of Linear Momentum is a principle that establishes the equality of two vectors: $\mathbf{p}_{\text{total}}$ before the collision and $\mathbf{p}_{\text{total}}$ after the collision. Writing this single vector equation as two equations, one for the x component and one for the y, we have

$$x \text{ component:} \quad mv = mv_1' \cos 30° + 2mv_2' \cos 45° \qquad (1)$$

$$y \text{ component:} \quad 0 = mv_1' \sin 30° - 2mv_2' \sin 45° \qquad (2)$$

Adding these equations eliminates v_2',

$$mv = mv_1'(\cos 30° + \sin 30°)$$

and lets us determine v_1':

$$v_1' = \frac{v}{\cos 30° + \sin 30°} = \frac{2v}{1+\sqrt{3}}$$

Substituting this result into Equation (2) gives us

$$0 = m\frac{2v}{1+\sqrt{3}}\sin 30^\circ - 2mv_2'\sin 45^\circ$$

$$2mv_2'\sin 45^\circ = m\frac{2v}{1+\sqrt{3}}\sin 30^\circ$$

$$v_2' = \frac{\frac{2v}{1+\sqrt{3}}\sin 30^\circ}{2\sin 45^\circ} = \frac{v}{\sqrt{2}(1+\sqrt{3})}$$

(b) The collision is elastic only if kinetic energy is conserved. The total kinetic energy after the collision, K', is calculated as follows:

$$\begin{aligned} K' &= \frac{1}{2}\cdot mv_1'^2 + \frac{1}{2}\cdot 2mv_2'^2 \\ &= \frac{1}{2}m\left(\frac{v}{1+\sqrt{3}}\right)^2 + m\left(\frac{v}{\sqrt{2}(1+\sqrt{3})}\right)^2 \\ &= mv^2\left[\frac{1}{2(1+\sqrt{3})^2} + \frac{1}{2(1+\sqrt{3})^2}\right] \\ &= \frac{mv^2}{(1+\sqrt{3})^2} \end{aligned}$$

However, the kinetic energy before the collision is just $K = \frac{1}{2}mv^2$, so the fact that

$$\frac{1}{(1+\sqrt{3})^2} < \frac{1}{2}$$

tells us that K' is less than K, so some kinetic energy is lost; the collision is inelastic.

CENTER OF MASS

The center of mass is the point where all of the mass of an object can be considered to be concentrated; it's the dot that represents the object of interest in a free-body diagram.

For a homogeneous body (that is, one for which the density is uniform throughout), the center of mass is where you intuitively expect it to be: at the geometric center. Thus, the center of mass of a uniform sphere or cube or box is at its geometric center.

If we have a collection of discrete particles, the center of mass of the system can be determined mathematically as follows. First consider the case where the particles all lie on a straight line. Call this the x axis. Select some point to be the origin ($x = 0$) and determine the positions of each particle on the axis. Multiply each position value by the mass of the particle at that location, and get the sum for all the particles. Divide this sum by the total mass, and the resulting x value is the center of mass:

$$x_{cm} = \frac{m_1x_1 + m_2x_2 + \cdots + m_nx_n}{m_1 + m_2 + \cdots + m_n}$$

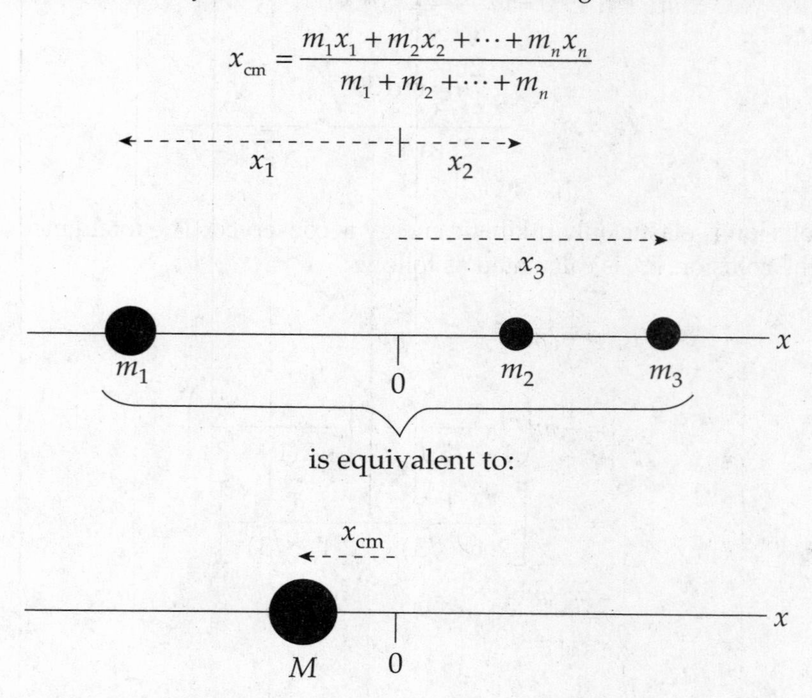

The system of particles behaves in many respects as if all its mass, $M = m_1 + m_2 + \cdots + m_n$, were concentrated at a single location, x_{cm}.

If the system consists of objects that are not confined to the same straight line, use the equation above to find the x coordinate of their center of mass, and the corresponding equation,

$$y_{cm} = \frac{m_1y_1 + m_2y_2 + \cdots + m_ny_n}{m_1 + m_2 + \cdots + m_n}$$

to find the y coordinate of their center of mass (and one more equation to calculate the z coordinate, if they are not confined to a single plane).

From the equation

$$x_{cm} = \frac{m_1x_1 + m_2x_2 + \cdots + m_nx_n}{M}$$

we can derive

$$Mv_{cm} = m_1v_1 + m_2v_2 + \cdots + m_nv_n$$

So, the total linear momentum of all the particles in the system ($m_1v_1 + m_2v_2 + \ldots + m_nv_n$) is the same as Mv_{cm}, the linear momentum of a single particle (whose mass is equal to the system's total mass) moving with the velocity of the center of mass.

We can also differentiate again and establish the following:

$$\mathbf{F}_{net} = M\mathbf{a}_{cm}$$

This says that the net (external) force acting on the system causes the center of mass to accelerate according to Newton's Second Law. In particular, *if the net external force on the system is zero, then the center of mass will not accelerate.*

Example 5.10 Two objects, one of mass m and one of mass $2m$, hang from light threads from the ends of a uniform bar of length $3L$ and mass $3m$. The masses m and $2m$ are at distances L and $2L$, respectively, below the bar. Find the center of mass of this system.

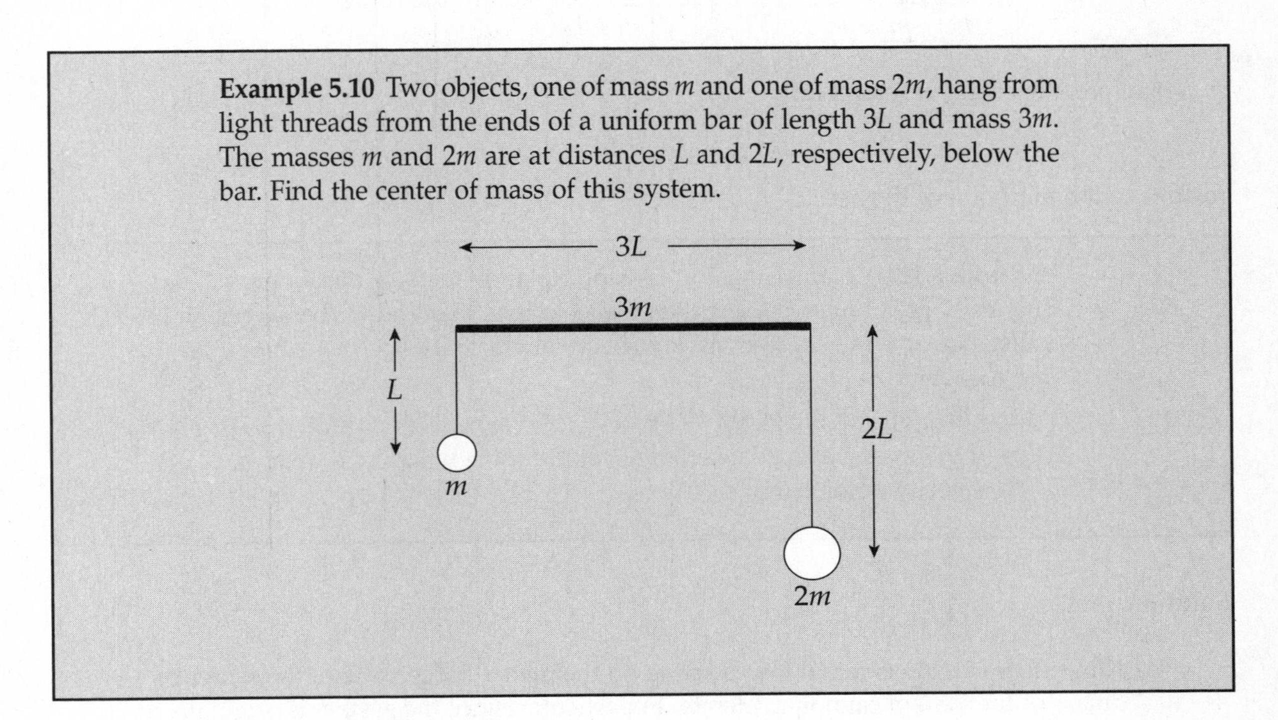

Solution. The center of mass of the bar alone is at its midpoint (because it is uniform), so we may treat the total mass of the bar as being concentrated at its midpoint. Constructing a coordinate system with this point as the origin, we now have three objects: one of mass m at $(-3L/2, -L)$, one of mass $2m$ at $(3L/2, -2L)$, and one of mass $3m$ at $(0, 0)$:

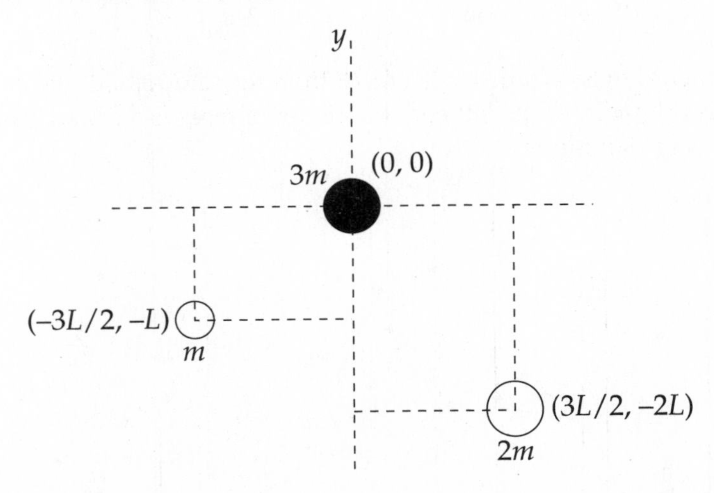

We figure out the x and y coordinates of the center of mass separately:

$$x_{cm} = \frac{m_1x_1 + m_2x_2 + m_3x_3}{m_1 + m_2 + m_3} = \frac{(m)(-3L/2) + (2m)(3L/2) + (3m)(0)}{m + 2m + 3m} = \frac{3mL/2}{6m} = \frac{L}{4}$$

$$y_{cm} = \frac{m_1y_1 + m_2y_2 + m_3y_3}{m_1 + m_2 + m_3} = \frac{(m)(-L) + (2m)(-2L) + (3m)(0)}{m + 2m + 3m} = \frac{-5mL}{6m} = -\frac{5L}{6}$$

Therefore, the center of mass is at

$$(x_{cm}, y_{cm}) = (L/4, -5L/6)$$

relative to the midpoint of the bar.

Example 5.11 A man of mass m is standing at one end of a stationary, floating barge of mass $3m$. He then walks to the other end of the barge, a distance of L meters. Ignore any frictional effects between the barge and the water.

(a) How far will the barge move?

(b) If the man walks at an average velocity of v, what is the average velocity of the barge?

Solution.

(a) Since there are no external forces acting on the man + barge system, the center of mass of the system cannot accelerate. In particular, since the system is originally at rest, the center of mass cannot move. Letting $x = 0$ denote the midpoint of the barge (which is its own center of mass, assuming it is uniform), we figure out the center of mass of the man + barge system:

$$x_{cm} = \frac{m_1x_1 + m_2x_2}{m_1 + m_2} = \frac{(m)(-L/2) + (3m)(0)}{m + 3m} = -\frac{L}{8}$$

So, the center of mass is a distance of $L/8$ from the midpoint of the barge, and since the mass is originally at the left end, the center of mass is a distance of $L/8$ to the left of the barge's midpoint.

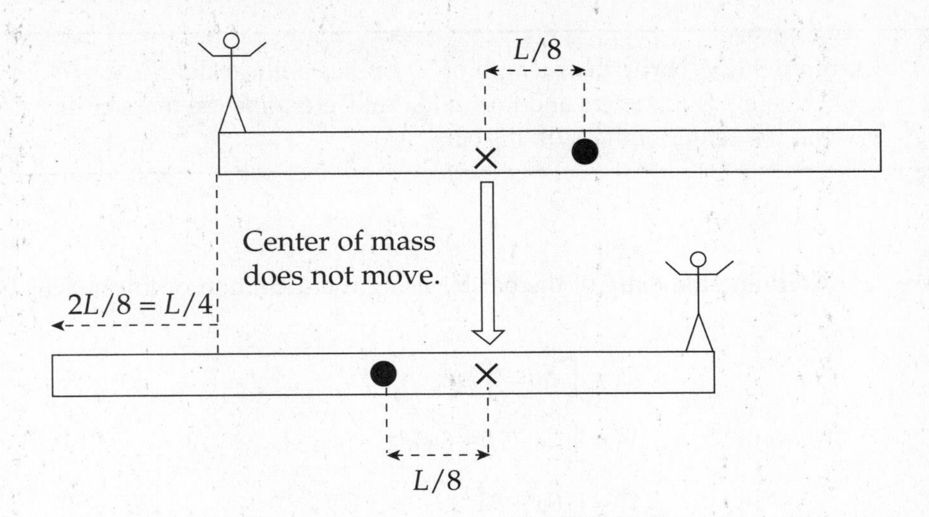

When the man reaches the other end of the barge, the center of mass will, by symmetry, be $L/8$ to the *right* of the midpoint of the barge. But, since the position of the center of mass cannot move, this means the barge itself must have moved a distance of

$$L/8 + L/8 = 2L/8 = L/4$$

to the left.

(b) Let the time it takes the man to walk across the barge be denoted by t; then $t = L/v$. In this amount of time, the barge moves a distance of $L/4$ in the *opposite* direction, so the velocity of the barge is

$$v_{\text{barge}} = \frac{-L/4}{t} = \frac{-L/4}{L/v} = -\frac{v}{4}$$

So far we have dealt with objects that can be considered point masses, or masses with uniform density. Now we will learn how to find the center of mass of objects with non-uniform density.

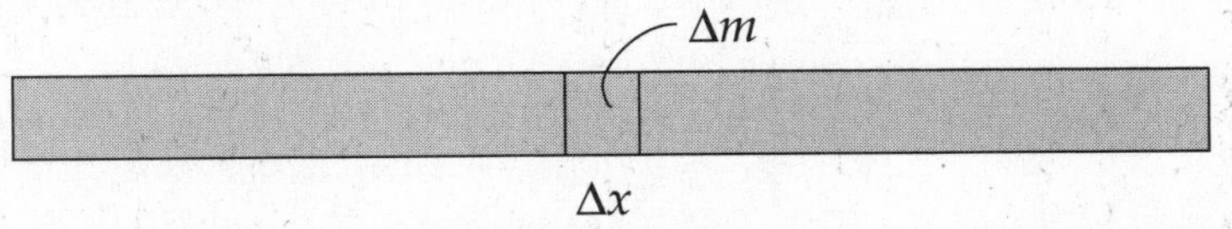

Take the example of a bar that becomes denser along its length. Here we will deal with the linear density λ as a function of x, $\lambda(x)$. Each small segment of the bar, Δx, has a different mass, Δm. We treat each Δm as a point mass and then take the limit as Δx approaches zero. Using the formula for calculating the center of mass of point masses, and replacing Δm with dm, we get the integral shown below.

$x_{cm} = \frac{1}{M}\int x\,dm$, where M is the total mass, x is the distance to each dm, and you can substitute for dm in terms of x. Linear density is mass per length, so the equation is $\lambda = \frac{dm}{dx}$, therefore $M = \int dm = \int \lambda dx$.

Example 5.12 A bar with of length of 30 cm has a linear density $\lambda = 10 + 6x$, where x is in meters and λ is in kg/m. Determine the mass of the bar and the center of mass of this bar.

Solution. We can determine the mass of the bar by using the definition of linear density, $\lambda = \frac{dm}{dx}$. Therefore

$$M = \int dm = \int \lambda dx$$

$$M = \int_0^{0.30} (10 + 6x)dx$$

$$M = (10x + 3x^2)\Big|_0^{0.30} = 3.27 \text{ kg}$$

To calculate the center of mass we will use the equation $x_{cm} = \frac{1}{M}\int xdm = \frac{1}{M}\int x\lambda dx.$

$$x_{cm} = \frac{1}{3.27}\int_0^{0.3} x(10 + 6x)dx$$

$$x_{cm} = \frac{1}{3.27}(5x^2 + 2x^3)\Big|_0^{0.3}$$

$$x_{cm} = \frac{1}{3.27}(0.504)$$

$$x_{cm} = 0.154 \text{ m}$$

This answer makes sense because the center of mass of the bar is beyond the midpoint.

CHAPTER 5 REVIEW QUESTIONS

Section I: Multiple Choice

1. An object of mass 2 kg has a linear momentum of magnitude 6 kg·m/s. What is this object's kinetic energy?

   (A) 3 J
   (B) 6 J
   (C) 9 J
   (D) 12 J
   (E) 18 J

2. A ball of mass 0.5 kg, initially at rest, acquires a speed of 4 m/s immediately after being kicked by a force of strength 20 N. For how long did this force act on the ball?

   (A) 0.01 s
   (B) 0.02 s
   (C) 0.1 s
   (D) 0.2 s
   (E) 1 s

3. A box with a mass of 2 kg accelerates in a straight line from 4 m/s to 8 m/s due to the application of a force whose duration is 0.5 s. Find the average strength of this force.

   (A) 2 N
   (B) 4 N
   (C) 8 N
   (D) 12 N
   (E) 16 N

4. A ball of mass m traveling horizontally with velocity $\mathbf{v}$ strikes a massive vertical wall and rebounds back along its original direction with no change in speed. What is the magnitude of the impulse delivered by the wall to the ball?

   (A 0
   (B) $\frac{1}{2}mv$
   (C) mv
   (D) $2mv$
   (E) $4mv$

5. Two objects, one of mass 3 kg and moving with a speed of 2 m/s and the other of mass 5 kg and speed 2 m/s, move toward each other and collide head-on. If the collision is perfectly inelastic, find the speed of the objects after the collision.

   (A) 0.25 m/s
   (B) 0.5 m/s
   (C) 0.75 m/s
   (D) 1 m/s
   (E) 2 m/s

6. Object 1 moves toward Object 2, whose mass is twice that of Object 1 and which is initially at rest. After their impact, the objects lock together and move with what fraction of Object 1's initial kinetic energy?

   (A) 1/18
   (B) 1/9
   (C) 1/6
   (D) 1/3
   (E) None of the above

7. Two objects move toward each other, collide, and separate. If there was no net external force acting on the objects, but some kinetic energy was lost, then

   (A) the collision was elastic and total linear momentum was conserved
   (B) the collision was elastic and total linear momentum was not conserved
   (C) the collision was not elastic and total linear momentum was conserved
   (D) the collision was not elastic and total linear momentum was not conserved
   (E) None of the above

8. Three thin, uniform rods each of length L are arranged in the shape of an inverted U:

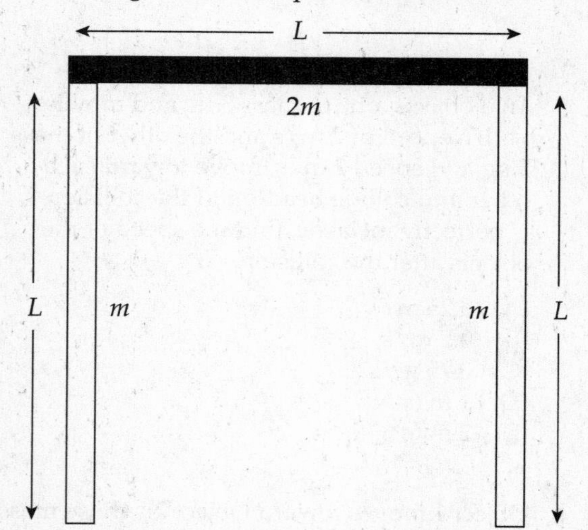

The two rods on the arms of the U each have mass m; the third rod has mass $2m$. How far below the midpoint of the horizontal rod is the center of mass of this assembly?

(A) $L/8$
(B) $L/4$
(C) $3L/8$
(D) $L/2$
(E) $3L/4$

9. A wooden block of mass M is moving at speed V in a straight line.

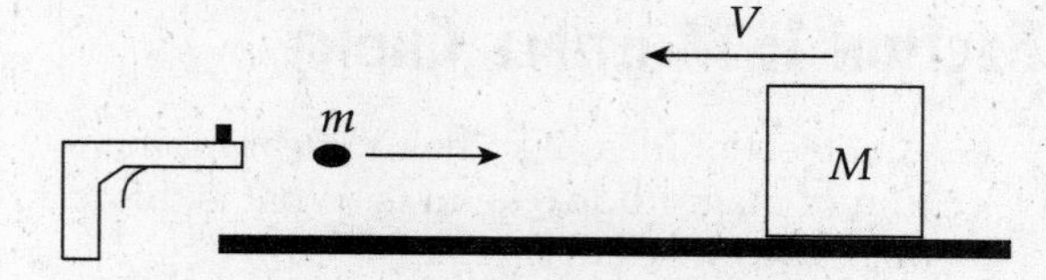

How fast would the bullet of mass m need to travel to stop the block (assuming that the bullet became embedded inside)?

(A) $mV/(m + M)$
(B) $MV/(m + M)$
(C) mV/M
(D) MV/m
(E) $(m + M)V/m$

10. Which of the following best describes a perfectly inelastic collision free of external forces?

(A) Total linear momentum is never conserved.
(B) Total linear momentum is sometimes conserved.
(C) Kinetic energy is never conserved.
(D) Kinetic energy is sometimes conserved.
(E) Kinetic energy is always conserved.

SECTION II: FREE RESPONSE

1. A steel ball of mass m is fastened to a light cord of length L and released when the cord is horizontal. At the bottom of its path, the ball strikes a hard plastic block of mass $M = 4m$, initially at rest on a frictionless surface. The collision is elastic.

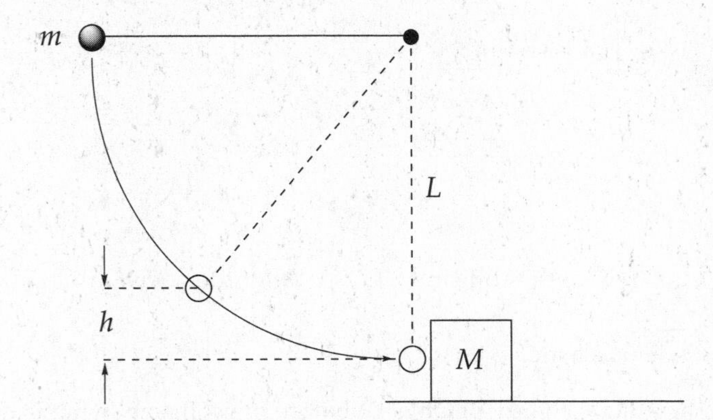

   (a) Find the tension in the cord when the ball's height above its lowest position is $\frac{1}{2}L$. Write your answer in terms of m and g.

   (b) Find the speed of the block immediately after the collision.

   (c) To what height h will the ball rebound after the collision?

2. A *ballistic pendulum* is a device that may be used to measure the muzzle speed of a bullet. It is composed of a wooden block suspended from a horizontal support by cords attached at each end. A bullet is shot into the block, and as a result of the perfectly inelastic impact, the block swings upward. Consider a bullet (mass m) with velocity v as it enters the block (mass M). The length of the cords supporting the block each have length L. The maximum height to which the block swings upward after impact is denoted by y, and the maximum horizontal displacement is denoted by x.

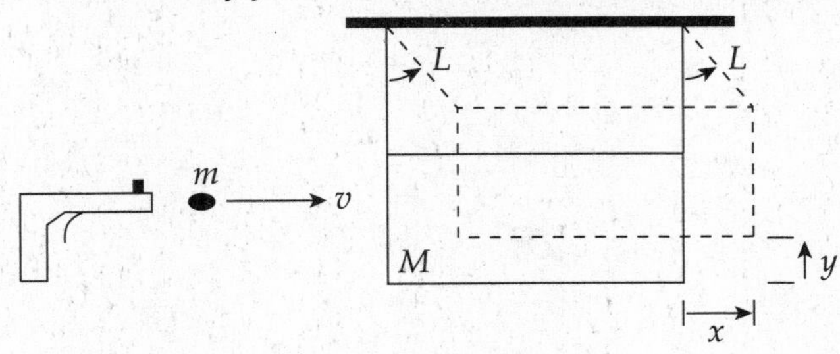

   (a) In terms of m, M, g, and y, determine the speed v of the bullet.

   (b) What fraction of the bullet's original kinetic energy is lost as a result of the collision? What happens to the lost kinetic energy?

   (c) If y is very small (so that y^2 can be neglected), determine the speed of the bullet in terms of m, M, g, x, and L.

   (d) Once the block begins to swing, does the momentum of the block remain constant? Why or why not?

3. An object of mass m moves with velocity $\mathbf{v}$ toward a stationary object of the same mass. After their impact, the objects move off in the directions shown in the following dagram:

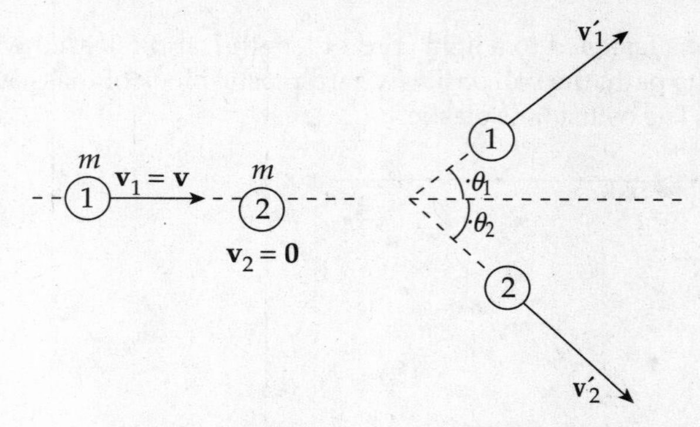

Before the collision After the collision

Assume that the collision is elastic.

(a) If K_1 denotes the kinetic energy of Object 1 before the collision, what is the kinetic energy of this object after the collision? Write your answer in terms of K_1 and θ_1.

(b) What is the kinetic energy of Object 2 after the collision? Write your answer in terms of K_1 and θ_1.

(c) What is the relationship between θ_1 and θ_2?

LINEAR MOMENTUM SUMMARY

MOMENTUM

- Linear momentum is given by the equation: $\mathbf{p} = m\mathbf{v}$
- Linear momentum is conserved when no external force acts on a system. This is known as the **Law of Conservation of Linear Momentum**. It can be written as

$$\text{Total } \mathbf{p}_{\text{before collision}} = \text{total } \mathbf{p}_{\text{after collission}}$$

- **Elastic collisions** conserve *kinetic energy* (in general, every collision conserves energy but not necessarily kinetic energy).
- **Inelastic collisions** do not conserve *kinetic energy.*
- When the objects stick together, the collision is known as **perfectly inelastic**.
- Anytime you are given a problem that involves a collision or separation, first consider whether you can use the Law of Conservation of Linear Momentum.

IMPULSE

- **Impulse** is given by the equation: $J = \overline{F}\Delta t$
- The **Impulse/Momentum Theorem** states that the impulse on an object is equal to the change in momentum of the object. The equation is: $J = \overline{F}\Delta t = \Delta p$

CENTER OF MASS

- Usually the motion of an object is describing the motion of the center of mass. When you use Newton's Second Law, F=ma, the acceleration you calculate is the acceleration of the center of mass.
- For point masses, $r_{cm} = \dfrac{\sum mr}{\sum m}$, where r is used for the position of each mass.
- For distributed mass (e.g. a bar with non-uniform density) the equation is:

  $r_{cm} = \int r^2 dm$
- You usually use linear density, $\lambda dr = dm$, for dm and then integrate to solve for the center of mass.

Rotational Motion

INTRODUCTION

So far we've studied only translational motion: objects sliding, falling, or rising, but in none of our examples have we considered spinning objects. We will now look at rotation, which will complete our study of motion. All motion is some combination of **translation** and **rotation**, which are illustrated in the figures below. Consider any two points in the object under study (on the left) and imagine connecting them by a straight line. If this line always remains parallel to itself while the object moves, then the object is translating only. However, if this line does not always remain parallel to itself while the object moves, then the object is rotating.

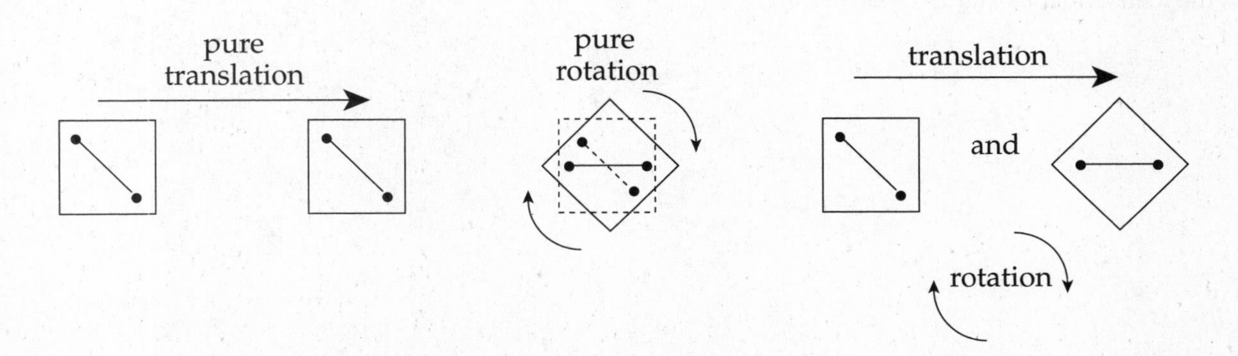

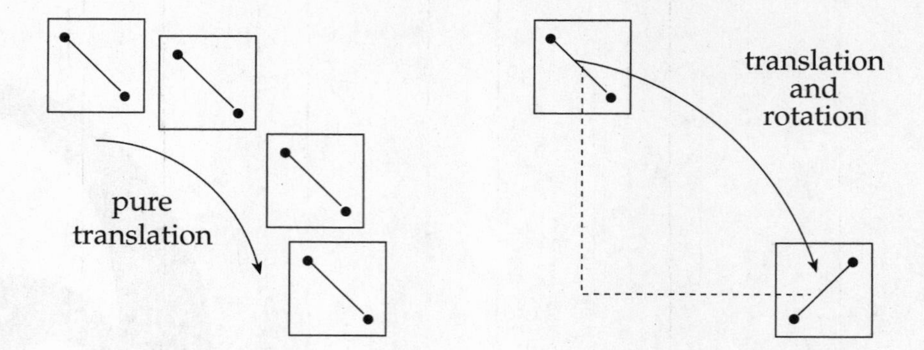

ROTATIONAL KINEMATICS

Mark several dots along a radius on a disk, and call this radius the *reference line*. If the disk rotates about its center, we can use the movement of these dots to talk about angular displacement, angular velocity, and angular acceleration.

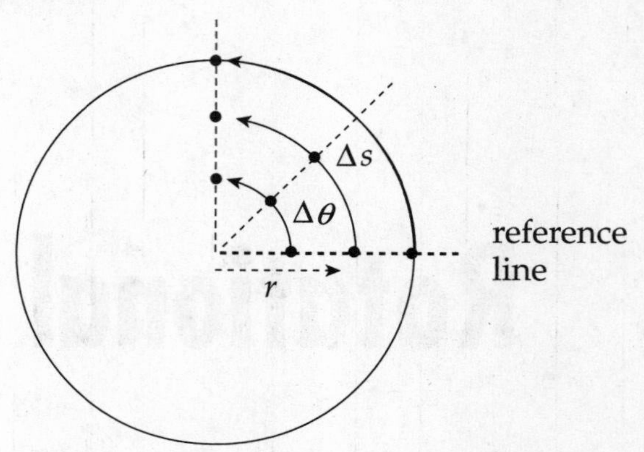

If the disk rotates as a rigid body, then all three dots shown have the same **angular displacement**, $\Delta\theta$. In fact, this is the definition of a **rigid body**: All points along a radial line always have the same angular displacement.

Just as the time rate-of-change of displacement gives velocity, the time rate-of-change of angular displacement gives angular velocity, denoted by ω (*omega*). The definition of the **average angular velocity** is:

$$\bar{\omega} = \frac{\Delta\theta}{\Delta t}$$

Note that if we let the time interval Δt approach 0, then the equation above leads to the definition of the instantaneous angular velocity:

$$\omega = \frac{d\theta}{dt}$$

And, finally, just as the time rate-of-change of velocity gives acceleration, the time rate-of-change of angular velocity gives angular acceleration, or α (*alpha*). The definition of the **average angular acceleration** is:

$$\bar{\alpha} = \frac{\Delta\omega}{\Delta t}$$

If we let the time interval Δt approach 0, then the equation above leads to the definition of the instantaneous angular velocity:

$$\alpha = \frac{d\omega}{dt}$$

On the rotating disk illustrated on the previous page, we said that all points undergo the same angular displacement in any given time interval; this means that all points on the disk have the same angular velocity, ω, but not all points have the same linear velocity, v. This follows from the definition of **radian** measure. Expressed in radians, the angular displacement, $\Delta\theta$, is related to the arc length, Δs, by the equation

$$\Delta\theta = \frac{\Delta s}{r}$$

Rearranging this equation and dividing by Δt, we find that

$$\Delta s = r\Delta\theta \quad \Rightarrow \quad \frac{\Delta s}{\Delta t} = r\frac{\Delta\theta}{\Delta t} \quad \Rightarrow \quad \bar{v} = r\bar{\omega}$$

Or, using the equations $v = ds/dt$ and $\omega = d\theta/dt$,

$$v = r\omega$$

Therefore, the greater the value of r, the greater the value of v. Points on the rotating body farther from the rotation axis move more quickly than those closer to the rotation axis.

From the equation $v = r\omega$, we can derive the relationship that connects angular acceleration and linear acceleration. Differentiating both sides with respect to t (holding r constant), gives us

$$\frac{dv}{dt} = r\frac{d\omega}{dt} \quad \Rightarrow \quad a = r\alpha$$

(It's important to realize that the acceleration a in this equation is *not* centripetal acceleration; it's tangential acceleration, which arises from a change in speed caused by an angular acceleration. By contrast, centripetal acceleration does not produce a change in speed.) Often, tangential acceleration is written as a_t to distinguish it from centripetal acceleration (a_c).

Example 6.1 A rotating, rigid body makes one complete revolution in 2s. What is its average angular velocity?

Solution. One complete revolution is equal to an angular displacement of 2π radians, so the body's average angular velocity is

$$\bar{\omega} = \frac{\Delta\theta}{\Delta t} = \frac{2\pi \text{ rad}}{2 \text{ s}} = \pi \text{ rad/s}$$

Example 6.2 The angular velocity of a rotating disk increases from 2 rad/s to 5 rad/s in 0.5 s. What's the disk's average angular acceleration?

Solution. By definition,

$$\bar{\alpha} = \frac{\Delta\omega}{\Delta t} = \frac{(5-2) \text{ rad/s}}{0.5 \text{ s}} = 6 \text{ rad/s}^2$$

Example 6.3 A disk of radius 20 cm rotates at a constant angular velocity of 6 rad/s. How fast does a point on the rim of this disk travel (in m/s)?

Solution. The linear speed, v, is related to the angular speed, ω, by the equation $v = r\omega$. Therefore,

$$v = r\omega = (0.20 \text{ m})(6 \text{ rad/s}) = 1.2 \text{ m/s}$$

Note that although we typically write the abbreviation *rad* when writing angular measurements, the radian is actually a dimensionless quantity, since, by definition, $\theta = s/r$. So $\Delta\theta = 6$ means the same thing as $\Delta\theta = 6$ rad.

Example 6.4 The angular velocity of a rotating disk of radius 50 cm increases from 2 rad/s to 5 rad/s in 0.5 s. What is the linear tangential acceleration of a point on the rim of the disk during this time interval?

Solution. The linear acceleration a is related to the angular acceleration α by the equation $a = r\alpha$. Since $\alpha = 6 \text{ rad/s}^2$ (as calculated in Example 6.2), we find that

$$a = r\alpha = (0.50 \text{ m})(6 \text{ rad/s}^2) = 3 \text{ m/s}^2$$

Example 6.5 Derive an expression for centripetal acceleration in terms of angular speed.

Solution. For an object revolving with linear speed v at a distance r from the center of rotation, the centripetal acceleration is given by the equation $a_c = v^2/r$. Using the fundamental equation $v = r\omega$, we find that

$$a_c = \frac{v^2}{r} = \frac{(r\omega)^2}{r} = \omega^2 r$$

THE BIG FIVE FOR ROTATIONAL MOTION

The simplest type of rotational motion to analyze is motion in which the angular acceleration is *constant* (possibly equal to zero). Another restriction that will make our analysis easier (and which doesn't diminish the power and applicability of our results too much) is to consider rotational motion around a *fixed* axis of rotation. In this case, there are only two possible directions for motion. One direction, counterclockwise, is called *positive* (+), and the opposite direction, clockwise, is called *negative* (–).

Let's review the quantities we've seen so far. The fundamental quantities for rotational motion are angular displacement ($\Delta\theta$), angular velocity (ω), and angular acceleration (α). Because we're dealing with angular acceleration, we know about changes in angular velocity, from initial velocity (ω_i or ω_0) to final velocity (ω_f or simply ω—with no subscript). And, finally, the motion takes place during some elapsed time interval, Δt. Therefore, we have five kinematics quantities: $\Delta\theta$, ω_0, ω, α, and Δt.

These five quantities are interrelated by a group of five equations which we call the *Big Five*. They work in cases in which the angular acceleration is uniform. These equations are identical to the Big Five we studied in Chapter 2 but, in these cases, the translational variables (s, v, or a) are replaced by the corresponding rotational variables (θ, ω, or α, respectively).

In Big Five #1, because angular acceleration is constant, the average angular velocity is simply the average of the initial angular velocity and the final angular velocity: $\bar{\omega} = \frac{1}{2}(\omega_0 + \omega)$. Also, if we decide that $t_i = 0$, then $\Delta t = t_f - t_i = t - 0 = t$, so we can just write "$t$" instead of "$\Delta t$" in the first four equations. This simplification in notation makes the equations a little easier to memorize.

		Variable that's missing
Big Five #1:	$\Delta\theta = \bar{\omega}t$	α
Big Five #2:	$\omega = \omega_0 + dt$	$\Delta\theta$
Big Five #3:	$\Delta\theta = \omega_0 t + \frac{1}{2}\alpha(t)^2$	ω
Big Five #4:	$\Delta\theta = \omega t - \frac{1}{2}\alpha(t)^2$	ω_0
Big Five #5:	$\omega^2 = \omega_0^2 + 2\alpha\Delta\theta$	Δt

Each of the Big Five equations is missing exactly one of the five kinematics quantities and, as with the other Big Five you learned, the way you decide which equation to use is to determine which of the kinematics quantities is missing from the problem, and use the equation that's also missing that quantity. For example, if the problem never mentions the final angular velocity—ω is neither given nor asked for—then the equation that will work is the one that's missing ω; that's Big Five #3.

Notice that Big Five #1 and #2 are simply the definitions of $\bar{\omega}$ and $\bar{\alpha}$ written in forms that don't involve fractions.

Example 6.6 An object with an initial angular velocity of 1 rad/s rotates with constant angular acceleration. Three seconds later, its angular velocity is 5 rad/s. Calculate its angular displacement during this time interval.

Solution. We're given ω_0, t, and ω, and asked for $\Delta\theta$. So α is missing, and we use Big Five #1:

$$\Delta\theta = \bar{\omega}\Delta t = \tfrac{1}{2}(\omega_0 + \omega)\Delta t = \tfrac{1}{2}(1 \text{ rad/s} + 5 \text{ rad/s})(3 \text{ s}) = 9 \text{ rad}$$

> **Example 6.7** Starting with zero initial angular velocity, a sphere begins to spin with constant angular acceleration about an axis through its center, achieving an angular velocity of 10 rad/s when its angular displacement is 20 rad. What is the value of the sphere's angular acceleration?

Solution. We're given ω_0, $\Delta\theta$, and ω, and asked for α. Since t is missing, we use Big Five #5:

$$\omega^2 = \omega_0^2 + 2\alpha\Delta\theta \quad \Rightarrow \quad \omega^2 = 2\alpha\Delta\theta \ (\text{since } \omega_0 = 0)$$

$$\alpha = \frac{\omega^2}{2\Delta\theta} = \frac{(10 \text{ rad/s})^2}{2(20 \text{ rad})} = 2.5 \text{ rad/s}^2$$

To summarize, here's a comparison of the fundamental quantities of translational and rotational motion and of the Big Five (assuming constant acceleration and a fixed axis of rotation):

	Translational	*Rotational*	*Connection*
displacement:	Δx	$\Delta\theta$	$\Delta x = r\Delta\theta$
velocity:	v	ω	$v = r\omega$
acceleration:	a	α	$a = r\alpha$
Big Five #1:	$\Delta x = x - x_0 = \bar{v}t$	$\Delta\theta = \bar{\omega}t$	
Big Five #2:	$v = v_0 + at$	$\omega = \omega_0 + dt$	
Big Five #3:	$x = x_0 + v_0 t + \frac{1}{2}at^2$	$\Delta\theta = \omega_0 t + \frac{1}{2}\alpha(t)^2$	
Big Five #4:	$x = x_0 + vt - \frac{1}{2}at^2$	$\Delta\theta = \omega t - \frac{1}{2}\alpha(t)^2$	
Big Five #5:	$v^2 = v_0^2 + 2a(x - x_0)$	$\omega^2 = \omega_0^2 + 2\alpha\Delta\theta$	

ROTATIONAL DYNAMICS

The dynamics of translational motion involve describing the acceleration of an object in terms of its mass (inertia) and the forces that act on it; $F_{net} = ma$. By analogy, the dynamics of rotational motion involve describing the angular (rotational) acceleration of an object in terms of its **rotational inertia** and the **torques** that act on it.

TORQUE

Intuitively, torque describes the effectiveness of a force in producing rotational acceleration. Consider a uniform rod that pivots around one of its ends, which is fixed. For simplicity, let's assume that the rod is at rest. What effect, if any, would each of the four forces in the figure below have on the potential rotation of the rod?

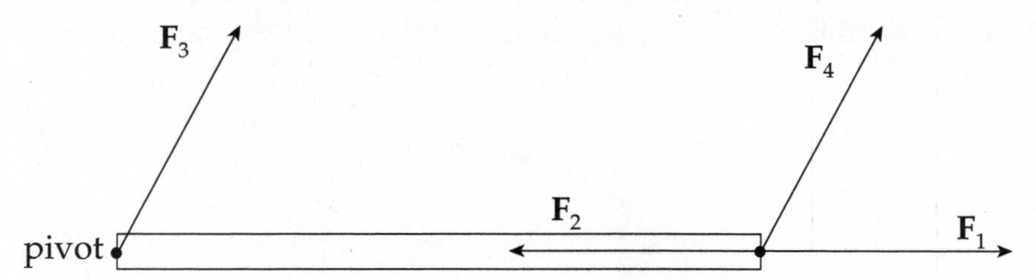

Our intuition tells us that $\mathbf{F}_1$, $\mathbf{F}_2$, and $\mathbf{F}_3$ would *not* cause the rod to rotate, but $\mathbf{F}_4$ would. What's different about $\mathbf{F}_4$? *It has torque.*

The torque of a force can be defined as follows. Let r be the distance from the pivot (axis of rotation) to the point of application of the force **F**, and let θ be the angle between vectors **r** and **F**.

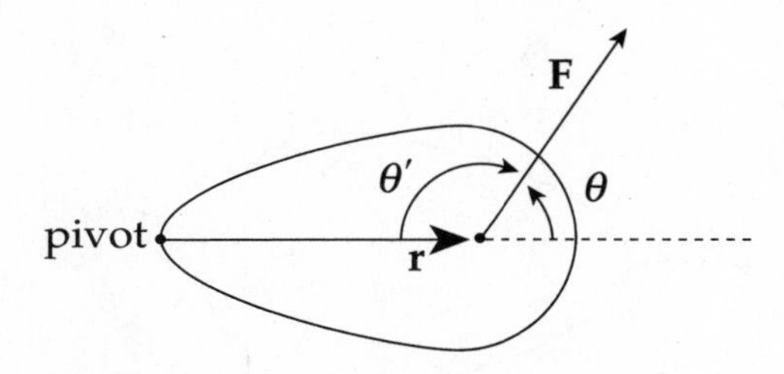

Then the torque of **F**, denoted by τ (*tau*), is defined as:

$$\tau = rF \sin \theta$$

In the figure above, the angle between the vectors **r** and **F** is θ. Imagine sliding **r** over so that its initial point is the same as that of **F**. *The angle between two vectors is the angle between them when they start at the same point.* However, the supplementary angle θ' can be used in place of θ in the definition of torque. This is because torque depends on sin θ, and the sine of an angle and the sine of its supplement are always equal. Therefore, when figuring out torque, use whichever of these angles is more convenient.

We will now see if this mathematical definition of torque supports our intuition about forces $\mathbf{F}_1$, $\mathbf{F}_2$, $\mathbf{F}_3$, and $\mathbf{F}_4$.

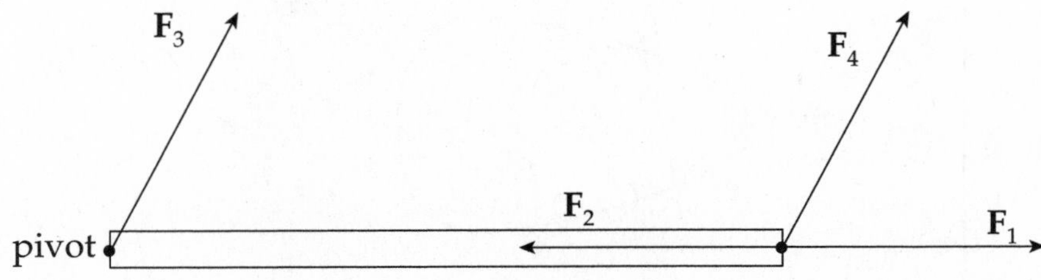

The angle between **r** and $\mathbf{F}_1$ is 0, and $\theta = 0$ implies $\sin\theta = 0$, so by the definition of torque, $\tau = 0$ as well. The angle between **r** and $\mathbf{F}_2$ is 180°, and $\theta = 180°$ gives us $\sin\theta = 0$, so $\tau = 0$. For $\mathbf{F}_3$, $r = 0$ (because $\mathbf{F}_3$ acts *at* the pivot, so the distance from the pivot to the point of application of $\mathbf{F}_3$ is zero); since $r = 0$, the torque is 0 as well. However, for $\mathbf{F}_4$, neither r nor $\sin\theta$ is zero, so $\mathbf{F}_4$ has a nonzero torque. Of the four forces shown in that figure, only $\mathbf{F}_4$ has torque and would produce rotational acceleration.

There's another way to determine the value of the torque. Of course, it gives the same result as the method given above, but this method is often easier to use. Look at the same object and force:

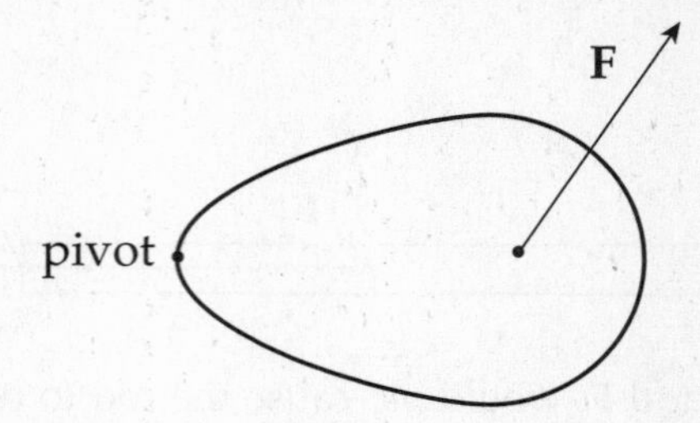

Instead of determining the distance from the pivot point to the point of application of the force, we will now determine the (perpendicular) distance from the pivot point to what's called the **line of action** of the force. This distance is the **lever arm** (or **moment arm**) of the force **F** relative to the pivot, and is denoted by l.

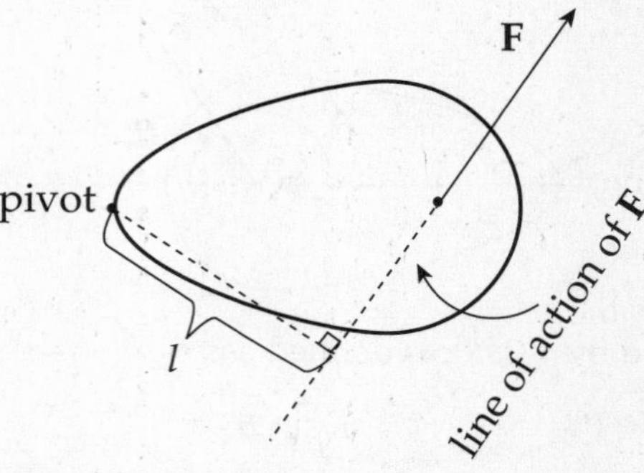

The torque of **F** is defined as the product

$$\tau = lF$$

(Just as the lever arm is often called the moment arm, the torque is called the **moment** of the force.) That these two definitions of torque, $\tau = rF\sin\theta$ and $\tau = lF$ are equivalent follows immediately from the fact that $l = r\sin\theta$:

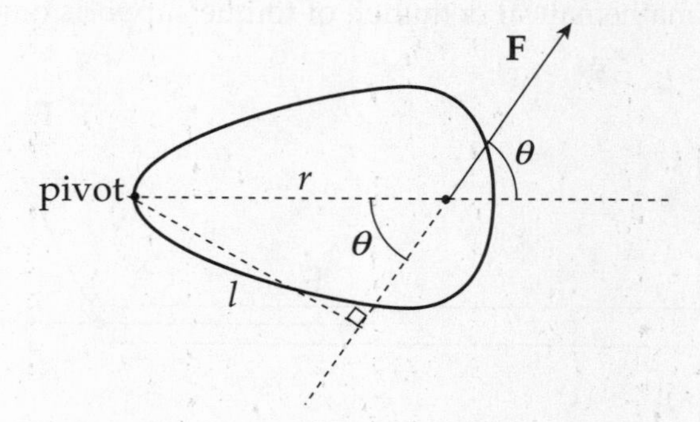

Since l is the component of **r** that's perpendicular to **F**, it is also denoted by $r_\perp$ ("r perp"). So the definition of torque can be written as $\tau = r_\perp F$.

These two equivalent definitions of torque make it clear that only the component of **F** that's perpendicular to **r** produces torque. The component of **F** that's parallel to **r** does not produce torque. Notice that $\tau = rF \sin \theta = rF_\perp$, where $F_\perp$ ("F perp") is the component of **F** that's perpendicular to **r**:

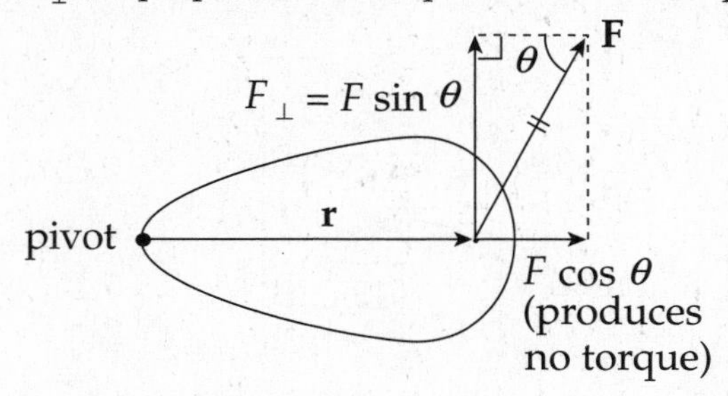

So the definition of torque can also be written as $\tau = rF_\perp$.

Remark. Because only the component of **F** (perpendicular to **r**) produces torque, it is not surprising that the torque can be written as the cross product of **r** and **F**:

$$\tau = \mathbf{r} \times \mathbf{F}$$

Example 6.8 A student pulls down with a force of 40 N on a rope that winds around a pulley of radius 5 cm.

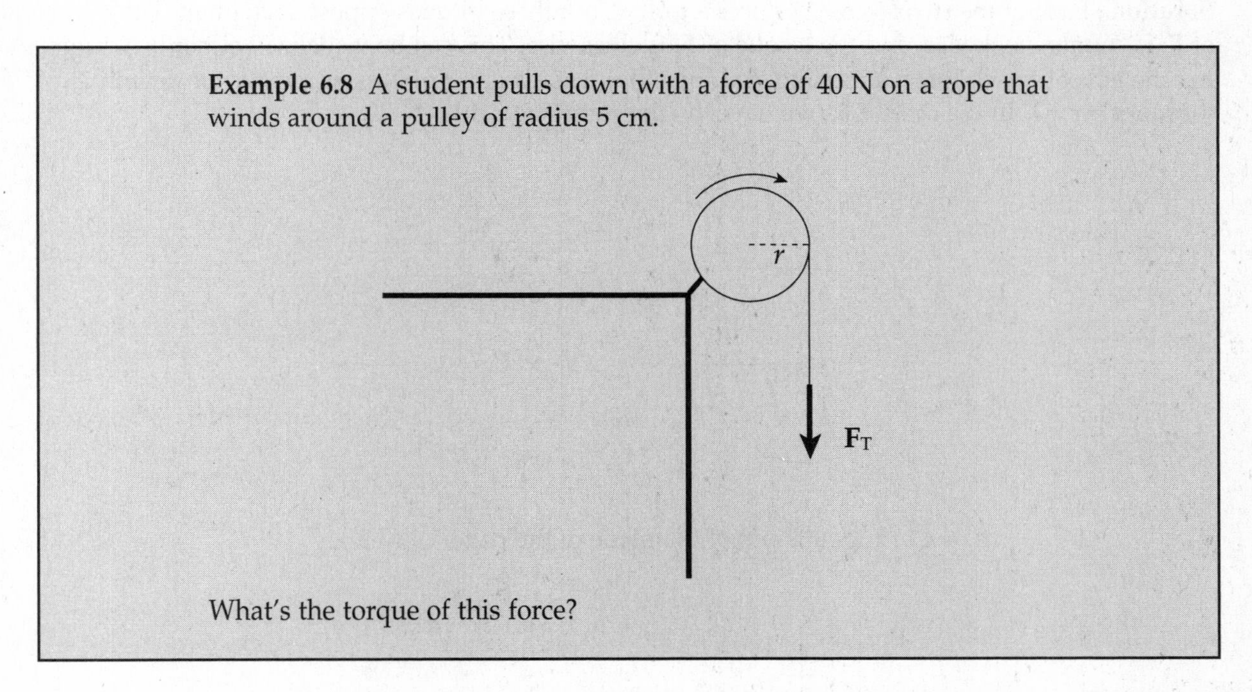

What's the torque of this force?

Solution. Since the tension force, $\mathbf{F}_T$, is tangent to the pulley, it is perpendicular to the radius vector **r** at the point of contact:

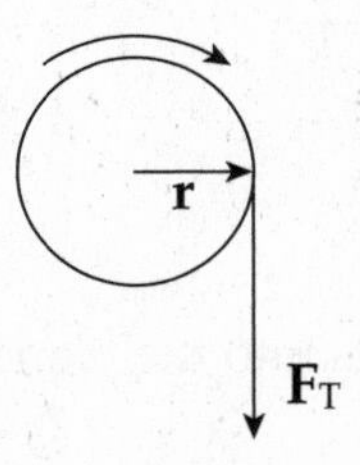

Therefore, the torque produced by this tension force is simply

$$\tau = rF_T = (0.05 \text{ m})(40 \text{ N}) = 2 \text{ N·m}$$

Example 6.9 What is the net torque on the cylinder shown below? The cylinder is pinned at its center.

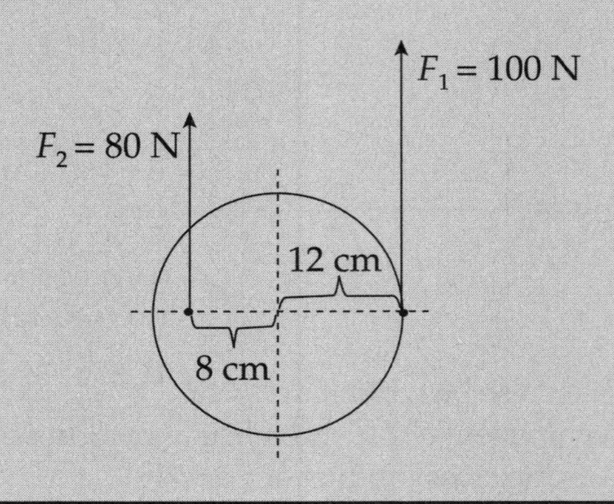

Solution. Each of the two forces produces a torque, but these torques oppose each other. The torque of $\mathbf{F}_1$ is counterclockwise, and the torque of $\mathbf{F}_2$ is clockwise. This can be visualized either by imagining the effect of each force, assuming that the other was absent, or by using the vector definition of torque, $\tau = \mathbf{r} \times \mathbf{F}$. In the case of $\mathbf{F}_1$, we have, by the right-hand rule,

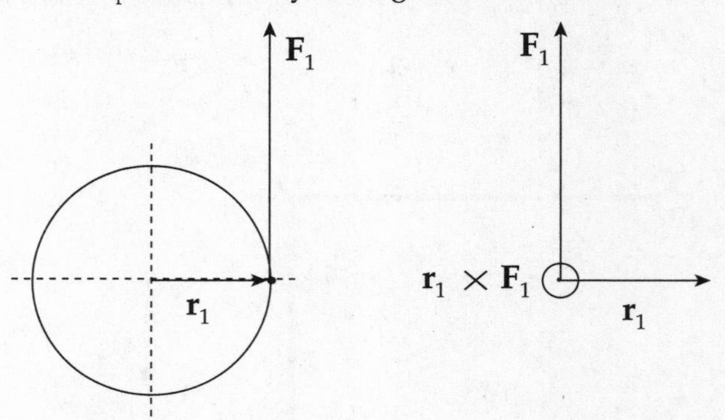

$\tau_1 = \mathbf{r}_1 \times \mathbf{F}_1$ points out of the plane of the page: ⊙, while

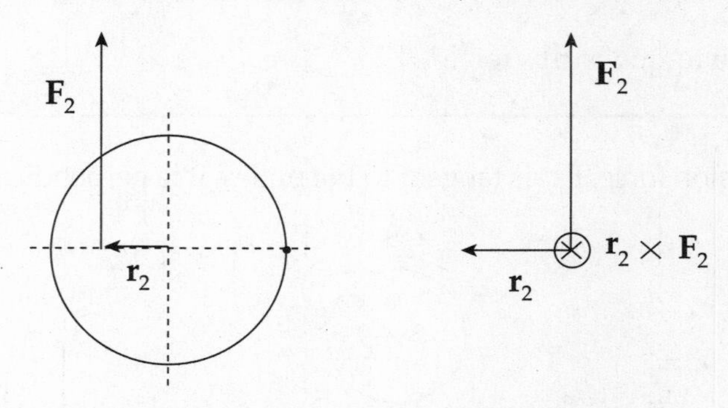

$\tau_2 = \mathbf{r}_2 \times \mathbf{F}_2$ points into the plane of the page: ⊗

These symbols are easy to remember if you think of a dart: ⊙ is the point of the dart coming at you and ⊗ is the pattern of feathers at the back of the dart you see as it flies away from you toward the dartboard.

If the torque vector points out of the plane of the page, this indicates a tendency to produce counterclockwise rotation and, if it points into the plane of the page, this indicates a tendency to produce clockwise rotation.

The **net torque** is the sum of all the torques. Counting a counterclockwise torque as positive and a clockwise torque as negative, we have

$$\tau_1 = +r_1F_1 = +(0.12 \text{ m})(100 \text{ N}) = +12 \text{ N}\cdot\text{m}$$

and

$$\tau_2 = -r_2F_2 = -(0.08 \text{ m})(80 \text{ N}) = -6.4 \text{ N}\cdot\text{m}$$

so

$$\tau_{net} = \Sigma\tau = \tau_1 + \tau_2 = (+12 \text{ N}\cdot\text{m}) + (-6.4 \text{ N}\cdot\text{m}) = +5.6 \text{ N}\cdot\text{m}$$

ROTATIONAL INERTIA

Our goal is to develop a rotational analog of Newton's Second Law, $F_{net} = ma$. We're almost there; torque is the rotational analog of force and, therefore, τ_{net} is the rotational analog of F_{net}. The rotational analog of translational acceleration, a, is rotational (or angular) acceleration, α. We will now look at the rotational analog of inertial mass, m.

Consider a small point mass m at a distance r from the axis of rotation, being acted upon by a tangential force **F**.

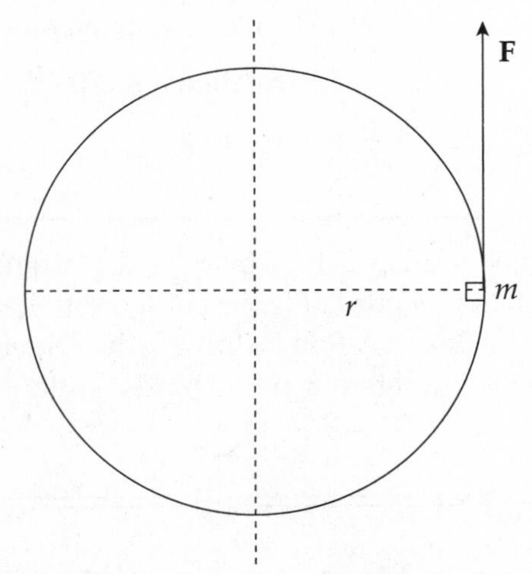

From Newton's Second Law, we have $F = ma$. Substituting $a = r\alpha$, this equation becomes $F = mr\alpha$. Now multiply both sides of this last equation by r to yield

$$rF = mr^2\alpha$$

or, since $rF = \tau$,

$$\tau = mr^2\alpha$$

In the equation $F = ma$, the quantity m is multiplied by the acceleration produced by the force F, while in the equation $\tau = mr^2\alpha$, the quantity mr^2 is multiplied by the rotational acceleration produced by the torque τ.

So for a point mass at a distance r from the axis of rotation, its **rotational inertia** (also called **moment of inertia**) is defined as mr^2. If we now take into account all the point masses that comprise the object under study, we can get the total rotational inertia, I, of the body by adding them:

$$I = \sum m_i r_i^2$$

For a continuous solid body, this sum becomes the integral

$$I = \int r^2 \, dm$$

This formula can be used to calculate expressions for the rotational inertia of cylinders (disks), spheres, slender rods, and hoops. Notice that the rotational inertia depends not only on m, but also on r; both the mass *and how it's distributed about the axis of rotation* determine I. By summing over all point masses and all external torques, the equation $\tau = mr^2\alpha$ becomes

$$\sum \tau_i = \left(\sum m_i r_i^2\right)\alpha \quad \text{or} \quad \tau_{\text{net}} = I\alpha$$

We've reached our goal:

Translational motion	*Rotational motion*
force, F	torque, τ
acceleration, a	rotational acceleration, α
mass, m	rotational inertia, I
$F_{\text{net}} = ma$	$\tau_{\text{net}} = I\alpha$

Example 6.10 Three beads, each of mass m, are arranged along a rod of negligible mass and length L. Figure out the rotational inertia of the assembly when the axis of rotation is through the center bead and when the axis of rotation is through one of the beads on the ends.

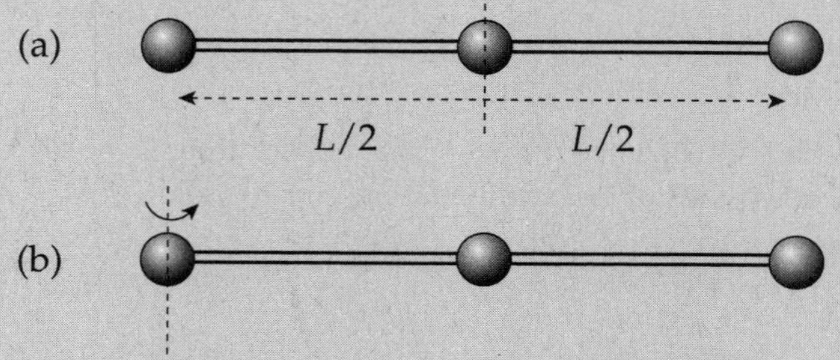

(a)

L/2 L/2

(b)

Solution.

(a) In the first case, both the left bead and the right bead are at a distance of $L/2$ from the axis of rotation, while the center bead is at distance zero from the axis of rotation. Therefore,

$$I = \sum m_i r_i^2 = m\left(\tfrac{L}{2}\right)^2 + m(0)^2 + m\left(\tfrac{L}{2}\right)^2 = \tfrac{1}{2}mL^2$$

(b) In the second case, the left bead is at distance zero from the rotation axis, the center bead is at distance $L/2$, and the right bead is at distance L. Therefore,

$$I' = \sum m_i r_i^2 = m(0)^2 + m\left(\tfrac{L}{2}\right)^2 + m(L)^2 = \tfrac{5}{4}mL^2$$

Note that, although both assemblies have the same mass (namely, $3m$), their rotational inertias are different, because of the different distribution of mass relative to the axis of rotation.

If the rotational inertia of a body is known relative to an axis that passes through the body's center of mass, then the rotational inertia, I', relative to any other rotation axis (parallel to the first one) can be calculated as follows. Let I_{cm} be the rotational inertia of a body relative to a rotation axis that passes through the body's center of mass, let the mass of the body be M, and let x be the distance from the axis through the center of mass to the rotation axis. Then

$$I' = I_{cm} + Mx^2$$

This is called the **parallel-axis theorem**. Let's use this result to calculate the rotational inertia of the three-bead assembly in part (b) from the value obtained in part (a), which is I_{cm}. Since $M = 3m$ and $x = L/2$, we have

$$I' = I_{cm} + Mx^2 = \tfrac{1}{2}mL^2 + (3m)\left(\tfrac{L}{2}\right)^2 = \tfrac{5}{4}mL^2$$

which agrees with the value calculated above.

Example 6.11 Show that the rotational inertia of a homogeneous cylinder of radius R and mass M, rotating about its central axis, is given by the equation $I = \frac{1}{2}MR^2$.

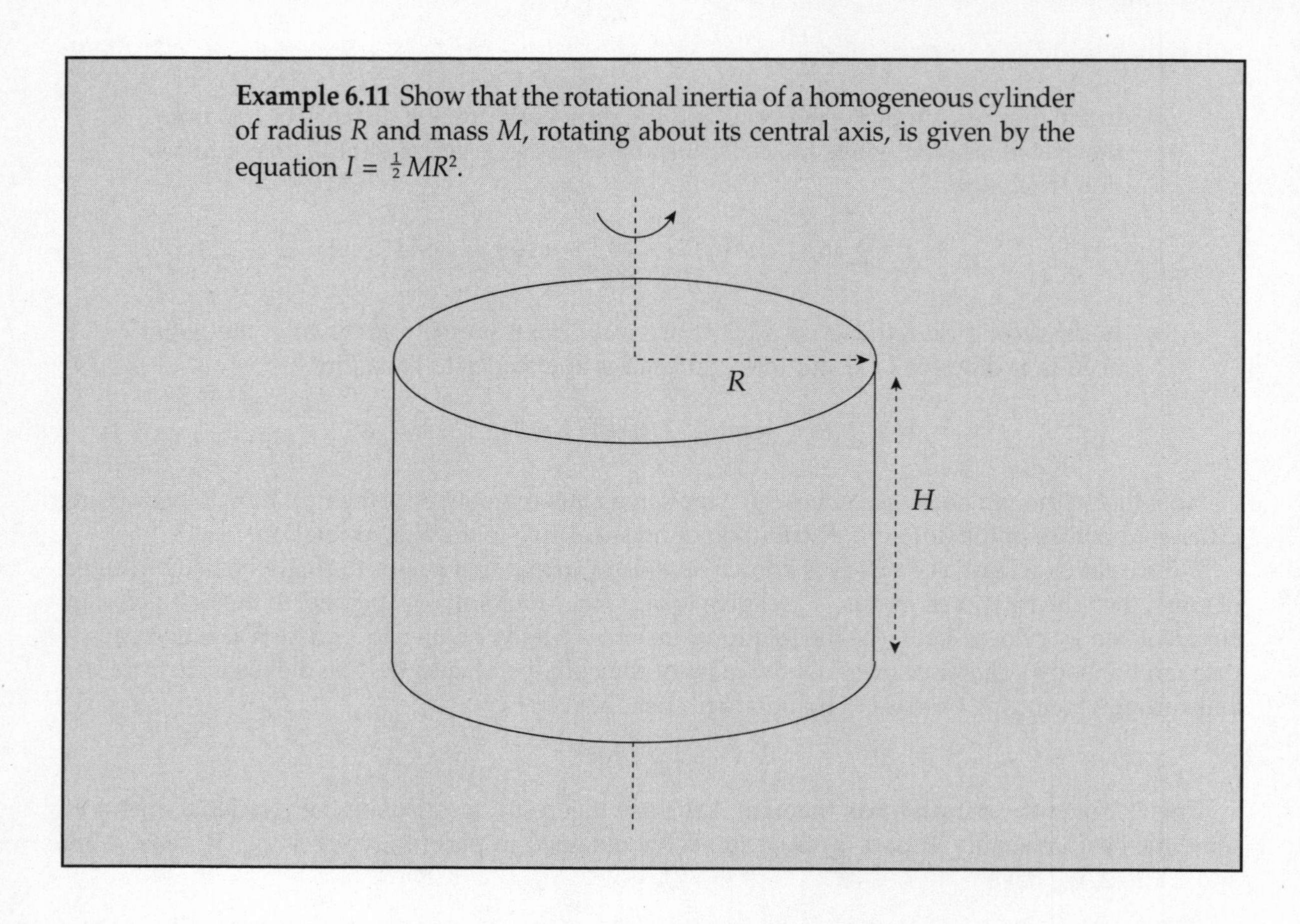

Solution. A solid is homogeneous if its density is constant throughout. Let ρ be the density of the cylinder. In order to compute I using the formula $I = \int r^2\,dm$, we choose our mass element to be an infinitesimally thin cylindrical ring of radius r:

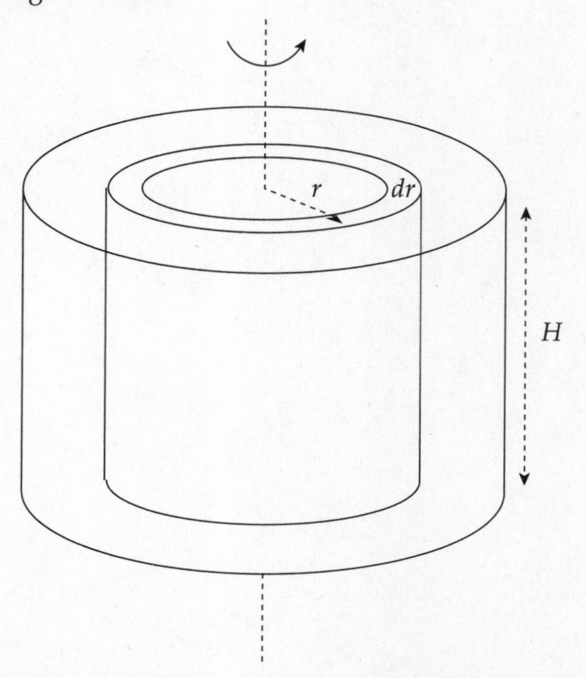

The mass of this ring is equal to its volume, $2\pi r\,dr \times H$, times the density; that is,

$$dm = \rho \times 2\pi r H\,dr$$

Therefore,

$$\begin{aligned} I = \int r^2\,dm &= \int_{r=0}^{r=R} r^2(\rho \cdot 2\pi r H\,dr) \\ &= 2\pi\rho H \int_0^R r^3\,dr \\ &= 2\pi\rho H\left[\tfrac{1}{4}r^4\right]_0^R \\ &= \tfrac{1}{2}\pi\rho H R^4 \end{aligned}$$

To eliminate ρ, use the fact that the total mass of the cylinder is

$$M = \rho V = \rho \cdot \pi R^2 H$$

Putting this into the expression derived for I, we find that

$$I = \tfrac{1}{2}\pi\rho H R^4 = \tfrac{1}{2}(\rho \cdot \pi R^2 H)R^2 = \tfrac{1}{2}MR^2$$

The height of the cylinder (the dimension parallel to the axis of rotation) is irrelevant. Therefore, the formula $I = \frac{1}{2}MR^2$ gives the rotational inertia of any homogeneous solid cylinder revolving around its central axis. This includes a disk (which is just a really short cylinder).

Example 6.12 Again, consider the system of Example 6.9:

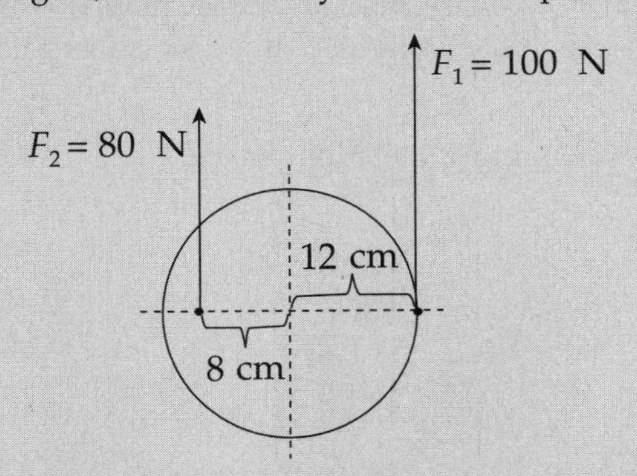

Assume that the mass of the cylinder is 50 kg. Given that the rotational inertia of a cylinder of radius R and mass M rotating about its central axis is given by the equation $I = \frac{1}{2}MR^2$, determine the rotational acceleration produced by the two forces shown.

Solution. In Example 6.9, we figured out that $\tau_{net} = +5.6$ N·m. The rotational inertia of the cylinder is

$$I = \tfrac{1}{2}MR^2 = \tfrac{1}{2}(50 \text{ kg})(0.12 \text{ m})^2 = 0.36 \text{ kg·m}^2$$

Therefore, from the equation $\tau_{net} = I\alpha$, we find that

$$\alpha = \frac{\tau_{net}}{I} = \frac{5.6 \text{ N} \cdot \text{m}}{0.36 \text{ kg} \cdot \text{m}^2} = 16 \text{ rad/s}^2$$

This angular acceleration will be counterclockwise, because τ_{net} is counterclockwise.

Example 6.13 A block of mass m is hung from a pulley of radius R and mass M and allowed to fall. What is the acceleration of the block?

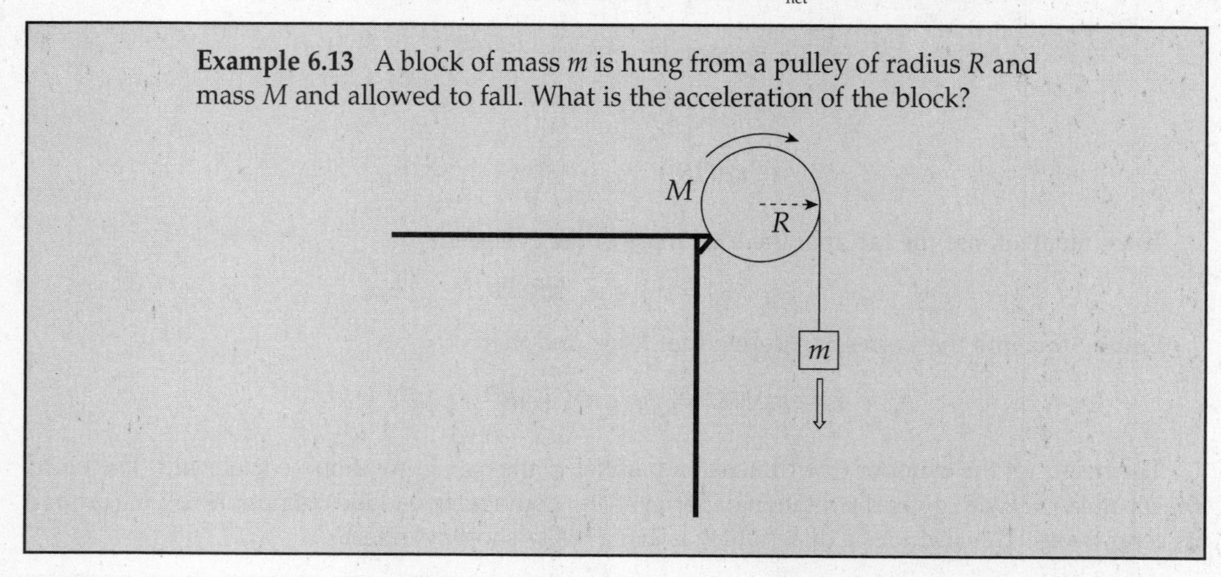

Solution. In Chapter 3, we treated pulleys as if they were massless, and no force was required to make them rotate. Now, however, we know how to take the mass of a pulley into account, by including its rotational inertia in our analysis. The pulley is a disk, so its rotational inertia is given by the formula $I = \frac{1}{2}MR^2$.

First we draw a free-body diagram for the falling block:

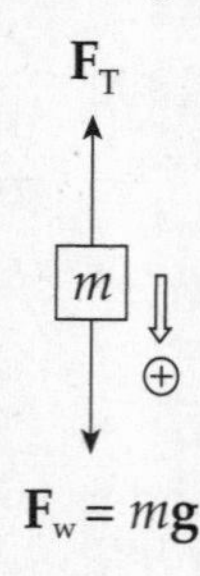

and apply Newton's Second Law:

$$mg - F_T = ma \quad (1)$$

Now the tension $\mathbf{F}_T$ in the cord produces a torque, $\tau = RF_T$, on the pulley:

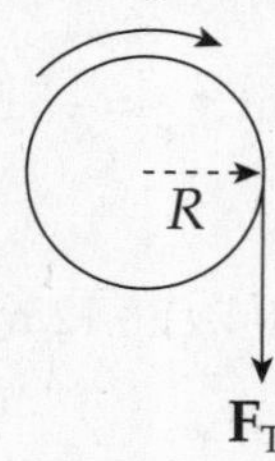

Since this is the only torque on the pulley, the equation $\tau_{net} = I\alpha$ becomes $RF_T = I\alpha$. But $I = \frac{1}{2}MR^2$ and $\alpha = a/R$. (This last equation says that the cord doesn't slip as it slides over the pulley; the linear acceleration of a point on the rim, $a = R\alpha$, is equal to the acceleration of the connected block.) Therefore,

$$\tau = I\alpha \quad \Rightarrow \quad RF_T = \tfrac{1}{2}MR^2 \cdot \frac{a}{R} = \tfrac{1}{2}MRa \ (1)$$

which tells us that

$$F_T = \tfrac{1}{2}Ma \qquad (2)$$

Substituting Equation (2) into Equation (1), we find that

$$mg - \tfrac{1}{2}Ma = ma \quad \Rightarrow \quad \left(\tfrac{1}{2}M + m\right)a = mg \quad \Rightarrow \quad a = \frac{m}{\tfrac{1}{2}M + m}g$$

KINETIC ENERGY OF ROTATION

A rotating object has rotational kinetic energy, just as a translating object has translational kinetic energy. The formula for kinetic energy is, of course, $K = \frac{1}{2}mv^2$, but this can't be directly used to calculate the kinetic energy of rotation because each point mass that makes up the body can have a different v. For this reason, we need a definition of $K_{rotational}$ that involves ω instead of v.

$$K_{rotational} = \sum K_i = \sum \tfrac{1}{2}(m_i r_i^2)\omega^2 = \tfrac{1}{2} \cdot \left(\sum m_i r_i^2\right) \cdot \omega^2 = \tfrac{1}{2}I\omega^2$$

Note that this expression for rotational kinetic energy follows the general pattern displayed by our previous results: I is the rotational analog of m and ω is the rotational analog of v. Therefore, the rotational analog of $\frac{1}{2}mv^2$ should be $\frac{1}{2}I\omega^2$.

ROLLING MOTION

One of the main types of motion associated with rotational motion is rolling motion. We will primarily deal with rolling without slipping.

Consider a disk rolling down an incline without slipping.

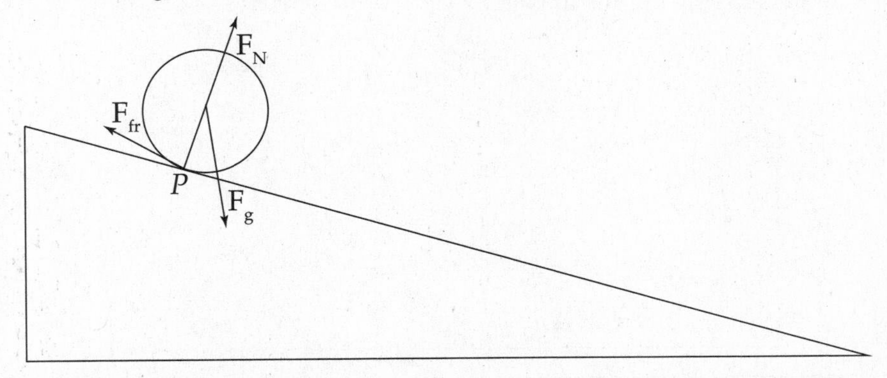

The point of contact of the object with the surface P is *instantaneously* at rest. If this were not the case then the disk would be slipping down the incline, so the contact point must not be moving relative to the surface. In this case, the velocity of the center of mass of the disk is equal to the radius times the angular velocity of the disk.

You can take the torque around any point to determine the acceleration of the disk. It is often easiest to take it around the contact point P because then only gravity provides a torque about this point, and if you know the mass of the object, then you know the force of gravity. Make sure to use the parallel axis theorem in this case since you are considering the rotational inertia about point P, not the center of mass. This will allow you to calculate the acceleration of the disk. Once you know the acceleration, you can calculate the necessary coefficient of friction to produce rolling without slipping using Newton's Second Law.

The total motion for an object that is rolling without slipping is the combined motion of the entire object translating with the velocity of the center of mass, and the object rotating about its center of mass, as shown below. This shows the object is instantaneously rotating about the contact point P.

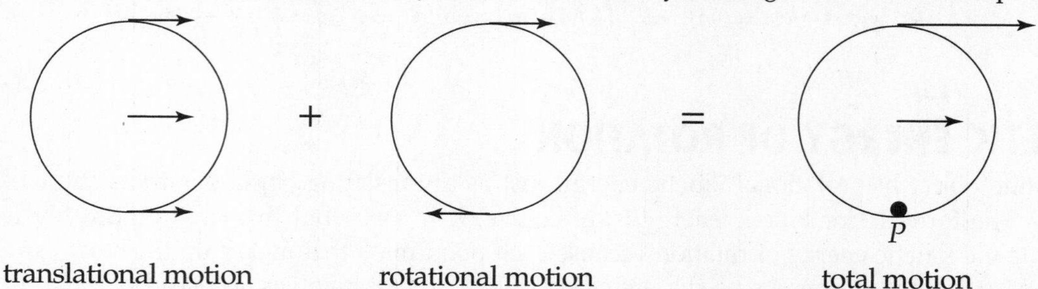

translational motion rotational motion total motion

For rolling motion the total kinetic energy is the translational kinetic energy and the rotational kinetic energy.

$$K_{\text{rolling}} = K_{\text{translation}} + K_{\text{rotation}}$$

$$K_{\text{rolling}} = \frac{1}{2}mv_{cm}^{2} + \frac{1}{2}I_{cm}\omega^{2}$$

Example 6.14 A cylinder of mass M and radius R rolls (without slipping) down an inclined plane whose incline angle with the horizontal is θ. Determine the acceleration of the cylinder's center of mass, and the minimum coefficient of friction that will allow the cylinder to roll without slipping on this incline.

Solution. First we draw a free-body diagram for the cylinder:

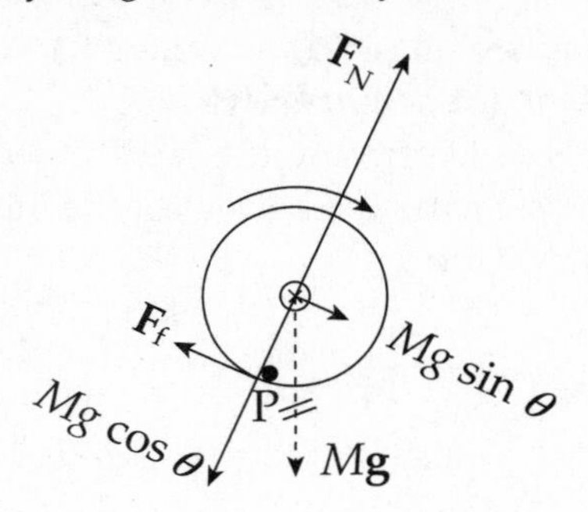

We know that the cylinder rolls without slipping, so the force of friction is not kinetic friction. Since the speed of the point on the cylinder in contact with the ramp is zero with respect to the ramp, *static* friction supplies the torque that allows the cylinder to roll smoothly.

Take the torque about the contact point to solve for the acceleration because the frictional force will not be part of the equation and we do not know it yet.

$\sum \tau_P = I\alpha$, use the parallel-axis theorem and $a = R\alpha$

$$(Mg \sin \theta)R = \left(\frac{1}{2}MR^2 + MR^2\right)\left(\frac{a}{R}\right)$$

$$g \sin \theta R = \frac{3}{2}Ra$$

$$a = \frac{2}{3}g \sin \theta$$

Now we will use Newton's Second Law to solve for μ since we know the acceleration.

$$\sum F = ma$$

$$Mg \sin \theta - F_f = Ma$$

$$Mg \sin \theta - \mu Mg \cos \theta = M\left(\frac{2}{3}g \sin \theta\right)$$, cancel M and g.

$$\sin \theta - \mu \cos \theta = \frac{2}{3}\sin \theta$$

$$\frac{1}{3}\sin \theta = \mu \cos \theta$$

$$\mu = \frac{1}{3}\tan \theta$$

Example 6.15 A cylinder of mass M and radius R rolls (without slipping) down an inclined plane (of height h and length L) whose incline angle with the horizontal is θ. Determine the linear speed of the cylinder's center of mass when it reaches the bottom of the incline (assuming that it started from rest at the top).

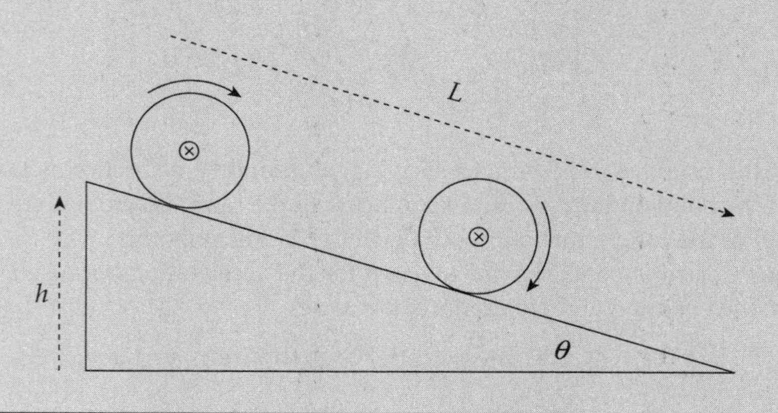

Solution. We will attack this problem using Conservation of Mechanical Energy. As the cylinder rolls down the ramp, its initial gravitational potential energy is converted into kinetic energy, which is a combination of translational kinetic energy (since the cylinder's center of mass is translating down the ramp) and rotational kinetic energy:

$$K_i + U_i = K_f + U_f$$
$$0 + Mgh = (\tfrac{1}{2}Mv_{cm}^2 + \tfrac{1}{2}I\omega^2) + 0$$

Since $I = \frac{1}{2}MR^2$ and $\omega = v_{cm}/R$, this equation becomes

$$Mgh = \tfrac{1}{2}Mv_{cm}^2 + \tfrac{1}{2}\left(\tfrac{1}{2}MR^2\right)\left(\frac{v_{cm}}{R}\right)^2 = \tfrac{1}{2}Mv_{cm}^2 + \tfrac{1}{4}Mv_{cm}^2 = \tfrac{3}{4}Mv_{cm}^2$$

Therefore,

$$v_{cm} = \sqrt{\tfrac{4}{3}gh}$$

We can verify this result using the result of the previous example. There we found that the acceleration of the cylinder's center of mass as it rolled down the ramp was $a = \frac{2}{3}g\sin\theta$. Applying Big Five #5 gives us:

$$v^2 = v_0^2 + 2a\Delta s = 2aL = 2a\frac{h}{\sin\theta} = 2 \cdot \tfrac{2}{3}g\sin\theta \cdot \frac{h}{\sin\theta} = \tfrac{4}{3}gh \quad \Rightarrow \quad v = \sqrt{\tfrac{4}{3}gh}$$

WORK AND POWER

Consider a small point mass m at distance r from the axis of rotation, acted upon by a tangential force **F**.

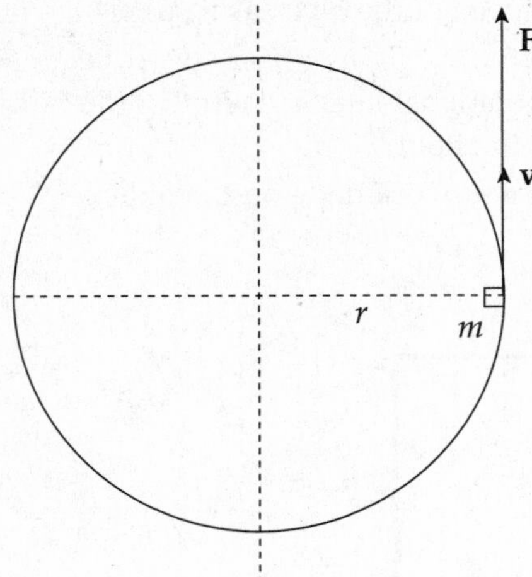

As it rotates through an angular displacement of $\Delta\theta$, the force does work on the point mass: $W = F\Delta s$, where $\Delta s = r\Delta\theta$. Therefore,

$$W = Fr\Delta\,\theta = \tau\,\Delta\theta$$

If the force is not purely tangential to the object's path, then only the tangential component of the force does work; the radial component does not (since it's perpendicular to the object's displacement). Therefore, for a general constant force F, the equation above would read $W = F_t r\Delta\theta = \tau\,\Delta\theta$, where F_t denotes the tangential component of F.

If we want to allow for a varying F—and a varying τ—then the work done is equal to the definite integral:

$$W = \int_{\theta_1}^{\theta_2} \tau\, d\theta$$

Again, notice the analogy between this equation and the one that defines work by F:

$$W = \int_{x_1}^{x_2} F\, dx$$

The work–energy theorem ($W = \Delta K$) also holds in the rotational case, where W is the work done by net torque and ΔK is the resulting change in the rotational kinetic energy.

The rate at which work is done, or the power (P), is defined by the equation

$$P = \frac{dW}{dt}$$

Over an infinitesimal angular displacement $d\theta$, the torque τ does an amount of work dW given by $dW = \tau\, d\theta$. This implies that

$$\frac{dW}{dt} = \tau\frac{d\theta}{dt} \quad \Rightarrow \quad P = \tau\omega$$

Once again, notice the parallel to $P = Fv$, the translation version of this equation.

Example 6.16 A block of mass $m = 5$ kg is hung from a pulley of radius $R = 15$ cm and mass $M = 8$ kg and then released from rest.

(a) What is the speed of the block as it strikes the floor, 2 m below its initial position?

(b) What is the rotational kinetic energy of the pulley just before the block strikes the floor?

(c) At what rate was work done on the pulley?

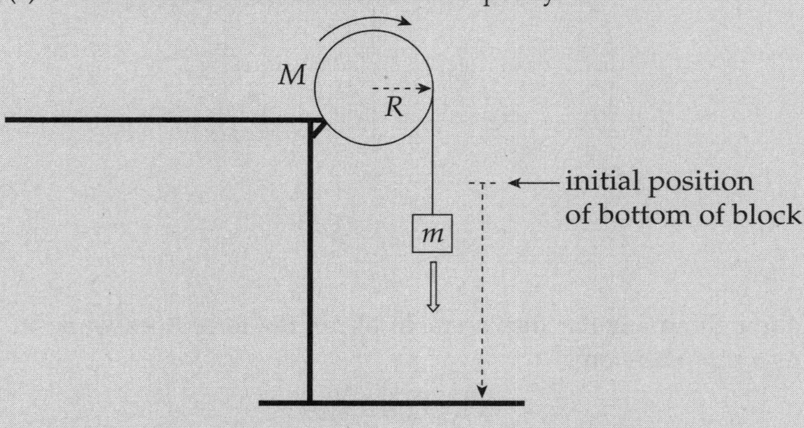

Solution.

(a) Apply Conservation of Mechanical Energy. The initial gravitational potential energy of the block is transformed into the purely rotational kinetic energy of the pulley and translational kinetic energy of the falling block:

$$
\begin{aligned}
K_i + U_i &= K_f + U_f \\
0 + mgh &= (\tfrac{1}{2}mv^2 + \tfrac{1}{2}I\omega^2) + 0 \\
&= \tfrac{1}{2}mv^2 + \tfrac{1}{2}\cdot\tfrac{1}{2}MR^2\cdot\left(\tfrac{v}{R}\right)^2 \\
&= \tfrac{1}{2}mv^2 + \tfrac{1}{4}Mv^2 \\
&= (\tfrac{1}{2}m + \tfrac{1}{4}M)v^2 \\
v &= \sqrt{\frac{mgh}{\tfrac{1}{2}m + \tfrac{1}{4}M}}
\end{aligned}
$$

Substituting in the given numerical values, we get

$$v = \sqrt{\frac{(5)(10)(2)}{\tfrac{1}{2}(5) + \tfrac{1}{4}(8)}} = 4.7 \text{ m/s}$$

(b) The rotational kinetic energy of the pulley as the block strikes the floor is

$$K = \tfrac{1}{2}I\omega^2 = \tfrac{1}{2}\cdot\tfrac{1}{2}MR^2\cdot\left(\tfrac{v}{R}\right)^2 = \tfrac{1}{4}Mv^2 = \tfrac{1}{4}(8)(4.7)^2 = 44 \text{ J}$$

(c) The rate at which work is done on the pulley is the power produced by the torque. One way to compute this is to first use the work–energy theorem to determine the work done by the torque and then divide this by the time during which the block fell. So,

$$W = \Delta K = K_f - K_i = K_f = 44 \text{ J}$$

The time during which this work was done is the time required for the block to drop to the ground. Using Big Five #1 and the result of part (a), we find that

$$\Delta s = \bar{v}t \quad \Rightarrow \quad t = \frac{\Delta s}{\bar{v}} = \frac{h}{\frac{1}{2}(v_0 + v)} = \frac{h}{\frac{1}{2}v} = \frac{2 \text{ m}}{\frac{1}{2}(4.7 \text{ m/s})} = 0.85 \text{ s}$$

Therefore,

$$P = \frac{W}{t} = \frac{44 \text{ J}}{0.85 \text{ s}} = 52 \text{ W}$$

ANGULAR MOMENTUM

So far we've developed rotational analogs for displacement, velocity, acceleration, force, mass, and kinetic energy. We will finish by developing a rotational analog for linear momentum; it's called **angular momentum**.

Consider a small point mass m at distance r from the axis of rotation, moving with velocity **v** and acted upon by a tangential force **F**.

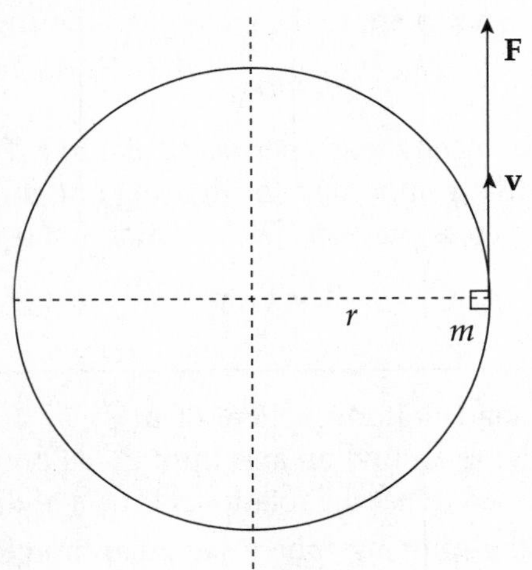

Then, by Newton's Second Law,

$$F = \frac{\Delta p}{\Delta t} = \frac{\Delta(mv)}{\Delta t}$$

If we multiply both sides of this equation by r and notice that $rF = \tau$, we get

$$\tau = \frac{\Delta(rmv)}{\Delta t}$$

Therefore, to form the analog of the law $F = \Delta p/\Delta t$ (force equals the rate-of-change of linear momentum), we say that torque equals the rate-of-change of angular momentum, and the angular momentum (denoted by L) of the point mass m is defined by the equation

$$L = rmv$$

If we now take into account all the point masses which comprise the object under study, we can get the angular momentum of the body by adding up all the individual contributions. This gives

$$L = I\omega$$

Note that this expression for angular momentum follows the general pattern we saw previously: I is the rotational analog of m, and ω is the rotational analog of v. Therefore, the rotational analog of mv should be $I\omega$.

If the point mass m does not move in a circular path, we can still define its angular momentum relative to any reference point.

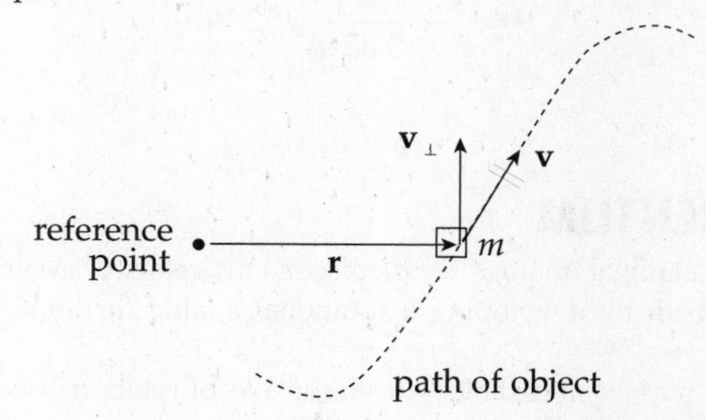

If **r** is the vector from the reference point to the mass, then the angular momentum is

$$L = rmv_{\perp}$$

where $v_{\perp}$ is the component of the velocity that's perpendicular to **r**. The fact that it's the perpendicular component of **v** relative to **r** that's important for figuring out angular momentum motivates the general vector definition with the cross product. The equation $L = (r)(mv_{\perp}) = (r)(p_{\perp})$ becomes

$$\mathbf{L} = \mathbf{r} \times \mathbf{p}$$

> **Example 6.17** A solid uniform sphere of mass $M = 8$ kg and radius $R = 50$ cm is revolving around an axis through its center at an angular speed of 10 rad/s. Given that the rotational inertia of the sphere is equal to $\frac{2}{5}MR^2$, what is the spinning sphere's angular momentum?

Solution. Apply the definition:

$$L = I\omega = \tfrac{2}{5}MR^2\omega = \tfrac{2}{5}(8 \text{ kg})(0.50 \text{ m})^2(10 \text{ rad/s}) = 8 \text{ kg} \cdot \text{m}^2/\text{s}$$

If you want to specify the direction of the angular momentum vector, **L**, use the right-hand rule. Let the fingers of your right hand curl in the direction of rotation of the body. Your thumb gives the direction of **L**, pointing along the rotation axis:

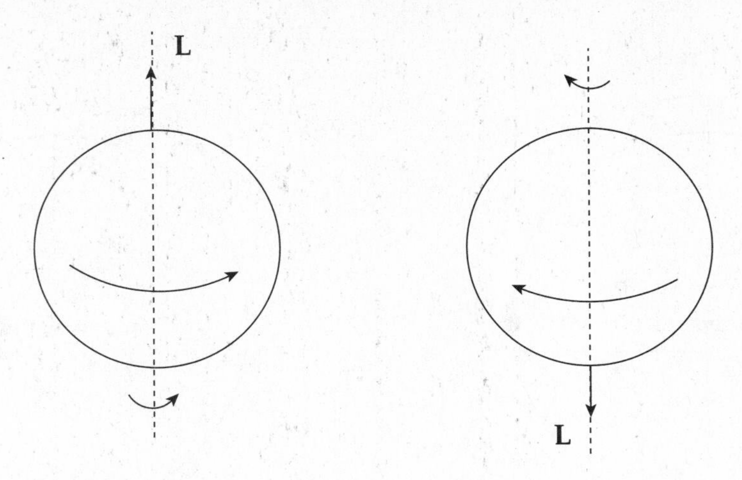

CONSERVATION OF ANGULAR MOMENTUM

Newton's Second Law says that

$$\mathbf{F}_{\text{net}} = \frac{d\mathbf{p}}{dt}$$

so if $F_{\text{net}} = 0$, then **p** is constant. This is Conservation of Linear Momentum.

The rotational analog of this is:

$$\tau_{\text{net}} = \frac{d\mathbf{L}}{dt}$$

so if $\tau_{\text{net}} = 0$, then **L** is constant. This is **Conservation of Angular Momentum**. Basically, this says that if the torques on a body balance so that the net torque is zero, then the body's angular momentum can't change.

An often cited example of this phenomenon is the spinning of a figure skater. As she pulls her arms inward, she moves more of her mass closer to the rotation axis and decreases her rotational inertia. Since the external torque on her is negligible, her angular momentum must be conserved. Since $L = I\omega$, a decrease in I causes an increase in ω, and she spins faster.

Example 6.18 A child of mass $m = 30$ kg stands at the edge of a small merry-go-round that's rotating at a rate of 1 rad/s. The merry-go-round is a disk of radius $R = 2.5$ m and mass $M = 100$ kg. If the child walks in, toward the center of the disk, and stops 0.5 m from the center, what will happen to the angular velocity of the merry-go-round (if friction can be ignored)?

Solution. The child walking toward the center of the merry-go-round does not provide an external torque to the child + disk system, so angular momentum is conserved. Let's denote the child as a point mass, and consider the following two views of the merry-go-round (looking down from above):

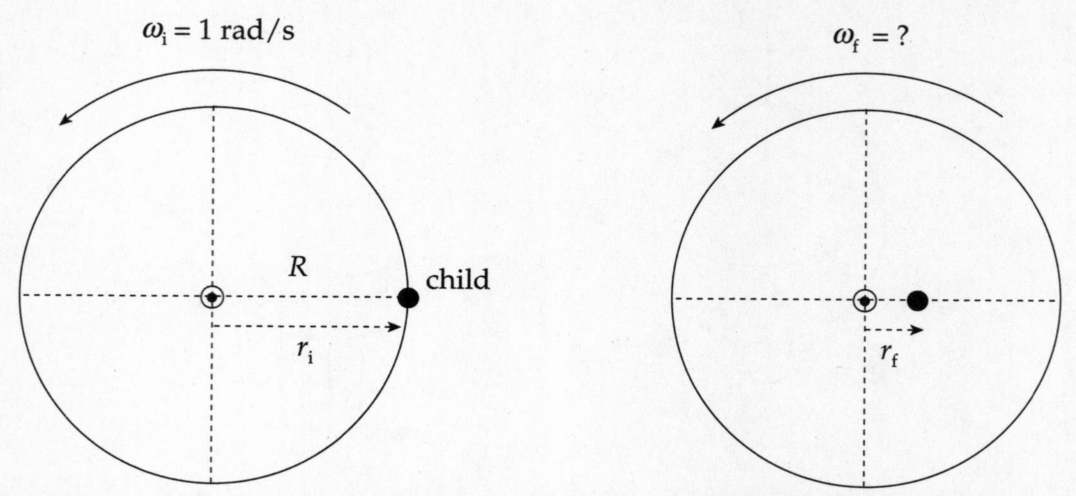

In the first picture, the total rotational inertia, I_i, is equal to the sum of the rotational inertia of the merry-go-round (MGR) and the child:

$$I_i = I_{MGR} + I_{child} = \tfrac{1}{2}MR^2 + mr_i^2 = \tfrac{1}{2}MR^2 + mR^2 = \left(\tfrac{1}{2}M + m\right)R^2$$

In the second picture, the total rotational inertia has decreased to

$$I_f = I_{MGR} + I'_{child} = \tfrac{1}{2}MR^2 + mr_f^2$$

So, by Conservation of Angular Momentum, we have

$$\begin{aligned} L_i &= L_f \\ I_i\omega_i &= I_f\omega_f \\ \left(\tfrac{1}{2}M + m\right)R^2\omega_i &= \left(\tfrac{1}{2}MR^2 + mr_f^2\right)\omega_f \\ \omega_f &= \frac{\left(\tfrac{1}{2}M + m\right)R^2}{\tfrac{1}{2}MR^2 + mr_f^2}\omega_i \end{aligned}$$

and substituting the given numerical values gives us

$$\begin{aligned} \omega_f &= \frac{\left(\tfrac{1}{2}M + m\right)R^2}{\tfrac{1}{2}MR^2 + mr_f^2}\omega_i \\ &= \frac{\left(\tfrac{1}{2}\cdot 100 + 30\right)(2.5)^2}{\tfrac{1}{2}\cdot 100\cdot (2.5)^2 + 30\cdot (0.5)^2}(1 \text{ rad/s}) \\ &= 1.6 \text{ rad/s} \end{aligned}$$

Notice that ω increased as I decreased, just as Conservation of Angular Momentum predicts.

EQUILIBRIUM

An object is said to be in **translational equilibrium** if the sum of the forces acting on it is zero; that is, if $F_{net} = 0$. Similarly, an object is said to be in **rotational equilibrium** if the sum of the torques acting on it is zero; that is, if $\tau_{net} = 0$. The term *equilibrium* by itself means both translational and rotational equilibrium. A body in equilibrium may be in motion; $F_{net} = 0$ does not mean that the velocity is zero; it only means that the velocity is constant. Similarly, $\tau_{net} = 0$ does not mean that the angular velocity is zero; it only means that it's constant. If an object is at rest, then it is said to be in **static equilibrium**.

Example 6.19 A uniform bar of mass m and length L extends horizontally from a wall. A supporting wire connects the wall to the bar's midpoint, making an angle of 55° with the bar. A sign of mass M hangs from the end of the bar.

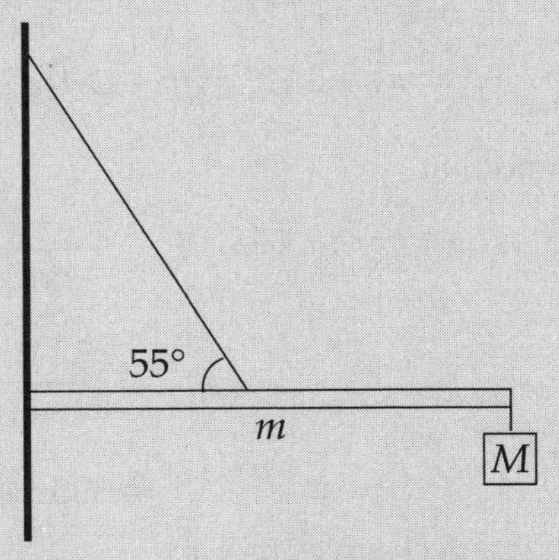

If the system is in static equilibrium, determine the tension in the wire and the strength of the force exerted on the bar by the wall if $m = 8$ kg and $M = 12$ kg.

Solution. Let $\mathbf{F}_C$ denote the (contact) force exerted by the wall on the bar. In order to simplify our work, we can write $\mathbf{F}_C$ in terms of its horizontal component, F_{Cx}, and its vertical component, F_{Cy}. Also, if $\mathbf{F}_T$ is the tension in the wire, then $F_{Tx} = F_T \cos 55°$ and $F_{Ty} = F_T \sin 55°$ are its components. This gives us the following force diagram:

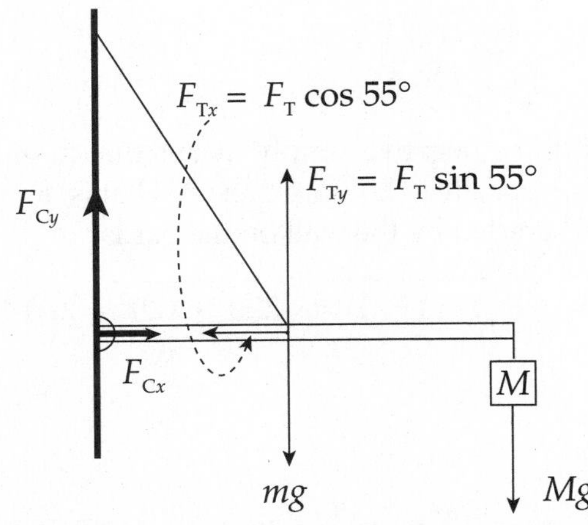

The first condition for equilibrium requires that the sum of the horizontal forces is zero and the sum of the vertical forces is zero:

$$\Sigma F_x = 0: \qquad F_{Cx} - F_T \cos 55° = 0 \qquad (1)$$

$$\Sigma F_y = 0: \qquad F_{Cy} + F_T \sin 55° - mg - Mg = 0 \qquad (2)$$

We notice immediately that we have more unknowns (F_{Cx}, F_{Cy}, F_T) than equations, so this system cannot be solved as is. The second condition for equilibrium requires that the sum of the torques about any point is equal to zero. Choosing the contact point between the bar and the wall as our pivot, only three of the forces in the diagram above produce torque: $\mathbf{F}_{Ty}$ produces a counterclockwise torque, and both $m\mathbf{g}$ and $M\mathbf{g}$ produce clockwise torques, which must balance. From the definition $\tau = lF$, and taking counterclockwise torque as positive and clockwise torque as negative, we have

$$\Sigma\tau = 0: \qquad (L/2)F_{Ty} - (L/2)(mg) - LMg = 0 \quad (3)$$

This equation contains only one unknown and can be solved immediately:

$$\tfrac{L}{2}F_{Ty} = \tfrac{L}{2}mg + LMg$$

$$F_{Ty} = mg + 2Mg = (m + 2M)g$$

Since $F_{Ty} = F_T \sin 55°$, we can find that

$$F_T \sin 55° = (m + 2M)g \quad \Rightarrow \quad F_T = \frac{(m+2M)g}{\sin 55°}$$

$$= \frac{(8 + 2 \cdot 12)(10)}{\sin 55°}$$

$$= 390 \text{ N}$$

Substituting this result into Equation (1) gives us F_{Cx}:

$$F_{Cx} = F_T \cos 55° = \frac{(m+2M)g}{\sin 55°}\cos 55° = (8 + 2 \cdot 12)(10)\cot 55° = 220 \text{ N}$$

And finally, from Equation (2), we get

$$F_{Cy} = mg + Mg - F_T \sin 55°$$

$$= mg + Mg - \frac{(m+2M)g}{\sin 55°}\sin 55°$$

$$= -Mg$$

$$= -(12)(10)$$

$$= -120 \text{ N}$$

The fact that F_{Cy} turned out to be negative simply means that in our original force diagram, the vector $\mathbf{F}_{Cy}$ points in the direction opposite to how we drew it. That is, $\mathbf{F}_{Cy}$ points downward. Therefore, the magnitude of the total force exerted by the wall on the bar is

$$F_C = \sqrt{(F_{Cx})^2 + (F_{Cy})^2} = \sqrt{220^2 + 120^2} = 250 \text{ N}$$

CHAPTER 6 REVIEW QUESTIONS

Section I: Multiple Choice

1. A compact disc has a radius of 6 cm. If the disc rotates about its central axis at an angular speed of 5 rev/s, what is the linear speed of a point on the rim of the disc?

   (A) 0.3 m/s
   (B) 1.9 m/s
   (C) 7.4 m/s
   (D) 52 m/s
   (E) 83 m/s

2. A compact disc has a radius of 6 cm. If the disc rotates about its central axis at a constant angular speed of 5 rev/s, what is the total distance traveled by a point on the rim of the disc in 40 min?

   (A) 180 m
   (B) 360 m
   (C) 540 m
   (D) 720 m
   (E) 4.5 km

3. An object of mass 0.5 kg, moving in a circular path of radius 0.25 m, experiences a centripetal acceleration of constant magnitude 9 m/s^2. What is the object's angular speed?

   (A) 2.3 rad/s
   (B) 4.5 rad/s
   (C) 6 rad/s
   (D) 12 rad/s
   (E) Cannot be determined from the information given

4. An object, originally at rest, begins spinning under uniform angular acceleration. In 10 s, it completes an angular displacement of 60 rad. What is the numerical value of the angular acceleration?

   (A) 0.3 rad/s^2
   (B) 0.6 rad/s^2
   (C) 1.2 rad/s^2
   (D) 2.4 rad/s^2
   (E) 3.6 rad/s^2

5. 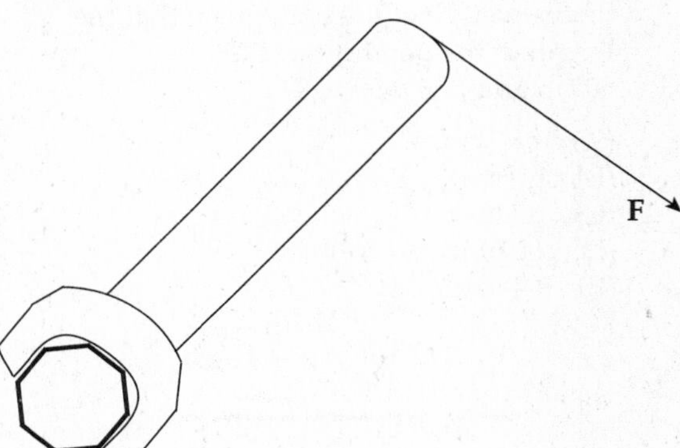

   In an effort to tighten a bolt, a force **F** is applied as shown in the figure above. If the distance from the end of the wrench to the center of the bolt is 20 cm and $F = 20$ N, what is the magnitude of the torque produced by **F**?

   (A) 0 N·m
   (B) 1 N·m
   (C) 2 N·m
   (D) 4 N·m
   (E) 10 N·m

6.

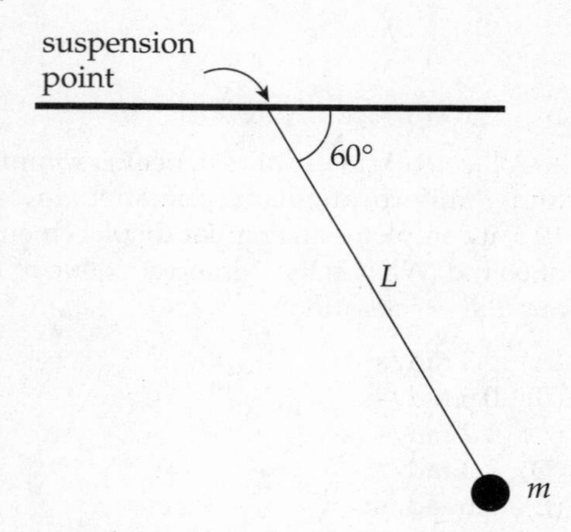

In the figure above, what is the torque about the pendulum's suspension point produced by the weight of the bob, given that the length of the pendulum, L, is 80 cm and $m = 0.50$ kg?

(A) 0.5 N·m
(B) 1.0 N·m
(C) 1.7 N·m
(D) 2.0 N·m
(E) 3.4 N·m

7.

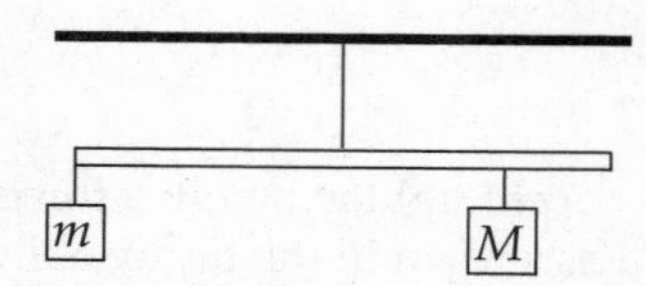

A uniform meter stick of mass 1 kg is hanging from a thread attached at the stick's midpoint. One block of mass $m = 3$ kg hangs from the left end of the stick, and another block, of unknown mass M, hangs below the 80 cm mark on the meter stick. If the stick remains at rest in the horizontal position shown above, what is M?

(A) 4 kg
(B) 5 kg
(C) 6 kg
(D) 8 kg
(E) 9 kg

8. What is the rotational inertia of the following body about the indicated rotation axis? (The masses of the connecting rods are negligible.)

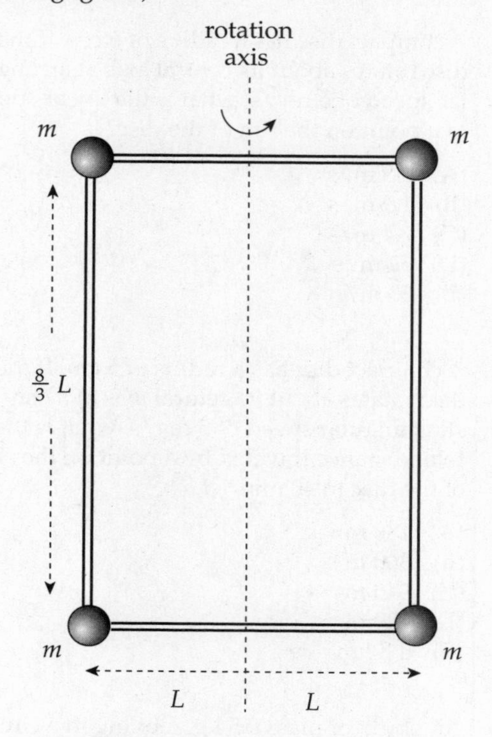

(A) $4mL^2$
(B) $\frac{32}{3}mL^2$
(C) $\frac{64}{9}mL^2$
(D) $\frac{128}{9}mL^2$
(E) $\frac{256}{9}mL^2$

9. The moment of inertia of a solid uniform sphere of mass M and radius R is given by the equation $I = \frac{2}{5}MR^2$. Such a sphere is released from rest at the top of an inclined plane of height h, length L, and incline angle θ. If the sphere rolls without slipping, find its speed at the bottom of the incline.

(A) $\sqrt{\frac{10}{7}gh}$
(B) $\sqrt{\frac{5}{2}gh}$
(C) $\sqrt{\frac{7}{2}gh}$
(D) $\sqrt{\frac{2}{7}gL\sin\theta}$
(E) $\sqrt{\frac{7}{10}gL\sin\theta}$

10. An object spins with angular velocity ω. If the object's moment of inertia increases by a factor of 2 without the application of an external torque, what will be the object's new angular velocity?

(A) $\omega/4$

(B) $\omega/2$

(C) $\omega/\sqrt{2}$

(D) $\sqrt{2}\,\omega$

(E) 2ω

SECTION II: FREE RESPONSE

1. In the figure below, the pulley is a solid disk of mass M and radius R, with rotational inertia $MR^2/2$. Two blocks, one of mass m_1 and one of mass m_2, hang from either side of the pulley by a light cord. Initially the system is at rest, with Block 1 on the floor and Block 2 held at height h above the floor. Block 2 is then released and allowed to fall.

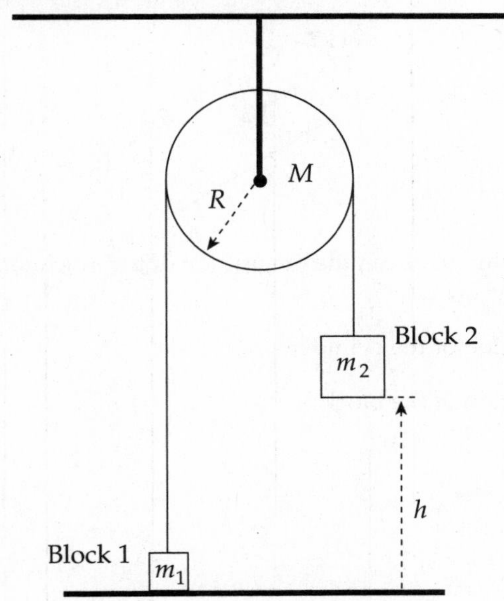

(a) What is the speed of Block 2 just before it strikes the ground?

(b) What is the angular speed of the pulley at this moment?

(c) What's the angular displacement of the pulley?

(d) How long does it take for Block 2 to fall to the floor?

2. The diagram below shows a solid uniform cylinder of radius R and mass M rolling (without slipping) down an inclined plane of incline angle θ. A thread wraps around the cylinder as it rolls down the plane and pulls upward on a block of mass m. Ignore the rotational inertia of the pulley.

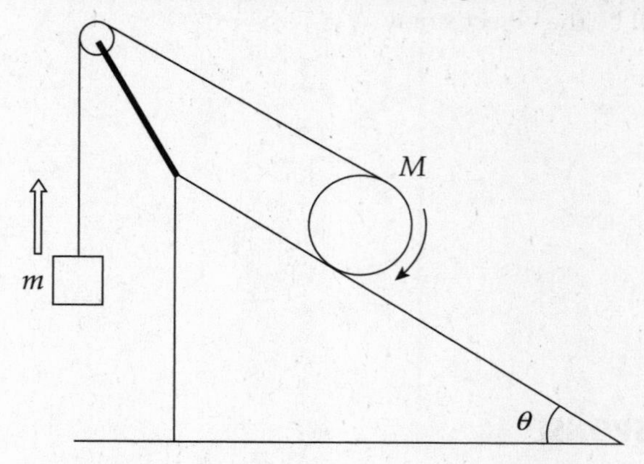

(a) Show that "rolling without slipping" means that the speed of the cylinder's center of mass, v_{cm}, is equal to $R\omega$, where ω is its angular speed.

(b) Show that, relative to P (the point of contact of the cylinder with the ramp), the speed of the top of the cylinder is $2v_{cm}$.

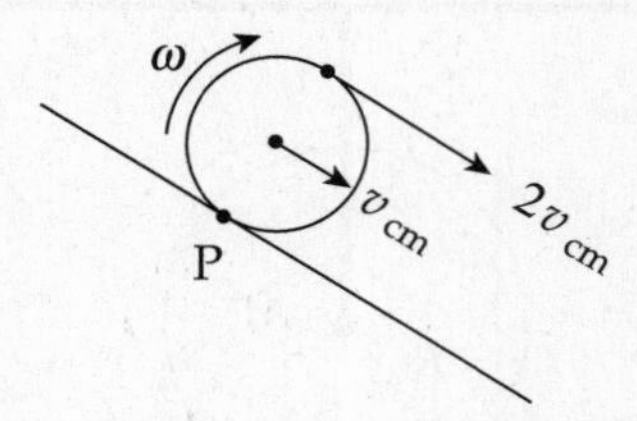

(c) What is the relationship between the magnitude of the acceleration of the block and the linear acceleration of the cylinder?

(d) What is the acceleration of the cylinder?

(e) What is the acceleration of the block?

3. Two slender uniform bars, each of mass M and length $2L$, meet at right angles at their midpoints to form a rigid assembly that's able to rotate freely about an axis through the intersection point, perpendicular to the page. Attached to each end of each rod is a solid ball of clay of mass m. A bullet of mass m_b is shot with velocity $\mathbf{v}$ as shown in the figure (which is a view from above of the assembly) and becomes embedded in the targeted clay ball.

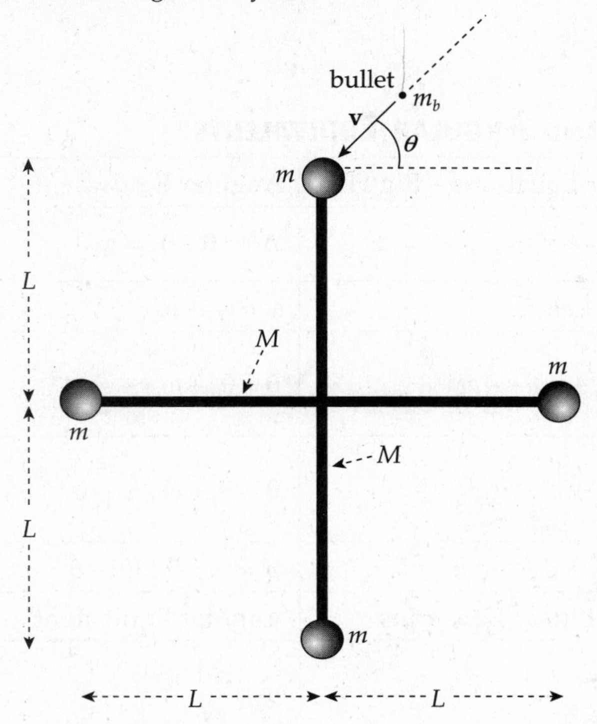

(a) Show that the moment of inertia of each slender rod about the given rotation axis, not including the clay balls, is $ML^2/3$.

(b) Determine the angular velocity of the assembly after the bullet has become lodged in the targeted clay ball.

(c) What is the resulting linear speed of each clay ball?

(d) Determine the ratio of the final kinetic energy of the assembly to the kinetic energy of the bullet before impact.

ROTATIONAL MOTION SUMMARY

RELATING LINEAR AND ANGULAR QUANTITIES

$s = r\theta$

$v = r\omega$

$a_{tan} = r\alpha$

LINEAR EQUATIONS AND ANGULAR EQUIVALENTS

Linear Equations – Big Five	**Angular Equivalent**
$\Delta x = x - x_0 = \bar{v}t$	$\Delta\theta = \theta - \theta_0 = \bar{\omega}t$
$v = v_0 + at$	$\omega = \omega_0 + \alpha t$
$x = x_0 + v_0 t + \frac{1}{2}at^2$	$\theta = \theta_0 + \omega_0 t + \frac{1}{2}\alpha t^2$
$x = x_0 + vt - \frac{1}{2}at^2$	$\theta = \theta_0 + \omega_f t - \frac{1}{2}\alpha t^2$
$v^2 = v_0^2 + 2a(x - x_0)$	$\omega^2 = \omega_0^2 + 2\alpha(\theta - \theta_0)$
Basic Linear Equations	**Angular Equivalent**
$v = \frac{dx}{dt}$	$\omega = \frac{d\theta}{dt}$
$a = \frac{dv}{dt}$	$\alpha = \frac{dw}{dt}$

BASIC ROTATION INFORMATION

- **Rotational inertia** is the rotational analog of inertia, essentially a measure of how difficult it is to change an object's rotational motion.

  For point masses, $I = mr^2$

  For distributed mass, $I = \int r^2 dm$
- **Parallel Axis Theorem**: $I = I_{cm} + md^2$
- **Torque** is a force's ability to cause an object to rotate. The equation for torque is

$$\tau = \mathbf{r} \times \mathbf{F} = I\alpha$$

- Rotating objects have **rotational kinetic energy** although the object does not necessarily translate. The equation for rotational kinetic energy is

$$K_{rotation} = \frac{1}{2}I\omega^2$$

Rolling:

$$K_{rolling} = K_{rotation} + K_{translation} = \frac{1}{2} I_{cm}\omega^2 + \frac{1}{2} m v_{cm}^{\ 2}$$

Angular Momentum

- **Angular momentum** for a point particle is given by the equation, $L = I\omega$.
- Angular momentum for a rigid object is given by the equation, $L = r \times p$.
- Angular momentum is conserved unless a net torque acts on the object. This is expressed by the equation, $\Sigma\tau = dL/dt$.

  For an object to be in static equilibrium the net force and the net torque must be zero: $\Sigma F = 0$, $\Sigma\tau = 0$.

Laws of Gravitation

KEPLER'S LAWS

Johannes Kepler spent years of exhaustive study distilling volumes of data collected by his mentor, Tycho Brahe, into three simple laws that describe the motion of planets.

KEPLER'S FIRST LAW

Every planet moves in an elliptical orbit, with the Sun at one focus.

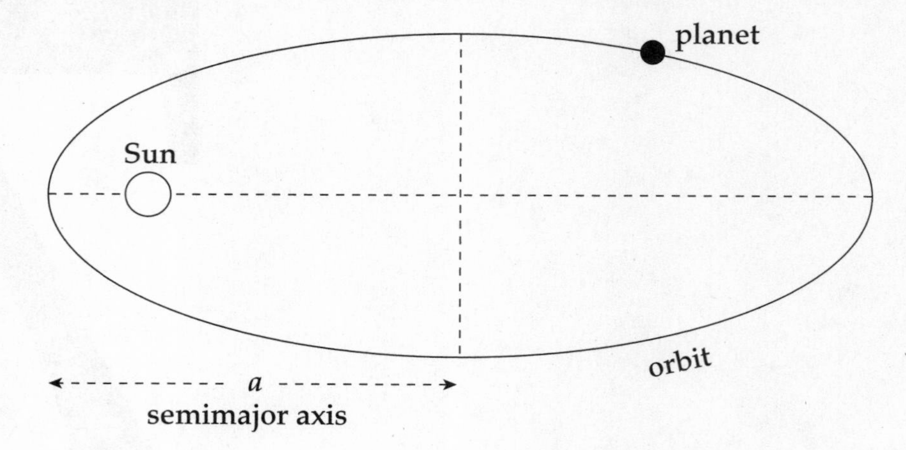

KEPLER'S SECOND LAW

As a planet moves in its orbit, a line drawn from the Sun to the planet sweeps out equal areas in equal time interals.

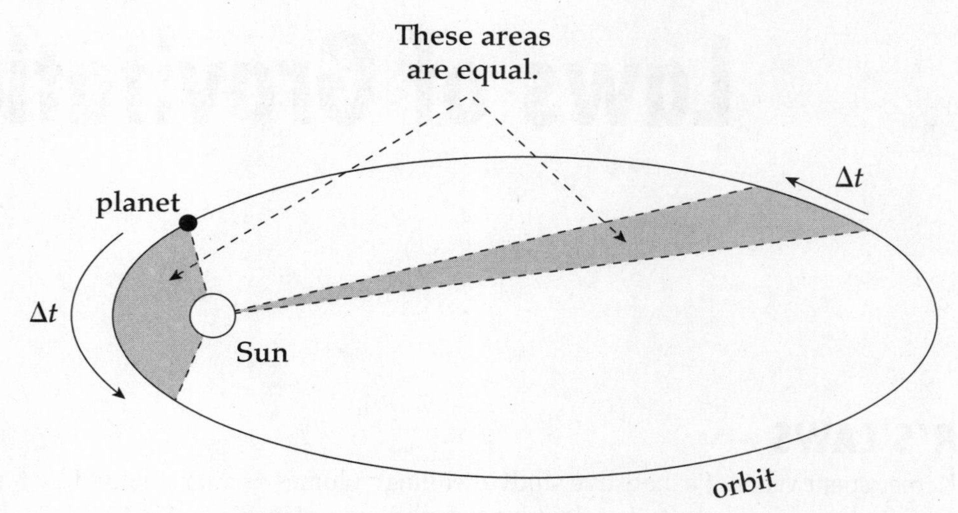

KEPLER'S THIRD LAW

If T is the period and a is the length of the semimajor axis of a planet's orbit, then the ratio T^2/a^3 is the same for all the planets.

NEWTON'S LAW OF GRAVITATION

Newton eventually proved that Kepler's first two laws imply a law of gravitation: Any two objects in the universe exert an attractive force on each other—called the **gravitational force**—whose strength is proportional to the product of the objects' masses and inversely proportional to the square of the center-to-center distance between them. If we let G be the **universal gravitational constant**, then the strength of the gravitational force is given by the equation:

$$F = G\frac{m_1 m_2}{r^2}$$

Consider a mass, m_1, close to the surface of the Earth. We will use Newton's Law of Gravitation and Newton's Second Law to show that the gravitational acceleration of m_1 is independent of the mass of the object, as shown below.

$$F_g = G\frac{m_1 M_E}{R_E^{\,2}} = m_1 a_g$$

$$\frac{GM_E}{R_E^{\,2}} = a_g = g$$

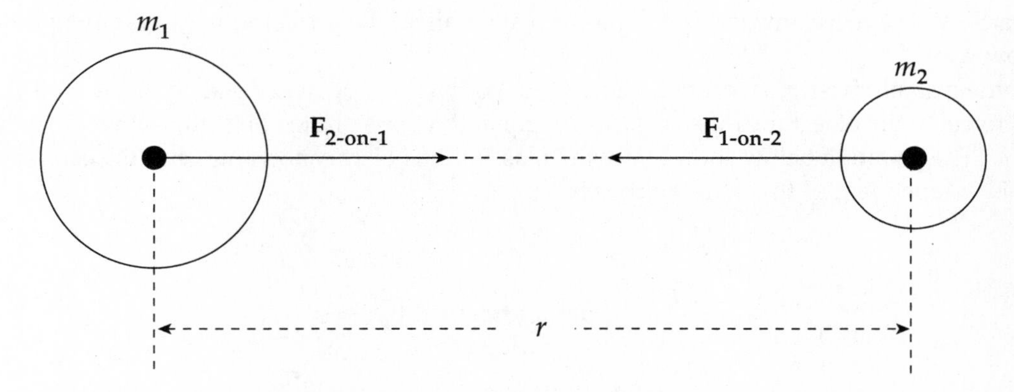

The forces $\mathbf{F}_{1\text{-on-}2}$ and $\mathbf{F}_{2\text{-on-}1}$ act along the line that joins the bodies and form an action/reaction pair.

The first reasonably accurate numerical value for G was determined by Cavendish more than one hundred years after Newton's Law was published. To three decimal places, the currently accepted value of G is

$$G = 6.67 \times 10^{-11}\ \text{N} \cdot \text{m}^2/\text{kg}^2$$

Kepler's Third Law then follows from Newton's Law of Gravitation. We'll show how this works, for the case of a circular orbit of radius R (which can be considered an elliptical orbit with eccentricity zero). If the orbit is circular, then Kepler's Second Law says that the planet's orbit speed, v, must be constant. Therefore, the planet executes uniform circular motion, and centripetal force is provided by the gravitational attraction of the Sun. If we let M be the mass of the Sun and m the mass of the planet, then this last statement can be expressed mathematically as:

$$\frac{mv^2}{R} = G\frac{Mm}{R^2} \quad (1)$$

The period of a planet's orbit is the time it requires to make one revolution around the Sun, so dividing the distance covered, $2\pi R$, by the planet's orbit speed, v, we have

$$T = \frac{2\pi R}{v} \quad (2)$$

Equation (1) implies that $v^2 = GM/R$. Squaring both sides of Equation (2) and then substituting $v^2 = GM/R$, we find that

$$T^2 = \frac{4\pi^2 R^2}{v^2} = \frac{4\pi^2 R^2}{GM/R} = \frac{4\pi^2}{GM}R^3$$

Therefore,

$$\frac{T^2}{R^3} = \frac{4\pi^2}{GM}, \text{ a constant}$$

which is Kepler's Third Law for a circular orbit of radius R.

Acceleration of Gravity Due to Large Bodies

We have been using the acceleration due to gravity $g = 9.8 \text{ m/s}^2$ for all objects falling near the surface of the earth. We have assumed that the mass does not affect the acceleration of gravity and now we will show why.

Assume a small mass m is located near a large body (i.e., a planet or star) of mass M. The gravitational force on the object near the surface will equal the mass of the object times the acceleration of gravity a_g. The equation below shows how the mass of the object cancels out, and the acceleration of gravity is independent of that mass of the object.

$$F_g = ma_g$$

$$\frac{GmM}{r^2} = ma_g, \text{cancel out the } m\text{'s}$$

$$\frac{GM}{r^2} = a_g$$

From this expression we can see that the acceleration due to gravity for any object on a planet would be related to the mass and radius of the planet, but not the mass of the object.

THE GRAVITATIONAL ATTRACTION DUE TO AN EXTENDED BODY

Newton's Law of Gravitation is really a statement about the force between two point particles: objects that are very small in comparison to the distance between them. Newton also proved that a uniform sphere attracts another body as if all of the sphere's mass were concentrated at its center.

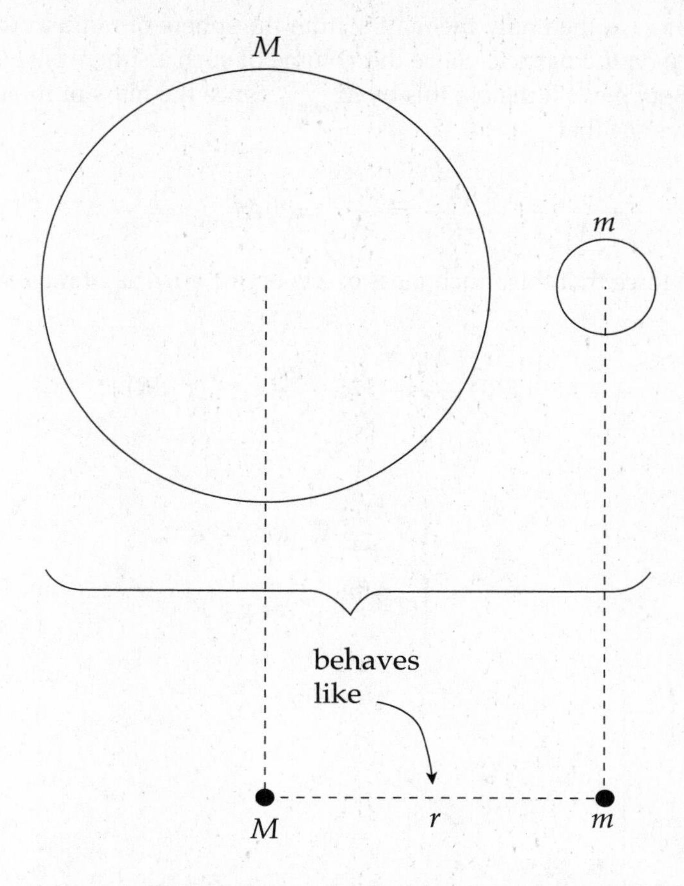

For this reason, we can apply Newton's Law of Gravitation to extended bodies, that is, to objects that are not small relative to the distance between them.

Additionally, a uniform shell of mass does not exert a gravitational force on a particle inside it. This means that if a spherical planet is uniform, then as we descend into it, only the mass of the sphere *underneath* us exerts a gravitational force; the shell *above* exerts no force because we're inside it.

Example 7.1 What is the gravitational force on a particle of mass m at a distance x from the center of a spherically-symmetric planet of uniform density ρ, total mass M, and radius R for

(a) $x \geq R$
(b) $x < R$

Solution.

(a) If $x \geq R$, then the planet can be treated as a point particle with all its mass concentrated at its center, and

$$F = G\frac{Mm}{x^2} \quad (x \geq R)$$

(b) However, if $x < R$, then only the mass within the sphere of radius x exerts a gravitational force on the particle. Since the volume of such a sphere is $(4/3)\pi x^3$, its mass is $(4/3)\pi x^3\rho$; we'll denote this by $M_{\text{within } x}$. Since the mass of the entire planet is $(4/3)\pi R^3\rho$, we see that

$$\frac{M_{\text{within } x}}{M} = \frac{\frac{4}{3}\pi x^3 \rho}{\frac{4}{3}\pi R^3 \rho} = \frac{x^3}{R^3} \quad \Rightarrow \quad M_{\text{within } x} = \frac{x^3}{R^3} M$$

Therefore, the force that this much mass exerts on the particle of mass m is

$$F = G\frac{\frac{x^3}{R^3}M \cdot m}{x^2} = G\frac{Mm}{R^3}x \qquad (x < R)$$

In summary then,

$$F_{\text{grav}} = \begin{cases} G\dfrac{Mm}{R^3}x & \text{for } x < R \\ G\dfrac{Mm}{x^2} & \text{for } x \geq R \end{cases}$$

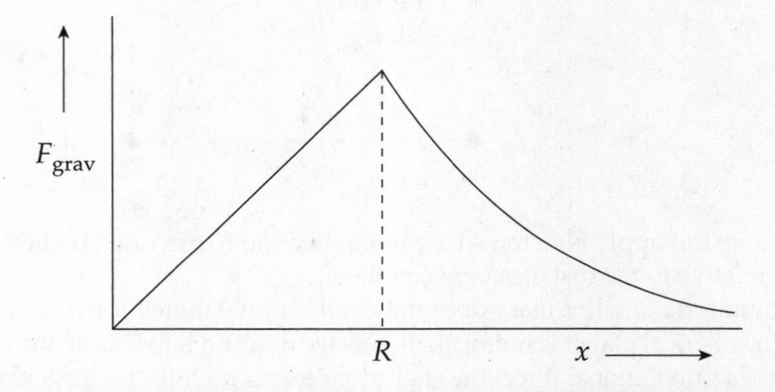

Example 7.2 Given that the radius of the earth is 6.37×10^6 m, determine the mass of the earth.

Solution. Consider a small object of mass m near the surface of the earth (mass M). Its weight is mg, but its weight is just the gravitational force it feels due to the earth, which is GMm/R^2. Therefore,

$$mg = G\frac{Mm}{R^2} \quad \Rightarrow \quad M = \frac{gR^2}{G}$$

Since we know that $g = 10 \text{ m/s}^2$ and $G = 6.67 \times 10^{-11} \text{ N·m}^2/\text{kg}^2$, we can substitute to find

$$M = \frac{gR^2}{G} = \frac{(10 \text{ m/s}^2)(6.37 \times 10^6 \text{ m})^2}{6.67 \times 10^{-11} \text{ N} \cdot \text{m}^2/\text{kg}^2} = 6.0 \times 10^{24} \text{ kg}$$

Example 7.3 We can derive the expression GM/R^2 by equating mg and GMm/R^2 (as we did in the previous example), and this gives the magnitude of the *absolute gravitational acceleration*, a quantity that's sometimes denoted g_0. The notation g is acceleration, but with the spinning of the earth taken into account. Show that if an object is at the equator, its *measured weight*, mg, is less than its *true weight*, mg_0, and compute the weight difference for a person of mass m = 60 kg.

Solution. Imagine looking down at the earth from above the North Pole.

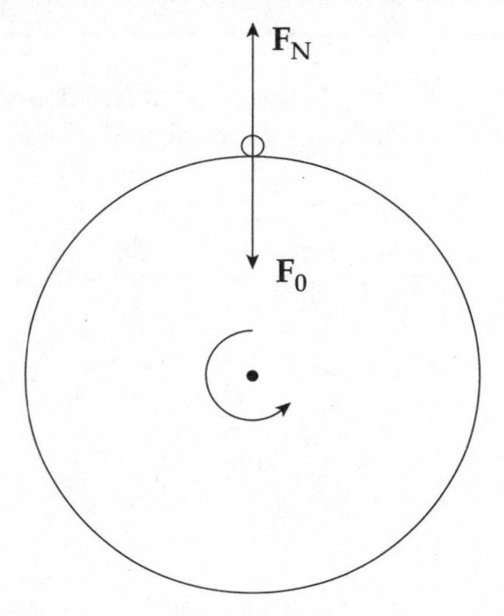

The net force toward the center of the earth is $\mathbf{F}_0 - \mathbf{F}_N$, which provides the centripetal force on the object. Therefore,

$$F_0 - F_N = \frac{mv^2}{R}$$

Since $v = 2\pi R/T$, where T is the earth's rotation period we have

$$F_0 - F_N = \frac{m}{R}\left(\frac{2\pi R}{T}\right)^2 = \frac{4\pi^2 mR}{T^2}$$

or, since $F_0 = mg_0$ and $F_N = mg$,

$$mg_0 - mg = \frac{4\pi^2 mR}{T^2}$$

Since the quantity $4\pi^2 mR/T^2$ is positive, mg must be less than mg_0. The difference between mg_0 and mg, for a person of mass m = 60 kg, is only:

$$\frac{4\pi^2 mR}{T^2} = \frac{4\pi^2(60\text{ kg})(6.37\times10^6\text{ m})}{\left(24\text{ hr}\times\frac{60\text{ min}}{\text{hr}}\times\frac{60\text{ s}}{\text{min}}\right)^2} = 2.0\text{ N}$$

and the difference between g_0 and g is

$$g_0 - g = \frac{mg_0 - mg}{m} = \frac{4\pi^2 R}{T^2} = \frac{4\pi^2(6.37\times10^6 \text{ m})}{\left(24 \text{ hr}\times\frac{60 \text{ min}}{\text{hr}}\times\frac{60 \text{ s}}{\text{min}}\right)^2} = 0.034 \text{ m/s}^2$$

Example 7.4 Communications satellites are often parked in geosynchronous orbits above Earth's surface. Such satellites have orbit periods that are equal to Earth's rotation period, so they remain above the same position on Earth's surface. Determine the altitude and the speed that a satellite must have to be in a geosynchronous orbit above a fixed point on Earth's equator. (The mass of the earth is 5.98×10^{24} kg.)

Solution. Let m be the mass of the satellite, M the mass of Earth, and R the distance from the center of Earth to the position of the satellite. The gravitational pull of Earth provides the centripetal force on the satellite, so

$$G\frac{Mm}{R^2} = \frac{mv^2}{R} \quad\Rightarrow\quad G\frac{M}{R} = v^2$$

The orbit speed of the satellite is $2\pi R/T$, so

$$G\frac{M}{R} = \left(\frac{2\pi R}{T}\right)^2$$

which implies that

$$G\frac{M}{R} = \frac{4\pi^2 R^2}{T^2} \quad\Rightarrow\quad 4\pi^2 R^3 = GMT^2 \quad\Rightarrow\quad R = \sqrt[3]{\frac{GMT^2}{4\pi^2}}$$

Now the key feature of a geosynchronous orbit is that its period matches Earth's rotation period, $T = 24$ hr. Substituting the numerical values of G, M, and T into this expression, we find that

$$R = \sqrt[3]{\frac{GMT^2}{4\pi^2}} = \sqrt[3]{\frac{(6.67\times10^{-11})(5.98\times10^{24})(24\cdot60\cdot60)^2}{4\pi^2}}$$
$$= 4.23\times10^7 \text{ m}$$

Therefore, if r_E is the radius of Earth, then the satellite's altitude above Earth's surface must be

$$h = R - r_E = (4.23\times10^7 \text{ m}) - (6.37\times10^6 \text{ m}) = 3.59\times10^7 \text{ m}$$

which is equal to $5.6r_E$. The speed of the satellite in this orbit is, from the first equation in our calculations,

$$v = \sqrt{G\frac{M}{R}} = \sqrt{(6.67\times10^{-11} \text{ N}\cdot\text{m}^2/\text{kg}^2)\frac{5.98\times10^{24} \text{ kg}}{4.23\times10^7 \text{ m}}} = 3070 \text{ m/s}$$

regardless of the mass of the satellite (well, as long as $m << M$).

Example 7.5 A uniform, slender bar of mass M has length L. Determine the gravitational force it exerts on the point particle of mass m shown below:

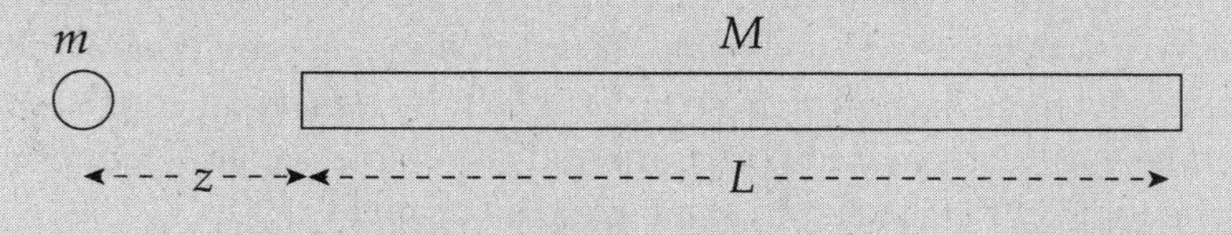

Solution. Since the bar is an extended body (and not spherically symmetric), we must calculate F using an integral. Select an arbitrary segment of length dx and mass dM in the bar, at a distance x from its left-hand end.

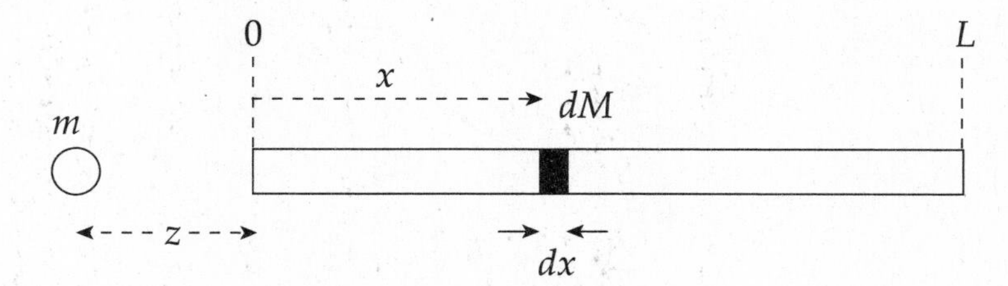

Then, since the bar is uniform, $dM = (M/L)dx$, so the gravitational force between m and dM is

$$dF = G\frac{m \cdot dM}{(z+x)^2} = G\frac{m \cdot \frac{M}{L}dx}{(z+x)^2} = G\frac{mM}{L(z+x)^2}dx$$

Now, by adding (that is, by integrating) all of the contributions dF, we get the total gravitational force, F:

$$\begin{aligned}
F &= \int dF \\
&= \int_{x=0}^{x=L} G\frac{mM}{L(z+x)^2}dx \\
&= G\frac{mM}{L}\int_{x=0}^{x=L} \frac{dx}{(z+x)^2} \\
&= G\frac{mM}{L}\left[\frac{-1}{z+x}\right]_{x=0}^{x=L} \\
&= G\frac{mM}{L}\left[\frac{-1}{z+L}+\frac{1}{z}\right] \\
&= G\frac{mM}{L}\frac{L}{z(z+L)} \\
&= G\frac{mM}{z(z+L)}
\end{aligned}$$

GRAVITATIONAL POTENTIAL ENERGY

When we developed the equation $U = mgh$ for the gravitational potential energy of an object of mass m at height h above the surface of the earth, we took the surface of the earth to be our $U = 0$ reference level and assumed that the height, h, was small compared to the earth's radius. In that case, the variation in g was negligible, so g was treated as constant. The work done by gravity as an object was raised to height h was then simply $-F_{grav} \times \Delta s = -mgh$, so U_{grav}, which by definition equals $-W_{by\ grav}$, was mgh.

But now we'll take variations in g into account and develop a general equation for gravitational potential energy, one that isn't restricted to small altitude changes.

Consider an object of mass m at a distance r_1 from the center of the earth (or any spherical body) moving by some means to a position r_2:

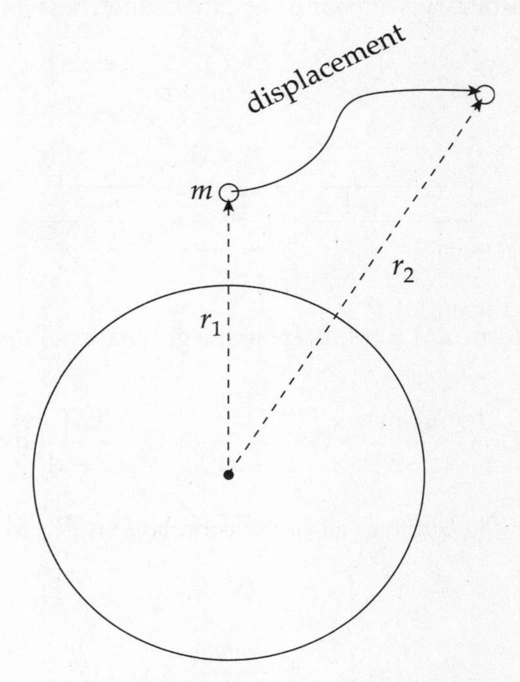

How much work did the gravitational force perform during this displacement? The answer is given by the equation:

$$W_{by\ grav} = GMm\left(\frac{1}{r_2} - \frac{1}{r_1}\right) \quad (*)$$

Therefore, since $\Delta U_{grav} = -W_{by\ grav}$, we get

$$U_2 - U_1 = -GMm\left(\frac{1}{r_2} - \frac{1}{r_1}\right)$$

Let's choose our $U = 0$ reference at infinity. That is, we decide to allow $U_2 \to 0$ as $r_2 \to \infty$. Then this equation becomes

$$U = -\frac{GMm}{r}$$

Notice that, according to this equation (and our choice of $U = 0$ when $r = \infty$), the gravitational potential energy is always negative. This just means that energy has to be added to bring an object (mass m) bound to the gravitational field of M to a point very far from M, at which $U = 0$.

A PROOF OF EQUATION (*)

Because the gravitational force is not constant over the displacement, the work done by this force must be calculated using a definite integral:

$$dW = \mathbf{F} \cdot d\mathbf{r} \quad \Rightarrow \quad W = \int_{r=r_1}^{r=r_2} \mathbf{F} \cdot d\mathbf{r}$$

Displacement can be broken into a series of infinitesimal steps of two types: Those that are at a constant distance from the earth's center and those that are along a radial line going away from the earth's center. The result is that the displacement is equivalent to the sum of two curves $C_1 + C_2$:

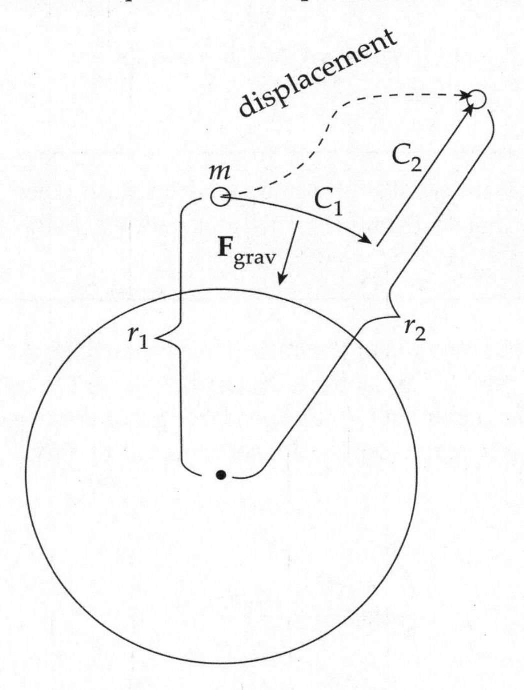

The gravitational force along C_1 does no work, because it's always perpendicular to the displacement. Along C_2, the inward gravitational force is in the opposite direction from the outward displacement, so along this ray,

$$\begin{aligned}\mathbf{F} \cdot d\mathbf{r} &= F \cdot dr \cdot \cos 180^\circ \\ &= -F\, dr \\ &= -\frac{GMm}{r^2} dr\end{aligned}$$

Therefore,

$$W_{\text{by grav}} = \int_{r=r_1}^{r=r_2} \mathbf{F} \cdot d\mathbf{r}$$

$$= \int_{C_2} \mathbf{F} \cdot d\mathbf{r}$$

$$= \int_{r_1}^{r_2} -\frac{GMm}{r^2}\,dr$$

$$= GMm \int_{r=r_1}^{r=r_2} -\frac{1}{r^2}\,dr$$

$$= GMm \left[\frac{1}{r}\right]_{r_1}^{r_2}$$

$$= GMm \left(\frac{1}{r_2} - \frac{1}{r_1}\right)$$

Example 7.6 With what minimum speed must an object of mass m be launched in order to escape Earth's gravitational field? (This is called **escape speed**, v_{esc}.)

Solution. When launched, the object is at the surface of the earth ($r_i = r_E$) and has an upward, initial velocity of magnitude v_i. To get it far away from the earth, we want to bring its gravitational potential energy to zero, but to find the *minimum* launch speed, we want the object's final speed to be zero by the time it gets to this distant location. So, by Conservation of Energy,

$$K_i + U_i = K_f + U_f$$

$$\frac{1}{2}mv_i^2 + \frac{-GM_Em}{r_E} = 0 + 0$$

which gives

$$\frac{1}{2}mv_i^2 = \frac{GM_Em}{r_E} \quad \Rightarrow \quad v_i = \sqrt{\frac{2GM_E}{r_E}}$$

Substituting the known numerical values for G, M_E, and r_E gives us:

$$v_i = v_{\text{esc}} = \sqrt{\frac{2GM_E}{r_E}} = \sqrt{\frac{2(6.67\times10^{-11})(5.98\times10^{24})}{6.37\times10^{6}}}$$

$$= 1.12\times10^{4}\ \text{m/s}$$

(Note: This is the minimum speed needed to escape *Earth's* gravitational field but, in this situation, a projectile would also have to contend with the *Sun's* gravitational field, etc.)

Example 7.7 A satellite of mass m is in a circular orbit of radius R around the earth (radius r_E, mass M).

(a) What is its total mechanical energy (where U_{grav} is considered zero as R approaches infinity)?

(b) How much work would be required to move the satellite into a new orbit, with radius $2R$?

Solution.

(a) The mechanical energy, E, is the sum of the kinetic energy, K, and potential energy, U. You can calculate the kinetic energy since you know that the centripetal force on the satellite is provided by the gravitational attraction of the earth:

$$\frac{mv^2}{R} = \frac{GMm}{R^2} \Rightarrow mv^2 = \frac{GMm}{R} \Rightarrow K = \tfrac{1}{2}mv^2 = \frac{GMm}{2R}$$

Therefore,

$$E = K + U = \frac{GMm}{2R} + \frac{-GMm}{R} = -\frac{GMm}{2R}$$

(b) From the equation $K_i + U_i + W = K_f + U_f$, we see that

$$\begin{aligned} W &= (K_f + U_f) - (K_i + U_i) \\ &= E_f - E_i \end{aligned}$$

Therefore, the amount of work necessary to effect the change in the satellite's orbit radius from R to $2R$ is

$$\begin{aligned} W &= E_{\text{at } 2R} - E_{\text{at } R} \\ &= -\frac{GMm}{2(2R)} - \left(\frac{-GMm}{2R}\right) \\ &= \frac{GMm}{4R} \end{aligned}$$

A Note on Elliptical Orbits

The expression for the total energy of a satellite in a circular orbit of radius R [derived in Example 7.7(a)] is:

$$E = -\frac{GMm}{2R} \quad \text{(circular orbit)}$$

and this also holds for a satellite traveling in an elliptical orbit, if the radius R is replaced by a, the length of the semimajor axis:

$$E = -\frac{GMm}{2a} \quad \text{(elliptical orbit)}$$

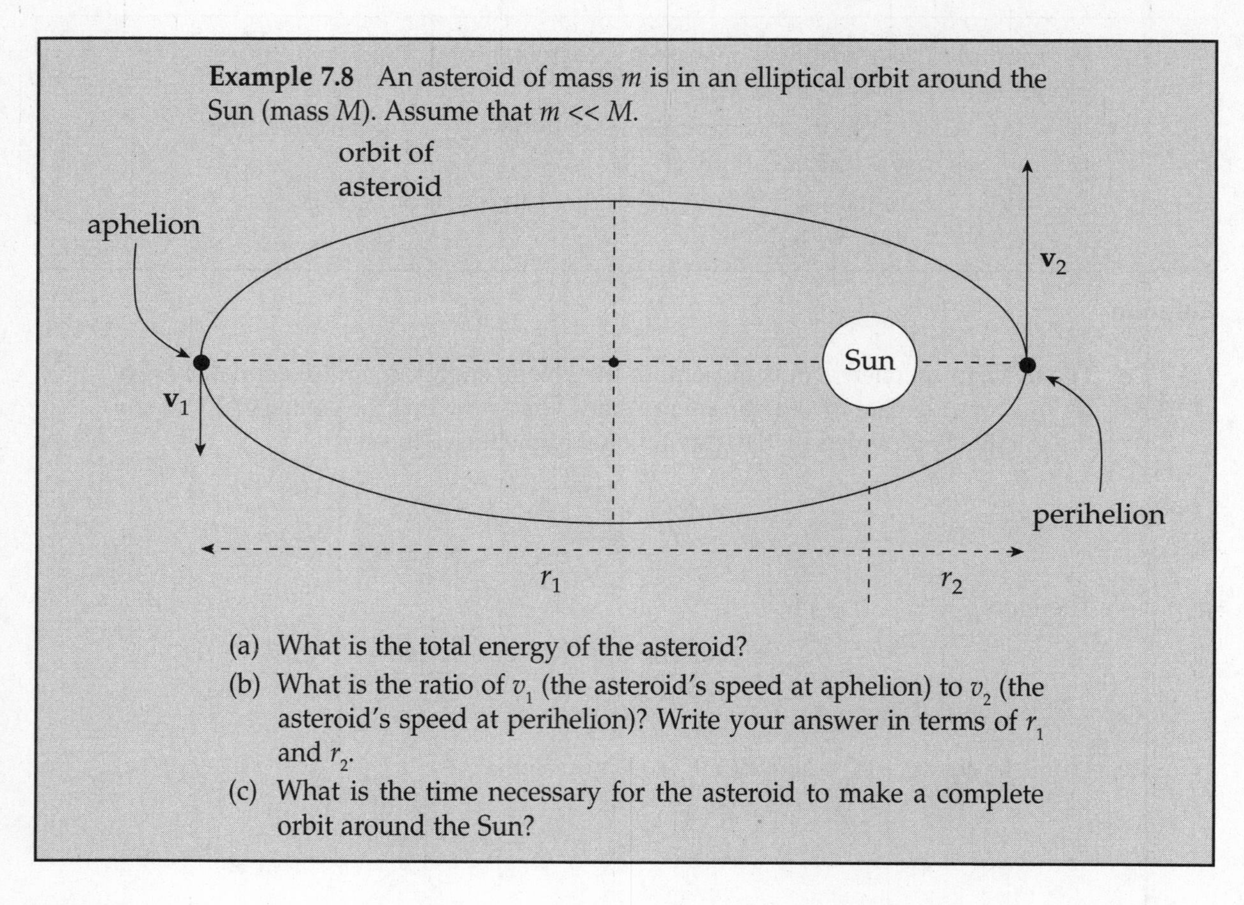

Example 7.8 An asteroid of mass m is in an elliptical orbit around the Sun (mass M). Assume that $m << M$.

(a) What is the total energy of the asteroid?

(b) What is the ratio of v_1 (the asteroid's speed at aphelion) to v_2 (the asteroid's speed at perihelion)? Write your answer in terms of r_1 and r_2.

(c) What is the time necessary for the asteroid to make a complete orbit around the Sun?

Solution.

(a) The total energy of the asteroid is equal to $-GMm/2a$, where a is the semimajor axis. However, notice in the figure above that $2a = r_1 + r_2$. Therefore, the total energy of the asteroid is

$$E = -\frac{GMm}{r_1 + r_2}$$

(b) One way to answer this question is to invoke Conservation of Angular Momentum. When the asteroid is at aphelion, its angular momentum (with respect to the center of the Sun) is $L_1 = r_1 m v_1$. When the asteroid is at perihelion, its angular momentum is $L_2 = r_2 m v_2$. Therefore,

$$L_1 = L_2 \quad \Rightarrow \quad r_1 m v_1 = r_2 m v_2 \quad \Rightarrow \quad \frac{v_1}{v_2} = \frac{r_2}{r_1}$$

This tells us that the asteroid's speed at aphelion is less than its speed at perihelion (because $r_2/r_1 < 1$), as implied by Kepler's Second Law (a line drawn from the Sun to the asteroid must sweep out equal areas in equal time intervals). The closer the asteroid is to the Sun, the faster it has to travel to make this true.

(c) As you know, the time necessary for the asteroid to make a complete orbit around the Sun is the orbit period, T. Using Kepler's Third Law, with $a = \frac{1}{2}(r_1 + r_2)$, we find

$$\frac{T^2}{a^3} = \frac{4\pi^2}{GM} \quad \Rightarrow \quad T = \sqrt{\frac{4\pi^2 a^3}{GM}} = \sqrt{\frac{4\pi^2[\frac{1}{2}(r_1 + r_2)]^3}{GM}} = \sqrt{\frac{\pi^2(r_1 + r_2)^3}{2GM}}$$

ORBITS OF THE PLANETS

Kepler's First Law states that the planets' orbits are ellipses, but the ellipses that the planets in our solar system travel are nearly circular. The deviation of an ellipse from a perfect circle is measured by a parameter called its **eccentricity**. The eccentricity, e, is the ratio of c (the distance between the center and either focus) to a, the length of the semimajor axis. For every point on the ellipse, the sum of the distances to the foci (plural of focus) is a constant (and is equal to $2a$ in the figure below).

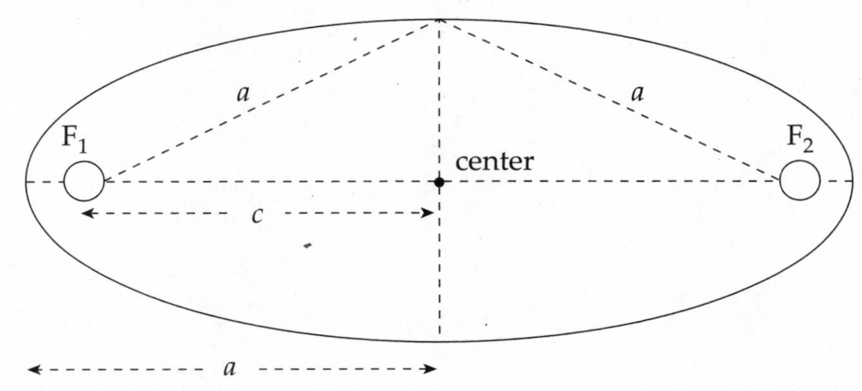

Kepler's First Law also states that one of the foci of a planet's elliptical orbit is located at the position of the Sun. Actually, one of the foci is at the center of mass of the Sun-planet system, because when one body orbits another, both bodies orbit around their center of mass, a point called the **barycenter**.

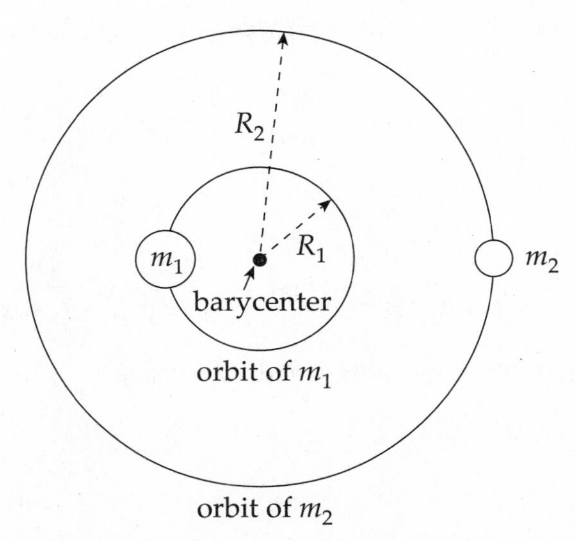

For most of the planets, which are much less massive than the Sun, this correction to Kepler's First Law has little significance, because the center of mass of the Sun and the planet system is close enough to the Sun's center. For example, let's figure out the center of mass of the Sun-Earth system. The mass of Earth is $m = 5.98 \times 10^{24}$ kg, the mass of the Sun is $M = 1.99 \times 10^{30}$ kg, and the Sun-Earth distance averages $R = 1.496 \times 10^{11}$ m. Therefore, letting $x = 0$ be at the Sun's center, we have

$$x_{\text{cm}} = \frac{Mx_{\text{Sun}} + mx_{\text{Earth}}}{M+m} = \frac{M(0) + (5.98\times10^{24}\text{ kg})(1.496\times10^{11}\text{ m})}{(1.99\times10^{30}\text{ kg}) + (5.98\times10^{24}\text{ kg})} = 450\text{ km}$$

So the center of mass of the Sun-Earth system is only 450 km from the center of the Sun, a distance of less than 0.1% of the Sun's radius.

Example 7.9 Derive a corrected version of Kepler's Third Law for the following orbiting system. Both bodies have orbit period T.

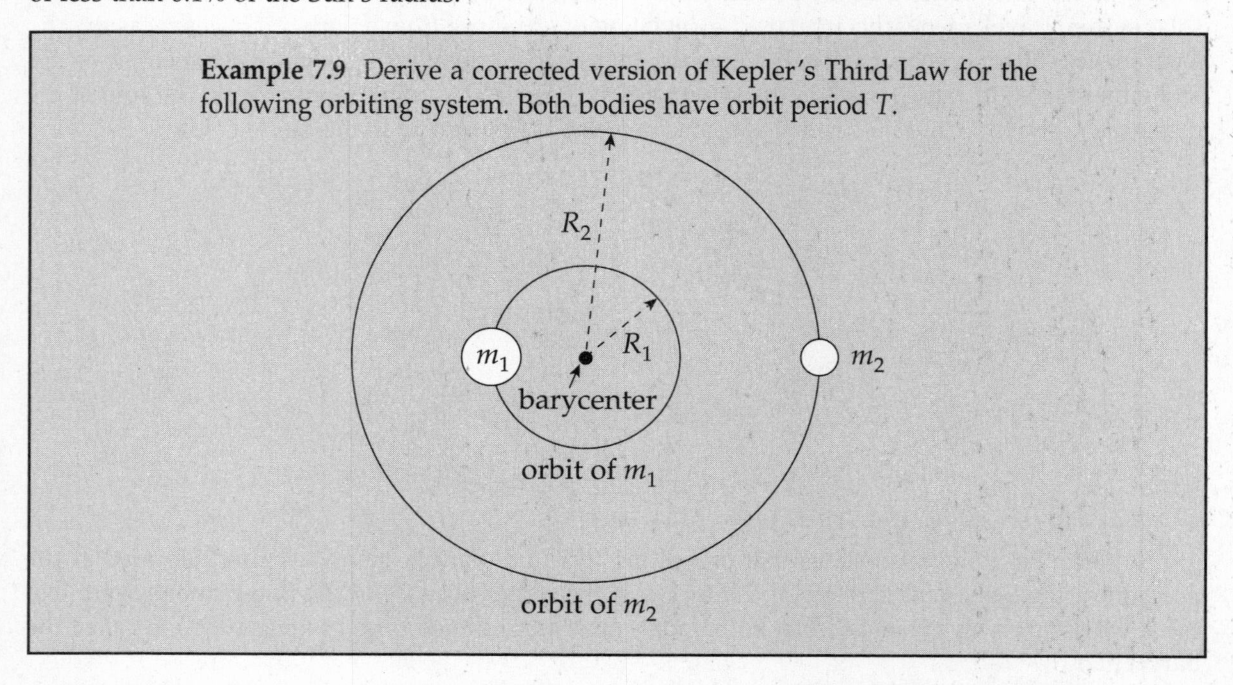

Solution. The centripetal force on each body is provided by the gravitational pull of the other body, so

$$\frac{m_1v_1^2}{R_1} = G\frac{m_1m_2}{(R_1+R_2)^2} \quad \text{and} \quad \frac{m_2v_2^2}{R_2} = G\frac{m_1m_2}{(R_1+R_2)^2}$$

which imply

$$\frac{v_1^2}{R_1} = G\frac{m_2}{(R_1+R_2)^2} \quad \text{and} \quad \frac{v_2^2}{R_2} = G\frac{m_1}{(R_1+R_2)^2}$$

But, since both bodies have the same orbit period, T, we have

$$v_1 = \frac{2\pi R_1}{T} \quad \text{and} \quad v_2 = \frac{2\pi R_2}{T}$$

Substituting these results into the preceding pair of equations gives us:

$$\frac{(2\pi R_1/T)^2}{R_1} = G\frac{m_2}{(R_1+R_2)^2} \quad \text{and} \quad \frac{(2\pi R_2/T)^2}{R_2} = G\frac{m_1}{(R_1+R_2)^2}$$

which simplify to

$$\frac{4\pi^2 R_1}{T^2} = G\frac{m_2}{(R_1+R_2)^2} \quad \text{and} \quad \frac{4\pi^2 R_2}{T^2} = G\frac{m_1}{(R_1+R_2)^2}$$

Adding this last pair of equations gives us the desired result:

$$\frac{4\pi^2(R_1+R_2)}{T^2} = G\frac{m_1+m_2}{(R_1+R_2)^2}$$

$$\frac{4\pi^2}{T^2} = \frac{G(m_1+m_2)}{(R_1+R_2)^3}$$

$$\frac{T^2}{(R_1+R_2)^3} = \frac{4\pi^2}{G(m_1+m_2)}$$

Note that this final equation is a general version of Kepler's Third Law for a circular orbit derived earlier, $T^2/R^3 = 4\pi^2/GM$, where it was assumed that the planet orbited at a distance R from the center of the Sun.

Example 7.10 An artificial satellite of mass m travels at a constant speed in a circular orbit of radius R around the Earth (mass M). What is the speed of the satellite?

Solution. The centripetal force on the satellite is provided by Earth's gravitational pull. Therefore,

$$\frac{mv^2}{R} = G\frac{Mm}{R^2}$$

Solving this equation for v yields

$$v = \sqrt{G\frac{M}{R}}$$

Notice that the satellite's speed doesn't depend on its mass; even if it were a baseball, if its orbit radius were R, then its orbit speed would still be $\sqrt{GM/R}$.

CHAPTER 7 REVIEW QUESTIONS

SECTION I: MULTIPLE CHOICE

1. If the distance between two point particles is doubled, then the gravitational force between them

   (A) decreases by a factor of 4
   (B) decreases by a factor of 2
   (C) increases by a factor of 2
   (D) increases by a factor of 4
   (E) Cannot be determined without knowing the masses

2. At the surface of the earth, an object of mass m has weight w. If this object is transported to a height above the surface that's twice the radius of the earth, then, at the new location,

   (A) its mass is $m/2$ and its weight is $w/2$
   (B) its mass is m and its weight is $w/2$
   (C) its mass is $m/2$ and its weight is $w/4$
   (D) its mass is m and its weight is $w/4$
   (E) its mass is m and its weight is $w/9$

3. A moon of mass m orbits a planet of mass $100m$. Let the strength of the gravitational force exerted by the planet on the moon be denoted by F_1, and let the strength of the gravitational force exerted by the moon on the planet be F_2. Which of the following is true?

   (A) $F_1 = 100F_2$
   (B) $F_1 = 10F_2$
   (C) $F_1 = F_2$
   (D) $F_2 = 10F_1$
   (E) $F_2 = 100F_1$

4. The dwarf planet Pluto has 1/500 the mass and 1/15 the radius of Earth. What is the value of g (in m/s²) on the surface of Pluto?

   (A) $\frac{25}{225}$
   (B) $\frac{50}{15}$
   (C) $\frac{15}{50}$
   (D) $\frac{225}{50}$
   (E) $\frac{50}{225}$

5. A satellite is currently orbiting Earth in a circular orbit of radius R; its kinetic energy is K_1. If the satellite is moved and enters a new circular orbit of radius $2R$, what will be its kinetic energy?

   (A) $K_1/4$
   (B) $K_1/2$
   (C) K_1
   (D) $2K_1$
   (E) $4K_1$

6. A moon of Jupiter has a nearly circular orbit of radius R and an orbit period of T. Which of the following expressions gives the mass of Jupiter?

   (A) $2\pi R/T$
   (B) $4\pi^2 R/T^2$
   (C) $2\pi R^3/(GT^2)$
   (D) $4\pi R^2/(GT^2)$
   (E) $4\pi^2 R^3/(GT^2)$

7. Two large bodies, Body A of mass m and Body B of mass $4m$, are separated by a distance R. At what distance from Body A, along the line joining the bodies, would the gravitational force on an object be equal to zero? (Ignore the presence of any other bodies.)

   (A) $R/16$
   (B) $R/8$
   (C) $R/5$
   (D) $R/4$
   (E) $R/3$

8. The mean distance from Saturn to the Sun is 9 times greater than the mean distance from Earth to the Sun. How long is a Saturn year?

   (A) 18 Earth years
   (B) 27 Earth years
   (C) 81 Earth years
   (D) 243 Earth years
   (E) 729 Earth years

9. The Moon has mass M and radius R. A small object is dropped from a distance of $3R$ from the Moon's center. The object's impact speed when it strikes the surface of the Moon is equal to

$$\sqrt{kGM/R} \text{ for } k =$$

(A) 1/3
(B) 2/3
(C) 3/4
(D) 4/3
(E) 3/2

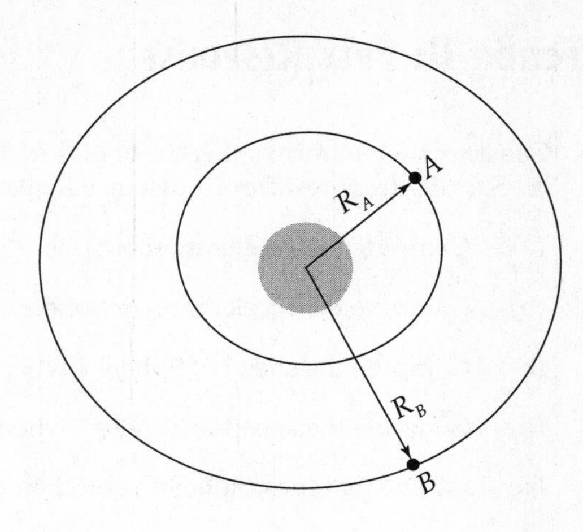

10. Two satellites, A and B, orbit a planet in circular orbits having radii R_A and R_B, respectively, as shown above. If $R_B = 3R_A$, the velocities v_A and v_B of the two satellites are related by which of the following?

(A) $v_B = v_A$
(B) $v_B = 3v_A$
(C) $v_B = 9v_A$
(D) $v_B = v_A\sqrt{3}$
(E) $v_B = \frac{v_A}{\sqrt{3}}$

SECTION II: FREE RESPONSE

1. Consider two uniform spherical bodies in deep space. Sphere 1 has mass m_1 and Sphere 2 has mass m_2. Starting from rest from a distance R apart, they are gravitationally attracted to each other.

   (a) Compute the acceleration of Sphere 1 when the spheres are a distance $R/2$ apart.

   (b) Compute the acceleration of Sphere 2 when the spheres are a distance $R/2$ apart.

   (c) Compute the speed of Sphere 1 when the spheres are a distance $R/2$ apart.

   (d) Compute the speed of Sphere 2 when the spheres are a distance $R/2$ apart.

   Now assume that these spheres orbit their center of mass with the same orbit period, T.

   (e) Determine the radii of their orbits. Write your answer in terms of m_1, m_2, T, and fundamental constants.

2. A satellite of mass m is in the elliptical orbit shown below around Earth (radius r_E, mass M). Assume that $m << M$.

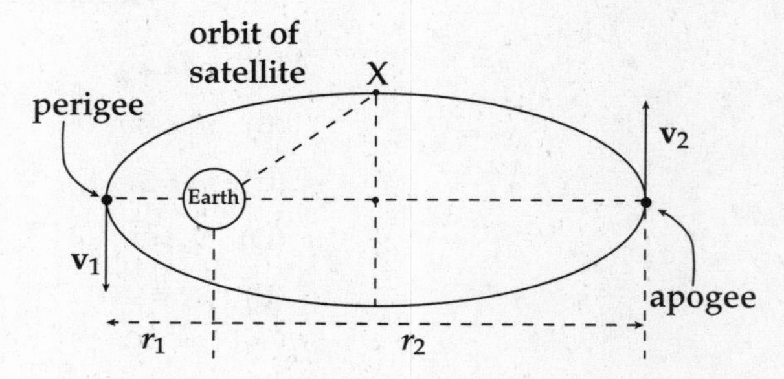

   (a) Determine v_1, the speed of the satellite at perigee (the point of the orbit closest to Earth). Write your answer in terms of r_1, r_2, M, and G.

   (b) Determine v_2, the speed of the satellite at apogee (the point of the orbit farthest from Earth). Write your answer in terms of r_1, r_2, M, and G.

   (c) Express the ratio v_1/v_2 in simplest terms.

   (d) What is the satellite's angular momentum (with respect to Earth's center) when it's at apogee?

   (e) Determine the speed of the satellite when it's at the point marked X in the figure.

   (f) Determine the period of the satellite's orbit. Write your answer in terms of r_1, r_2, M, and fundamental constants.

LAWS OF GRAVITATION SUMMARY

NEWTON'S LAW OF GRAVITATION

Newton's Law of Gravitation gives the force between any two point masses, regardless of their mass or location. This uniform circular motion can also be described by an angular velocity and a centripetal acceleration.

$$F_g = \frac{GM_1M_2}{r^2}$$

$$a_g = \frac{GM_1}{r^2}$$

CIRCULAR ORBITS

$$F_g = F_c$$

$$\frac{GM_1M_2}{r^2} = \frac{M_2v^2}{r}$$

$$\frac{GM_1}{r} = v^2$$

$$2\pi r = vT$$

GENERAL ORBITS

$$U_g = -\frac{GM_1M_2}{r}$$

Both *mechanical energy* and *angular momentum* are conserved for orbits. For the AP Exam, consider using conservation of angular momentum because its equation is simpler than the equation for gravitational potential energy.

Systems are bound when the total mechanical energy is less than zero. This means that the gravitational potential energy is greater than the kinetic energy of the system.

Therefore, for an object to escape, its kinetic energy must be greater than or equal to its gravitational potential energy.

$$E_{total} = 0$$

$$\frac{1}{2}M_1v_{esc}{}^2 - \frac{GM_1M_2}{d} = 0$$

$$v_{esc} = \sqrt{\frac{2GM_2}{d}}$$

Gravity of Spheres and Shells

Gravity due to a spherical shell of mass M and radius R. A small mass, *m*, is at different locations, a distance *r*, away from the center of the shell.

Outside the shell:

$$F_g = \frac{GMm}{r^2}$$

Inside the shell:

$$F_g = 0$$

Gravity due to a uniform, solid sphere of mass M and radius R. A small mass, *m*, is at different locations, a distance *r*, away from the center of the sphere.

Outside the sphere:

$$F_g = \frac{GMm}{r^2}$$

Inside the sphere:

$$F_g = \frac{GMm}{R^3}r$$

Gravity of a spherically symmetric sphere of mass M and radius R (i.e. $\rho(r)$).

$$F_g = -Gm\int \frac{dM}{r^2}\hat{r} \text{ where } dM = \rho(r)dV$$

8

Oscillations

INTRODUCTION

In this chapter, we'll concentrate on a kind of periodic motion that's straightforward and that, fortunately, actually describes many real-life systems. This type of motion is called *simple harmonic motion.* The prototypical example of simple harmonic motion is a block that's oscillating on the end of a spring, and what we learn about this simple system, we can apply to many other oscillating systems.

SIMPLE HARMONIC MOTION (SHM): THE SPRING-BLOCK OSCILLATOR

When a spring is compressed or stretched from its natural length, a force is created. If the spring is displaced by $\mathbf{x}$ from its natural length, the force it exerts in response is given by the equation

$$\mathbf{F}_S = -k\mathbf{x}$$

This is known as **Hooke's Law**. Many, but not all, springs obey Hooke's Law. The proportionality constant, k, is a positive number called the **spring** (or **force**) **constant** that indicates how stiff the spring is. The stiffer the spring, the greater the value of k. The minus sign in Hooke's Law tells us that $\mathbf{F}_S$ and **x** always point in opposite directions. For example, referring to the figure below, when the spring is stretched (**x** is to the right), the spring pulls back (**F** is to the left); when the spring is compressed (**x** is to the left), the spring pushes outward (**F** is to the right). In all cases, the spring wants to return to its original length. As a result, the spring tries to restore the attached block to the **equilibrium position**, which is the position at which the net force on the block is zero. For this reason, we say that the spring provides a **restoring force**.

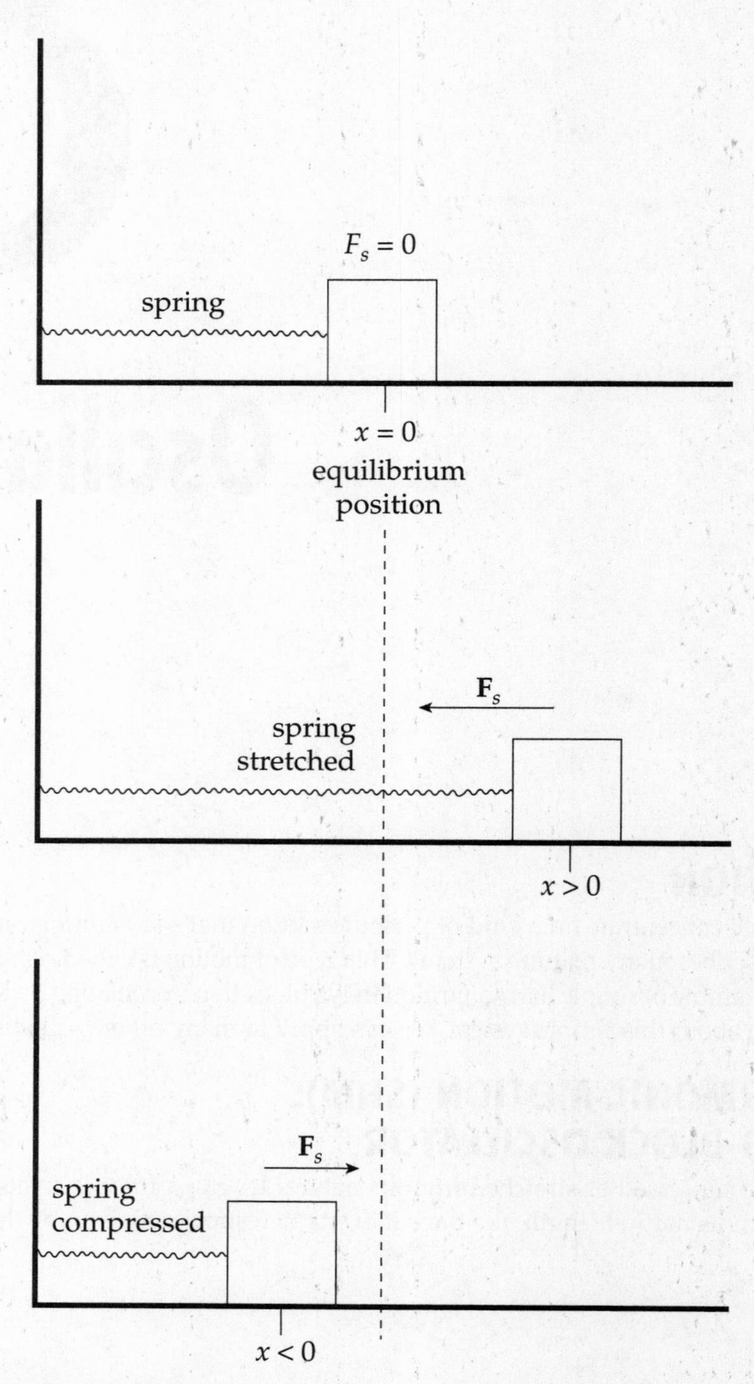

> **Example 8.1** A 12 cm-long spring has a force constant (k) of 400 N/m. How much force is required to stretch the spring to a length of 14 cm?

Solution. The displacement of the spring has a magnitude of 14 – 12 = 2 cm = 0.02 m so, according to Hooke's Law, the spring exerts a force of magnitude $F = kx$ = (400 N/m)(0.02 m) = 8 N. Therefore, we'd have to exert this much force to keep the spring in this stretched state.

Springs that obey Hooke's Law (called **ideal** or **linear** springs) provide an ideal mechanism for defining the most important kind of vibrational motion: simple harmonic motion.

Consider a spring with force constant k, attached to a vertical wall, with a block of mass m on a frictionless table attached to the other end.

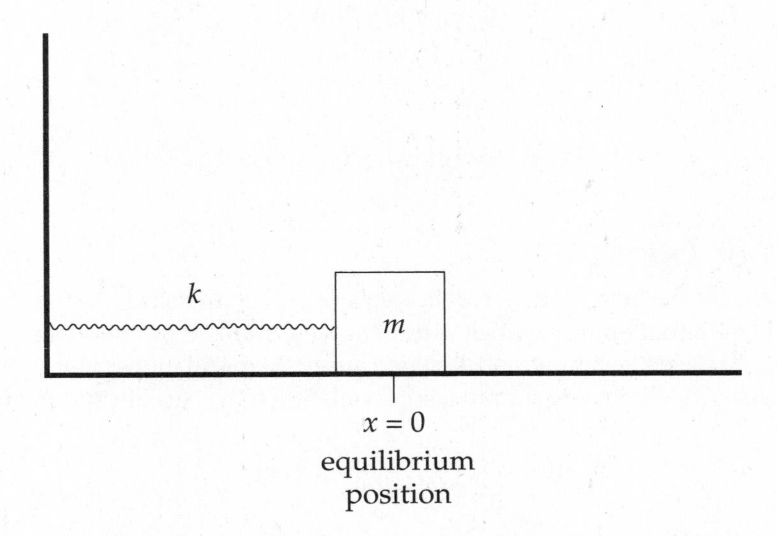

Grab the block, pull it some distance from its original position, and release it. The spring will pull the block back toward equilibrium. Of course, because of its momentum, the block will pass through the equilibrium position and compress the spring. At some point, the block will stop, and the compressed spring will push the block back. In other words, the block will oscillate.

During the oscillation, the force on the block is zero when the block is at equilibrium (the point we designate as $x = 0$). This is because Hooke's Law says that the strength of the spring's restoring force is given by the equation $F_S = kx$, so $F_S = 0$ at equilibrium. The acceleration of the block is also equal to zero at $x = 0$, since $F_S = 0$ at $x = 0$ and $a = F_S/m$. At the endpoints of the oscillation region, where the block's displacement, x, has the greatest magnitude, the restoring force and the magnitude of the acceleration are both at their maximum.

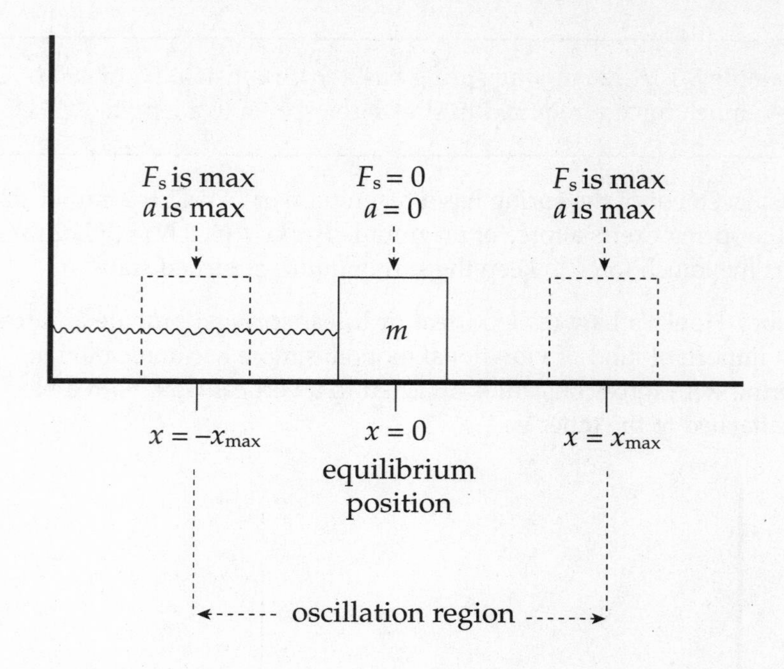

SHM IN TERMS OF ENERGY

Another way to describe the block's motion is in terms of energy transfers. A stretched or compressed spring stores **elastic potential energy**, which is transformed into kinetic energy (and back again); this shuttling of energy between potential and kinetic causes the oscillations. For a spring with spring constant k, the elastic potential energy it possesses—relative to its equilibrium position—is given by the equation

$$U_s = \frac{1}{2}kx^2$$

Notice that the farther you stretch or compress a spring, the more work you have to do, and, as a result, the more potential energy that's stored.

In terms of energy transfers, we can describe the block's oscillations as follows. When you initially pull the block out, you increase the elastic potential energy of the system. Upon releasing the block, this potential energy turns into kinetic energy, and the block moves. As it passes through equilibrium, $U_S = 0$, so all the potential energy is kinetic. Then, as the block continues through equilibrium, it compresses the spring and the kinetic energy is transformed back into elastic potential energy.

By Conservation of Mechanical Energy, the sum $K + U_S$ is a constant. Therefore, when the block reaches the endpoints of the oscillation region (that is, when $x = \pm x_{max}$), U_S is maximized, so K must be minimized; in fact, $K = 0$ at the endpoints. As the block is passing through equilibrium, $x = 0$, so $U_S = 0$ and K is maximized.

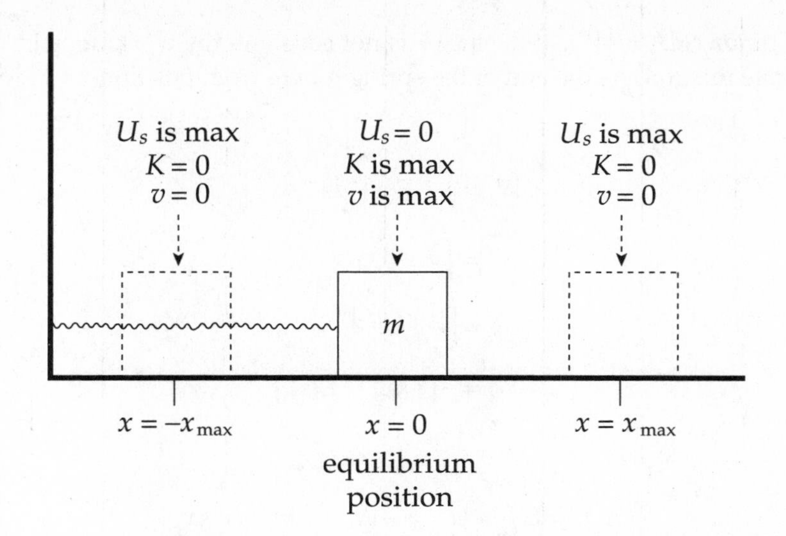

The maximum displacement from equilibrium is called the **amplitude** of oscillation, and is denoted by A. So instead of writing $x = x_{max}$, we write $x = A$ (and $x = -x_{max}$ will be written as $x = -A$).

Example 8.2 A block of mass $m = 0.05$ kg oscillates on a spring whose force constant k is 500 N/m. The amplitude of the oscillations is 4.0 cm. Calculate the maximum speed of the block.

Solution. First let's get an expression for the maximum elastic potential energy of the system:

$$U_S = \tfrac{1}{2}kx^2 \quad \Rightarrow \quad U_{S,\,max} = \tfrac{1}{2}kx_{max}^2 = \tfrac{1}{2}kA^2$$

When all this energy has been transformed into kinetic energy—which, as we discussed earlier, occurs just as the block is passing through equilibrium—the block will have maximum kinetic energy and maximum speed

$$U_{S,\,max} \to K_{max} \quad \Rightarrow \quad \tfrac{1}{2}kA^2 = \tfrac{1}{2}mv_{max}^2$$

$$v_{max} = \sqrt{\frac{kA^2}{m}}$$

$$= \sqrt{\frac{(500 \text{ N/m})(0.04 \text{ m})^2}{.05 \text{ kg}}}$$

$$= 4 \text{ m/s}$$

Example 8.3 Show why $U_s = \frac{1}{2}kx^2$ if $\mathbf{F}_S = -k\mathbf{x}$.

Solution. By definition, $\Delta U_S = -W_S$, and, since **F** is not constant, the work done by **F** must be calculated using a definite integral. As the end of the spring moves from position $x = x_1$ to $x = x_2$, the work it does is equal to

$$\begin{aligned} W_S &= \int_{x_1}^{x_2} \mathbf{F}(x) \cdot d\mathbf{x} \\ &= \int_{x_1}^{x_2} -kx\,dx \\ &= \left[-\tfrac{1}{2}kx^2\right]_{x_1}^{x_2} \\ &= -\left(\tfrac{1}{2}kx_2^2 - \tfrac{1}{2}kx_1^2\right) \end{aligned}$$

Therefore,

$$\Delta U_S = U_{S2} - U_{S1} = -W_S = \tfrac{1}{2}kx_2^2 - \tfrac{1}{2}kx_1^2$$

So if we designate $U_{S1} = 0$ at $x_1 = 0$, the equation above yields $U_s = \frac{1}{2}kx^2$, as desired. (Even if a spring does not obey Hooke's Law, this method can still be used to find the work done and the potential energy stored.)

> **Example 8.4** A block of mass $m = 2.0$ kg is attached to an ideal spring of force constant $k = 500$ N/m. The amplitude of the resulting oscillations is 8.0 cm. Determine the total energy of the oscillator and the speed of the block when it's 4.0 cm from equilibrium.

Solution. The total energy of the oscillator is the sum of its kinetic and potential energies. By Conservation of Mechanical Energy, the sum $K + U_S$ is a constant, so if we can determine what this sum is at some point in the oscillation region, we'll know the sum at every point. When the block is at its amplitude position, $x = 8$ cm, its speed is zero; so at this position, E is easy to figure out:

$$E = K + U_S = 0 + \tfrac{1}{2}kA^2 = \tfrac{1}{2}(500\text{ N/m})(0.08\text{ m})^2 = 1.6\text{ J}$$

This gives the total energy of the oscillator at *every* position. At any position x, we have

$$\tfrac{1}{2}mv^2 + \tfrac{1}{2}kx^2 = E$$

$$v = \sqrt{\frac{E - \frac{1}{2}kx^2}{\frac{1}{2}m}}$$

so when we substitute in the numbers, we get

$$v = \sqrt{\frac{E - \frac{1}{2}kx^2}{\frac{1}{2}m}} = \sqrt{\frac{(1.6\text{ J}) - \frac{1}{2}(500\text{ N/m})(0.04\text{ m})^2}{\frac{1}{2}(2.0\text{ kg})}}$$
$$= 1.1\text{ m/s}$$

Example 8.5 A block of mass $m = 3.0$ kg is attached to an ideal spring of force constant $k = 500$ N/m. The block is at rest at its equilibrium position. An impulsive force acts on the block, giving it an initial speed of 2.0 m/s. Find the amplitude of the resulting oscillations.

Solution. The block will come to rest when all of its initial kinetic energy has been transformed into the spring's potential energy. At this point, the block is at its maximum displacement from equilibrium, that is, it's at one of its amplitude positions, and

$$\begin{aligned} K_i + U_i &= K_f + U_f \\ \tfrac{1}{2}mv_i^2 + 0 &= 0 + \tfrac{1}{2}kA^2 \\ A &= \sqrt{\frac{mv_i^2}{k}} \\ &= \sqrt{\frac{(3.0\text{ kg})(2.0\text{ m/s})^2}{500\text{ N/m}}} \\ &= 0.15\text{ m} \end{aligned}$$

THE KINEMATICS OF SHM

Now that we've explored the dynamics of the block's oscillations in terms of force and energy, let's talk about motion—or kinematics. As you watch the block oscillate, you should notice that it repeats each **cycle** of oscillation in the same amount of time. A cycle is a *round-trip*: for example, from position $x = A$ over to $x = -A$ and back again to $x = A$. The amount of time it takes to complete a cycle is called the **period** of the oscillations, or T. If T is short, the block is oscillating rapidly, and if T is long, the block is oscillating slowly.

Another way of indicating the rapidity of the oscillations is to count the number of cycles that can be completed in a given time interval; the more completed cycles, the more rapid the oscillations. The number of cycles that can be completed per unit time is called the **frequency** of the oscillations, or f, and frequency is expressed in cycles per second. One cycle per second is one **hertz** (abbreviated **Hz**).

One of the most basic equations of oscillatory motion expresses the fact that the period and frequency are reciprocals of each other:

$$\text{period} = \frac{\text{\# seconds}}{\text{cycle}} \qquad \text{while} \qquad \text{frequency} = \frac{\text{\# cycles}}{\text{second}}$$

Therefore

$$T = \frac{1}{f} \quad \text{and} \quad f = \frac{1}{T}$$

Example 8.6 A block oscillating on the end of a spring moves from its position of maximum spring stretch to maximum spring compression in 0.25 s. Determine the period and frequency of this motion.

Solution. The period is defined as the time required for one full cycle. Moving from one end of the oscillation region to the other is only half a cycle. Therefore, if the block moves from its position of maximum spring stretch to maximum spring compression in 0.25 s, the time required for a full cycle is twice as much; $T = 0.5$ s. Because frequency is the reciprocal of period, the frequency of the oscillations is $f = 1/T = 1/(0.5 \text{ s}) = 2$ Hz.

Example 8.7 A student observing an oscillating block counts 45.5 cycles of oscillation in one minute. Determine its frequency (in hertz) and period (in seconds).

Solution. The frequency of the oscillations, in hertz (which is the number of cycles per second), is

$$f = \frac{45.5 \text{ cycles}}{\text{min}} \times \frac{1 \text{ min}}{60 \text{ s}} = \frac{0.758 \text{ cycles}}{\text{s}} = 0.758 \text{ Hz}$$

Therefore,

$$T = \frac{1}{f} = \frac{1}{0.758 \text{ Hz}} = 1.32 \text{ s}$$

One of the defining properties of the spring-block oscillator is that the frequency and period can be determined from the mass of the block and the force constant of the spring. The equations are as follows:

$$f = \frac{1}{2\pi}\sqrt{\frac{k}{m}} \quad \text{and} \quad T = 2\pi\sqrt{\frac{m}{k}}$$

Let's analyze these equations. Suppose we had a small mass on a very stiff spring; then intuitively, we would expect that this strong spring would make the small mass oscillate rapidly, with high frequency and short period. Both of these predictions are substantiated by the equations above, because if m is small and k is large, then the ratio k/m is large (high frequency) and the ratio m/k is small (short period).

Example 8.8 A block of mass $m = 2.0$ kg is attached to a spring whose force constant, k, is 300 N/m. Calculate the frequency and period of the oscillations of this spring–block system.

Solution. According to the equations above,

$$f = \frac{1}{2\pi}\sqrt{\frac{k}{m}} = \frac{1}{2\pi}\sqrt{\frac{300\text{ N/m}}{2.0\text{ kg}}} = 1.9\text{ Hz}$$

$$T = 2\pi\sqrt{\frac{m}{k}} = 2\pi\sqrt{\frac{2.0\text{ kg}}{300\text{ N/m}}} = 0.51\text{ s}$$

Notice that $f \approx 2$ Hz and $T \approx 0.5$ s, and that these values satisfy the basic equation $T = 1/f$.

> **Example 8.9** A block is attached to a spring and set into oscillatory motion, and its frequency is measured. If this block were removed and replaced by a second block with 1/4 the mass of the first block, how would the frequency of the oscillations compare to that of the first block?

Solution. Since the same spring is used, k remains the same. According to the equation given above, f is inversely proportional to the square root of the mass of the block: $f \propto 1/\sqrt{m}$. Therefore, if m decreases by a factor of 4, then f increases by a factor of $\sqrt{4} = 2$.

The equations we saw above for the frequency and period of the spring-block oscillator do not contain A, the amplitude of the motion. In simple harmonic motion, *both the frequency and the period are independent of the amplitude*. The reason that the frequency and period of the spring-block oscillator are independent of amplitude is that F, the strength of the restoring force, is proportional to x, the displacement from equilibrium, as given by Hooke's Law: $F_S = -kx$.

> **Example 8.10** A student performs an experiment with a spring-block simple harmonic oscillator. In the first trial, the amplitude of the oscillations is 3.0 cm, while in the second trial (using the same spring and block), the amplitude of the oscillations is 6.0 cm. Compare the values of the period, frequency, and maximum speed of the block between these two trials.

Solution. If the system exhibits simple harmonic motion, then the period and frequency are independent of amplitude. This is because the same spring and block were used in the two trials, so the period and frequency will have the same values in the second trial as they had in the first. But the maximum speed of the block will be greater in the second trial than in the first. Since the amplitude is greater in the second trial, the system possesses more total energy ($E = \frac{1}{2}kA^2$). So when the block is passing through equilibrium (its position of greatest speed), the second system has more energy to convert to kinetic, meaning that the block will have a greater speed. In fact, from Example 8.2, we know that $v_{max} = A\sqrt{k/m}$ so, since A is twice as great in the second trial than in the first, v_{max} will be twice as great in the second trial than in the first.

Example 8.11 For each of the following arrangements of two springs, determine the **effective spring constant**, k_{eff}. This is the force constant of a single spring that would produce the same force on the block as the pair of springs shown in each case.

(a)

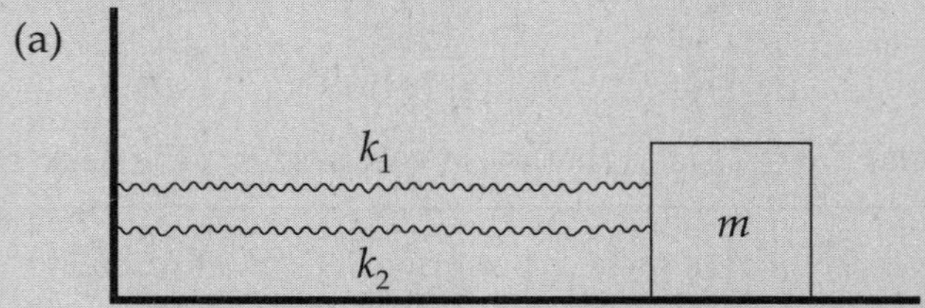

(b)

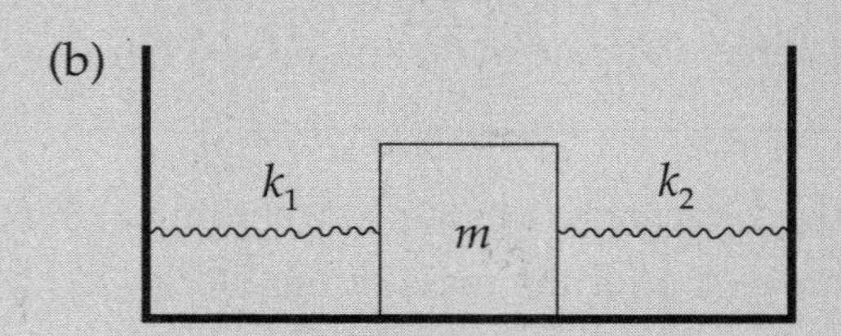

(c)

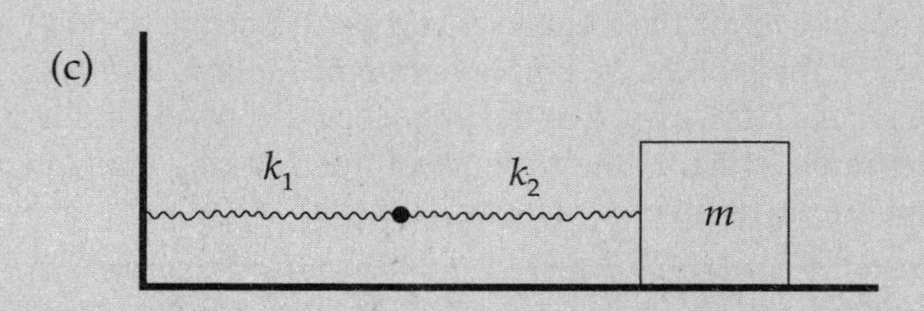

(d) Determine k_{eff} in each of these cases if $k_1 = k_2 = k$.

Solution.

(a) Imagine that the block was displaced a distance x to the right of its equilibrium position. Then the force exerted by the first spring would be $F_1 = -k_1x$ and the force exerted by the second spring would be $F_2 = -k_2x$. The net force exerted by the springs would be

$$F_1 + F_2 = -k_1x + -k_2x = -(k_1 + k_2)x$$

Since $F_{eff} = -(k_1 + k_2)x$, we see that $k_{eff} = k_1 + k_2$.

(b) Imagine that the block was displaced a distance x to the right of its equilibrium position. Then the force exerted by the first spring would be $F_1 = -k_1x$ and the force exerted by the second spring would be $F_2 = -k_2x$. The net force exerted by the springs would be

$$F_1 + F_2 = -k_1x + -k_2x = -(k_1 + k_2)x$$

As in part (a), we see that, since $F_{eff} = -(k_1 + k_2)x$, we get $k_{eff} = k_1 + k_2$.

(c) Imagine that the block was displaced a distance x to the right of its equilibrium position. Let x_1 be the distance that the first spring is stretched, and let x_2 be the distance that the second spring is stretched. Then $x = x_1 + x_2$. But $x_1 = -F/k_1$ and $x_2 = -F/k_2$, so

$$\frac{-F}{k_1} + \frac{-F}{k_2} = x$$

$$-F\left(\frac{1}{k_1} + \frac{1}{k_2}\right) = x$$

$$F = -\left(\frac{1}{\frac{1}{k_1} + \frac{1}{k_2}}\right)x$$

$$F = -\frac{k_1 k_2}{k_1 + k_2}x$$

Therefore,

$$k_{\text{eff}} = \frac{k_1 k_2}{k_1 + k_2}$$

(d) If the two springs have the same force constant, that is, if $k_1 = k_2 = k$, then in the first two cases, the pairs of springs are equivalent to one spring that has twice their force constant: $k_{\text{eff}} = k_1 + k_2 = k + k = 2k$. In (c), the pair of springs is equivalent to a single spring with half their force constant:

$$k_{\text{eff}} = \frac{k_1 k_2}{k_1 + k_2} = \frac{kk}{k + k} = \frac{k^2}{2k} = \frac{k}{2}$$

THE SPRING-BLOCK OSCILLATOR: VERTICAL MOTION

So far we've looked at a block sliding back and forth on a horizontal table, but the block could also oscillate vertically. The only difference would be that gravity would cause the block to move downward, to an equilibrium position at which, in contrast with the horizontal SHM we've examined, the spring would not be at its natural length.

Consider a spring of negligible mass hanging from a stationary support. A block of mass m is attached to its end and allowed to come to rest, stretching the spring a distance d. At this point, the block is in equilibrium; the upward force of the spring is balanced by the downward force of gravity. Therefore,

$$kd = mg \quad \Rightarrow \quad d = \frac{mg}{k}$$

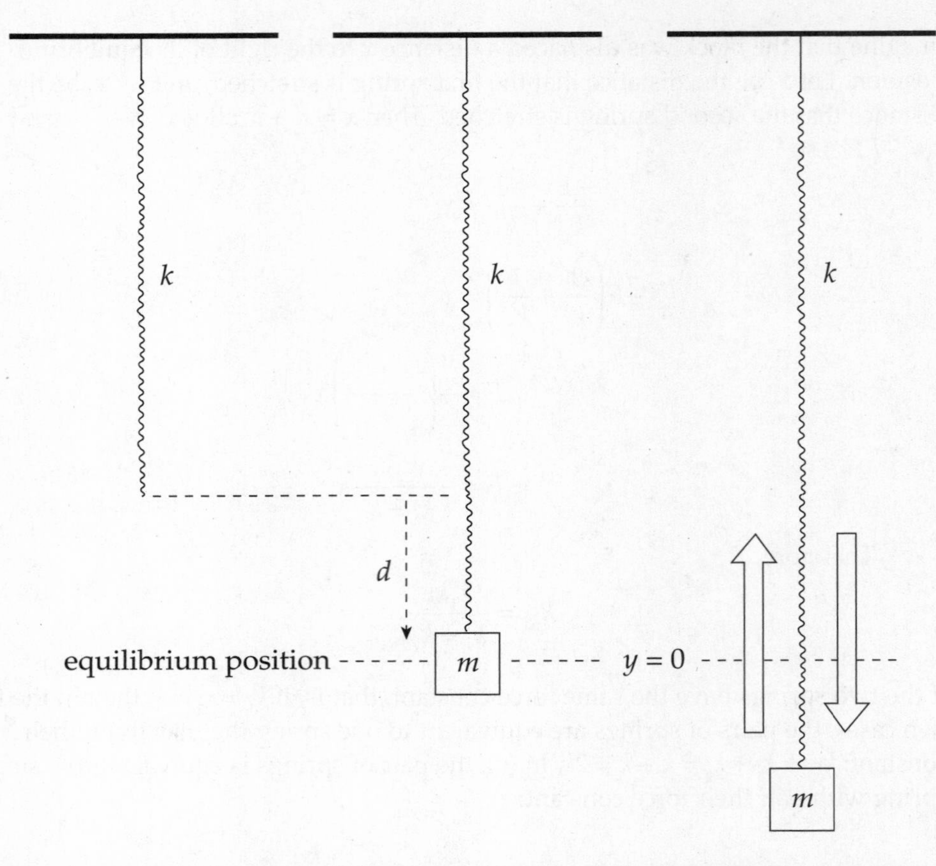

Next, imagine that the block is pulled down a distance A and released. The spring force increases (because the spring was stretched farther); it's stronger than the block's weight, and, as a result, the block accelerates upward. As the block's momentum carries it up, through the equilibrium position, the spring becomes less stretched than it was at equilibrium, so F_S is less than the block's weight. As a result, the block decelerates, stops, and accelerates downward again, and the up-and-down motion repeats.

When the block is at a distance y below its equilibrium position, the spring is stretched a total distance of $d + y$, so the upward spring force is equal to $k(d + y)$, while the downward force stays the same, mg. The net force on the block is

$$F = k(d + y) - mg$$

but this equation becomes $F = ky$, because $kd = mg$ (as we saw above). Since the resulting force on the block, $F = ky$, has the form of Hooke's Law, we know that the vertical simple harmonic oscillations of the block have the same characteristics as do horizontal oscillations, with the equilibrium position, $y = 0$, not at the spring's natural length, but at the point where the hanging block is in equilibrium.

Example 8.12 A block of mass m = 1.5 kg is attached to the end of a vertical spring of force constant k = 300 N/m. After the block comes to rest, it is pulled down a distance of 2.0 cm and released.

(a) What is the frequency of the resulting oscillations?

(b) What are the minimum and maximum amounts of stretch of the spring during the oscillations of the block?

Solution.

(a) The frequency is given by

$$f = \frac{1}{2\pi}\sqrt{\frac{k}{m}} = \frac{1}{2\pi}\sqrt{\frac{300\text{ N/m}}{1.5\text{ kg}}} = 2.3\text{ Hz}$$

(b) Before the block is pulled down, to begin the oscillations, it stretches the spring by a distance

$$d = \frac{mg}{k} = \frac{(1.5\text{ kg})(10\text{ N/kg})}{300\text{ N/m}} = .05\text{ m} = 5\text{ cm}$$

Since the amplitude of the motion is 2.0 cm, the spring is stretched a maximum of 5 cm + 2.0 cm = 7 cm when the block is at the lowest position in its cycle, and a minimum of 5 cm – 2.0 cm = 3 cm when the block is at its highest position.

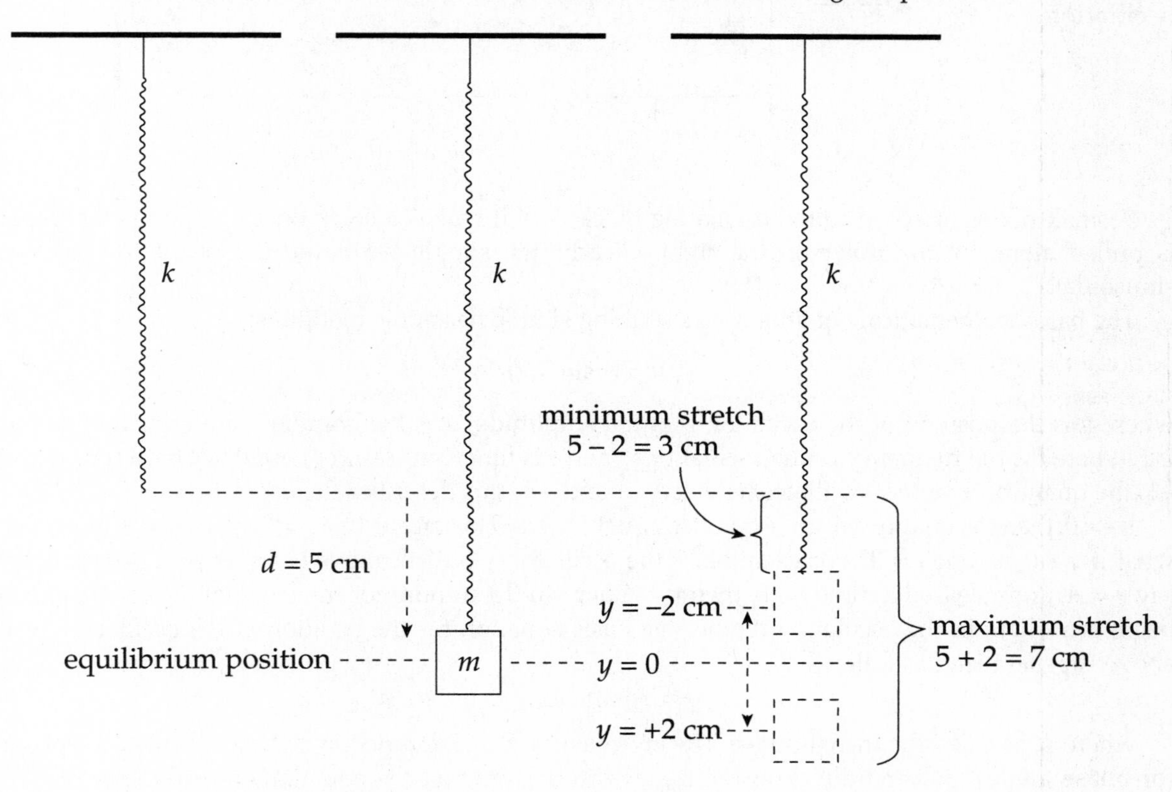

THE SINUSOIDAL DESCRIPTION OF SHM

The position of the block during its oscillation can be written as a function of time. Take a look at the experimental set-up below.

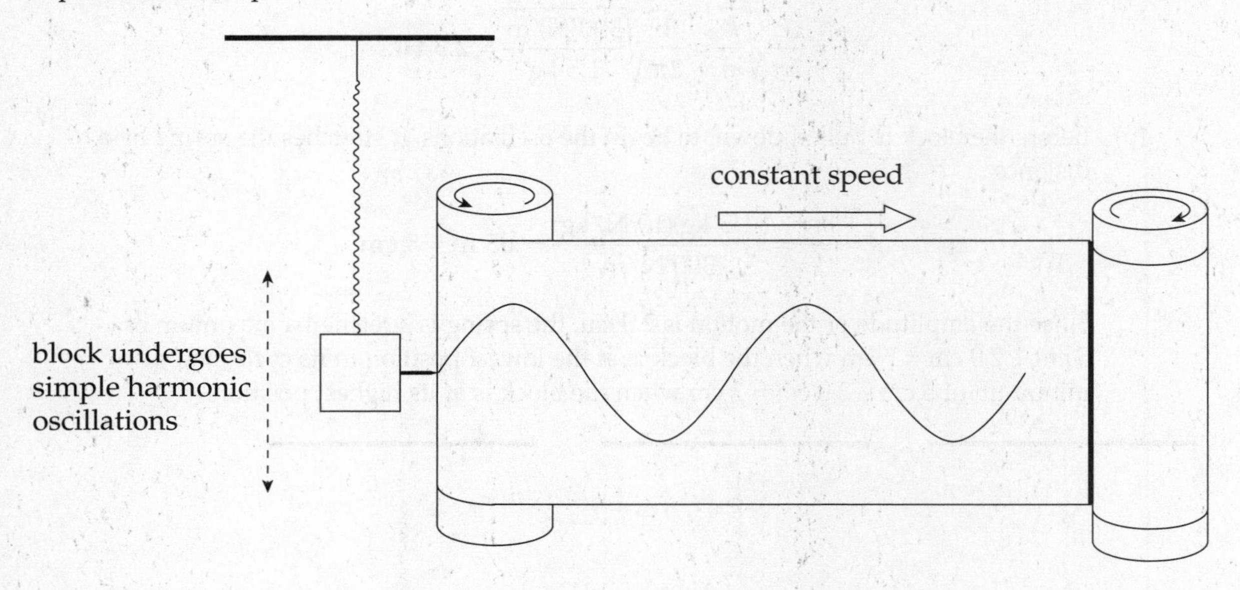

A small pen is attached to the oscillating block, and it makes a mark on the paper as the paper is pulled along by the roller on the right. Clearly, the simple harmonic motion of the block is sinusoidal.

The basic mathematical equation for describing simple harmonic motion is:

$$y = A \sin (\omega t)$$

where y is the position of the oscillator, A is the amplitude, ω is the **angular frequency** (defined as $2\pi f$, where f is the frequency of the oscillations), and t is time. Since $\sin(\omega t)$ oscillates between –1 and +1, the quantity $A \sin(\omega t)$ oscillates between $-A$ and $+A$; this describes the oscillation region.

If $t = 0$, then the quantity $A \sin (\omega t)$ is also equal to zero. This means that $y = 0$ at time $t = 0$. However, what if $y \neq 0$ at time $t = 0$? For example, if the oscillator is pulled to one of its amplitude positions, say $y = A$, and released at time $t = 0$, then $y = A$ at $t = 0$. To account for the fact that the oscillator can begin anywhere in the oscillation region, the basic equation for the position of the oscillator given above is generalized as follows:

$$y = A \sin (\omega t + \phi_0)$$

where ϕ_0 is called the **initial phase**. The argument of the sine function, $\omega t + \phi_0$, is called the **phase** (or **phase angle**). By carefully choosing ϕ_0, we can be sure that the equation correctly specifies the oscillator's position no matter where it may have been at time $t = 0$. The value of ϕ_0 can be calculated from the equation

$$\phi_0 = \sin^{-1}\left(\frac{y_{\text{at } t=0}}{A}\right)$$

Example 8.13 A simple harmonic oscillator has an amplitude of 3.0 cm and a frequency of 4.0 Hz. At time $t = 0$, its position is $y = 3.0$ cm. Where is it at time $t = 0.3$ s?

Solution. First, A = 3 cm and $\omega = 2\pi f = 2\pi(4.0\ \text{s}^{-1}) = 8\pi\ \text{s}^{-1}$. The value of the initial phase is

$$\phi_0 = \sin^{-1}\left(\frac{y_{\text{at }t=0}}{A}\right) = \sin^{-1}\frac{3\ \text{cm}}{3\ \text{cm}} = \sin^{-1} 1 = \tfrac{1}{2}\pi$$

Therefore, the position of the oscillator at any time t is given by the equation

$$y = (3\ \text{cm}) \cdot \sin[(8\pi\ \text{s}^{-1})t + \tfrac{1}{2}\pi]$$

So, at time t = 0.3 s, we find that $y = (3\ \text{cm}) \cdot \sin[(8\pi\ \text{s}^{-1})(0.3\ \text{s}) + \tfrac{1}{2}\pi] = 0.93$. (Make sure your calculator is in *radian mode* when you evaluate this expression!)

Example 8.14 The position of a simple harmonic oscillator is given by the equation

$$y = (4\ \text{cm}) \cdot \sin[(6\pi\ \text{s}^{-1})t - \tfrac{1}{2}\pi]$$

(a) Where is the oscillator at time t = 0?
(b) What is the amplitude of the motion?
(c) What is the frequency?
(d) What is the period?

Solution.

(a) To determine y at time t = 0, we simply substitute t = 0 into the given equation and evaluate:

$$y_{\text{at }t=0} = (4\ \text{cm}) \cdot \sin(-\tfrac{1}{2}\pi) = -4\ \text{cm}$$

(b) The amplitude of the motion, A, is the coefficient in front of the $\sin(\omega t + \phi_0)$ expression. In this case, we read directly from the given equation that A = 4 cm.

(c) The coefficient of t is the angular frequency, ω. In this case, we see that $\omega = 6\pi\ \text{s}^{-1}$. Since $\omega = 2\pi f$ by definition, we have $f = \omega/(2\pi)$ = 3 Hz.

(d) Since $T = 1/f$, we calculate that $T = 1/f = 1/(3\ \text{Hz})$ = 0.33 s.

INSTANTANEOUS VELOCITY AND ACCELERATION

If the position of a simple harmonic oscillator is given by the equation $y = A\sin(\omega t + \phi_0)$, its velocity and acceleration can be found by differentiation:

$$v(t) = \dot{y}(t) = \frac{d}{dt}[A\sin(\omega t + \phi_0)] = A\omega\cos(\omega t + \phi_0)$$

and

$$a(t) = \dot{v}(t) = \frac{d}{dt}[A\omega\cos(\omega t + \phi_0)] = -A\omega^2\sin(\omega t + \phi_0)$$

Note that both the velocity and acceleration vary with time.

DIFFERENTIAL EQUATION FOR SHM

The differential equation, $\frac{d^2y}{dt^2} = -\omega^2 y$ has a solution of $y(t) = A \sin(\omega t + \phi_0)$. To check that this function, $y(t)$, is a solution, put in the equation for $y(t)$ and the equation for the $a(t)$, the second derivative of $y(t)$, and see if both sides of the equation are the same.

$$\frac{d^2y}{dt^2} = -\omega^2 y$$

$$-A\omega^2 \sin(\omega t + \phi_0) = -\omega^2 (A \sin(\omega t + \phi_0))$$

check

$$-A\omega^2 \sin(\omega t + \phi_0) = -\omega^2 (A \sin(\omega t + \phi_0))$$

Yes, the two sides of the equation are the same so $y(t) = A \sin(\omega t + \phi_0)$ is a solution to the differential equation $\frac{d^2y}{dt^2} = -\omega^2 y$. Therefore, if we can derive a differential equation that takes this form we will know the solution to the equation and that the object is undergoing SHM.

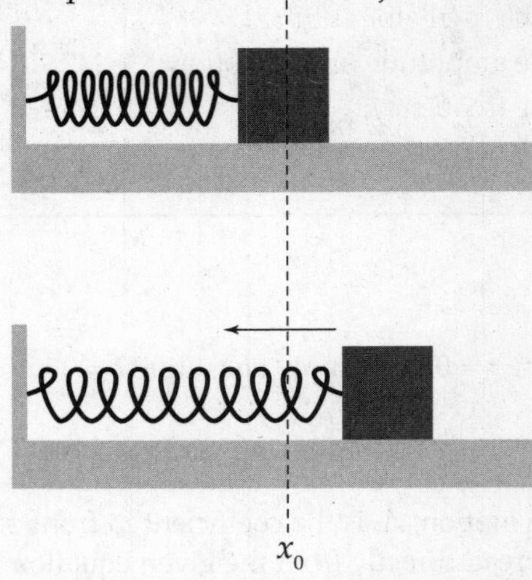

Let's do this for a spring/mass system sliding along a frictionless, horizontal floor, as shown above.

$$F = ma$$

$$-kx = m\frac{d^2x}{dt^2}$$

$$-\frac{k}{m}x = \frac{d^2x}{dt^2}$$

This is the same basic form of the differential equation from above, and for the spring mass system:

$$\omega = \sqrt{\frac{k}{m}} \text{ and } x(t) = A\sin(\omega t)$$

> **Example 8.15** The position of a simple harmonic oscillator is given by the equation
>
> $$y = 4\sin(6\pi t - \tfrac{1}{2}\pi)$$
>
> where y is in cm and t is in seconds. What are the maximum speed and maximum acceleration of the oscillator?

Solution. Since the maximum value of cos $(\omega t + \phi_0)$ is 1, the maximum value of v is $v_{max} = A\omega$; similarly, since the minimum value of $(\omega t + \phi_0)$ is –1, the maximum value of a is $a_{max} = A\omega^2$. From the equation for y, we see that $A = 4$ cm $= 0.04$ m and $\omega = 6\pi$ s^{-1}. Therefore, $v_{max} = A\omega = (0.04\text{ m})\cdot(6\pi\text{ s}^{-1}) = 0.75$ m/s and $a_{max} = A\omega^2 = (0.04\text{ m})(6\pi\text{ s}^{-1})^2 = 14$ m/s^2.

PENDULUMS

A **simple pendulum** consists of a weight of mass m attached to a massless rod that swings, without friction, about the vertical equilibrium position. The restoring force is provided by gravity and, as the figure below shows, the magnitude of the restoring force when the bob is θ to an angle to the vertical is given by the equation:

$$F_{restoring} = mg\sin\theta$$

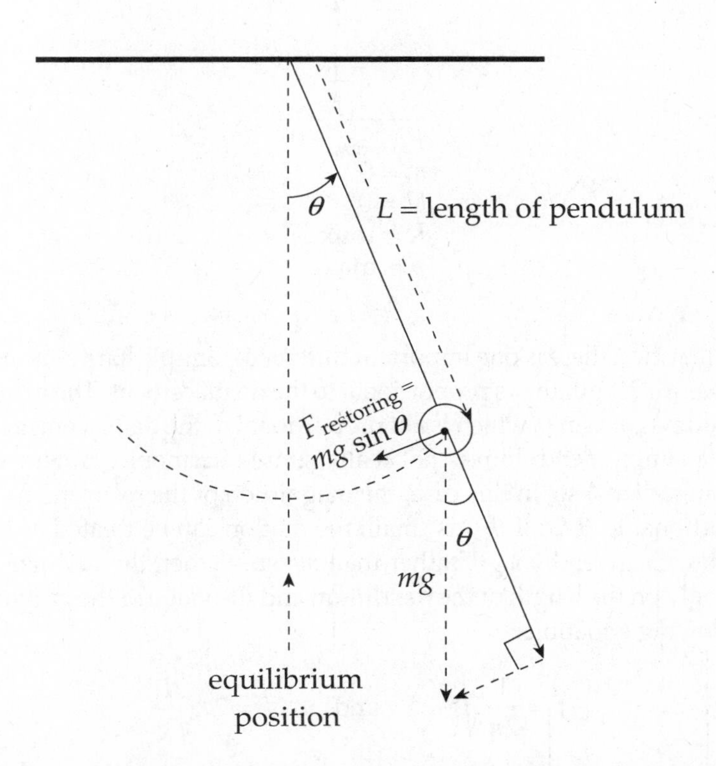

Although the displacement of the pendulum is measured by the angle that it makes with the vertical, rather than by its linear distance from the equilibrium position (as was the case for the spring–block oscillator), the simple pendulum shares many of the important features of the spring–block oscillator. For example,

- Displacement is zero at the equilibrium position.
- At the endpoints of the oscillation region (where $\theta = \pm\theta_{max}$), the restoring force and the tangential acceleration (a_t) have their greatest magnitudes, the speed of the pendulum is zero, and the potential energy is maximized.
- As the pendulum passes through the equilibrium position, its kinetic energy and speed are maximized.

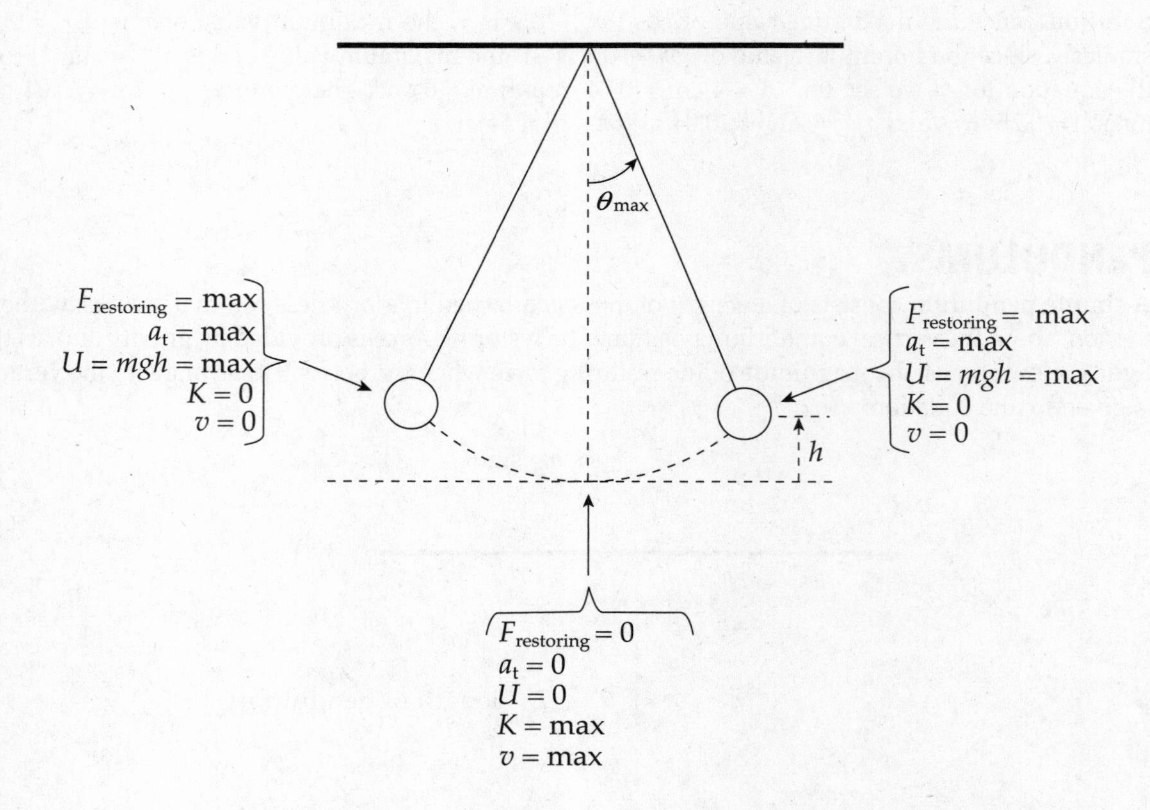

Despite these similarities, there is one important difference. Simple harmonic motion results from a restoring force that has a strength that's proportional to the displacement. The magnitude of the restoring force on a pendulum is $mg \sin \theta$, which is *not* proportional to the displacement θ. Strictly speaking, then, the motion of a simple pendulum is not really simple harmonic. However, if θ is small, then $\sin \theta \approx \theta$ (measured in radians) so, in this case, the magnitude of the restoring force is approximately $mg\theta$, which *is* proportional to θ. So if θ_{max} is small, the motion can be treated as simple harmonic.

If the restoring force is given by $mg\theta$, rather than $mg \sin \theta$, then the frequency and period of the oscillations depend only on the length of the pendulum and the value of the gravitational acceleration, according to the following equations:

$$f = \frac{1}{2\pi}\sqrt{\frac{g}{L}} \quad \text{and} \quad T = 2\pi\sqrt{\frac{L}{g}}$$

DIFFERENTIAL EQUATION FOR A PENDULUM

We have derived a differential equation for a spring-mass system that has a solution of $y(t) = A\sin(\omega t + \phi_o)$. Now we will do the same for a pendulum with small amplitude oscillations, which allows us to approximate $\sin\theta = \theta$. We will use s for the arc length displacement.

Consider a pendulum of length L and mass m undergoing small amplitude oscillations.

$$F = ma$$

$$-mg\sin\theta = m\frac{d^2s}{dt^2}$$

$$-g\theta = \frac{d^2s}{dt^2}$$

$$-g\theta = L\frac{d^2\theta}{dt^2} \quad \text{because } s = L\theta$$

$$-\frac{g}{L}\theta = \frac{d^2\theta}{dt^2}$$

This is the same basic form of the differential equation for the spring-mass system. Therefore, for a pendulum: $\omega = \sqrt{\frac{g}{L}}$ and $\theta(t) = \theta_{max}\sin(\omega t + \phi_0)$.

Note that neither frequency nor period depends on the amplitude (the maximum angular displacement, θ_{max}); this is a characteristic feature of simple harmonic motion. Also notice that neither depends on the mass of the weight.

Example 8.16 A simple pendulum has a period of 1 s on Earth. What would its period be on the Moon (where g is one-sixth of its value here)?

Solution. The equation $T = 2\pi\sqrt{L/g}$ shows that T is inversely proportional to $\sqrt{g}$, so if g decreases by a factor of 6, then T increases by factor of $\sqrt{6}$. That is,

$$T_{\text{on Moon}} = \sqrt{6} \times T_{\text{on Earth}} = \sqrt{6}(1\text{ s}) = 2.4\text{ s}$$

CHAPTER 8 REVIEW QUESTIONS

SECTION I: MULTIPLE CHOICE

1. Which of the following is/are characteristics of simple harmonic motion?

   I. The acceleration is constant.
   II. The restoring force is proportional to the displacement.
   III. The frequency is independent of the amplitude.

   (A) II only
   (B) I and II only
   (C) I and III only
   (D) II and III only
   (E) I, II, and III

2. A block attached to an ideal spring undergoes simple harmonic motion. The acceleration of the block has its maximum magnitude at the point where

   (A) the speed is the maximum
   (B) the potential energy is the minimum
   (C) the speed is the minimum
   (D) the restoring force is the minimum
   (E) the kinetic energy is the maximum

3. A block attached to an ideal spring undergoes simple harmonic motion about its equilibrium position ($x = 0$) with amplitude A. What fraction of the total energy is in the form of kinetic energy when the block is at position $x = \frac{1}{2}A$?

   (A) 1/3
   (B) 3/8
   (C) 1/2
   (D) 2/3
   (E) 3/4

4. A student measures the maximum speed of a block undergoing simple harmonic oscillations of amplitude A on the end of an ideal spring. If the block is replaced by one with twice the mass but the amplitude of its oscillations remains the same, then the maximum speed of the block will

   (A) decrease by a factor of 4
   (B) decrease by a factor of 2
   (C) decrease by a factor of $\sqrt{2}$
   (D) remain the same
   (E) increase by a factor of 2

5. A spring–block simple harmonic oscillator is set up so that the oscillations are vertical. The period of the motion is T. If the spring and block are taken to the surface of the Moon, where the gravitational acceleration is 1/6 of its value here, then the vertical oscillations will have a period of

   (A) $T/6$
   (B) $T/3$
   (C) $T/\sqrt{6}$
   (D) T
   (E) $T\sqrt{6}$

6. A linear spring of force constant k is used in a physics lab experiment. A block of mass m is attached to the spring and the resulting frequency, f, of the simple harmonic oscillations is measured. Blocks of various masses are used in different trials, and in each case, the corresponding frequency is measured and recorded. If f^2 is plotted versus $1/m$, the graph will be a straight line with slope

   (A) $4\pi^2/k^2$
   (B) $4\pi^2/k$
   (C) $4\pi^2 k$
   (D) $k/(4\pi^2)$
   (E) $k^2/(4\pi^2)$

7. A block of mass $m = 4$ kg on a frictionless, horizontal table is attached to one end of a spring of force constant $k = 400$ N/m and undergoes simple harmonic oscillations about its equilibrium position ($x = 0$) with amplitude $A = 6$ cm. If the block is at $x = 6$ cm at time $t = 0$, then which of the following equations (with x in centimeters and t in seconds) gives the block's position as a function of time?

(A) $x = 6 \sin(10t + \frac{1}{2}\pi)$

(B) $x = 6 \sin(10\pi t + \frac{1}{2}\pi)$

(C) $x = 6 \sin(10\pi t - \frac{1}{2}\pi)$

(D) $x = 6 \sin(10t)$

(E) $x = 6 \sin(10t - \frac{1}{2}\pi)$

8. A block attached to an ideal spring undergoes simple harmonic motion about its equilibrium position with amplitude A and angular frequency ω. What is the maximum magnitude of the block's velocity?

(A) $A\omega$
(B) $A^2\omega$
(C) $A\omega^2$
(D) A/ω
(E) A/ω^2

9. A simple pendulum swings about the vertical equilibrium position with a maximum angular displacement of 5° and period T. If the same pendulum is given a maximum angular displacement of 10°, then which of the following best gives the period of the oscillations?

(A) $T/2$
(B) $T/\sqrt{2}$
(C) T
(D) $T\sqrt{2}$
(E) $2T$

10. A simple pendulum of length L and mass m swings about the vertical equilibrium position ($\theta = 0$) with a maximum angular dispacement of θ_{max}. What is the tension in the connecting rod when the pendulum's angular displceent is $\theta = \theta_{max}$?

(A) $mg \sin \theta_{max}$
(B) $mg \cos \theta_{max}$
(C) $mgL \sin \theta_{max}$
D) $mgL \cos \theta_{max}$
(E) $mgL(1 - \cos \theta_{max})$

SECTION II: FREE RESPONSE

1. The figure below shows a block of mass m (Block 1) that's attached to one end of an ideal spring of force constant k and natural length L. The block is pushed so that it compresses the spring to 3/4 of its natural length and then released from rest. Just as the spring has extended to its natural length L, the attached block collides with another block (also of mass m) at rest on the edge of the frictionless table. When Block 1 collides with Block 2, half of its kinetic energy is lost to heat; the other half of Block 1's kinetic energy at impact is divided between Block 1 and Block 2. The collision sends Block 2 over the edge of the table, where it falls a vertical distance H, landing at a horizontal distance R from the edge.

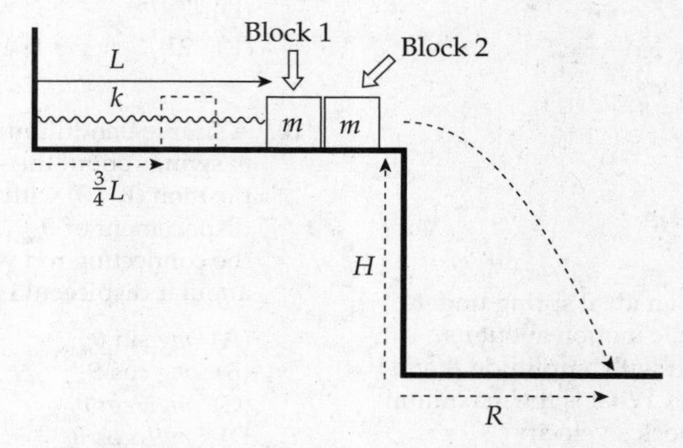

(a) What is the acceleration of Block 1 at the moment it's released from rest from its initial position? Write your answer in terms of k, L, and m.

(b) If v_1 is the velocity of Block 1 just before impact, show that the velocity of Block 1 just after impact is $\frac{1}{2}v_1$.

(c) Determine the amplitude of the oscillations of Block 1 after Block 2 has left the table. Write your answer in terms of L only.

(d) Determine the period of the oscillations of Block 1 after the collision, writing your answer in terms of T_0, the period of the oscillations that Block 1 would have had if it did not collide with Block 2.

(e) Find an expression for R in terms of H, k, L, m, and g.

2. A bullet of mass m is fired horizontally with speed v into a block of mass M initially at rest, at the end of an ideal spring on a frictionless table. At the moment the bullet hits, the spring is at its natural length, L. The bullet becomes embedded in the block, and simple harmonic oscillations result.

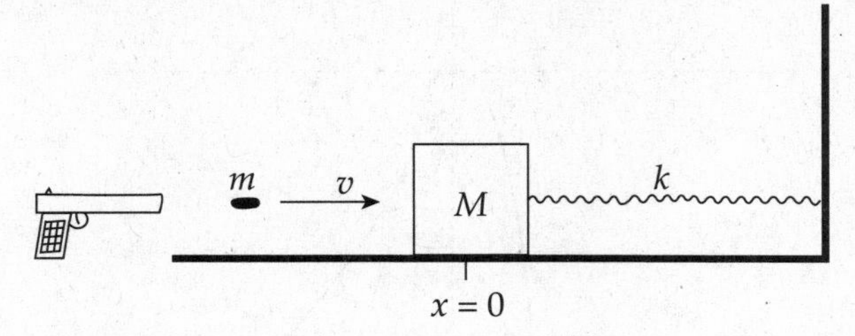

(a) Determine the speed of the block immediately after the impact by the bullet.

(b) Determine the amplitude of the resulting oscillations of the block.

(c) Compute the frequency of the resulting oscillations.

(d) Derive an equation which gives the position of the block as a function of time (relative to $x = 0$ at time $t = 0$).

3. A block of mass M oscillates with amplitude A on a frictionless horizontal table, connected to an ideal spring of force constant k. The period of its oscillations is T. At the moment when the block is at position $x = \frac{1}{2}A$ and moving to the right, a ball of clay of mass m dropped from above lands on the block.

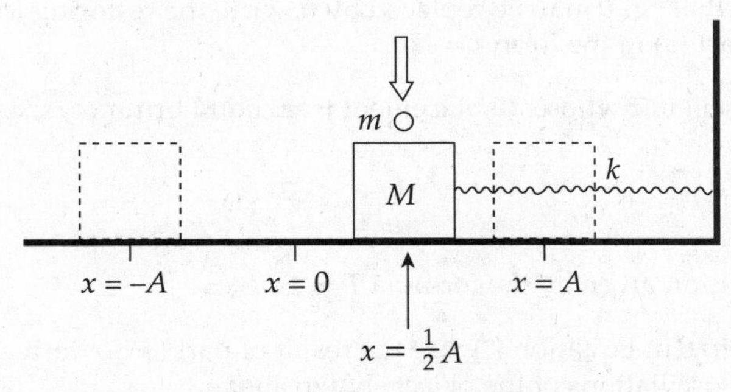

(a) What is the velocity of the block just before the clay hits?

(b) What is the velocity of the block just after the clay hits?

(c) What is the new period of the oscillations of the block?

(d) What is the new amplitude of the oscillations? Write your answer in terms of A, k, M, and m.

(e) Would the answer to part (c) be different if the clay had landed on the block when it was at a different position? Support your answer briefly.

(f) Would the answer to part (d) be different if the clay had landed on the block when it was at a different position? Support your answer briefly.

4. An object of total mass M is allowed to swing around a fixed suspension point P. The object's moment of inertia with respect to the rotation axis perpendicular to the page through P is denoted by I. The distance between P and the object's center of mass C, is d.

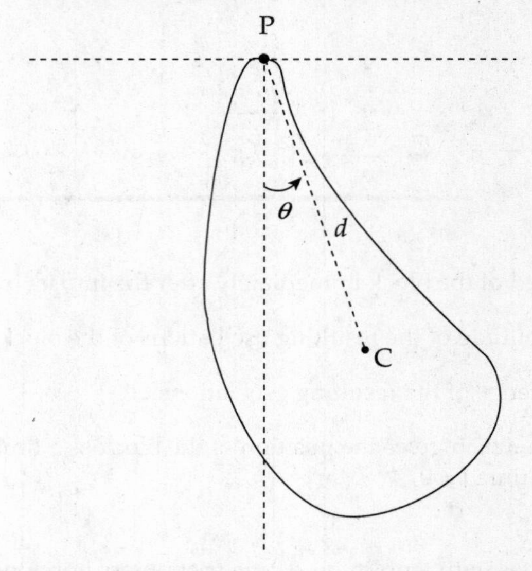

(a) Compute the torque τ produced by the weight of the object when the line PC makes an angle θ with the vertical. (Take the counterclockwise direction as positive for both θ and τ.)

(b) If θ is small, so that $\sin\theta$ may be replaced by θ, write the restoring torque τ computed in part (a) in the form $\tau = -k\theta$.

A simple harmonic oscillator whose displacement from equilibrium, z, satisfies an equation of the form

$$\frac{d^2z}{dt^2} = -bz \quad (*)$$

has a period of oscillation given by the formula $T = 2\pi / \sqrt{b}$.

(c) Taking z equal to θ in Equation (*), use the result of part (b) to derive an expression for the period of small oscillations of the object shown above.

(d) Answer the question posed in part (c) if the object were a uniform bar of mass M and length L (whose moment of inertia about one of its ends is given by the equation $I = \frac{1}{3}ML^2$.)

OSCILLATION SUMMARY

Simple Harmonic Motion

- Simple harmonic motion will occur when there is a restoring force on an object that is proportional to the displacement from equilibrium.
- The mathematical equation of an object undergoing SHM is $y(t) = A \sin(\omega t + \phi)$
- This equation is a solution to the differential equation $\frac{d^2y}{dt^2} = -\omega^2 y$
- The period is how long the object takes to complete one cycle: $T = \frac{time}{cycles}$
- The frequency is how many cycles the object completes in one unit of time:

  $f = \frac{1}{T} = \frac{cycles}{time}$
- An object undergoing SHM typically transforms potential energy to kinetic energy and back to potential energy in a repeating cycle.

9

Electric Forces and Fields

ELECTRIC CHARGE

The basic components of atoms are protons, neutrons, and electrons. Protons and neutrons form the nucleus (and are referred to collectively as *nucleons*), while the electrons keep their distance, swarming around the nucleus. Most of an atom consists of empty space. In fact, if a nucleus were the size of the period at the end of this sentence, then the electrons would be 5 meters away. So what holds such an apparently tenuous structure together? One of the most powerful forces in nature: the *electromagnetic force*. Protons and electrons have a quality called **electric charge** that gives them an attractive force. Electric charge comes in two varieties; positive and negative. A positive particle always attracts a negative particle, and particles of the same charge always repel each other. Protons are positively charged, and electrons are negatively charged.

Protons and electrons are intrinsically charged, but bulk matter is not. This is because the amount of charge on a proton exactly balances the charge on an electron, which is quite remarkable in light of the fact that protons and electrons are very different particles. Since most atoms contain an equal number of protons and electrons, their overall electric charge is 0, because the negative charges cancel out the positive charges. Therefore, in order for matter to be **charged**, an imbalance between the numbers of protons and electrons must exist. This can be accomplished by either the removal or addition of electrons (that is, by the **ionization** of some of the object's atoms). If you remove electrons, then the object becomes positively-charged, while if you add electrons, then it becomes negatively-charged. Furthermore, charge is **conserved**. For example, if you rub a glass rod with a piece of silk, then the silk will acquire a negative charge and the glass will be left with an *equal* positive charge. *Net charge cannot be created or destroyed.* (*Charge* can be created or destroyed—it happens all the time—but *net* charge cannot.)

The magnitude of charge on an electron (and therefore on a proton) is denoted e. This stands for **elementary charge**, because it's the basic unit of electric charge. The charge of an ionized atom must be a whole number times e, because charge can be added or subtracted only in lumps of size e. For this reason we say that charge is **quantized**. To remind us of the quantized nature of electric charge, the charge of a particle (or object) is denoted by the letter q. In the SI system of units, charge is expressed in **coulombs** (abbreviated **C**). One coulomb is a tremendous amount of charge; the value of e is about 1.6×10^{-19} C.

COULOMB'S LAW

The electric force between two charged particles obeys the same mathematical law as the gravitational force between two masses, that is, it's an inverse-square law. The **electric force** between two particles with charges of q_1 and q_2, separated by a distance r, is given by the equation

$$F_E = k\frac{q_1 q_2}{r^2}$$

This is **Coulomb's Law**. We interpret a negative F_E as an attraction between the charges and a positive F_E as a repulsion. The value of the proportionality constant, k, depends on the material between the charged particles. In empty space (vacuum)—or air, for all practical purposes—it is called **Coulomb's constant** and has the approximate value $k_0 = 9 \times 10^9\ \text{N} \cdot \text{m}^2/\text{C}^2$. For reasons that will become clear later in this chapter, k_0 is usually written in terms of a fundamental constant known as the **permittivity of free space**, denoted ε_0, whose numerical value is approximately $8.85 \times 10^{-12}\ \text{C}^2/\text{N}\cdot\text{m}^2$. The equation that gives k_0 in terms of ε_0 is:

$$k_0 = \frac{1}{4\pi\varepsilon_0}$$

Coulomb's Law for the force between two point charges is then written as

$$F_E = \frac{1}{4\pi\varepsilon_0}\frac{q_1 q_2}{r^2}$$

Let's compare the electric force to the gravitational force between two protons, a distance r apart.

$$F_E = k\frac{ee}{r^2} \;\&\; F_g = G\frac{m_p m_p}{r^2}$$

$$\frac{F_E}{F_g} = \frac{ke^2}{Gm_p^{\,2}} = 1.29x10^{36}$$

> **Example 9.1** Consider two small spheres, one carrying a charge of +1.5 nC and the other a charge of –2.0 nC, separated by a distance of 1.5 cm. Find the electric force between them. ("n" is the abbreviation for "nano," which means 10^{-9}.)

Solution. The electric force between the spheres is given by Coulomb's Law:

$$F_E = \frac{1}{4\pi\varepsilon_0}\frac{q_1 q_2}{r^2} = (9\times10^9\ \text{N}\cdot\text{m}^2/\text{C}^2)\frac{(1.5\times10^{-9}\ \text{C})(-2.0\times10^{-9}\ \text{C})}{(1.5\times10^{-2}\ \text{m})^2} = -1.2\times10^{-4}\ \text{N}$$

The fact that F_E is negative means that the force is one of *attraction*, which we naturally expect, since one charge is positive and the other is negative. The force between the spheres is along the line that joins the charges, as we've illustrated below. The two forces shown form an action/reaction pair.

$q_1 \oplus \longrightarrow \mathbf{F_E}$ $\qquad$ $\mathbf{F_E} \longleftarrow \ominus q_2$

Superposition

Consider three point charges: q_1, q_2, and q_3. The total electric force acting on, say, q_2 is simply the sum of $\mathbf{F}_{\text{1-on-2}}$, the electric force on q_2 due to q_1, and $\mathbf{F}_{\text{3-on-2}}$, the electric force on q_2 due to q_3:

$$\mathbf{F}_{\text{on 2}} = \mathbf{F}_{\text{1-on-2}} + \mathbf{F}_{\text{3-on-2}}$$

The fact that electric forces can be added in this way is known as **superposition**.

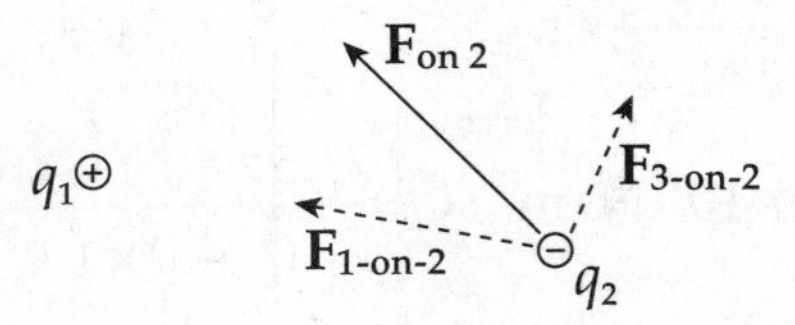

> **Example 9.2** Consider four equal, positive point charges that are situated at the vertices of a square. Find the net electric force on a negative point charge placed at the square's center.

Solution. Refer to the diagram on the next page. The attractive forces due to the two charges on each diagonal cancel out: $\mathbf{F}_1 + \mathbf{F}_3 = \mathbf{0}$, and $\mathbf{F}_2 + \mathbf{F}_4 = \mathbf{0}$, because the distances between the negative charge and the positive charges are all the same and the positive charges are all equivalent. Therefore, by symmetry, the net force on the center charge is zero.

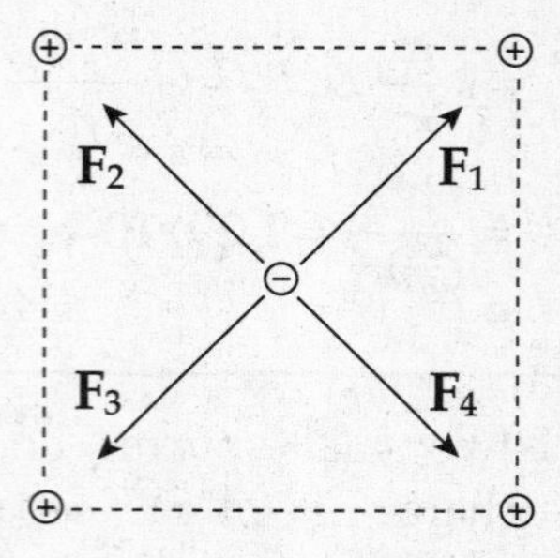

> **Example 9.3** If the two positive charges on the bottom side of the square in the previous example were removed, what would be the net electric force on the negative charge? Assume that each side of the square is 4.0 cm, each positive charge is 1.5 μC, and the negative charge is –6.2 nC. (μ is the symbol for "micro-," which equals 10^{-6}.)

Solution. If we break down $\mathbf{F}_1$ and $\mathbf{F}_2$ into horizontal and vertical components, then by symmetry the two horizontal components will cancel each other out, and the two vertical components will add:

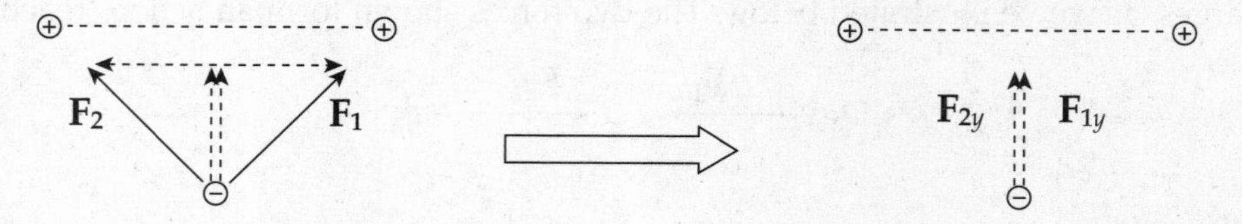

It is clear from the diagram on the left that $F_{1y} = F_1 \sin 45°$ and $F_{2y} = F_2 \sin 45°$. Also, the magnitude of $\mathbf{F}_1$ equals that of $\mathbf{F}_2$. So the net electric force on the negative charge is $F_{1y} + F_{2y} = 2F \sin 45°$, where F is the strength of the force between the negative charge and each of the positive charges. If s is the length of each side of the square, then the distance r between each positive charge and the negative charge is $r = \frac{1}{2}s\sqrt{2}$ and

$$F_E = 2F \sin 45° = 2\frac{1}{4\pi\varepsilon_0}\frac{qQ}{r^2}\sin 45°$$

$$= 2(9\times10^9 \text{ N}\cdot\text{m}^2/\text{C}^2)\frac{(1.5\times10^{-6}\text{ C})(6.2\times10^{-9}\text{ C})}{(\frac{1}{2}\cdot4.0\times10^{-2}\cdot\sqrt{2}\text{ m})^2}\sin 45°$$

$$= 0.15 \text{ N}$$

The direction of the net force is straight upward, toward the center of the line that joins the two positive charges.

> **Example 9.4** Two pith balls of mass m are each given a charge of $+q$. They are hung side-by-side from two threads each of length L, and move apart as a result of their electrical repulsion. Find the equilibrium separation distance x in terms of m, q, and L. (Use the fact that if θ is small, then $\tan\theta \approx \sin\theta$.)

Solution. Three forces act on each ball: weight, tension, and electrical repulsion:

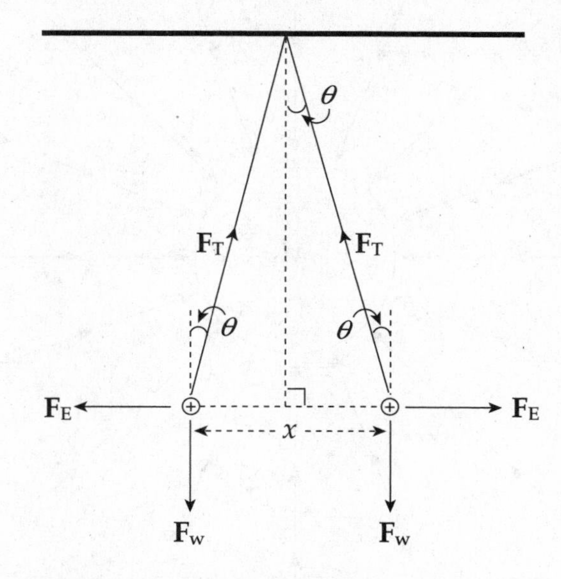

When the balls are in equilibrium, the net force each feels is zero. Therefore, the vertical component of $\mathbf{F}_T$ must cancel out $\mathbf{F}_w$ and the horizontal component of $\mathbf{F}_T$ must cancel out $\mathbf{F}_E$:

$$F_T \cos\theta = F_w \qquad \text{and} \qquad F_T \sin\theta = F_E$$

Dividing the second equation by the first, we get $\tan\theta = F_E/F_w$. Therefore,

$$\tan\theta = \frac{k\frac{q^2}{x^2}}{mg} = \frac{kq^2}{mgx^2}$$

Now, to approximate: If θ is small, the $\tan\theta \approx \sin\theta$ and, from the diagram $\sin\theta = \frac{1}{2}x/L$. Therefore, the equation above becomes

$$\frac{\frac{1}{2}x}{L} = \frac{kq^2}{mgx^2} \quad \Rightarrow \quad \tfrac{1}{2}mgx^3 = kq^2L \quad \Rightarrow \quad x = \sqrt[3]{\frac{2kq^2L}{mg}}$$

THE ELECTRIC FIELD

The presence of a massive body such as the earth causes objects to experience a gravitational force directed toward the earth's center. For objects located outside the earth, this force varies inversely with the square of the distance and directly with the mass of the gravitational source. A vector diagram of the gravitational field surrounding the earth looks like this:

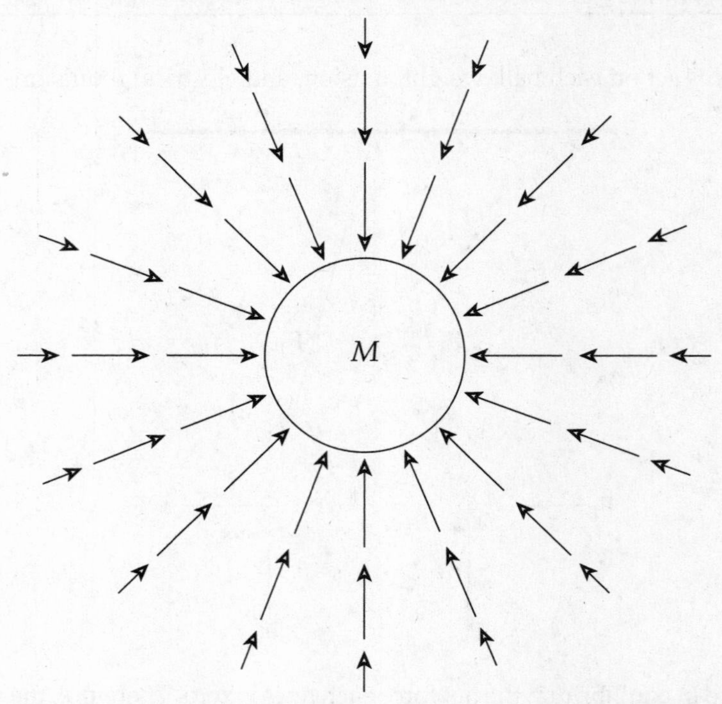

We can think of the space surrounding the earth as permeated by a **gravitational field** that's created by the earth. Any mass that's placed in this field then experiences a gravitational force due to this field.

The same process is used to describe the electric force. Rather than having two charges reach out across empty space to each other to produce a force, we can instead interpret the interaction in the following way: The presence of a charge creates an **electric field** in the space that surrounds it. Another charge placed in the field created by the first will experience a force due to the field.

Consider a point charge Q in a fixed position and assume that it's positive. Now imagine moving a tiny positive test charge q around to various locations near Q. At each location, measure the force that the test charge experiences, and call it $\mathbf{F}_{\text{on } q}$. Divide this force by the test charge q; the resulting vector is the **electric field vector, E**, at that location:

$$\mathbf{E} = \frac{\mathbf{F}_{\text{on } q}}{q}$$

The reason for dividing by the test charge is simple. If we were to use a different test charge with, say, twice the charge of the first one, then each of the forces **F** we'd measure would be twice as much as before. But when we divided this new, stronger force by the new, greater test charge, the factors of 2 would cancel, leaving the same ratio as before. So this ratio tells us the intrinsic strength of the field due to the source charge, independent of whatever test charge we may use to measure it.

What would the electric field of a positive charge Q look like? Since the test charge used to measure the field is positive, every electric field vector would point radially away from the source charge. *If the source charge is positive, the electric field vectors point away from it; if the source charge is negative, then the field vectors point toward it.* And, since the force decreases as we get farther away from the charge (as $1/r^2$), so does the electric field. This is why the electric field vectors farther from the source charge are shorter than those that are closer.

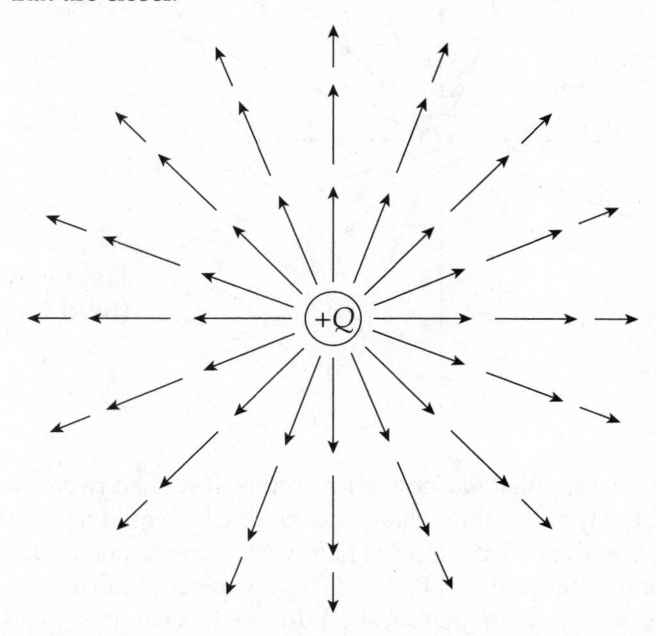

Since the force on the test charge q has a strength of $qQ/4\pi\varepsilon_0 r^2$, when we divide this by q, we get the expression for the strength of the electric field created by a point-charge source of magnitude Q:

$$E = \frac{1}{4\pi\varepsilon_0}\frac{Q}{r^2}$$

To make it easier to sketch an electric field, lines are drawn through the vectors such that the electric field vector is tangent to the line everywhere it's drawn.

Now, your first thought might be that obliterating the individual field vectors deprives us of information, since the length of the field vectors told us how strong the field was. Well, although the individual field vectors are gone, the strength of the field can be figured out by looking at the density of the field lines. Where the field lines are denser, the field is stronger.

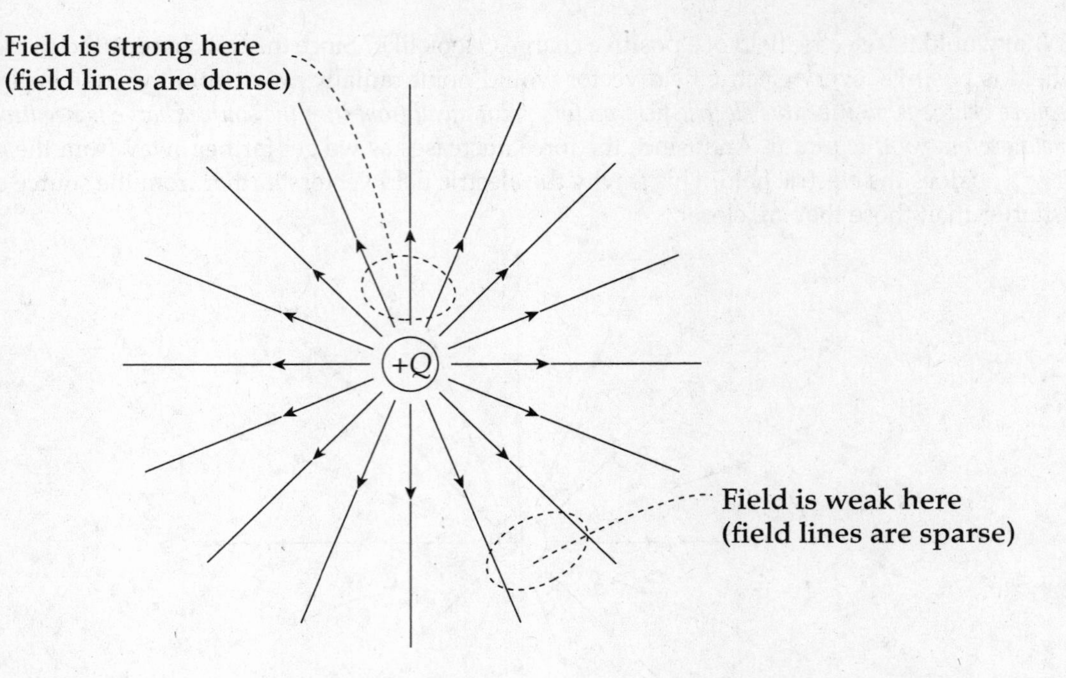

Electric field vectors can be added like any other vectors. If we had two source charges, their fields would overlap and effectively add; a third charge wandering by would feel the effect of the combined field. At each position in space, add the electric field vector due to one of the charges to the electric field vector due to the other charge: $\mathbf{E}_{total} = \mathbf{E}_1 + \mathbf{E}_2$. (This is superposition again.) In the diagram below, $\mathbf{E}_1$ is the electric field vector at a particular location due to the charge $+Q$, and $\mathbf{E}_2$ is the electric field vector at that same location due to the other charge, $-Q$. Adding these vectors gives the overall field vector $\mathbf{E}_{total}$ at that location.

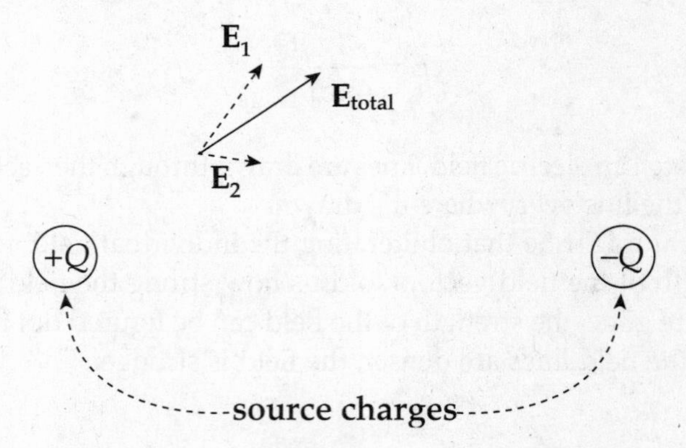

If this is done at enough locations, the electric field lines can be sketched.

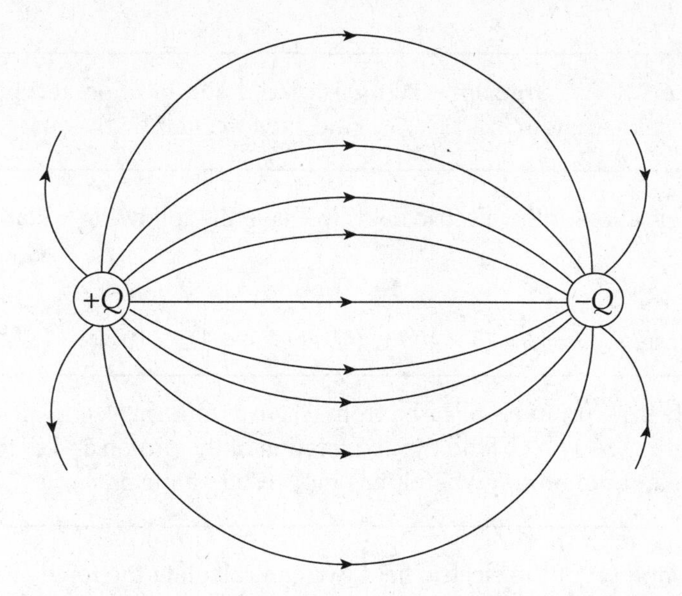

Note that, like electric field vectors, electric field lines always point away from positive source charges and toward negative ones. Two equal but opposite charges, like the ones shown in the diagram above, form a pair called an **electric dipole**.

If a positive charge $+q$ were placed in the electric field above, it would experience a force that is tangent to, and in the same direction as, the field line passing through $+q$'s location. After all, electric fields are sketched from the point of view of what a positive test charge would do. On the other hand, if a negative charge $-q$ were placed in the electric field, it would experience a force that is tangent to, but in the direction opposite from, the field line passing through $-q$'s location.

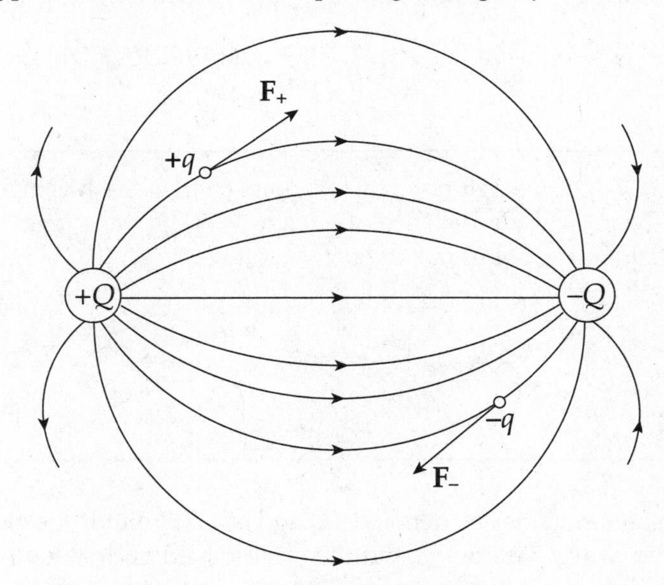

Finally, notice that electric field lines never cross.

Example 9.5 A charge $q = +3.0$ nC is placed at a location at which the electric field strength is 400 N/C. Find the force felt by the charge q.

Solution. From the definition of the electric field, we have the following equation:

$$\mathbf{F}_{\text{on } q} = q\mathbf{E}$$

Therefore, in this case, $F_{\text{on } q} = qE = (3 \times 10^{-9}\text{ C})(400\text{ N/C}) = 1.2 \times 10^{-6}\text{ N}$.

Example 9.6 A proton is released from rest in a uniform electric field with a strength of 500 N/C. Find the acceleration of the proton and determine the speed of the proton when it has moved half a meter.

Solution. From the definition of the electric field, we can calculate the force and then use Newton's Second Law to determine the acceleration.

$$F_{\text{on } q} = qE = ma$$

$$(1.6 \times 10^{-19})(500) = (1.67 \times 10^{-27})a$$

$$a = 4.79 \times 10^{10}\text{ m/s}^2$$

We can use the constant acceleration equations to solve for the speed of the proton after traveling 1 m.

$$v_2^2 = v_1^2 + 2a\Delta x \quad \Rightarrow \quad v_2^2 = 2(4.79 \times 10^{10})(0.5)$$

$$v_2 = 2.19 \times 10^5\text{ m/s}$$

Example 9.7 A dipole is formed by two point charges, each of magnitude 4.0 nC, separated by a distance of 6.0 cm. What is the strength of the electric field at the point midway between them?

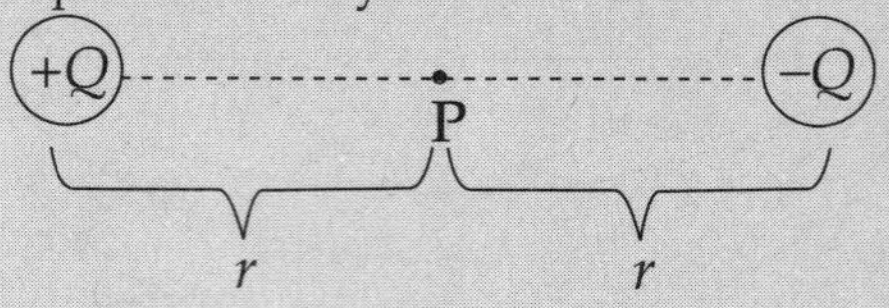

Solution. Let the two source charges be denoted $+Q$ and $-Q$. At Point P, the electric field vector due to $+Q$ would point directly away from $+Q$, and the electric field vector due to $-Q$ would point directly toward $-Q$. The refore, these two vectors point in the same direction (from $+Q$ to $-Q$), so their magnitudes would add.

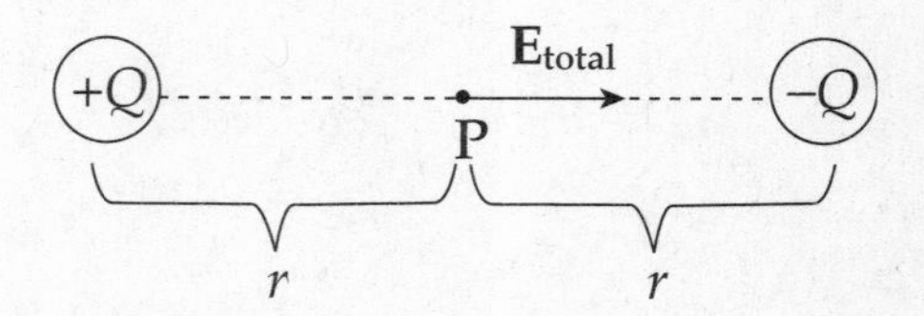

Using the equation for the electric field strength due to a single point charge, we find that

$$E_{\text{total}} = \frac{1}{4\pi\varepsilon_0}\frac{Q}{r^2} + \frac{1}{4\pi\varepsilon_0}\frac{Q}{r^2} = 2\frac{1}{4\pi\varepsilon_0}\frac{Q}{r^2}$$

$$= 2(9\times10^9\ \text{N}\cdot\text{m}^2/\text{C}^2)\frac{4.0\times10^{-9}\ \text{C}}{[\frac{1}{2}(6.0\times10^{-2}\ \text{m})]^2}$$

$$= 8.0\times10^4\ \text{N/C}$$

Example 9.8 If a charge $q = -5.0$ pC were placed at the midway point described in the previous example, describe the force it would feel. ("p" is the abbreviation for "pico-," which means 10^{-12}.)

Solution. Since the field **E** at this location is known, the force felt by q is easy to calculate:

$$\mathbf{F}_{\text{on } q} = q\mathbf{E} = (-5.0\times10^{-1}\ \text{C})(8.0\times10^4\ \text{N/C to the right}) = 4.0\times10^{-7}\ \text{N to the } \textit{left}$$

Example 9.9 What can you say about the electric force that a charge would feel if it were placed at a location at which the electric field was zero?

Solution. Remember that $\mathbf{F}_{\text{on } q} = q\mathbf{E}$. So if $\mathbf{E} = \mathbf{0}$, then $\mathbf{F}_{\text{on } q} = \mathbf{0}$. (Zero field means zero force.)

Example 9.10 Positive charge is distributed uniformly over a large, horizontal plate, which then acts as the source of a vertical electric field. An object of mass 5 g is placed at a distance of 2 cm above the plate. If the strength of the electric field at this location is 10^6 N/C, how much charge would the object need to have in order for the electrical repulsion to balance the gravitational pull?

Solution. Clearly, since the plate is positively charged, the object would also have to carry a positive charge so that the electric force would be repulsive.

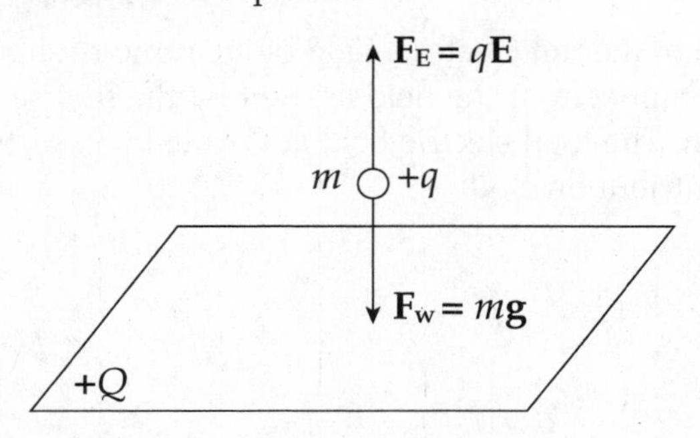

Let q be the charge on the object. Then, in order for F_E to balance mg, we must have

$$qE = mg \quad \Rightarrow \quad q = \frac{mg}{E} = \frac{(5\times10^{-3}\text{ kg})(10\text{ N/kg})}{10^6\text{ N/C}} = 5\times10^{-8}\text{ C} = 50\text{ nC}$$

Example 9.11 A thin, nonconducting rod that carries a uniform linear charge density λ is bent into a semicircle of radius R. Find the electric field at the center of curvature of the semicircle.

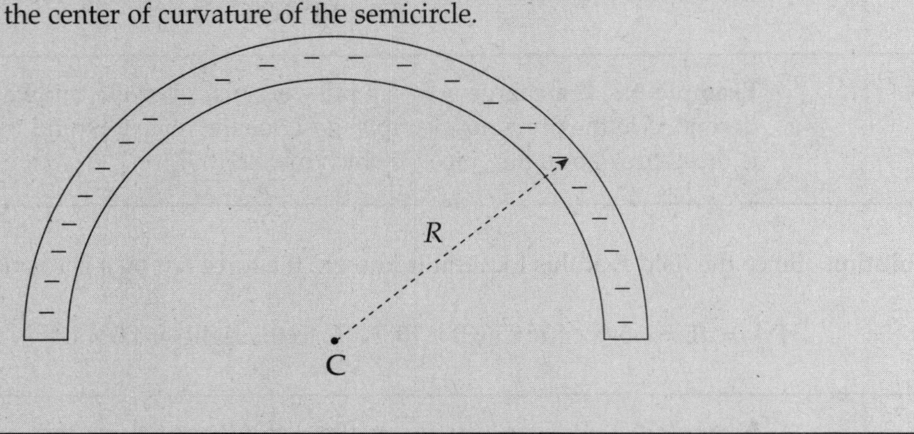

Solution. Refer to the figure below. Consider a small section of the rod subtended by an angle $d\theta$. The length of this section is (by the definition of radian measure) $R\,d\theta$, so the charge it carries is equal to $\lambda R\,d\theta$. The field it creates at C is $(1/4\pi\varepsilon_0)(\lambda R\,d\theta/R^2)$, pointing directly toward this section.

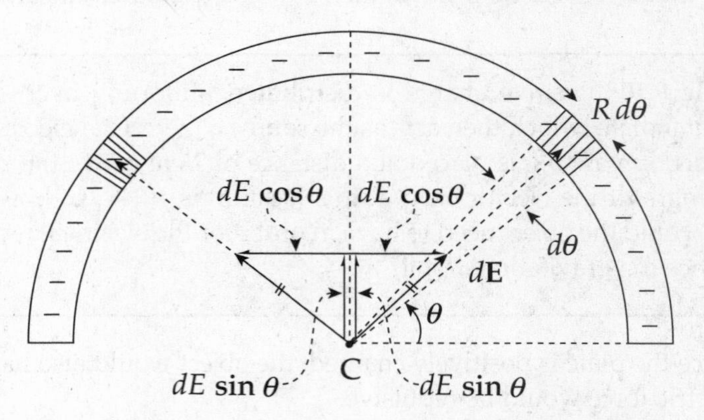

However, another section of the rod creates a field of the same strength and, as the figure shows, cancels out the horizontal component of the field created by the first section. Only the two vertical components, $dE \sin\theta$, remain. The total electric field at C, due to all sections of the rod, is found by adding up the individual contributions:

$$
\begin{aligned}
E &= \int 2\,dE \sin\theta \\
&= \int_{\theta=0}^{\theta=\pi/2} 2\left(\frac{1}{4\pi\varepsilon_0}\cdot\frac{\lambda R\,d\theta}{R^2}\right)\sin\theta \\
&= \frac{\lambda}{2\pi\varepsilon_0 R}\int_0^{\pi/2}\sin\theta\,d\theta \\
&= \frac{\lambda}{2\pi\varepsilon_0 R}\Big[-\cos\theta\Big]_0^{\pi/2} \\
&= \frac{\lambda}{2\pi\varepsilon_0 R}
\end{aligned}
$$

Since the rod is negatively charged, the field **E**, at C, points upward, toward the rod.

CONDUCTORS AND INSULATORS

Materials can be classified into broad categories based on their ability to permit the flow of charge. If electrons were placed on a metal sphere, they would quickly spread out and cover the outside of the sphere uniformly. These electrons would be free to flow through the metal and redistribute themselves, moving to get as far away from each other as they could. Materials that permit the flow of excess charge are called **conductors**; they conduct electricity. Metals are the best examples of conductors, but other conductors are aqueous solutions that contain dissolved electrolytes (such as salt water). Metals conduct electricity because the structure of a typical metal consists of a lattice of nuclei and inner-shell electrons, with about one electron per atom not bound to its nucleus. Electrons are free to move about the lattice, creating a sort of sea of mobile (or conduction) electrons. This freedom allows excess charge to flow freely.

Insulators, on the other hand, closely guard their electrons—and even extra ones that might be added. Electrons are not free to roam throughout the atomic lattice. Examples of insulators are glass, wood, rubber, and plastic. If excess charge is placed on an insulator, it stays put.

Midway between conductors and insulators is a class of materials known as **semiconductors**. As the name indicates, they're more or less conductors. That is, they are less conducting than most metals, but more conducting than most insulators. Examples of semiconducting materials are silicon and germanium.

An extreme example of a conductor is the **superconductor**. This is a material that offers absolutely no resistance to the flow of charge; it is a *perfect* conductor of electric charge. Many metals and ceramics become superconducting when they are brought to extremely low temperatures.

> **Example 9.12** A solid sphere of copper is given a negative charge. Discuss the electric field inside and outside the sphere.

Solution. The excess electrons that are deposited on the sphere move quickly to the outer surface (copper is a great conductor). *Any excess charge on a conductor resides entirely on the outer surface.*

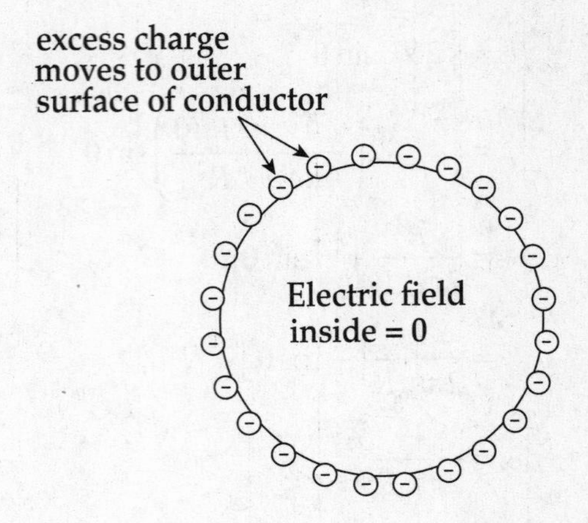

Once these excess electrons establish a uniform distribution on the outer surface of the sphere, there will be no net electric field within the sphere. Why not? Since there is no additional excess charge inside the conductor, there are no excess charges to serve as a source or sink of an electric field line cutting down into the sphere, because field lines begin or end on excess charges.

There can be no electrostatic field within the body of a conductor.

In fact, you can shield yourself from electric fields simply by surrounding yourself with metal. Charges may move around on the outer surface of your cage, but within the cage, the electric field will be zero. For points outside the sphere, it can be shown that the sphere behaves as if all its excess charge were concentrated at its center. (Remember that this is just like the gravitational field due to a uniform spherical mass.) Also, *the electric field is always perpendicular to the surface, no matter what shape the surface may be.* See the diagram below.

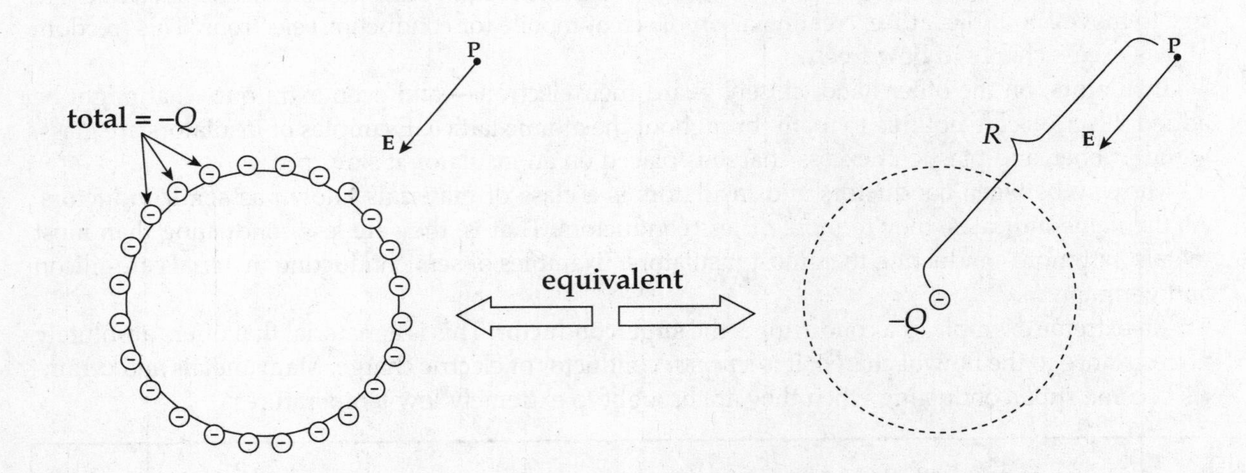

GAUSS'S LAW

For individual point charges, we can compute the electric field they produce by using the equation $E = Q/4\pi\varepsilon_0 r^2$ for each source charge, Q, then add up the resulting electric field vectors to obtain the net field. However, if the source charge is spread over a wire or a plate or throughout a cylinder or a sphere, then we need another method for determining the resulting electric field.

First, let's discuss the concept of **flux**. Consider the flow of a fluid through a pipe and imagine a small rectangular surface perpendicular to the flow. Multiplying the area of this rectangle (A) by the speed of the flow (v) gives the volume of fluid flowing through the surface per unit time. This is called the **fluid-volume flux**, symbolized Φ (the uppercase Greek letter *phi*):

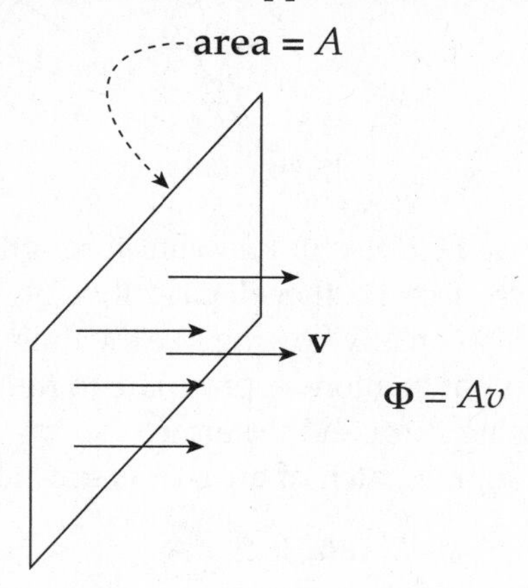

$$\Phi = Av$$

In general, the fluid-volume flux through the surface is given by the equation $\Phi = Av \cos \theta$, where θ is the angle between the velocity **v** of the flow and the normal to the surface. If this rectangular surface were tilted at an angle θ to the flow, then less fluid would flow through it per unit time. In fact, if θ were 90°, then no fluid would flow through the surface. (You might also recognize this as the dot product, $\Phi = \mathbf{A} \cdot \mathbf{v}$, where **A** is the vector whose magnitude is A and whose direction is normal to the surface.)

All the fluid that used to cross the surface down here (below the dotted line) now no longer passes through the tilted surface. So, the flux through the surface has been decreased. It's now the same as the flux through the rectangular portion perpendicular to v that lies above the dotted line. This area is $A \cos \theta$.

$$\Phi = \mathbf{A} \cdot \mathbf{v} = Av \cos \theta$$

Let's now apply this same idea to an electric field. Consider a source charge $+Q$ and imagine a tiny patch (area = A) of a spherical surface of radius r. The diagram on the next page shows a cross-section of the situation.

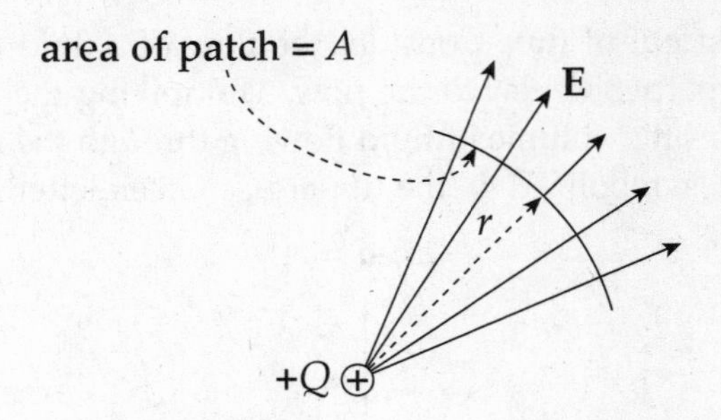

Then we can think of the electric field vectors flowing through the area A and, since **E** is everywhere perpendicular to this patch area A, the **electric flux** (Φ_E) would simply be the product: $\Phi_E = EA$. Although the electric field isn't really flowing like the fluid in the previous discussion, we can still use the same terminology. It might be more appropriate to think of the electric flux as a measure of the amount of electric field passing through the surface.

In general, the electric flux through a patch of area dA is given by the equation:

$$d\Phi_E = \mathbf{E} \cdot d\mathbf{A}$$

The total electric flux is then found by adding (that is, integrating) all of the contributions:

$$\Phi_E = \int d\Phi_E = \int \mathbf{E} \cdot d\mathbf{A}$$

Let's now imagine a *closed* surface—a sphere, for simplicity—surrounding the charge $+Q$. Place this closed spherical surface at a distance r from $+Q$.

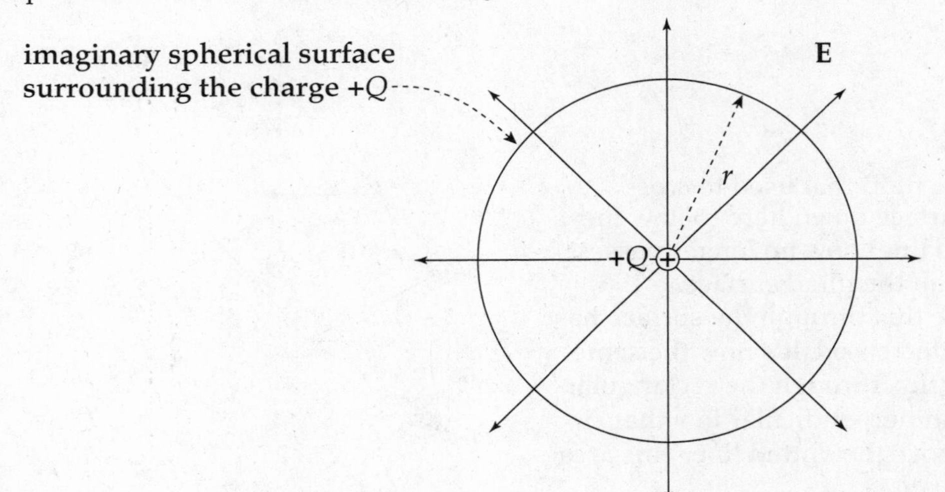

The total flux outward through this closed surface is

$\Phi_E = EA = E(4\pi r^2)$, where $4\pi^2$ is the surface area of a sphere.

We know what the electric field is at the position of this imaginary sphere; it's $E = Q/4_0\pi\varepsilon_0 r^2$. Therefore, the previous equation becomes

$$\Phi_E = \frac{1}{4\pi\varepsilon_0}\frac{Q}{r^2} \cdot 4\pi r^2 = \frac{Q}{\varepsilon_0}$$

(Now you can finally see why we write Coulomb's constant, k_0, as $1/4\pi\varepsilon_0$. It makes the 4π's cancel out in this equation.) It can be shown that the previous equation holds true for *any closed surface*; the total electric flux through the surface is equal to $1/\varepsilon_0$ times the charge enclosed:

$$\Phi_E = \frac{Q_{\text{enclosed}}}{\varepsilon_0}$$

This is **Gauss's Law**. The quantity Q_{enclosed} is the algebraic sum of the charges inside. For example, if the imaginary closed surface we construct encloses, say, charges of $+Q$ and $-3Q$, then the total charge enclosed is $-2Q$, the algebraic sum, so this is Q_{enclosed} in this case. In fact, many people place a Σ in front of Q_{enclosed} to remind them that they need to take the *sum* of the charges enclosed when using Gauss's Law. The power of Gauss's Law to determine electric fields is evident if we can write Φ_E in terms of E—and this requires symmetry—because then the equation can be solved for E. The imaginary closed surface that we will construct when using Gauss's Law to determine an electric field is known as a **Gaussian surface**.

Example 9.13 Consider a long, straight wire carrying a total charge of $+Q$, distributed uniformly along its entire length, L. Use Gauss's Law to find an expression for the electric field created by the wire.

Solution. By symmetry, the electric field must point radially away from the wire. To take advantage of this symmetry, we will construct a cylindrical Gaussian surface—of length x and radius r—around the wire:

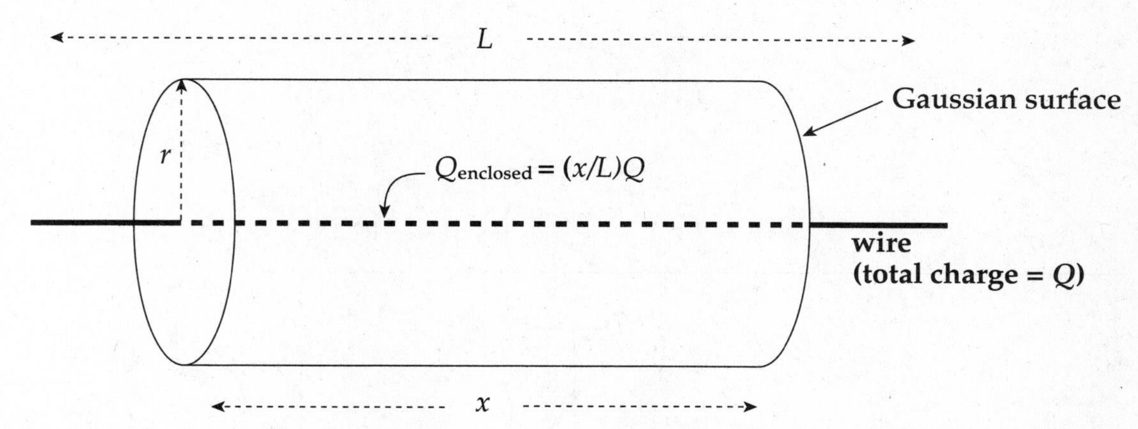

The electric field is always perpendicular to the lateral surface area, A, of the cylinder. Therefore, the electric flux through the cylinder is equal to $EA = E(2\pi rx)$. The flux through each of the two lids of the cylinder is zero, because there **E** is parallel to the end surfaces (so $\mathbf{E} \cdot d\mathbf{A} = E\,dA \cos 90° = 0$). So the total electric flux through the Gaussian surface is $\Phi_E = 0 + E(2\pi rx) + 0 = E(2\pi rx)$. Next we need to know the charge enclosed by the Gaussian surface. Since a total charge of Q is distributed uniformly along the entire length L of the wire, the amount of charge on a section of length x is $(x/L)Q$. Therefore, Gauss's Law becomes

$$\Phi_E = \frac{Q_{\text{enclosed}}}{\varepsilon_0}$$

$$E \cdot 2\pi rx = \frac{(x/L)Q}{\varepsilon_0}$$

$$E = \frac{Q/L}{2\pi\varepsilon_0 r}$$

The ratio Q/L (charge per unit length) is called the *linear charge density*, and is denoted by λ; using this notation, the electric field at a distance r from the wire is given by the equation $E = \lambda/(2\pi\varepsilon_0 r)$. (If the charge on the wire were $-Q$, then the field would point radially inward.)

Example 9.14 A very large rectangular plate has a surface charge density of $+\sigma$ (this is the charge per unit area). Use Gauss's Law to determine an expression for the electric field it creates.

Solution. By symmetry, the electric field lines must be perpendicular to the plate. We will construct a cylindrical Gaussian surface, with cross-sectional area A and total length $2r$, perpendicular to the plate:

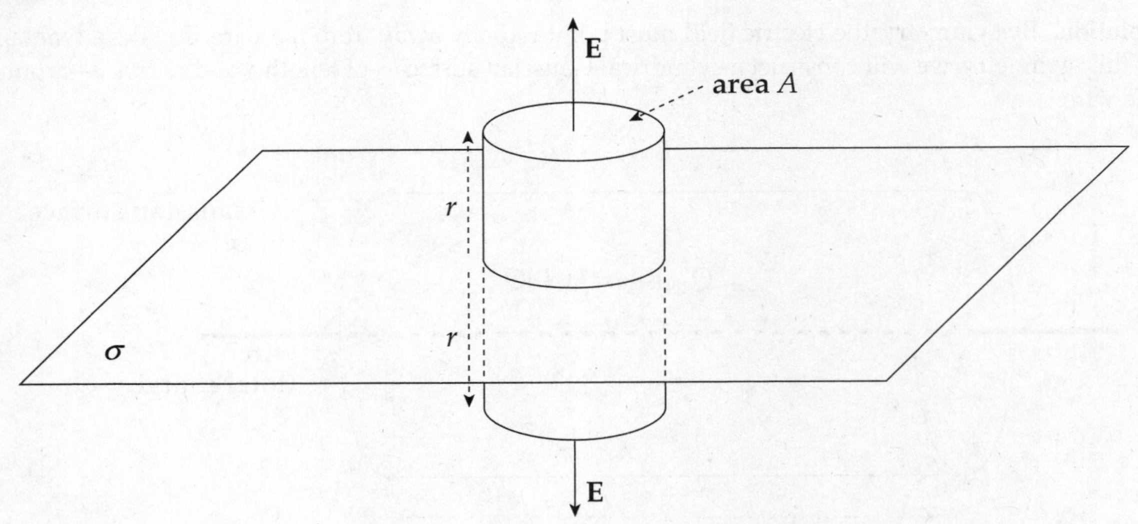

The electric flux through the lateral surface area is now zero, since **E** is parallel to the cylindrical surface here. However, there is flux through the end caps: $EA + EA = 2EA$. (Flux outward from the closed surface is counted as positive; flux inward is counted as negative.) Therefore,

$$\Phi_E = \frac{Q_{\text{enclosed}}}{\varepsilon_0}$$

$$2EA = \frac{\sigma A}{\varepsilon_0}$$

$$E = \frac{\sigma}{2\varepsilon_0}$$

The electric field does not depend on the distance r from the plate. This would only be true if the plate had infinite area. In practical terms, this result means that if we are close to a very large plate (and away from its edges), then the plate looks infinite, and moving a small amount toward or away from the plate won't make much difference; the electric field will remain essentially constant.

Example 9.15 A nonconducting sphere of radius a has excess charge distributed throughout its volume so that the volume charge density ρ as a function of r (the distance from the sphere's center) is given by the equation $\rho(r) = \rho_0(r/a)^2$, where ρ_0 is a constant. Determine the electric field at points inside and outside the sphere.

Solution. In order to apply Gauss's Law, we need to find the total charge enclosed by a Gaussian surface of radius r. Since the charge density depends on r only, consider a thin spherical shell of radius R and thickness dR.

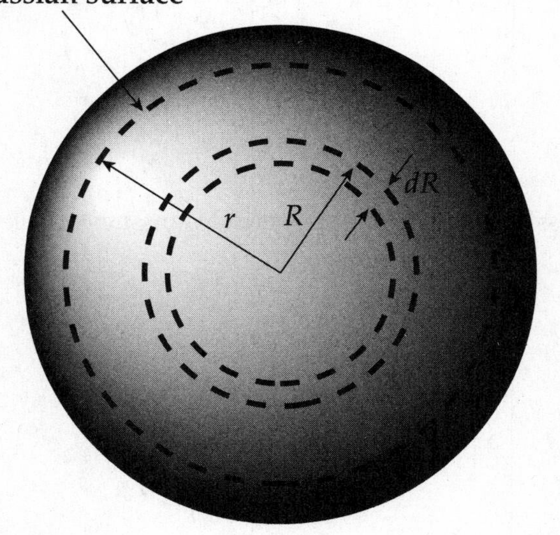

The shell's volume is $dV = 4\pi R^2\,dR$, so the charge contained in the shell is

$$dQ = \rho\,dV = \rho_0(R/a)^2 \cdot 4\pi R^2\,dR$$

The total charge enclosed within a sphere of radius r is found by adding up all the charges on these shells, from $R = 0$ to $R = r$:

$$\begin{aligned}Q(\text{within } r) &= \int dQ = \int_{R=0}^{R=r} \rho_0 (R/a)^2 \cdot 4\pi R^2 \, dR \qquad (*) \\ &= (4\pi\rho_0 / a^2) \int_{R=0}^{R=r} R^4 \, dR \\ &= \frac{4\pi\rho_0}{5a^2} r^5\end{aligned}$$

The electric flux through the spherical Gaussian surface of radius r is

$$\Phi_E = EA = E(4\pi r^2)$$

so, by Gauss's Law,

$$\begin{aligned}\Phi_E &= \frac{Q_{\text{encl}}}{\varepsilon_0} \\ E(4\pi r^2) &= \frac{1}{\varepsilon_0} \frac{4\pi\rho_0}{5a^2} r^5 \\ E &= \frac{\rho_0}{5\varepsilon_0 a^2} r^3\end{aligned}$$

This gives the electric field at all points within the given sphere. For points outside, we first figure out the charge on the entire sphere, which from the integral (*) above with $r = a$ is

$$\begin{aligned}Q(\text{within } a) &= \int dQ = \int_{R=0}^{R=a} \rho_0 (R/a)^2 \cdot 4\pi R^2 \, dR \\ &= (4\pi\rho_0 / a^2) \int_{R=0}^{R=a} R^4 \, dR \\ &= \frac{4\pi\rho_0 a^3}{5}\end{aligned}$$

Then, applying Gauss's Law again on a spherical Gaussian surface of radius $r > a$, we get

$$\begin{aligned}\Phi_E &= \frac{Q_{\text{encl}}}{\varepsilon_0} \\ E(4\pi r^2) &= \frac{1}{\varepsilon_0} \frac{4\pi\rho_0 a^3}{5} \\ E &= \frac{\rho_0 a^3}{5\varepsilon_0 r^2}\end{aligned}$$

To summarize then we have found

$$E(r) = \begin{cases} \dfrac{\rho_0}{5\varepsilon_0 a^2} r^3 & \text{for } r \le a \\[2ex] \dfrac{\rho_0 a^3}{5\varepsilon_0 r^2} & \text{for } r \ge a \end{cases}$$

Example 9.16 Use Gauss's Law to show that the electric field inside a charged hollow metal sphere is zero.

Solution. Recall that for a conductor, any excess charge resides on its surface. Therefore, if a Gaussian surface is constructed anywhere *within* the metal, it contains no net charge, so $\mathbf{E} = \mathbf{0}$ everywhere inside the sphere.

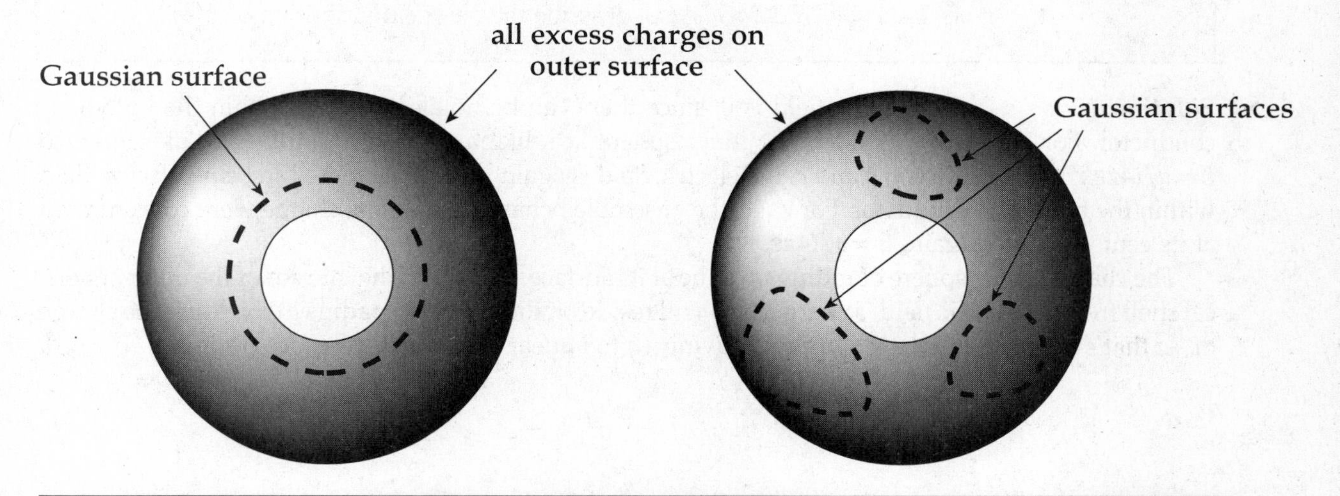

Example 9.17 Consider a neutral, hollow metal sphere. Imagine that a charge of $+q$ could be introduced into the cavity, insulated from the inner wall. What will happen?

Solution. *A static electric field cannot be sustained within the body of a conductor.* Therefore, enough electrons will migrate to the inner wall of the cavity to form a *shield* of charge $-q$ around the inside charge $+q$.

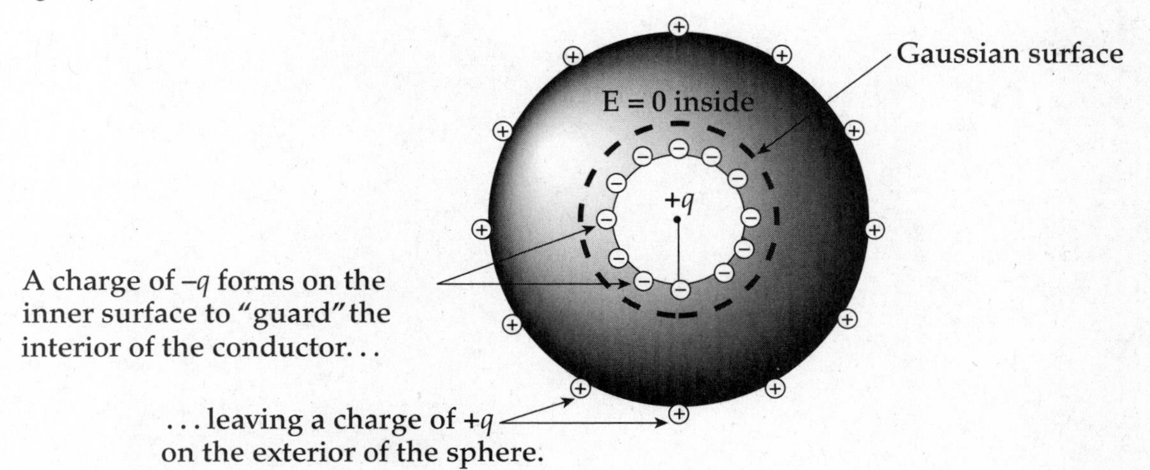

To see why this must be so, if any Gaussian surface surrounding the cavity is drawn within the conductor, then, because E *must* be 0 inside, the electric flux must be 0, so Q_{enclosed} must be 0. The only way to ensure that $Q_{\text{enclosed}} = 0$ is to have a charge of $-q$ on the inner wall of the cavity. Since the sphere was originally neutral, an excess charge of $+q$ will then appear on the outer surface.

> **Example 9.18** A solid metal sphere of radius a, carrying a charge of $+q$, is placed inside and concentric with a neutral hollow metal sphere of inner radius b and outer radius c. Determine the electric field for $r < a$, $a < r < b$, $b < r < c$, and $r > c$. Also, describe the charge distributions.

Solution. For $r < a$, the electric field is 0, since there can be no electric field within the body of a conductor. For r between a and b, the inner sphere acts like a point charge of $+q$ at its center, so $E = q/(4\pi\varepsilon_0 r^2)$. For r between b and c, the electric field is again zero, since there can be no electric field within the body of a conductor. For $r > c$, the ensemble behaves as if all its charge were concentrated at its center, so once again, $E = q/(4\pi\varepsilon_0 r^2)$.

The charge on the sphere of radius a will be at its surface. To protect the interior of the outer spherical shell from an electric field, a charge of $-q$ will reside on the surface of radius b (shielding the charge of $+q$ that's introduced into the cavity), leaving $+q$ to appear on the outer surface (radius c).

CHAPTER 9 REVIEW QUESTIONS

SECTION I: MULTIPLE CHOICE

1. If the distance between two positive point charges is tripled, then the strength of the electrostatic repulsion between them will decrease by a factor of

   (A) 3
   (B) 6
   (C) 8
   (D) 9
   (E) 12

2. Two 1 kg spheres each carry a charge of magnitude 1 C. How does F_E, the strength of the electric force between the spheres, compare to F_G, the strength of their gravitational attraction?

   (A) $F_E < F_G$
   (B) $F_E = F_G$
   (C) $F_E > F_G$
   (D) If the charges on the spheres are of the same sign, then $F_E > F_G$; but if the charges on the spheres are of opposite sign, then $F_E < F_G$.
   (E) Cannot be determined without knowing the distance between the spheres

3. The figure below shows three point charges, all positive. If the net electric force on the center charge is zero, what is the value of y/x?

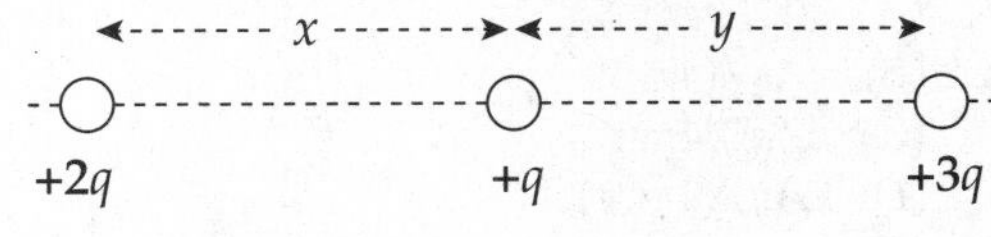

   (A) $4/9$
   (B) $\sqrt{2/3}$
   (C) $\sqrt{3/2}$
   (D) $3/2$
   (E) $9/4$

4.

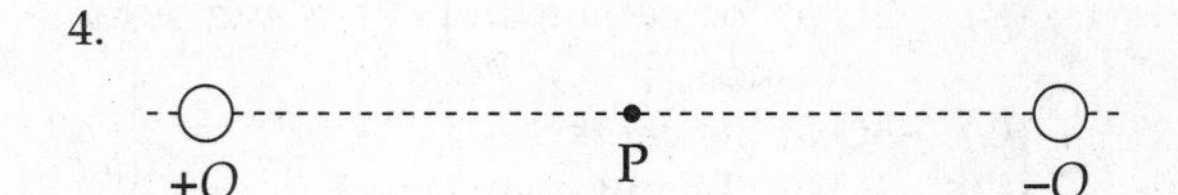

The figure above shows two point charges, $+Q$ and $-Q$. If the negative charge were absent, the electric field at Point P due to $+Q$ would have strength E. With $-Q$ in place, what is the strength of the total electric field at P, which lies at the midpoint of the line segment joining the charges?

   (A) 0
   (B) $E/4$
   (C) $E/2$
   (D) E
   (E) $2E$

5. A sphere of charge $+Q$ is fixed in position. A smaller sphere of charge $+q$ is placed near the larger sphere and released from rest. The small sphere will move away from the large sphere with

   (A) decreasing velocity and decreasing acceleration
   (B) decreasing velocity and increasing acceleration
   (C) decreasing velocity and constant acceleration
   (D) increasing velocity and decreasing acceleration
   (E) increasing velocity and increasing acceleration

6. An object of charge $+q$ feels an electric force $\mathbf{F}_E$ when placed at a particular location in an electric field, $\mathbf{E}$. Therefore, if an object of charge $-2q$ were placed at the same location where the first charge was, it would feel an electric force of

   (A) $-\mathbf{F}_E/2$
   (B) $-2\mathbf{F}_E$
   (C) $-2q\mathbf{F}_E$
   (D) $-2\mathbf{F}_E/q$
   (E) $-\mathbf{F}_E/(2q)$

7. A charge of $-3Q$ is transferred to a solid metal sphere of radius r. Where will this excess charge reside?

   (A) $-Q$ at the center, and $-2Q$ on the outer surface
   (B) $-2Q$ at the center, and $-Q$ on the outer surface
   (C) $-3Q$ at the center
   (D) $-3Q$ on the outer surface
   (E) $-Q$ at the center, $-Q$ in a ring of radius $\frac{1}{2}r$, and $-Q$ on the outer surface

8. The figure below shows four point charges and the cross section of a Gaussian surface:

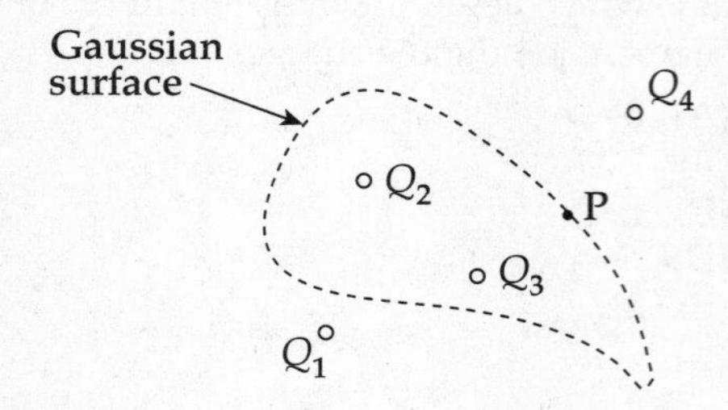

Which of the following statements is true concerning the situation depicted?

   (A) The net electric flux through the Gaussian surface depends on all four charges shown, but the electric field at point P depends only on charges Q_2 and Q_3.
   (B) The net electric flux through the Gaussian surface depends only on charges Q_2 and Q_3, but the electric field at point P depends on all four charges.
   (C) The net electric flux through the Gaussian surface depends only on charges Q_2 and Q_3, and the electric field at point P depends only on charges Q_2, Q_3, and Q_4.
   (D) The net electric flux through the Gaussian surface depends only on charges Q_1 and Q_4, and the electric field at point P depends only on charges Q_2 and Q_3.
   (E) Both the net electric flux through the Gaussian surface and the electric field at point P depend on all four charges.

9. A nonconducting sphere of radius R contains a total charge of $-Q$ distributed uniformly throughout its volume (that is, the volume charge density, ρ is constant).

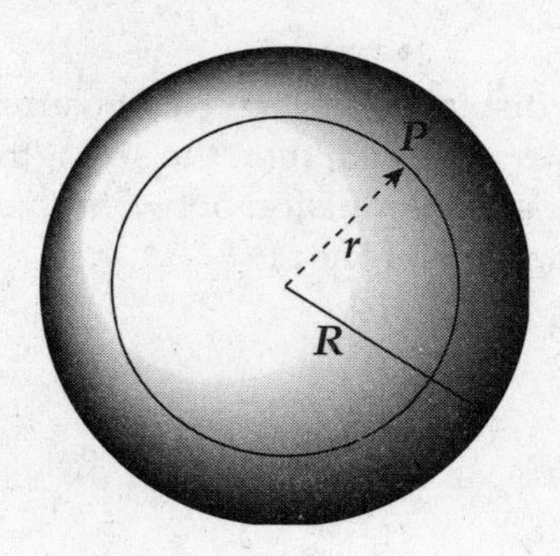

The magnitude of the electric field at Point P, at a distance $r < R$ from the sphere's center, is equal to

   (A) $\frac{1}{4\pi\varepsilon_0}\frac{Q}{R^3}r$
   (B) $\frac{1}{4\pi\varepsilon_0}\frac{Q}{R^2}r^2$
   (C) $\frac{1}{4\pi\varepsilon_0}\frac{Q}{R^3}r^3$
   (D) $\frac{1}{4\pi\varepsilon_0}\frac{Q}{R^3r^2}$
   (E) $\frac{1}{4\pi\varepsilon_0}\frac{Q}{r^2}$

10. Calculate the electric flux through a Gaussian surface of area A enclosing an electric dipole where each charge has magnitude q.

   (A) 0
   (B) $Aq/(4\pi\varepsilon_0)$
   (C) $Aq^2/4\pi\varepsilon_0)$
   (D) $Aq/(4\pi\varepsilon_0 r)$
   (E) $Aq/(4\pi\varepsilon_0 r^2)$

Section II: Free Response

1. In the figure shown, all four charges ($+Q$, $+Q$, $-q$, and $-q$) are situated at the corners of a square. The net electric force on each charge $+Q$ is zero.

   (a) Express the magnitude of q in terms of Q.

   (b) Is the net electric force on each charge $-q$ also equal to zero? Justify your answer.

   (c) Determine the electric field at the centerof the square.

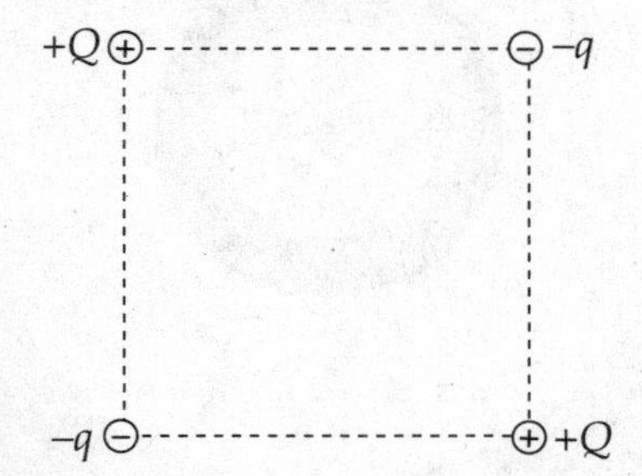

2. Two charges, $+Q$ and $+2Q$, are fixed in place along the y axis of an x-y coordinate system as shown in the figure below. Charge 1 is at the point $(0, a)$, and Charge 2 is at the point $(0, -2a)$.

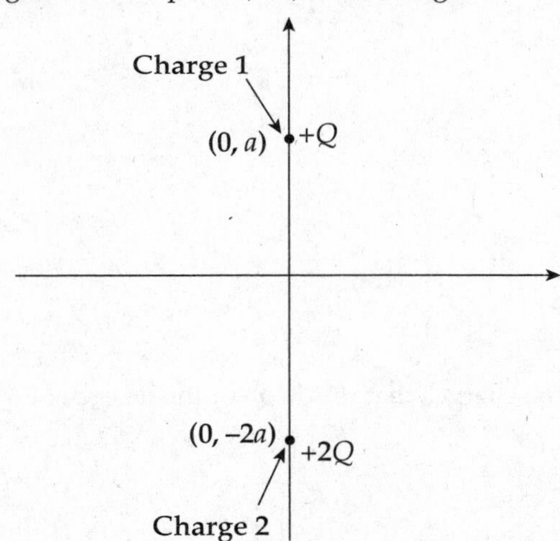

   (a) Find the electric force (magnitude and direction) felt by Charge 1 due to Charge 2.

   (b) Find the electric field (magnitude and direction) at the origin created by both Charges 1 and 2.

   (c) Is there a point on the x axis where the total electric field is zero? If so, where? If not, explain briefly.

   (d) Is there a point on the y axis where the total electric field is zero? If so, where? If not, explain briefly.

   (e) If a small negative charge, $-q$, of mass m were placed at the origin, determine its initial acceleration (magnitude and direction).

3. A conducting spherical shell of inner radius a and outer radius b is inside (and concentric with) a larger conducting spherical shell of inner radius c and outer radius d. The inner shell carries a net charge of $+2q$, and the outer shell has a net charge of $+3q$.

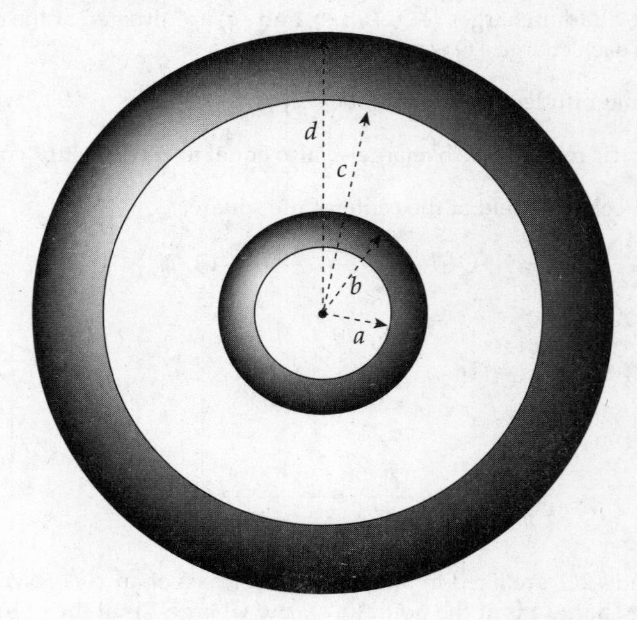

(a) Determine the electric field for

(i) $r < a$

(ii) $a < r < b$

(iii) $b < r < c$

(iv) $c < r < d$

(v) $r > d$

(b) Show in the figure the charges that reside on or inside each of the two shells.

4. A positively-charged, thin nonconducting rod of length ℓ lies along the y axis with its midpoint at the origin. The linear charge density within the rod is uniform and denoted by λ. Points P_1 and P_2 lie on the positive x axis, at distances x_1 and x_2, respectively from the rod.

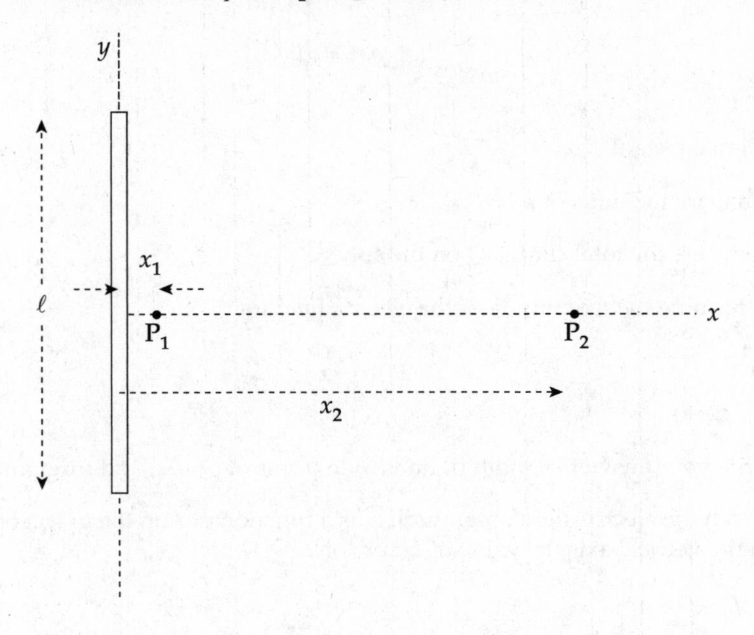

(a) Use Gauss's Law to approximate the electric field at Point P_1, given that x_1 is very small compared to ℓ. Write your answer in terms of λ, x_1, and fundamental constants.

(b) What is the total charge Q on the rod?

(c) Compute the electric field at Point P_2, given that x_2 is not small compared to ℓ. For $x_2 = \ell$, write your answer in terms of Q, ℓ, and fundamental constants. You may use the fact that

$$\int (x^2 + y^2)^{-3/2}\, dy = \frac{y}{x^2\sqrt{x^2 + y^2}} + c$$

5. A solid glass sphere of radius a contains excess charge distributed throughout its volume such that the volume charge density depends on the distance r from the sphere's center according to the equation

$$\rho(r) = \rho_s(r/a)$$

where ρ_s is a constant.

(a) What are the units of ρ_s?

(b) Compute the total charge Q on the sphere.

(c) Determine the magnitude of the electric field for

(i) $r < a$

(ii) $r \geq a$

Write your answers to both (i) and (ii) in terms of Q, a, r, and fundamental constants.

(d) Sketch the electric field magnitude E as a function of r on the graph below. Be sure to indicate on the vertical axis the value of E at $r = a$.

ELECTRIC FORCES AND FIELDS SUMMARY

General Information

- There are positive and negative electric charges. Opposite charges attract and like charges repel.
- Conductors allow excess charge to flow and insulators do not allow charge to flow.

Electric Forces

- The force between two charges is given by Coulomb's Law: $F = k\frac{q_1q_2}{r^2}$

Electric Fields

- The electric field indicates the direction of the force on a positive test charge, q.
- The strength of the electric field is given by: $E = \frac{F_{on\,q}}{q}$
- For a point charge, Q, the electric field is: $E = k\frac{Q}{r^2}$
- For distributed charge (see example 10.10): $E = k\int\frac{dq}{r^2}\hat{r}$

Electric Flux and Gauss's Law

- Electric flux is given by the number of electric field lines that pass through a surface.
- It is given by the dot product of the electric field and area vector: $\Phi_E = \mathbf{E} \bullet \mathbf{A}$
- For a variable electric field the flux is given by: $\Phi_E = \int \mathbf{E} \bullet d\mathbf{A}$
- Gauss's Law is given by the equation: $\Phi_E = \int E \bullet dA = \frac{q_{in}}{\varepsilon_0}$
- Gauss's Law is generally used to solve for the electric field created by symmetric charge distributions (e.g., spheres, spherical shells, lines of charge, and a sheet of charge).

10

Electric Potential and Capacitance

INTRODUCTION

When an object moves in a gravitational field, it usually experiences a change in kinetic energy and in gravitational potential energy due to the work done on the object by gravity. Similarly, when a charge moves in an electric field, it generally experiences a change in kinetic energy and in electrical potential energy due to the work done on it by the electric field. By exploring the idea of electric potential, we can simplify our calculations of work and energy changes within electric fields.

ELECTRICAL POTENTIAL ENERGY

When a charge moves in an electric field, then unless its displacement is always perpendicular to the field, the electric force does work on the charge. If W_E is the work done by the electric force, then the change in the charge's **electrical potential energy** is defined by

$$\Delta U_E = -W_E$$

Notice that this is the same equation that defined the change in the gravitational potential energy of an object of mass m undergoing a displacement in a gravitational field ($\Delta U_G = -W_G$).

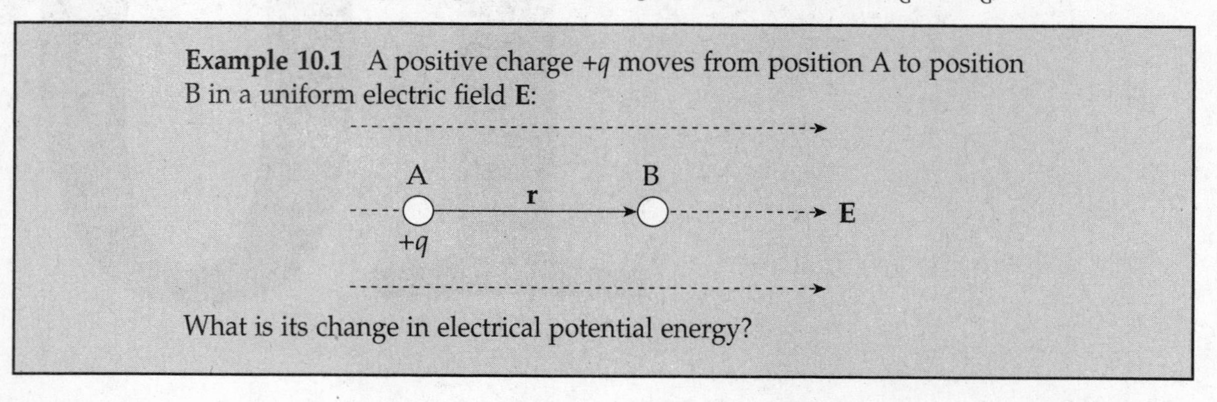

Example 10.1 A positive charge $+q$ moves from position A to position B in a uniform electric field **E**:

What is its change in electrical potential energy?

Solution. Since the field is uniform, the electric force that the charge feels, $\mathbf{F}_E = q\mathbf{E}$, is constant. Since q is positive, $\mathbf{F}_E$ points in the same direction as **E**, and, as the figure shows, they point in the same direction as the displacement, **r**. This makes the work done by the electric field equal to $W_E = F_E r = qEr$, so the change in the electrical potential energy is

$$\Delta U_E = -qEr$$

Note that the change in potential energy is negative, which means that potential energy has decreased; this always happens when the field does positive work. It's just like when you drop a rock to the ground: Gravity does positive work, and the rock loses gravitational potential energy.

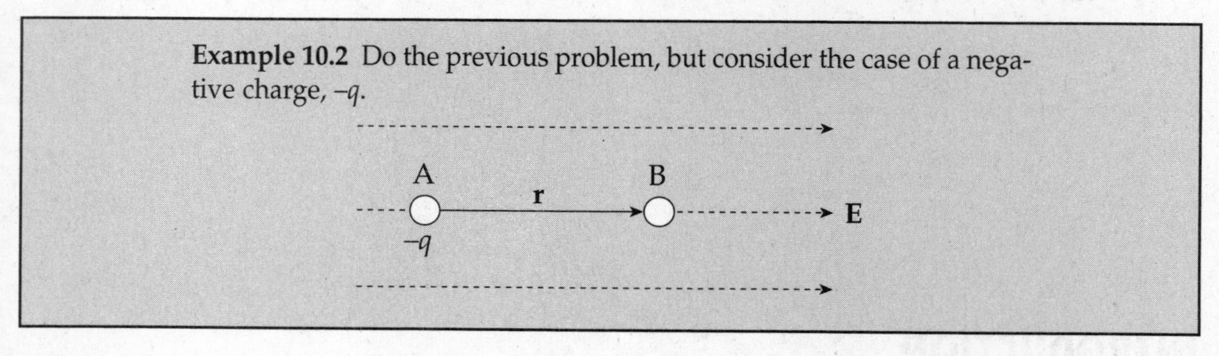

Example 10.2 Do the previous problem, but consider the case of a negative charge, $-q$.

Solution. In this case, an outside agent must be pushing the charge to make it move, because the electric force *naturally* pushes negative charges against field lines. Therefore, we expect that the work done by the electric field is negative. The electric force, $\mathbf{F}_E = (-q)\mathbf{E}$, points in the direction opposite to the displacement, so the work it does is $W_E = -F_E r = -qEr = -qEr$, so the change in electrical potential energy is positive: $\Delta U_E = -W_E = -(-qEr) = qEr$. Since the change in potential energy is positive, the potential energy increased; this always happens when the field does negative work. It's like when you lift a rock off the ground: Gravity does negative work, and the rock gains gravitational potential energy.

Example 10.3 A positive charge $+q$ moves from position A to position B in a uniform electric field E:

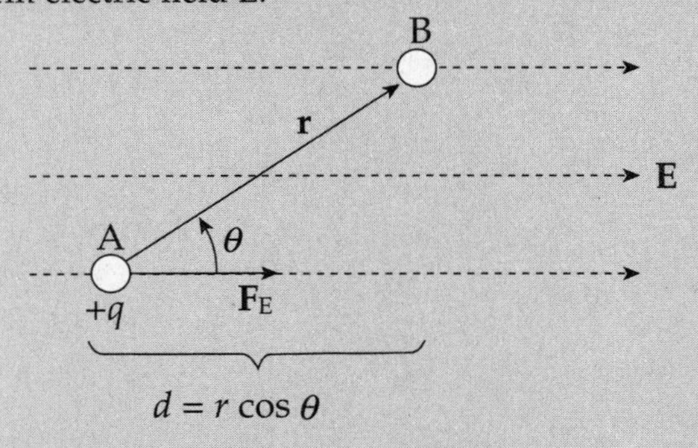

What is its change in electrical potential energy?

Solution. The electric force felt by the charge q is $\mathbf{F}_E = q\mathbf{E}$ and this force is parallel to **E,** because q is positive. In this case, because $\mathbf{F}_E$ is not parallel to **r** (as it was in Example 10.1), we will use the more general definition of work:

$$W_E = \mathbf{F}_E \cdot \mathbf{r} = F_E r \cos \theta = qEr \cos \theta$$

But $r \cos \theta = d$, so

$$W_E = qEd$$

and

$$\Delta U_E = -W_E = -qEd$$

Because the electric force is a conservative force, which means that the work done does not depend on the path that connects the positions A and B, the work calculated above could have been figured out by considering the path from A to B composed of the segments $\mathbf{r}_1$ and $\mathbf{r}_2$:

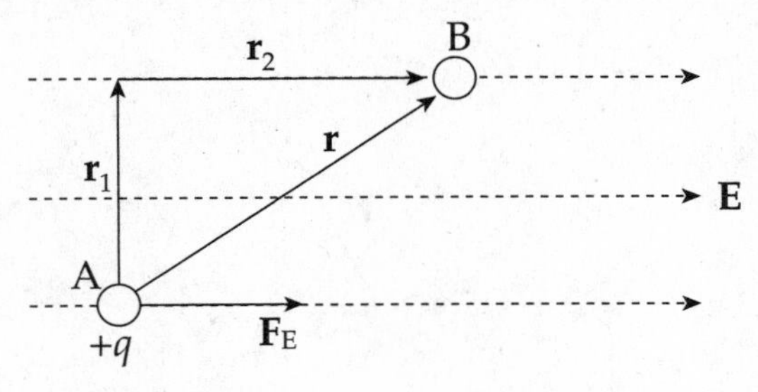

Along $\mathbf{r}_1$, the electric force does no work since this displacement is perpendicular to the force. Thus, the work done by the electric field as q moves from A to B is simply equal to the work it does along $\mathbf{r}_2$. And since the length of $\mathbf{r}_2$ is $d = r \cos \theta$, we have $W_E = F_E d = qEd$, just as before.

Example 10.4 A positive charge +q moves from position A to B in the electric field, E, created by the source charge +Q (only a portion of the electric field is drawn):

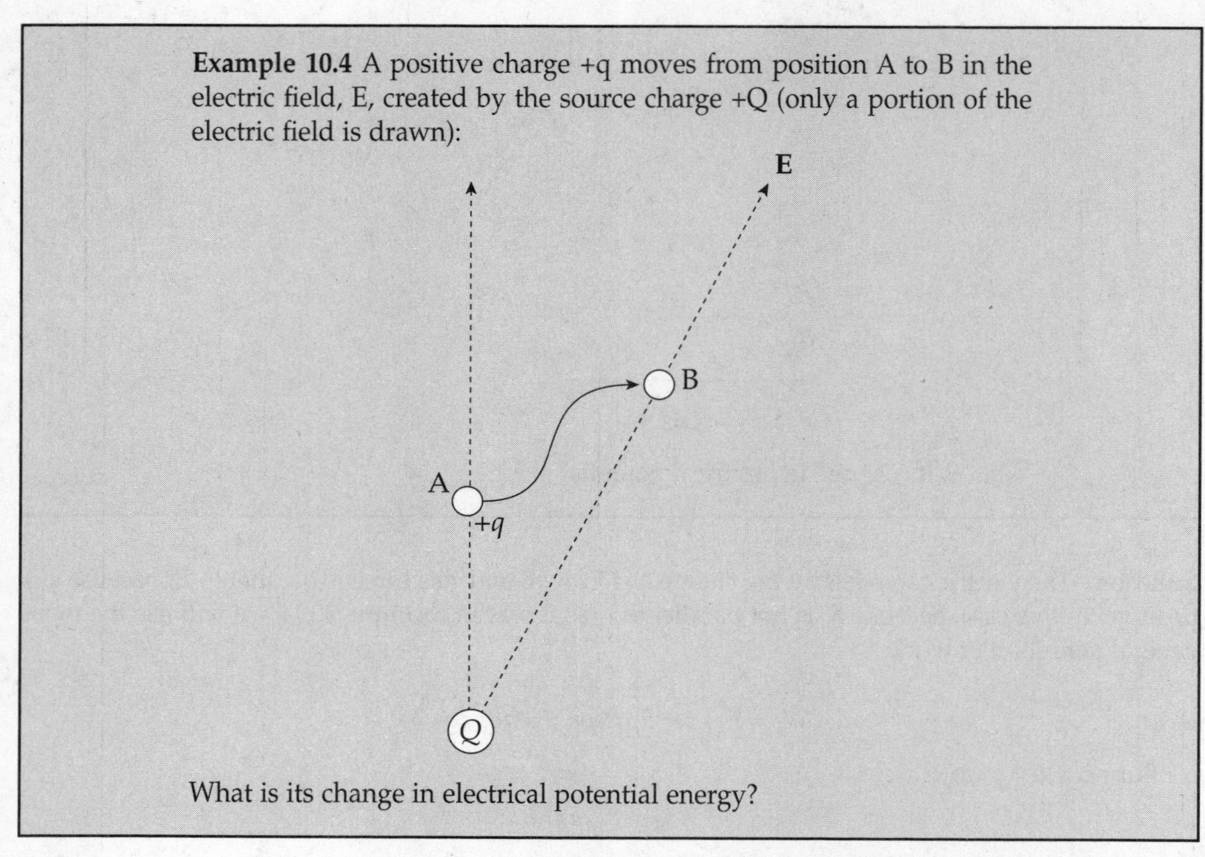

What is its change in electrical potential energy?

Solution. As you know, movement that's perpendicular to the field lines causes the electric field to do no work. So replace the path pictured by another path, composed of $\mathbf{r}_1$ and $\mathbf{r}_2$:

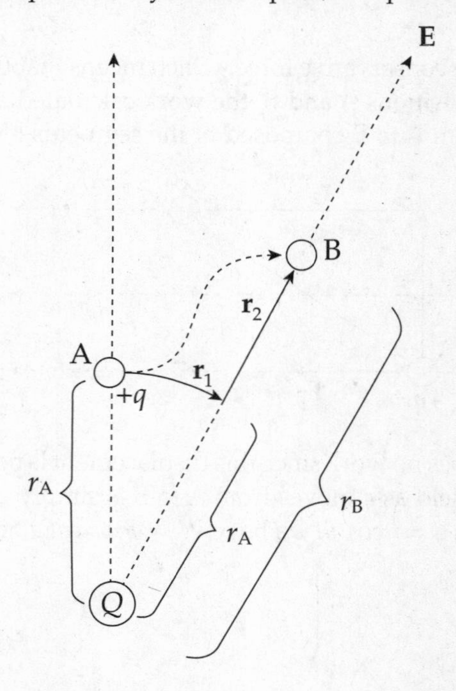

At every point along the arc r_1, the electric force is radial and thus perpendicular to the displacement, so $\mathbf{F}_E$ does no work on the charge as it moves along $\mathbf{r}_1$. But along radial line $\mathbf{r}_2$, the electric force is parallel to the displacement. The work done by the electric force is

$$W_E = \int \mathbf{F}_E \cdot d\mathbf{r} = \int_{r=r_A}^{r=r_B} \frac{1}{4\pi\varepsilon_0} \frac{Qq}{r^2} dr$$

$$= \frac{Qq}{4\pi\varepsilon_0} \int_{r=r_A}^{r=r_B} \frac{1}{r^2} dr$$

$$= \frac{Qq}{4\pi\varepsilon_0} \left[-\frac{1}{r} \right]_{r_A}^{r_B}$$

$$= \frac{Qq}{4\pi\varepsilon_0} \left(\frac{1}{r_A} - \frac{1}{r_B} \right)$$

and the change in electrical potential energy is

$$\Delta U_E = -W_E = \frac{Qq}{4\pi\varepsilon_0} \left(\frac{1}{r_B} - \frac{1}{r_A} \right)$$

The equation above holds true regardless of the signs of Q and q. Our illustration had both Q and q positive, but this procedure can be used with any combination of signs and the result will be the same. Since the equation can be written in the form

$$U_B - U_A = \frac{1}{4\pi\varepsilon_0} \frac{Qq}{r_B} - \frac{1}{4\pi\varepsilon_0} \frac{Qq}{r_A}$$

we can define the potential energy of two charges (let's just call them q_1 and q_2), separated by a distance r to be

$$U = \frac{1}{4\pi\varepsilon_0} \frac{q_1 q_2}{r}$$

This definition says that when the charges are infinitely far apart, their potential energy is zero.

Example 10.5 A positive charge $q_1 = +2 \times 10^{-6}$ C is held stationary, while a negative charge, $q_2 = -1 \times 10^{-8}$ C, is released from rest at a distance of 10 cm from q_1. Find the kinetic energy of charge q_2 when it's 1 cm from q_1.

Solution. The gain in kinetic energy is equal to the loss in potential energy; you know this from Conservation of Energy. The change in electrical potential energy is

$$\Delta U = \frac{q_1 q_2}{4\pi\varepsilon_0} \left(\frac{1}{r_B} - \frac{1}{r_A} \right)$$

$$= (9 \times 10^9 \text{ N} \cdot \text{m}^2 / \text{C}^2)(+2 \times 10^{-6} \text{ C})(-1 \times 10^{-8} \text{ C}) \left(\frac{1}{0.01 \text{ m}} - \frac{1}{0.10 \text{ m}} \right)$$

$$= -0.016 \text{ J}$$

So the gain in kinetic energy is +0.016 J. Since q_2 started from rest (with no kinetic energy), this is the kinetic energy of q_2 when it's 1 cm from q_1.

Example 10.6 Two positive charges, q_1 and q_2, are held in the positions shown below. How much work would be required to bring (from infinity) a third positive charge, q_3, and place it so that the three charges form the corners of an equilateral triangle of side length s?

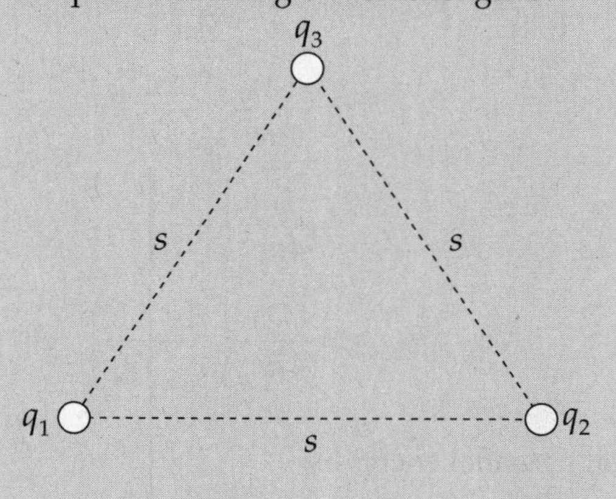

Solution. An external agent would need to do positive work, equal in magnitude to the negative work done by the electric force on q_3 as it is brought into place, so let's first compute this quantity. Let's first compute the work done *by the electric force* as q_3 is brought in. Since q_3 is fighting against both q_1's and q_2's electric field, the total work done on q_3 by the electric force, W_E, is equal to the work done on q_3 by q_1 ($W_{1\text{-}3}$) plus the work done on q_3 by q_2 ($W_{2\text{-}3}$). Using the equation $W_E = -\Delta U_E$ and the one we derived above for ΔU_E, we have

$$\begin{aligned} W_{1\text{-}3} + W_{2\text{-}3} &= -\Delta U_{1\text{-}3} + -\Delta U_{2\text{-}3} \\ &= -\frac{q_1 q_3}{4\pi\varepsilon_0}\left(\frac{1}{s} - 0\right) + -\frac{q_2 q_3}{4\pi\varepsilon_0}\left(\frac{1}{s} - 0\right) \\ &= -\left(\frac{1}{4\pi\varepsilon_0}\frac{q_1 q_3}{s} + \frac{1}{4\pi\varepsilon_0}\frac{q_2 q_3}{s}\right) \end{aligned}$$

Therefore, the work that an external agent must do to bring q_3 into position is

$$-W_E = \frac{1}{4\pi\varepsilon_0}\frac{q_1 q_3}{s} + \frac{1}{4\pi\varepsilon_0}\frac{q_2 q_3}{s}$$

In general, the work required by an external agent to assemble a collection of point charges $q_1, q_2, \ldots, q_n$ (bringing each one from infinity), such that the final fixed distance between q_i and q_j is r_{ij}, is equal to the total electrical potential energy of the arrangement:

$$W_{\text{by external agent}} = U_{\text{total}} = \frac{1}{4\pi\varepsilon_0}\sum_{i<j}\frac{q_i q_j}{r_{ij}}$$

ELECTRIC POTENTIAL

Let W_E be the work done by the electric field on a charge q as it undergoes a displacement. If another charge, say $2q$, were to undergo the same displacement, the electric force would be twice as great on this second charge, and the work done by the electric field would be twice as much, $2W_E$. Since the work would be twice as much in the second case, the change in electrical potential energy would be twice as great as well, but the ratio of the change in potential energy to the charge would be the same: $U_E/q = U_E/2q$. This ratio says something about the *field* and the *displacement*, but not the charge that made the move. The change in **electric potential**, ΔV, is defined as this ratio:

$$\Delta V = \frac{\Delta U_E}{q}$$

Electric potential is electrical potential energy *per unit charge*; the units of electric potential are joules per coulomb. One joule per coulomb is called one **volt** (abbreviated V); so 1 J/C = 1 V.

Consider the electric field that's created by a point source charge Q. If a charge q moves from a distance r_A to a distance r_B from Q, then the change in the potential energy is

$$U_B - U_A = \frac{Qq}{4\pi\varepsilon_0}\left(\frac{1}{r_B} - \frac{1}{r_A}\right)$$

The difference in electric potential between positions A and B in the field created by Q is

$$V_B - V_A = \frac{U_B - U_A}{q} = \frac{Q}{4\pi\varepsilon_0}\left(\frac{1}{r_B} - \frac{1}{r_A}\right)$$

If we designate $V_A \to 0$ as $r_A \to \infty$ (an assumption that's stated on the AP Physics Exam), then the electric potential at a distance r from Q is

$$V = \frac{1}{4\pi\varepsilon_0}\frac{Q}{r}$$

Note that the potential depends on the source charge making the field and the distance from it.

Example 10.7 Let $Q = 2 \times 10^{-9}$ C. What is the potential at a Point P that is 2 cm from Q?

Solution. Relative to $V = 0$ at infinity, we have

$$V = \frac{1}{4\pi\varepsilon_0}\frac{Q}{r} = (9\times10^9\ \text{N}\cdot\text{m}^2/\text{C}^2)\frac{2\times10^{-9}\ \text{C}}{0.02\ \text{m}} = 900\ \text{V}$$

This means that the work done by the electric field on a charge of q coulombs brought to a point 2 cm from Q would be $-900q$ joules.

Note that, like potential energy, potential is a *scalar*. In the preceding example, we didn't have to specify the direction of the vector from the position of Q to the Point P, because it didn't matter. At *any* point on a sphere that's 2 cm from Q, the potential will be 900 V. These spheres around Q are called **equipotential surfaces**, and they're surfaces of constant potential. Their cross sections in any plane are circles and are (therefore) perpendicular to the electric field lines. The equipotentials are always perpendicular to the electric field line.

Example 10.8 How much work is done by the electric field as a charge moves along an equipotential surface?

Solution. If the charge always remains on a single equipotential, then, by definition, the potential, V, never changes. Therefore, $\Delta V = 0$, so $\Delta U_E = 0$. Since $W_E = -\Delta U_E$, the work done by the electric field is zero.

Example 10.9 If charges $q_1 = 4 \times 10^{-9}$ C and $q_2 = -6 \times 10^{-9}$ C are stationary, calculate the potential at Point A in the figure below:

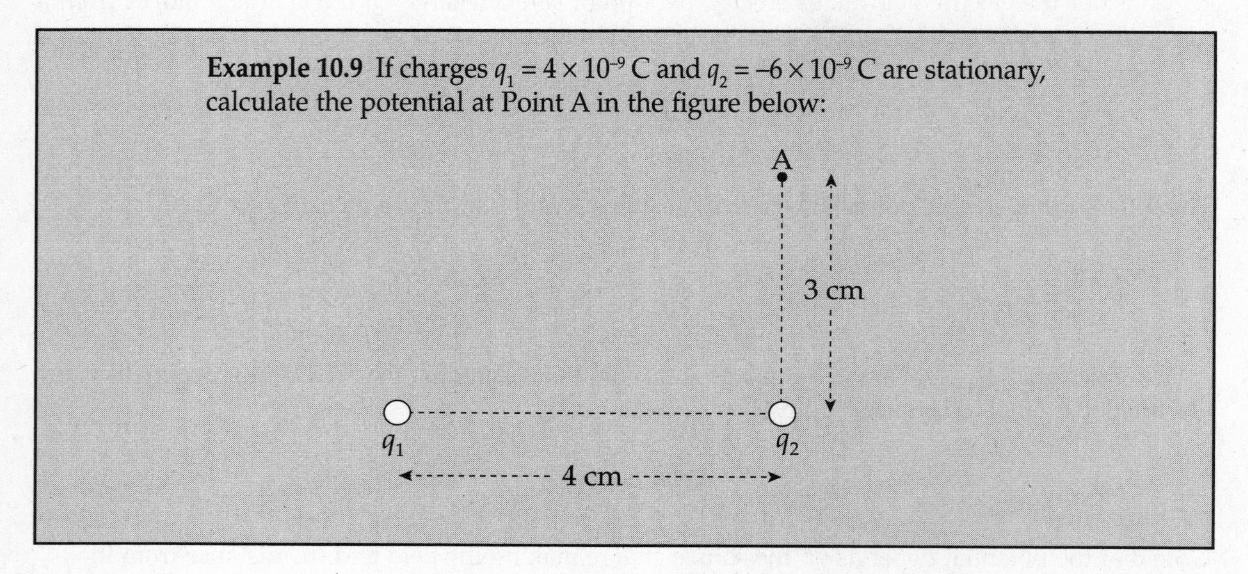

Solution. Potentials add like ordinary numbers. Therefore, the potential at A is just the sum of the potentials at A due to q_1 and q_2. Note that the distance from q_1 to A is 5 cm.

$$
\begin{aligned}
V &= \frac{1}{4\pi\varepsilon_0}\frac{q_1}{r_{1A}} + \frac{1}{4\pi\varepsilon_0}\frac{q_2}{r_{2A}} \\
&= \frac{1}{4\pi\varepsilon_0}\left(\frac{q_1}{r_{1A}} + \frac{q_2}{r_{2A}}\right) \\
&= (9\times10^9\ \text{N}\cdot\text{m}^2/\text{C}^2)\left(\frac{4\times10^{-9}\ \text{C}}{0.05\ \text{m}} + \frac{-6\times10^{-9}\ \text{C}}{0.03\ \text{m}}\right) \\
&= -1080\ \text{V}
\end{aligned}
$$

Example 10.10 How much work would it take to move a charge $q = +1 \times 10^{-2}$ C from Point A to Point B (the point midway between q_1 and q_2?

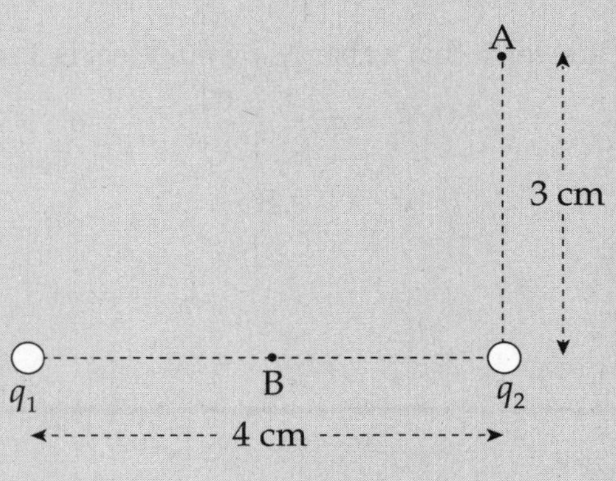

Solution. $\Delta U_E = q\Delta V$, so if we calculate the potential difference between Points A and B and multiply by q, we will have found the change in the electrical potential energy: $\Delta U_{A\to B} = q\Delta V_{A\to B}$. Then, since the work by the electric field is $-\Delta U$, the work required by an external agent is ΔU. In this case, the potential at Point B is

$$\begin{aligned} V_B &= \frac{1}{4\pi\varepsilon_0}\left(\frac{q_1}{r_{1B}} + \frac{q_2}{r_{2B}}\right) \\ &= (9\times 10^9\ \text{N}\cdot\text{m}^2/\text{C}^2)\left(\frac{4\times 10^{-9}\ \text{C}}{0.02\ \text{m}} + \frac{-6\times 10^{-9}\ \text{C}}{0.02\ \text{m}}\right) \\ &= -900\ \text{V} \end{aligned}$$

In the preceding example, we calculated the potential at Point A: $V_A = -1080$ V, so $\Delta V_{A\to B} = V_B - V_A = (-900\text{ V}) - (-1080\text{ V}) = +180$ V. This means that the change in electrical potential energy as q moves from A to B as

$$\Delta U_{A\to B} = qV_{A\to B} = (+1 \times 10^{-2}\text{ C})(+180\text{ V}) = 1.8\text{ J}$$

This is the work required by an external agent to move q from A to B.

The Potential in a Uniform Field

Example 10.11 Consider a very large, flat plate that contains a uniform surface charge density σ. At points that are not too far from the plate, the electric field is uniform and given by the equation

$$E = \frac{\sigma}{2\varepsilon_0}$$

What is the potential at a point which is a distance d from the sheet, relative to the potential of the sheet itself?

Solution. Let A be a point on the plate and let B be a point a distance d from the sheet. Then

$$V_B - V_A = \frac{-W_{E,\,A\to B} \text{ on } q}{q}$$

Since the field is constant, the force that a charge q would feel is also constant, and is equal to

$$F_E = qE = q\frac{\sigma}{2\varepsilon_0}$$

B
d
q
σ
charged plate
A

Therefore,

$$W_{E,\,A\to B} = F_E d = \frac{q\sigma}{2\varepsilon_0}d$$

so applying the definition gives us

$$V_B - V_A = \frac{-W_{E,\,A\to B}}{q} = -\frac{\sigma}{2\varepsilon_0}d$$

This says that for a positive σ, the potential decreases linearly as we move away from the plate.

> **Example 10.12** Two large flat plates—one carrying a charge of $+Q$, the other $-Q$—are separated by a distance d. The electric field between the plates, **E**, is uniform. Determine the potential difference between the plates.

Solution. Imagine a positive charge q moving from the positive plate to the negative plate:

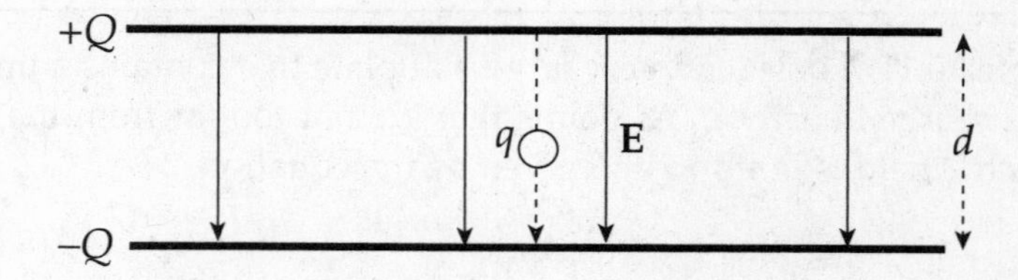

Since the work done by the electric field is

$$W_{E,+\to-} = F_E d = qEd$$

the potential difference between the plates is

$$V_- - V_+ = \frac{-W_{E,+\to-}}{q} = \frac{-qEd}{q} = -Ed$$

This tells us that the potential of the positive plate is greater than the potential of the negative plate, by the amount Ed. This equation can also be written as

$$E = -\frac{V_- - V_+}{d}$$

Therefore, if the potential difference and the distance between the plates are known, then the magnitude of the electric field can be determined quickly.

THE POTENTIAL OF A SPHERE

Think about a conducting spherical shell of radius R. If the shell carries an excess charge Q, we know that this excess charge will be found on the outer surface and that the electric field inside the sphere will be zero. What is the potential due to the shell? At points outside the shell, the field exists as if all of the charge of the sphere were concentrated at the center. That is, for points outside the shell, the potential is the same as it would be due to a single point charge Q:

$$V = \frac{1}{4\pi\varepsilon_0}\frac{Q}{r} \quad \text{for } r > R$$

What about at points inside the shell? Since $\mathbf{E} = \mathbf{0}$ everywhere inside, there would be no electric force and no work done on a charge moving inside the shell. *The potential is constant within the sphere.* So if we moved a charge q from the surface of the spherical shell to its inside, the potential wouldn't change. The potential everywhere inside the shell is equal to the potential on the surface, which is $(1/4\pi\varepsilon_0)(Q/R)$. Therefore,

$$V = \begin{cases} \dfrac{1}{4\pi\varepsilon_0}\dfrac{Q}{R} & \text{for } r \le R \\[2ex] \dfrac{1}{4\pi\varepsilon_0}\dfrac{Q}{r} & \text{for } r > R \end{cases}$$

The same is also true for a solid conducting sphere of charge Q.

Example 10.13 The figure below shows two concentric, conducting, thin spherical shells. The inner shell has a radius of a and carries a charge of q. The outer shell has a radius of b and carries a charge of Q. The inner shell is supported on an insulating stand.

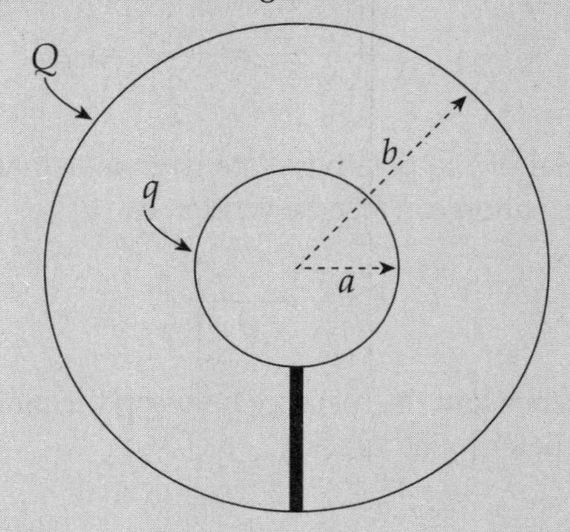

(a) What would the potential of the inner shell be if the outer shell were absent?

(b) What is the potential of the inner shell with the outer shell present?

(c) Show that the potential difference between the inner shell and the outer shell, $V_a - V_b$, does not depend on the charge on the outer shell.

Solution.

(a) The potential on (and within) a sphere of radius a containing a charge q is given by the equation $V_a = (1/4\pi\varepsilon_0)(q/a)$.

(b) The potential inside the outer sphere (if the inner sphere were absent) is equal to $(1/4\pi\varepsilon_0)(Q/b)$. The potential of the inner sphere if the outer sphere were absent is $(1/4\pi\varepsilon_0)(q/a)$. So the potential of the inner sphere with the outer sphere present is the sum: $(1/4\pi\varepsilon_0)(Q/b) + (1/4\pi\varepsilon_0)(q/a)$.

Here's another way to arrive at this result: Outside (and on) the outer sphere, the potential is the same as if the total charge were concentrated at the center, $(1/4\pi\varepsilon_0)(q + Q)/r$. So at a point *on* the outer sphere, the potential is found by substituting b for r in this expression; this gives $V_b = (1/4\pi\varepsilon_0)(q + Q)/b$. Now, the electric field between the spheres is, by Gauss's Law, simply equal to $(1/4\pi\varepsilon_0)(q/r^2)$. Therefore, the work required *by an external agent* to move a charge (let's call it q_0) from infinity to the inner sphere is equal to:

$$
\begin{aligned}
W_{\text{from } \infty \text{ to } a} &= W_{\text{from } \infty \text{ to } b} + W_{\text{from } b \text{ to } a} \\
&= q_0 V_b + -\int_{r=b}^{r=a} q_0 E\, dr \\
&= q_0 V_b - q_0 \int_{r=b}^{r=a} \frac{1}{4\pi\varepsilon_0} \frac{q}{r^2}\, dr \\
&= q_0 V_b - q_0 \frac{q}{4\pi\varepsilon_0} \left[-\frac{1}{r} \right]_b^a \\
&= q_0 V_b - q_0 \frac{q}{4\pi\varepsilon_0} \left(\frac{1}{b} - \frac{1}{a} \right)
\end{aligned}
$$

Therefore,

$$
\begin{aligned}
V_a &= \frac{W_{\text{from } \infty \text{ to } a}}{q_0} \\
&= V_b - \frac{q}{4\pi\varepsilon_0} \left(\frac{1}{b} - \frac{1}{a} \right) \\
&= \left[\frac{1}{4\pi\varepsilon_0} \frac{q+Q}{b} \right] - \frac{q}{4\pi\varepsilon_0} \left(\frac{1}{b} - \frac{1}{a} \right) \\
&= \frac{1}{4\pi\varepsilon_0} \left(\frac{Q}{b} + \frac{q}{a} \right)
\end{aligned}
$$

(c) At points outside the outer sphere, the two spheres behave as if all their charge was concentrated at the center. So at points outside the sphere ($r > b$) the potential is equal to $(1/4\pi\varepsilon_0)(q + Q)/r$. This must also give the potential at the surface of the outer sphere when $r = b$. Therefore, the potential difference between the inner sphere and the outer sphere is

$$
\begin{aligned}
V_a - V_b &= \left[\frac{1}{4\pi\varepsilon_0} \frac{q}{a} + \frac{1}{4\pi\varepsilon_0} \frac{Q}{b} \right] - \frac{1}{4\pi\varepsilon_0} \frac{q+Q}{b} \\
&= \frac{1}{4\pi\varepsilon_0} \frac{q}{a} - \frac{1}{4\pi\varepsilon_0} \frac{q}{b} \\
&= \frac{q}{4\pi\varepsilon_0} \left(\frac{1}{a} - \frac{1}{b} \right)
\end{aligned}
$$

This expression does not depend on Q, which is what we were asked to show.

Example 10.14 A *nonconducting* sphere of radius R has an excess charge of Q distributed uniformly throughout its volume (that is, the volume charge density, ρ, is a constant). The electric field at a distance $r < R$ from the center is given by the equation

$$E = \frac{1}{4\pi\varepsilon_0}\frac{Q}{R^3}r$$

What is the potential inside the sphere? That is, what is the potential at a distance r from the sphere's center?

Solution. The potential at some distance, let's call it r_0, from the center is equal to the negative of the work done by the electric field as a charge q is brought to r_0 from infinity, divided by q. (By definition, $V = U_E/q = -W_E/q$, where we take $V = 0$ at infinity.) So our first step (the big one) is to figure out the work done by the electric field in bringing a charge q in, from infinity. At points outside the sphere, the electric field is simply $(1/4\pi\varepsilon_0)(Q/r^2)$, since the sphere behaves as if all of its charge were concentrated at its center. Once we get inside, however, the electric field is given by the formula in the question. Therefore,

$$\begin{aligned}
W_E &= W_{E,\,\infty \text{ to } R} + W_{E,\,R \text{ to } r_0} \\
&= \int_\infty^R q\cdot\frac{1}{4\pi\varepsilon_0}\frac{Q}{r^2}\,dr + \int_R^{r_0} q\cdot\frac{1}{4\pi\varepsilon_0}\frac{Q}{R^3}r\,dr \\
&= \frac{qQ}{4\pi\varepsilon_0}\int_\infty^R \frac{1}{r^2}\,dr + \frac{qQ}{4\pi\varepsilon_0 R^3}\int_R^{r_0} r\,dr \\
&= \frac{qQ}{4\pi\varepsilon_0}\left[-\frac{1}{r}\right]_\infty^R + \frac{qQ}{4\pi\varepsilon_0 R^3}\left[\frac{r^2}{2}\right]_R^{r_0} \\
&= -\frac{qQ}{4\pi\varepsilon_0}\frac{1}{R} + \frac{qQ}{8\pi\varepsilon_0 R^3}(r_0^2 - R^2) \\
&= \frac{qQ}{4\pi\varepsilon_0}\left(\frac{r_0^2 - 3R^2}{2R^3}\right)
\end{aligned}$$

Taking the negative of this result, dividing by q, and replacing r_0 by r gives us our answer:

$$V = \frac{-W_E}{q} = \frac{Q}{4\pi\varepsilon_0}\left(\frac{3R^2 - r^2}{2R^3}\right)$$

Note how this result differs from the potential of a *conducting* sphere. In that case, the potential was constant throughout the interior of the sphere (and was equal to the value of the potential on the surface). In this case the potential is not constant; it depends on r.

THE POTENTIAL OF A CYLINDER

Example 10.15 Consider a very long conducting cylinder of radius R, which carries a uniform linear charge density λ. The electric field at a distance r, where $r > R$, from the center of the cylinder is given by the equation

$$E = \frac{\lambda}{2\pi\varepsilon_0 r}$$

Determine a formula for the potential at a point outside the cylinder, relative to the potential on the cylinder.

Solution. Let A be a point on the cylinder (at $r = R$) and let B be a point outside the cylinder (at $r = r_0 > R$). Then, by definition,

$$V_B - V_A = \frac{-W_{E,\,A\to B} \text{ on } q}{q}$$

So we figure out the work done by the electric field as a charge q moves from A to B:

$$\begin{aligned}
W_{E,\,A\to B} \text{ on } q &= \int_A^B \mathbf{F}_E \cdot d\mathbf{r} \\
&= \int_{r=R}^{r=r_0} q \cdot \frac{\lambda}{2\pi\varepsilon_0 r}\, dr \\
&= \frac{q\lambda}{2\pi\varepsilon_0} \int_{r=R}^{r=r_0} \frac{dr}{r} \\
&= \frac{q\lambda}{2\pi\varepsilon_0} \left[\ln r\right]_R^{r_0} \\
&= \frac{q\lambda}{2\pi\varepsilon_0} (\ln r_0 - \ln R) \\
&= \frac{q\lambda}{2\pi\varepsilon_0} \ln \frac{r_0}{R}
\end{aligned}$$

Replacing r_0 with r, we get

$$\begin{aligned}
V_B - V_A &= \frac{-W_{E,\,A\to B} \text{ on } q}{q} \\
&= -\frac{\lambda}{2\pi\varepsilon_0} \ln \frac{r}{R} \\
&= \frac{\lambda}{2\pi\varepsilon_0} \ln \frac{R}{r}
\end{aligned}$$

DERIVING THE FIELD FROM THE POTENTIAL

From the definition of potential,

$$\begin{aligned} V_{\mathrm{B}} - V_{\mathrm{A}} &= \frac{-W_{\mathrm{E, A \to B}} \text{ on } q}{q} \\ &= -\frac{1}{q}\int_{\mathrm{A}}^{\mathrm{B}} F_{\mathrm{E}}\, dr \\ &= -\frac{1}{q}\int_{\mathrm{A}}^{\mathrm{B}} qE\, dr \\ &= \int_{\mathrm{A}}^{\mathrm{B}} (-E)\, dr \end{aligned}$$

If A and B are separated by an infinitesimal distance dr, then

$$V_{\mathrm{B}} - V_{\mathrm{A}} = \int_{\mathrm{A}}^{\mathrm{B}} dV$$

and this gives us

$$\int_{\mathrm{A}}^{\mathrm{B}} (-E)\, dr = \int_{\mathrm{A}}^{\mathrm{B}} dV$$

from which we can conclude that

$$\begin{aligned} (-E)\, dr &= dV \\ E &= -\frac{dV}{dr} \end{aligned}$$

So, if we know how the potential varies as a function of r, we can determine the electric field variation with r.

Example 10.16 If the potential at a distance r from a source point charge Q is given by the equation $V(r) = (1/4\pi\varepsilon_0)(Q/r)$, determine a formula for the electric field.

Solution. Using the relationship derived above,

$$E = -\frac{dV}{dr} = -\frac{d}{dr}\left(\frac{1}{4\pi\varepsilon_0}\frac{Q}{r}\right) = -\frac{Q}{4\pi\varepsilon_0}\frac{d}{dr}\left(\frac{1}{r}\right) = -\frac{Q}{4\pi\varepsilon_0}\left(-\frac{1}{r^2}\right) = \frac{1}{4\pi\varepsilon_0}\frac{Q}{r^2}$$

a result we know well.

CAPACITANCE

Consider two conductors, separated by some distance, that carry equal but opposite charges, $+Q$ and $-Q$. Such a pair of conductors comprise a system called a **capacitor**. Work must be done to create this separation of charge, and, as a result, potential energy is stored. Capacitors are basically storage devices for electrical potential energy.

The conductors may have any shape, but the most common conductors are parallel metal plates or sheets. These types of capacitors are called **parallel-plate capacitors**. We'll assume that the distance d between the plates is small compared to the dimensions of the plates since, in this case, the electric field between the plates is uniform. The electric field due to *one* such plate, if its surface charge density is $\sigma = Q/A$, is given by the equation $E = \sigma/(2\varepsilon_0)$, with **E** pointing away from the sheet if σ is positive and toward the plate if σ is negative.

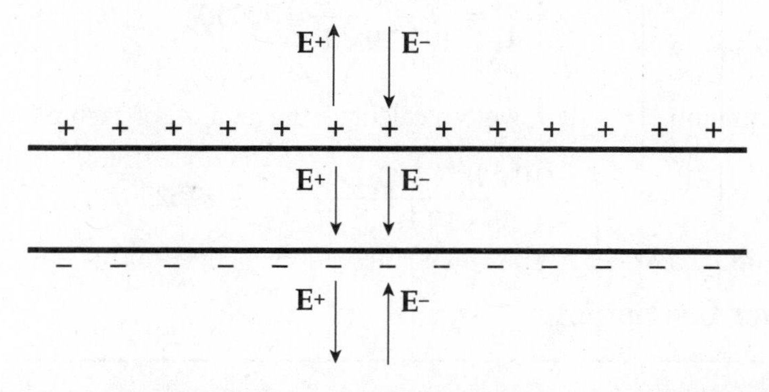

Therefore, with two plates, one with surface charge density $+\sigma$ and the other, $-\sigma$, the electric fields combine to give a field that's zero outside the plates and that has the magnitude

$$E_{\text{total}} = \frac{\sigma}{2\varepsilon_0} + \frac{\sigma}{2\varepsilon_0} = \frac{\sigma}{\varepsilon_0}$$

in between. In Example 10.12, we learned that the magnitude of the potential difference, ΔV, between the plates satisfies the relationship $\Delta V = Ed$, so combining this with the previous equation, we get

$$E = \frac{\sigma}{\varepsilon_0} \quad \Rightarrow \quad \frac{\Delta V}{d} = \frac{\sigma}{\varepsilon_0} \quad \Rightarrow \quad \frac{\Delta V}{d} = \frac{Q/A}{\varepsilon_0} \quad \Rightarrow \quad \frac{Q}{\Delta V} = \frac{\varepsilon_0 A}{d}$$

The ratio of Q to ΔV, for *any* capacitor, is called its **capacitance** (C),

$$C = Q/\Delta V$$

so for a parallel-plate capacitor, we get

$$C = \frac{\varepsilon_0 A}{d}$$

The capacitance measures the capacity for holding charge. The greater the capacitance, the more charge can be stored on the plates at a given potential difference. The capacitance of any capacitor depends only on the size, shape, and separation of the conductors. From the definition, $C = Q/\Delta V$, the units of C are coulombs per volt. One coulomb per volt is renamed one **farad** (abbreviated F): $1\ \text{C/V} = 1\ \text{F}$.

Example 10.17 A 10-nanofarad parallel-plate capacitor holds a charge of magnitude 50 μC on each plate.

(a) What is the potential difference between the plates?

(b) If the plates are separated by a distance of 0.2 mm, what is the area of each plate?

Solution.

(a) From the definition, $C = Q/\Delta V$, we find that

$$\Delta V = \frac{Q}{C} = \frac{50 \times 10^{-6}\ \text{C}}{10 \times 10^{-9}\ \text{F}} = 5000\ \text{V}$$

(b) From the equation $C = \varepsilon_0 A/d$, we can calculate the area, A, of each plate:

$$A = \frac{Cd}{\varepsilon_0} = \frac{(10 \times 10^{-9}\ \text{F})(0.2 \times 10^{-3}\ \text{m})}{8.85 \times 10^{-12}\ \text{C}^2/\text{N} \cdot \text{m}^2} = 0.23\ \text{m}^2$$

Capacitors of Other Geometries

Example 10.18 A long cable consists of a solid conducting cylinder of radius a, which carries a linear charge density of $+\lambda$, concentric with an outer cylindrical shell of radius b, which carries a linear charge density of $-\lambda$. This is a **coaxial** cable. Determine the capacitance of the cable.

Solution. We first apply the definition, $C = Q/\Delta V$. In Example 9.13, we saw that the electric field outside of a conducting cylinder is given by the equation

$$E = \frac{\lambda}{2\pi\varepsilon_0 r}$$

Therefore, we can get the potential difference between the inner cylinder and the outer cylindrical shell by integrating the electric field. Since

$$V_B - V_A = \Delta V_{A\to B} = \frac{-W_{E,\,A\to B}\ \text{on}\ q}{q} = -\frac{1}{q}\int_A^B q\mathbf{E} \cdot d\mathbf{r} = \int_B^A \mathbf{E} \cdot d\mathbf{r}$$

we have

$$V_{\text{inner}} - V_{\text{outer}} = \int_{r=a}^{r=b} \frac{\lambda}{2\pi\varepsilon_0 r}\,dr = \frac{\lambda}{2\pi\varepsilon_0}[\ln r]_a^b = \frac{\lambda}{2\pi\varepsilon_0}\ln\frac{b}{a}$$

Therefore, over a length ℓ of cable, the magnitude of the charge Q on each cylinder is $\lambda\ell$, so by the definition of capacitance, we get

$$C = \frac{Q}{\Delta V} = \frac{\lambda\ell}{\frac{\lambda}{2\pi\varepsilon_0}\ln\frac{b}{a}} = \frac{2\pi\varepsilon_0\ell}{\ln\frac{b}{a}}$$

Example 10.19 A spherical conducting shell of radius a, which carries a charge of $+Q$, is concentric with an outer spherical shell of radius b, which carries a charge of $-Q$. What is the capacitance of this spherical capacitor?

Solution. We first apply the definition, $C = Q/\Delta V$. We know that, in the region between the shells, the field is due to the inner shell alone:

$$E = \frac{1}{4\pi\varepsilon_0}\frac{Q}{r^2}$$

Therefore, the potential difference between the inner sphere and the outer sphere shell is obtained by integrating the electric field as follows. Since

$$V_B - V_A = \Delta V_{A\to B} = \frac{-W_{E,\,A\to B}\text{ on }q}{q} = -\frac{1}{q}\int_A^B q\mathbf{E}\cdot d\mathbf{r} = \int_B^A \mathbf{E}\cdot d\mathbf{r}$$

we have

$$V_{\text{inner}} - V_{\text{outer}} = \int_{r=a}^{r=b}\frac{1}{4\pi\varepsilon_0}\frac{Q}{r^2}dr = \frac{Q}{4\pi\varepsilon_0}\int_{r=a}^{r=b}\frac{1}{r^2}dr = \frac{Q}{4\pi\varepsilon_0}\left[-\frac{1}{r}\right]_a^b = \frac{Q}{4\pi\varepsilon_0}\left(\frac{1}{a}-\frac{1}{b}\right)$$

So, by definition of capacitance, we get

$$C = \frac{Q}{\Delta V} = \frac{Q}{\frac{Q}{4\pi\varepsilon_0}\left(\frac{1}{a}-\frac{1}{b}\right)} = \frac{4\pi\varepsilon_0}{\frac{1}{a}-\frac{1}{b}} = 4\pi\varepsilon_0\frac{ab}{b-a}$$

The capacitance of a single, isolated conductor can also be defined; we can just think of the other conductor as being infinitely far away. In this case, if a conductor holds a charge Q at a potential of V, then its capacitance is defined as $C = Q/V$. For a single sphere of radius a, we determined earlier in this chapter that the potential at its surface is $V\ (1/4\pi\varepsilon_0)Q/a$, so its capacitance is

$$C = \frac{Q}{V} = \frac{Q}{\frac{1}{4\pi\varepsilon_0}\frac{Q}{a}} = 4\pi\varepsilon_0 a$$

which says that C is proportional to the radius of the sphere. This makes sense; a larger sphere would have the capacity to hold more charge and should therefore have a greater capacitance. This same result for the capacitance of a single sphere can be obtained from the equation we derived above for the spherical shell capacitor simply by letting the radius b of the outer sphere go to infinity:

$$\lim_{b\to\infty}\frac{4\pi\varepsilon_0}{\frac{1}{a}-\frac{1}{b}} = \frac{4\pi\varepsilon_0}{\frac{1}{a}-0} = 4\pi\varepsilon_0 a$$

COMBINATIONS OF CAPACITORS

Capacitors are often arranged in combination in electric circuits. Here we'll look at two types of arrangements, the parallel combination and the series combination.

A collection of capacitors are said to be in **parallel** if they all share the same potential difference. The following diagram shows two capacitors wired in parallel:

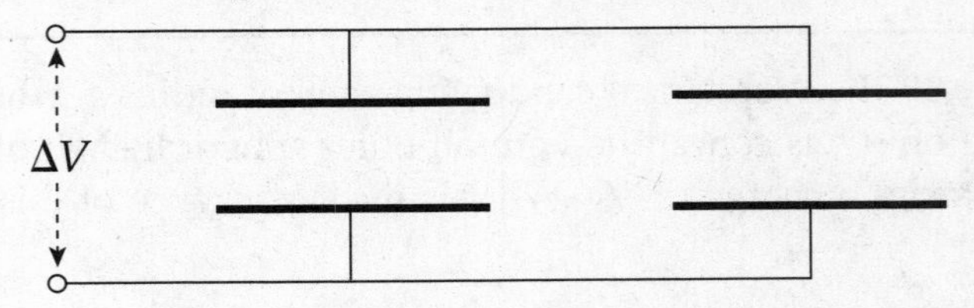

The top plates are connected by a wire and form a single equipotential; the same is true for the bottom plates. Therefore, the potential difference across one capacitor is the same as the potential difference across the other capacitor.

We want to find the capacitance of a *single* capacitor that would perform the same function as this combination. If the capacitances are C_1 and C_2, then the charge on the first capacitor is $Q_1 = C_1\Delta V$ and the charge on the second capacitor is $Q_2 = C_2\Delta V$. The total charge on the combination is $Q_1 + Q_2$, so the equivalent capacitance, C_P, must be

$$C_P = \frac{Q}{\Delta V} = \frac{Q_1 + Q_2}{\Delta V} = \frac{Q_1}{\Delta V} + \frac{Q_2}{\Delta V} = C_1 + C_2$$

So the **equivalent** capacitance of a collection of capacitors in parallel is found by adding the individual capacitances.

A collection of capacitors are said to be in **series** if they all share the same charge magnitude. The following diagram shows two capacitors wired in series:

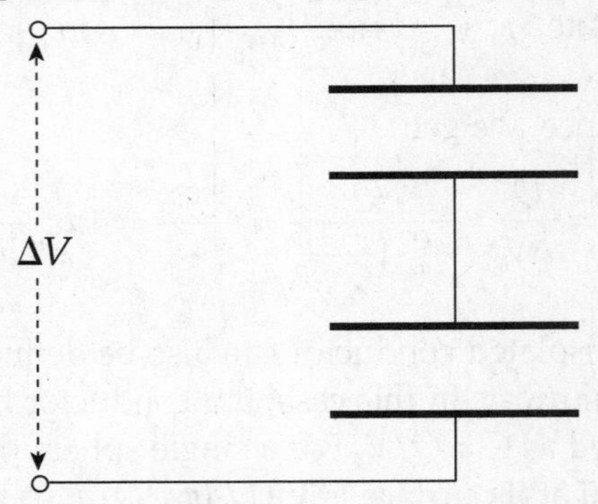

When a potential difference is applied, as shown, negative charge will be deposited on the bottom plate of the bottom capacitor; this will push an equal amount of negative charge away from the top plate of the bottom capacitor toward the bottom plate of the top capacitor. When the system has reached equilibrium, the charges on all the plates will have the same magnitude:

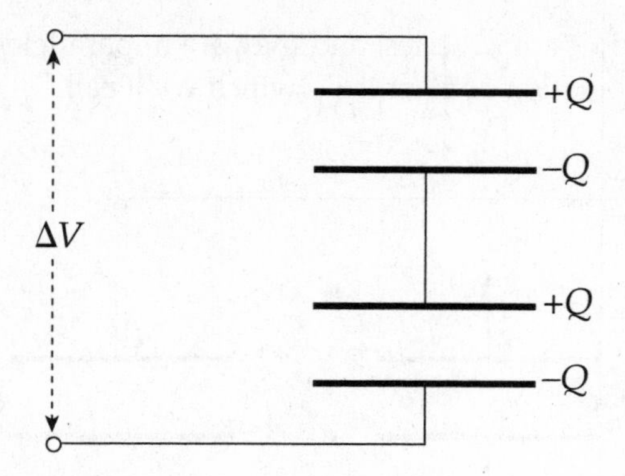

If the top and bottom capacitors have capacitances of C_1 and C_2, respectively, then the potential difference across the top capacitor is $\Delta V_1 = Q/C_1$, and the potential difference across the bottom capacitor is $\Delta V_2 = Q/C_2$. The total potential difference across the combination is $\Delta V_1 + \Delta V_2$, which must equal ΔV. Therefore, the equivalent capacitance, C_S, must be

$$C_S = \frac{Q}{\Delta V} = \frac{Q}{\Delta V_1 + \Delta V_2} = \frac{Q}{\frac{Q}{C_1} + \frac{Q}{C_2}} = \frac{1}{\frac{1}{C_1} + \frac{1}{C_2}}$$

We can write this in another form:

$$\frac{1}{C_S} = \frac{1}{C_1} + \frac{1}{C_2}$$

In words, the *reciprocal* of the capacitance of a collection of capacitors in series is found by adding the reciprocals of the individual capacitances.

Example 10.20 Given that $C_1 = 2$ μF, $C_2 = 4$ μF, an $C_3 = 6$ μF, calculate the equivalent capacitance for the following combination:

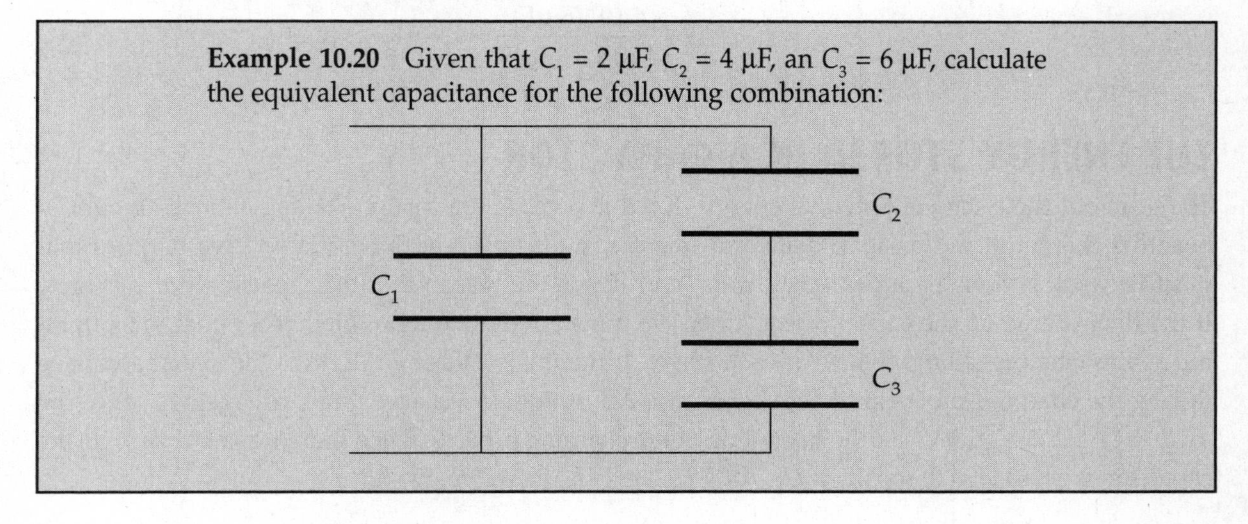

Solution. Notice that C_2 and C_3 are in series, and they are in parallel with C_1. That is, the capacitor equivalent to the series combination of C_2 and C_3 (which we'll call $C_{2\text{-}3}$) is in parallel with C_1. We can represent this as follows:

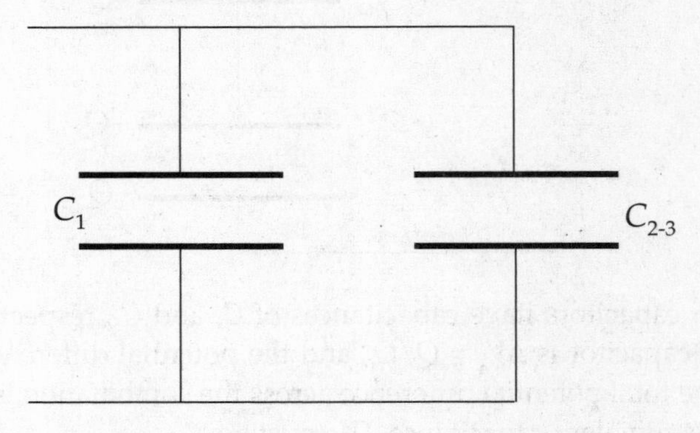

So, the first step is to find $C_{2\text{-}3}$:

$$\frac{1}{C_{2\text{-}3}} = \frac{1}{C_2} + \frac{1}{C_3} \quad \Rightarrow \quad C_{2\text{-}3} = \frac{C_2 C_3}{C_2 + C_3}$$

Now this is in parallel with C_1, so the overall equivalent capacitance ($C_{1\text{-}2\text{-}3}$) is

$$C_{1\text{-}2\text{-}3} = C_1 + C_{2\text{-}3} = C_1 + \frac{C_2 C_3}{C_2 + C_3}$$

Substituting in the given numerical values, we get

$$C_{1\text{-}2\text{-}3} = (2\ \mu\text{F}) + \frac{(4\ \mu\text{F})(6\ \mu\text{F})}{(4\ \mu\text{F}) + (6\ \mu\text{F})} = 4.4\ \mu\text{F}$$

THE ENERGY STORED IN A CAPACITOR

To figure out the electrical potential energy stored in a capacitor, imagine taking a small amount of negative charge off the positive plate and transferring it to the negative plate. This requires that positive work is done by an external agent, and this is the reason that the capacitor stores energy. If the final charge on the capacitor is Q, then we transferred an amount of charge equal to Q, fighting against the prevailing voltage at each stage. If the final voltage is ΔV, then the average voltage during the charging process is $\frac{1}{2}\Delta V$; so, because ΔU_E is equal to charge times voltage, we can write $\Delta U_E = Q \cdot \frac{1}{2}\Delta V = \frac{1}{2}Q\Delta V$. At the beginning of the charging process, when there was no charge on the capacitor, we had $U_i = 0$, so $\Delta U_E = U_f - U_i = U_f - 0 = U_f$; therefore, we have

$$U_E = \tfrac{1}{2}Q\Delta V$$

This is the electrical potential energy stored in a capacitor. Because of the definition $C = Q / \Delta V$, the equation for the stored potential energy can be written in two other forms:

$$U_E = \tfrac{1}{2}(C\Delta V)\cdot \Delta V = \tfrac{1}{2}C(\Delta V)^2 \text{—or—} U_E = \tfrac{1}{2}Q\cdot\frac{Q}{C} = \frac{Q^2}{2C}$$

DIELECTRICS

One method of keeping the plates of a capacitor apart, which is necessary to maintain charge separation and store potential energy, is to insert an insulator (called a **dielectric**) between the plates.

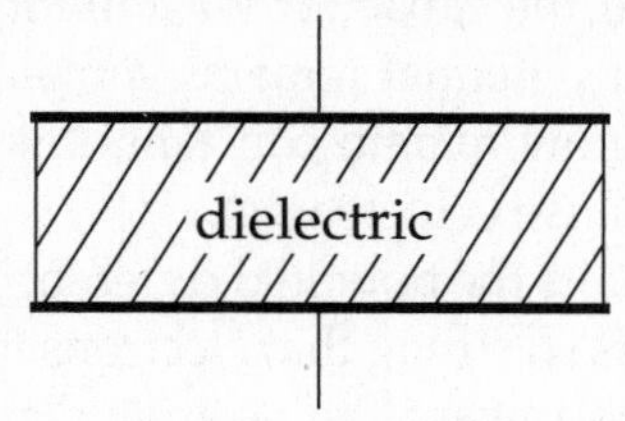

A dielectric always increases the capacitance of a capacitor. Let's see why this happens. Imagine charging a capacitor to a potential difference of V with charge $+Q$ on one plate and $-Q$ on the other. Now disconnect the capacitor from the charging source and insert a dielectric. What happens? Although the dielectric is not a conductor, the electric field that existed between the plates causes the molecules within the dielectric material to polarize; there is more electron density on the side of the molecule near the positive plate.

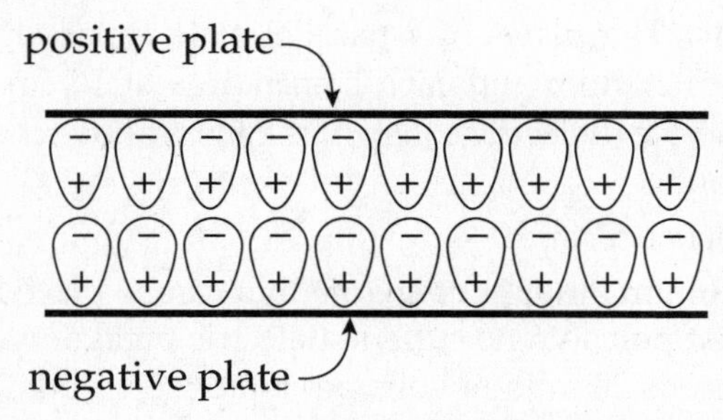

The effect of this is to form a layer of negative charge along the top surface of the dielectric and a layer of positive charge along the bottom surface; this separation of charge induces its own electric field ($\mathbf{E}_i$), within the dielectric, that opposes the original electric field, **E**, within the capacitor:

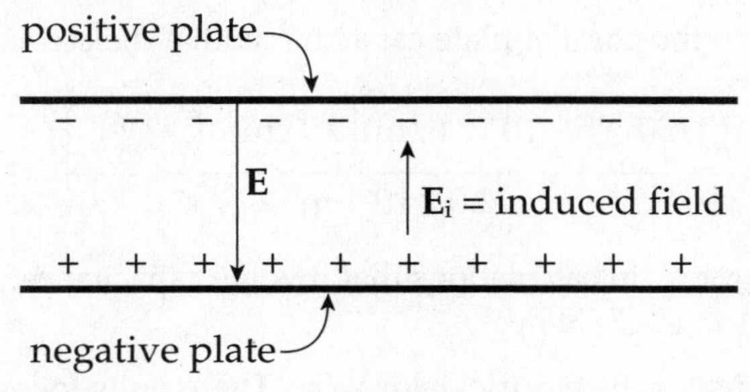

So the overall electric field has been reduced from its previous value: $\mathbf{E}_{total} = \mathbf{E} + \mathbf{E}_i$, and $E_{total} = E - E_i$. Let's say that the electric field has been reduced by a factor of κ (the Greek letter *kappa*) from its original value

$$E_{\text{with dielectric}} = E_{\text{without dielectric}} - E_i = \frac{E}{\kappa}$$

Since $\Delta V = Ed$ for a parallel-plate capacitor, we see that ΔV must have decreased by a factor of κ. But $C = Q/\Delta V$, so if ΔV decreases by a factor of κ, then C increases by a factor of κ:

$$C_{\text{with dielectric}} = \kappa C_{\text{without dielectric}}$$

The value of κ, called the **dielectric constant**, varies from material to material, but it's always greater than 1.

Although the description for why the capacitance increases with the insertion of a dielectric assumed that the source that charged the capacitor was disconnected (so that Q remains constant), the same result holds if the source of potential remains connected to the capacitor. The capacitance still increases, but because ΔV must now remain constant, the equation $Q = C\Delta V$ tells us that more charge will appear on the plates (because C increases).

The presence of a dielectric also limits the potential difference that can be applied across the plates. If ΔV gets too high, then $E = \Delta V/d$ gets so strong that electrons in the dielectric material can be ripped right out of their atoms and propelled toward the positive plate. This discharges the capacitor (and typically burns a hole through the dielectric). This event is called **dielectric breakdown**.

The capacitance formulas derived in this chapter have assumed that no dielectric was present; the permittivity constant that appears in the formulas is ε_0, the permittivity of *free space* (vacuum). If a dielectric is present, then the permittivity increases to $\varepsilon = \kappa\varepsilon_0$, so the occurrence of ε_0 in each formula is simply replaced by $\varepsilon = \kappa\varepsilon_0$.

Example 10.21 The plates of a parallel-plate capacitor are separated by a distance of 2.0 mm and each has an area of 10 cm^2. If a layer of polystyrene (whose dielectric constant is 2.6) is sandwiched between the plates, figure out

(a) the capacitance, and

(b) the maximum amount of charge that can be placed on the plates, given that polystyrene suffers dielectric breakdown if the electric field exceeds 20 million volts per meter.

Solution.

(a) The capacitance of the parallel-plate capacitor, with a dielectric, is

$$C = \frac{\kappa\varepsilon_0 A}{d} = \frac{(2.6)(8.85\times10^{-12}\ \text{F/m})\left[10\ \text{cm}^2\cdot\left(\frac{1\ \text{m}}{100\ \text{cm}}\right)^2\right]}{2.0\times10^{-3}\ \text{m}} = 1.2\times10^{-11}\ \text{F}$$

Notice the units for ε_0; in calculations that involve capacitance, it is usually easier to write F/m rather than C^2/N·m^2.

(b) First note the units for the electric field: V/m. These units follow from the equation $\Delta V = Ed$. In the past, we've written the units of E as N/C. These two units are actually equivalent: 1 V/m = 1 N/C. We were asked to determine Q_{max} if E_{max} is 20×10^6 V/m and, from the equations $\Delta V = Ed$ and $Q = C\,\Delta\,V$, we get

$$\begin{aligned} Q_{max} &= C\Delta V_{max} = CE_{max}d \\ &= (1.2\times10^{-11}\ \text{F})(20\times10^6\ \text{V/m})(2\times10^{-3}\ \text{m}) \\ &= 4.8\times10^{-7}\ \text{C} \end{aligned}$$

CHAPTER 10 REVIEW QUESTIONS

SECTION I: MULTIPLE CHOICE

1. Which of the following statements is/are true?

   I. If the electric field at a certain point is zero, then the electric potential at the same point is also zero.
   II. If the electric potential at a certain point is zero, then the electric field at the same point is also zero.
   III. The electric potential is inversely proportional to the strength of the electric field.

   (A) I only
   (B) II only
   (C) I and II only
   (D) I and III only
   (E) None are true

2. If the electric field does negative work on a negative charge as the charge undergoes a displacement from Position A to Position B within an electric field, then the electrical potential energy

   (A) is negative
   (B) is positive
   (C) increases
   (D) decreases
   (E) Cannot be determined from the information given

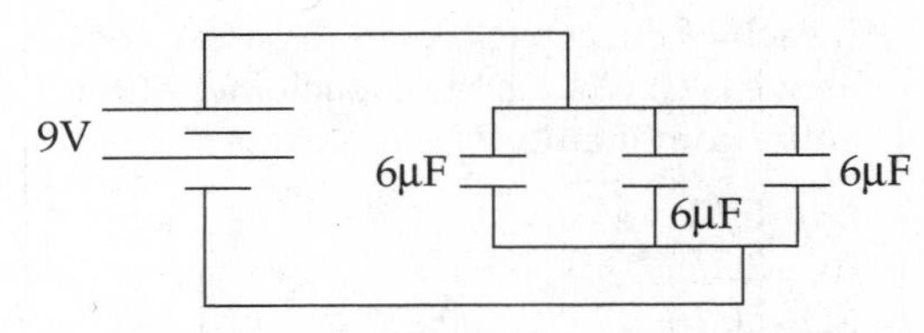

3. Three 6 μF capacitors are connected in parallel to a 9 V battery as shown above. Determine the energy stored in each capacitor.

   (A) 243 J
   (B) 7.29×10^{-4} J
   (C) 8.10×10^{-5} J
   (D) 2.43×10^{-4} J
   (E) 27 J

4. Negative charges are accelerated by electric fields toward points

   (A) at lower electric potential
   (B) at higher electric potential
   (C) where the electric field is zero
   (D) where the electric field is weaker
   (E) where the electric field is stronger

5. A charge q experiences a displacement within an electric field from Position A to Position B. The change in the electrical potential energy is ΔU_E, and the work done by the electric field during this displacement is W_E. Then

   (A) $V_B - V_A = W_E/q$
   (B) $V_A - V_B = qW_E$
   (C) $V_B - V_A = qW_E$
   (D) $V_A - V_B = \Delta U_E/q$
   (E) $V_B - V_A = \Delta_E/q$

6.

+Q

1 •

4 •   • 2

• 3

–Q

Which points in this uniform electric field (between the plates of the capacitor) shown above lie on the same equipotential?

   (A) 1 and 2 only
   (B) 1 and 3 only
   (C) 2 and 4 only
   (D) 3 and 4 only
   (E) 1, 2, 3 and 4 all lie on the same equipotential since the electric field is uniform.

7. Two isolated and widely separated conducting spheres each carry a charge of $-Q$. Sphere 1 has a radius of a and Sphere 2 has a radius of $4a$. If the spheres are now connected by a conducting wire, what will be the final charge on each sphere?

	Sphere 1	Sphere 2
(A)	$-Q$	$-Q$
(B)	$-2Q/3$	$-4Q/3$
(C)	$-4Q/3$	$-2Q/3$
(D)	$-2Q/5$	$-8Q/5$
(E)	$-8Q/5$	$-2Q/5$

8. A parallel-plate capacitor is charged to a potential diffrence of ΔV; this results in a charge of $+Q$ on one plate and a charge of $-Q$ on the other. The capacitor is disconnected from the charging source, and a dielectric is then inserted. What happens to the potential difference and the stored electrical potential energy?

(A) The potential difference decreases, and the stored electrical potential energy decreases.
(B) The potential difference decreases, and the stored electrical potential energy increases.
(C) The potential difference increases, and the stored electrical potential energy decreases.
(D) The potential difference increases, and the stored electrical potential energy increases.
(E) The potential difference decreases, and the stored electrical potential energy remains unchanged.

9.

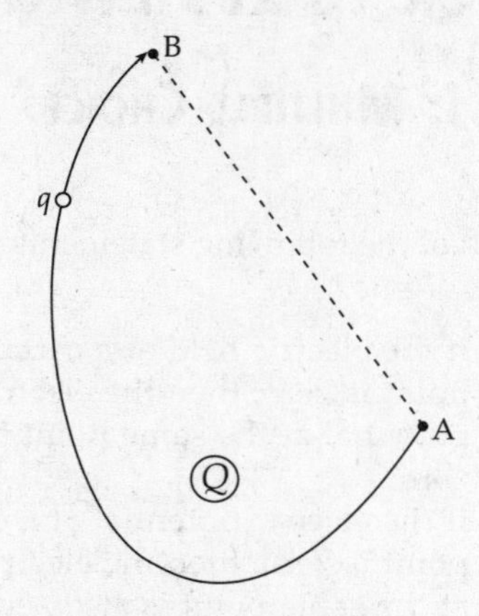

How much work would the electric field (created by the stationary charge Q) perform as a charge q is moved from Point A to B along the curved path shown?
$V_A = 200$ V, $V_B = 100$ V, $q = -0.05$ C, length of line segment AB = 10 cm, length of curved path = 20 cm.

(A) −10 J
(B) −5 J
(C) +5 J
(D) +10 J
(E) +2 J

10.

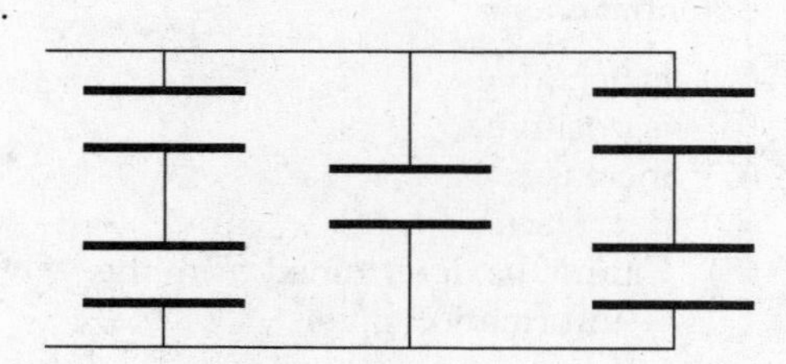

If each of the capacitors in the array shown above is C, what is the capacitance of the entire combination?

(A) C/2
(B) 2C/3
(C) 5C/6
(D) 2C
(E) 5C/3

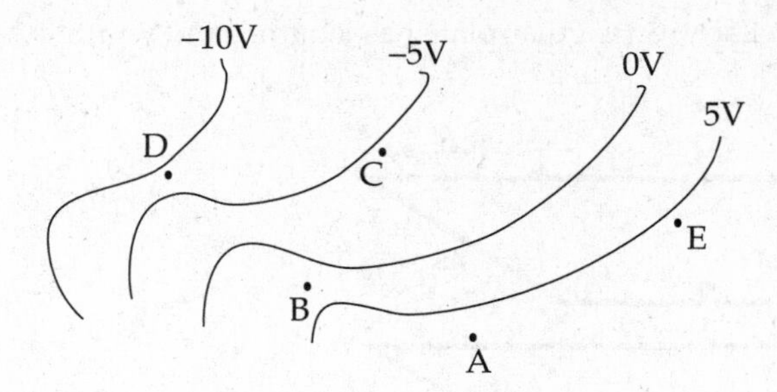

11. The diagram above shows equipotential lines produced by a charge distribution. *A*, *B*, *C*, *D*, and *E* are points in the plane. An electron begins at point *A*. The electron is then moved to point *E* and then from point *E* to point *C*. Which of the following correctly describes the work done *by the field* for each part of the movement?

	Movement from A to E	Movement from E to C
(A)	Negative	Positive
(B)	Zero	Positive
(C)	Zero	Negative
(D)	Negative	Zero
(E)	Positive	Positive

SECTION II: FREE RESPONSE

1. In the figure shown, all four charges are situated at the corners of a square with sides *s*.

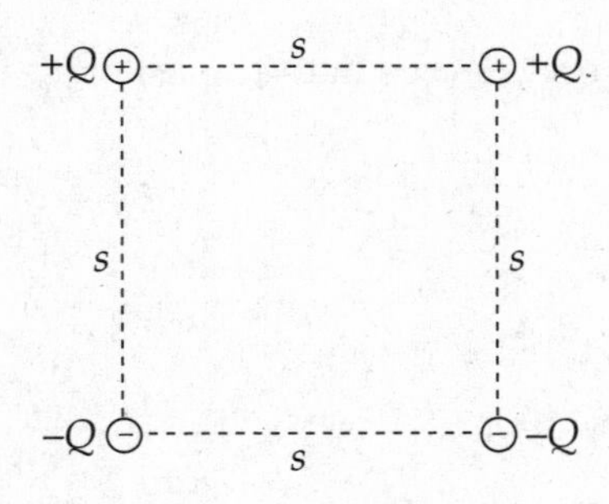

(a) What is the total electrical potential energy of this array of fixed charges?

(b) What is the electric field at the center of the square?

(c) What is the electric potential at the center of the square?

(d) Sketch (on the diagram) the portion of the equipotential surface that lies in the plane of the figure and passes through the center of the square.

(e) How much work would the electric field perform on a charge *q* as it moved from the midpoint of the right side of the square to the midpoint of the top of the square?

2. The figure below shows a parallel-plate capacitor. Each rectangular plate has length L and width w, and the plates are separated by a distance d.

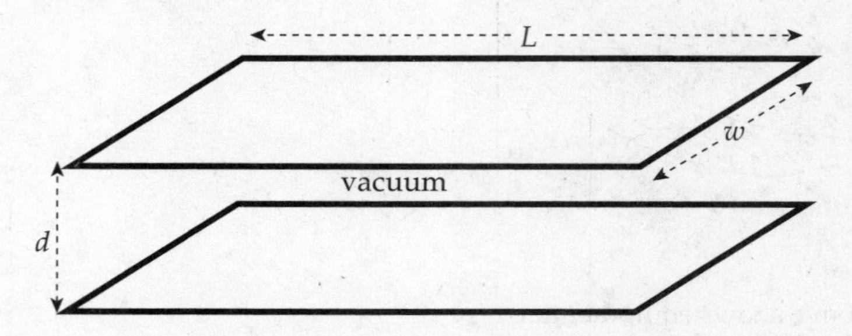

(a) Determine the capacitance.

An electron (mass m, charge $-e$) is shot horizontally into the empty space between the plates, midway between them, with an initial velocity of magnitude v_0. The electron just barely misses hitting the end of the top plate as it exits. (Ignore gravity.)

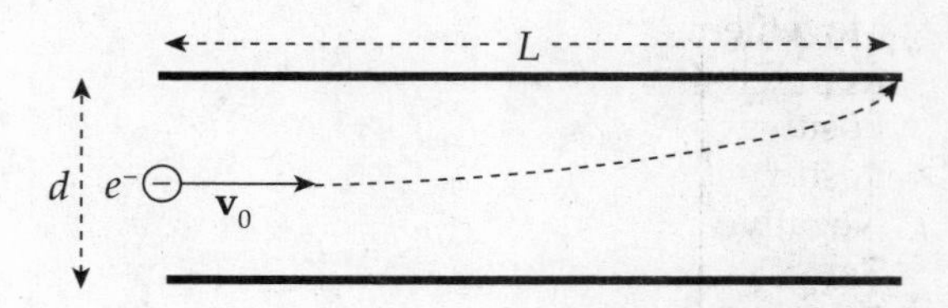

(b) In the diagram, sketch the electric field vector at the position of the electron when it has traveled a horizontal distance of $L/2$.

(c) In the diagram, sketch the electric force vector on the electron at the same position as in part (b).

(d) Determine the strength of the electric field between the plates. Write your answer in terms of L, d, m, e, and v_0.

(e) Determine the charge on the top plate.

(f) How much potential energy is stored in the capacitor?

3. A solid conducting sphere of radius a carries an excess charge of Q.

   (a) Determine the electric field magnitude, $E(r)$, as a function of r, the distance from the sphere's center.

   (b) Determine the potential, $V(r)$, as a function of r. Take the zero of potentil at $r = \infty$.

   (c) On the diagrams below, sketch $E(r)$ and $V(r)$. (Cover at least the range $0 < r < 2a$.)

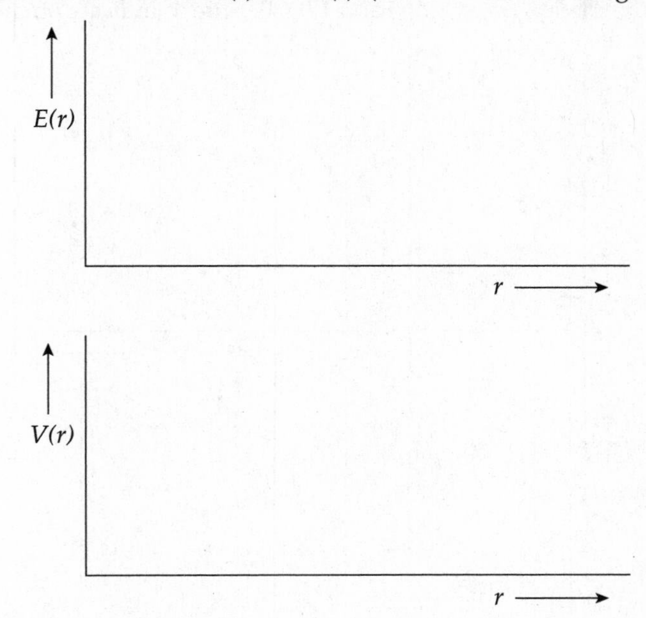

4. A solid, nonconducting sphere of radius a has a volume charge density given by the equation $\rho(r) = \rho_0(r/a)^3$, where r is the distance from the sphere's center.

(a) Determine the electric field magnitude, $E(r)$, as a function of r.

(b) Determine the potential, $V(r)$, as a function of r. Take the zero of potential at $r = \infty$.

(c) On the diagrams below, sketch $E(r)$ and $V(r)$. Be sure to indicate on the vertical axis in each plot the value at $r = a$.

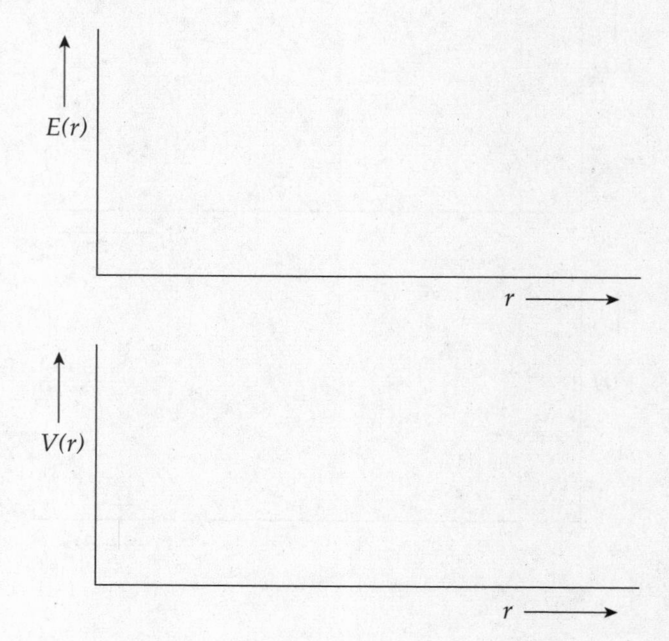

ELECTRIC POTENTIAL AND CAPACITANCE SUMMARY

Electric Potential

- When a positive test charge q_0 moves through an electric field, its **electric potential energy** changes. The change in electric potential energy is: $\Delta U = -q_0 \int E \bullet ds$. Note that this is a dot product between the electric field and the path, *ds*.
- When the electric field is uniform then $\Delta U = -q_0 Ed$, where *d* is the distance along the electric field that the charge moved.
- The **electric potential** is the electric potential energy divided by the test charge. Therefore $\Delta V = \frac{\Delta U}{q_0} = -\int \mathbf{E} \bullet d\mathbf{s}$. This can also be rearranged as: $-\frac{dV}{dr} = E$. Either one of these equations can be used to relate the electric potential to the electric field strength. The units of electric potential are volts (V), and it is often called voltage.
- When the electric field is uniform, the electric potential is given by the strength of the electric field and the distance the object moves along the electric field: $\Delta V = -Ed$
- The electric potential due to a point charge, *Q*, is $V = k\frac{Q}{r}$.
- The electric potential energy between a pair of charges is: $U = k\frac{q_1 q_2}{r}$

Capacitance

- Capacitors store electric potential energy by separating opposite charges. Capacitance is measure by the ratio of the charge stored to the potential difference across the capacitor. $C = \frac{Q}{\Delta V}$. The units are farads (F).
- The capacitance of a parallel plate capacitor is: $C = \frac{\varepsilon_0 A}{d}$
- When capacitors are added in parallel the equivalent capacitance is the sum of the capacitors: $C_{eq} = C_1 + C_2 + \ldots = \sum_i C_i$
- When capacitors are added in series the reciprocal of the equivalent capacitance is the sum of the reciprocals of each capacitor: $\frac{1}{C_{eq}} = \frac{1}{C_1} + \frac{1}{C_2} + \ldots = \sum_i \frac{1}{Ci}$
- The energy stored in a capacitor is: $U = \frac{1}{2} Q\Delta V^2$
- Dielectrics are used in capacitors to induce an opposing electric field that decreases the total electric field between the plates. This decreases the electric potential and therefore increases the capacitance.

11

Direct Current Circuits

INTRODUCTION

In the previous chapter, when we studied electrostatic fields, we learned that, within a conductor, an electrostatic field cannot be sustained; the source charges move to the surface and the conductor forms a single equipotential. We will now look at conductors within which an electric field can be sustained because a power source maintains a potential difference across the conductor, allowing charges to continually move through it. This ordered motion of charge through a conductor is called electric current.

ELECTRIC CURRENT

Picture a piece of metal wire. Within the metal, electrons are zooming around at speeds of about a million m/s in random directions, colliding with other electrons and positive ions in the lattice. This constitutes charge in motion, but it doesn't constitute *net* movement of charge, because the electrons move randomly. If there's no net motion of charge, there's no current. However, if we were to create a potential difference between the ends of the wire, meaning if we set up an electric field,

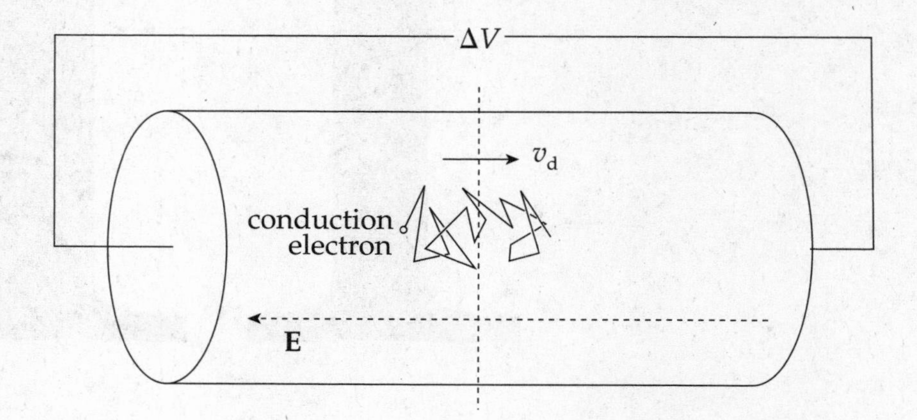

the electrons would experience an electric force, and they would start to drift through the wire. This is current. Although the electric field would travel through the wire at nearly the speed of light, the electrons themselves would still have to make their way through a crowd of atoms and other free electrons, so their **drift speed**, v_d, would be relatively slow: on the order of a millimeter per second.

To measure the current, we have to measure how much charge crosses a plane per unit time. If an amount of charge of magnitude ΔQ crosses an imaginary plane in a time interval Δt, then the **current** is

$$I = \frac{\Delta Q}{\Delta t}$$

If the amount of current is changing during the time interval Δt, then the equation above gives the average current. We can define the instantaneous current as

$$I = \frac{dQ}{dt}$$

Because current is charge per unit time, it's expressed in coulombs per second. One coulomb per second is an **ampere** (abbreviated **A**), or amp. So 1 C/s = 1 A.

Although the charge carriers that constitute the current within a metal are electrons, the direction of the current is taken to be the direction that *positive* charge carriers would move. (This is explicitly stated on the AP Physics Exam.) So, if the conduction electrons drift to the right, we'd say the current points toward the left.

RESISTANCE

Let's say we had a copper wire and a glass fiber that had the same length and cross-sectional area, and that we hooked up the ends of the metal wire to a source of potential difference and measured the resulting current. If we were to do the same thing with the glass fiber, the current would probably be too small to measure, but why? Well, the glass provided more resistance to the flow of charge. If the potential difference is ΔV and the current is I, then the **resistance** is

$$R = \frac{\Delta V}{I}$$

This is known as **Ohm's Law**. Not all electrical devices are ohmic, but many are. Notice that if the current is large, the resistance is low, and if the current is small, then resistance is high. The Δ in the equation above is often omitted, but you should always assume that, in this context, $V = \Delta V$ = potential difference, also called voltage.

Because resistance is voltage divided by current, it is expressed in volts per amp. One volt per amp is one **ohm** (Ω, *omega*). So, 1 V/A = 1 Ω.

Resistivity

The resistance of an object depends on two things: the material it's made of and its shape. For example, again think of the copper wire and glass fiber of the same length and area. They have the same shape, but their resistances are different because they're made of different materials. Glass has a much greater intrinsic resistance than copper does; it has a greater **resistivity**. Each material has its own characteristic resistivity, and resistance depends on how the material is shaped. For a wire of length L and cross-sectional area A made of a material with resistivity ρ, resistance is given by:

$$R = \frac{\rho L}{A}$$

The resistivity of copper is around 10^{-8} Ω·m, while the resistivity of glass is *much* greater, around 10^{12} Ω·m.

> **Example 11.1** A cylindrical wire of radius 1 mm and length 2 m is made of platinum (resistivity = 1×10^{-7} Ω·m). If a voltage of 9 V is applied between the ends of the wire, what will be the resulting current?

Solution. First, the resistance of the wire is given by the equation

$$R = \frac{\rho L}{A} = \frac{\rho L}{\pi r^2} = \frac{(1 \times 10^{-7}\ \Omega \cdot \text{m})(2\ \text{m})}{\pi (0.001\ \text{m})^2} = 0.064\ \Omega$$

Then, from $I = V/R$, we get

$$I = \frac{V}{R} = \frac{9\ \text{V}}{0.064\ \Omega} = 140\ \text{A}$$

ELECTRIC CIRCUITS

An electric current is maintained when the terminals of a voltage source (a battery, for example) are connected by a conducting pathway, in what's called a **circuit**. If the current always travels in the same direction through the pathway, it's called a **direct current**.

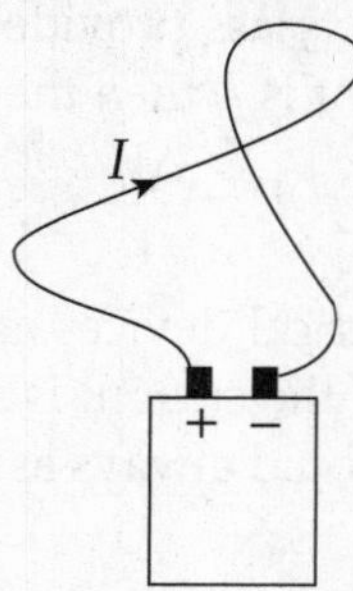

The job of the voltage source is to provide a potential difference called an **emf**, or **electromotive force**, which drives the flow of charge. The emf isn't really a force, it's the work done per unit charge, and it's measured in volts.

To try to imagine what's happening in a circuit in which a steady-state current is maintained, let's follow one of the charge carriers that's drifting through the pathway. (Remember we're pretending that the charge carriers are positive.) The charge is introduced by the positive terminal of the battery and enters the wire, where it's pushed by the electric field. It encounters resistance, bumping into the relatively stationary atoms that make up the metal's lattice and setting them into greater motion. So the electrical potential energy that the charge had when it left the battery is turning into heat. By the time the charge reaches the negative terminal, all of its original electrical potential energy is lost. In order to keep the current going, the voltage source must do positive work on the charge, forcing it to move from the negative terminal toward the positive terminal. The charge is now ready to make another journey around the circuit.

Energy and Power

When a carrier of positive charge q drops by an amount V in potential, it loses potential energy in the amount qV. If this happens in time t, then the rate at which this energy is transformed is equal to $(qV)/t = (q/t)V$. But q/t is equal to the current, I, so the rate at which electrical energy is transferred is given by the equation

$$P = IV$$

This equation works for the power delivered by a battery to the circuit as well as for resistors. The power dissipated in a resistor, as electrical potential energy is turned into heat, is given by $P = IV$, but, because of the relationship $V = IR$, we can express this in two other ways:

$$P = IV = I(IR) = I^2R$$

or

$$P = IV = \frac{V}{R} \cdot V = \frac{V^2}{R}$$

Resistors become hot when current passes through them; the thermal energy generated is called **joule heat**.

CIRCUIT ANALYSIS

We will now develop a way of specifying the current, voltage, and power associated which each element in a circuit. Our circuits will contain three basic elements: batteries, resistors, and connecting wires. As we've seen, the resistance of an ordinary metal wire is negligible; resistance is provided by devices that control the current: **resistors**. All the resistance of the system is concentrated in the resistors, which are symbolized in a circuit diagram by this symbol:

Batteries are denoted by the symbol:

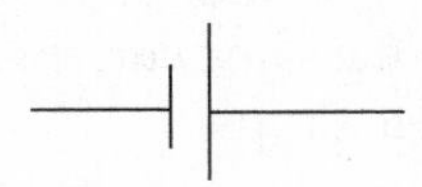

where the longer line represents the **positive** (higher potential) terminal, and the shorter line is the **negative** (lower potential) terminal. Sometimes a battery is denoted by more than one pair of such lines:

Here's a simple circuit diagram:

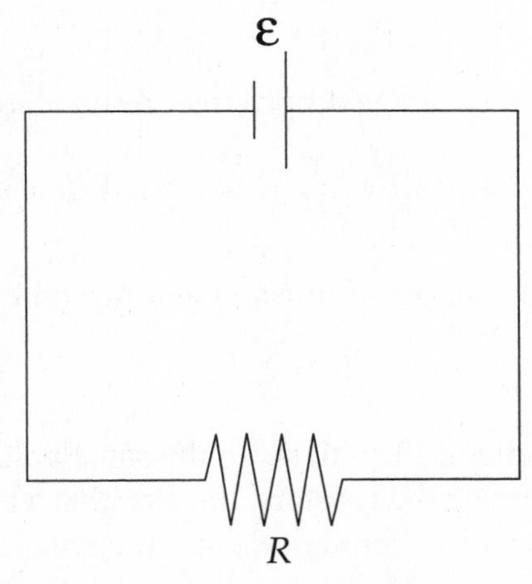

The emf (ε) of the battery is indicated, as is the resistance (R) of the resistor. Determining the current in this case is straightforward, because there's only one resistor. The equation $V = IR$, with given by ε, gives us

$$I = \frac{\varepsilon}{R}$$

Combinations of Resistors

Two common ways of combining resistors within a circuit is to place them either in **series** (one after the other),

R_1 R_2

or in **parallel** (that is, side-by-side):

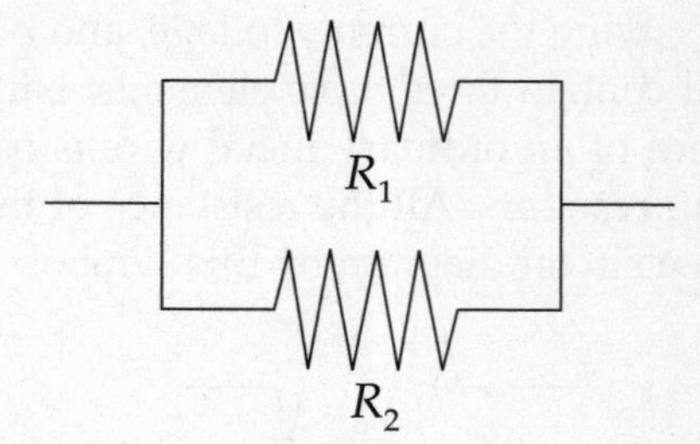

In order to simplify the circuit, our goal is to find the equivalent resistance of combinations. Resistors are said to be in series if they all share the same current and if the total voltage drop across them is equal to the sum of the individual voltage drops.

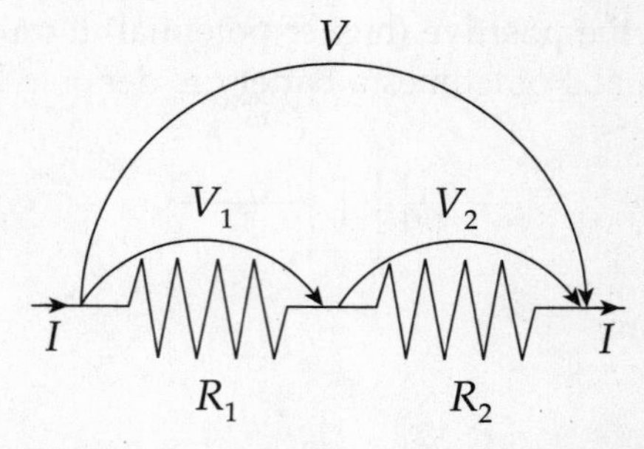

In this case, then, if V denotes the voltage drop across the combination, we have

$$R_{\text{equiv}} = \frac{V}{I} = \frac{V_1 + V_2}{I} = \frac{V_1}{I} + \frac{V_2}{I} = R_1 + R_2$$

This idea can be applied to any number of resistors in series (not just two):

$$R_S = \sum_i R_i$$

Resistors are said to be in parallel if they all share the same voltage drop, and the total current entering the combination is split among the resistors. Imagine that a current I enters the combination. It splits; some of the current, I_1, would go through R_1, and the remainder, I_2, would go through R_2.

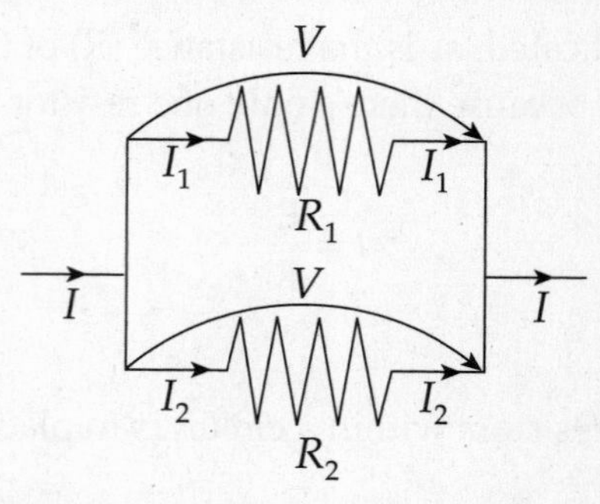

So if V is the voltage drop across the combination, we have

$$I = I_1 + I_2 \quad \Rightarrow \quad \frac{V}{R_{\text{equiv}}} = \frac{V}{R_1} + \frac{V}{R_2} \quad \Rightarrow \quad \frac{1}{R_{\text{equiv}}} = \frac{1}{R_1} + \frac{1}{R_2}$$

This idea can be applied to any number of resistors in parallel (not just two): The reciprocal of the equivalent resistance for resistors in parallel is equal to the sum of the reciprocals of the individual resistances:

$$\frac{1}{R_P} = \sum_i \frac{1}{R_i}$$

Example 11.2 Calculate the equivalent resistance for the following circuit:

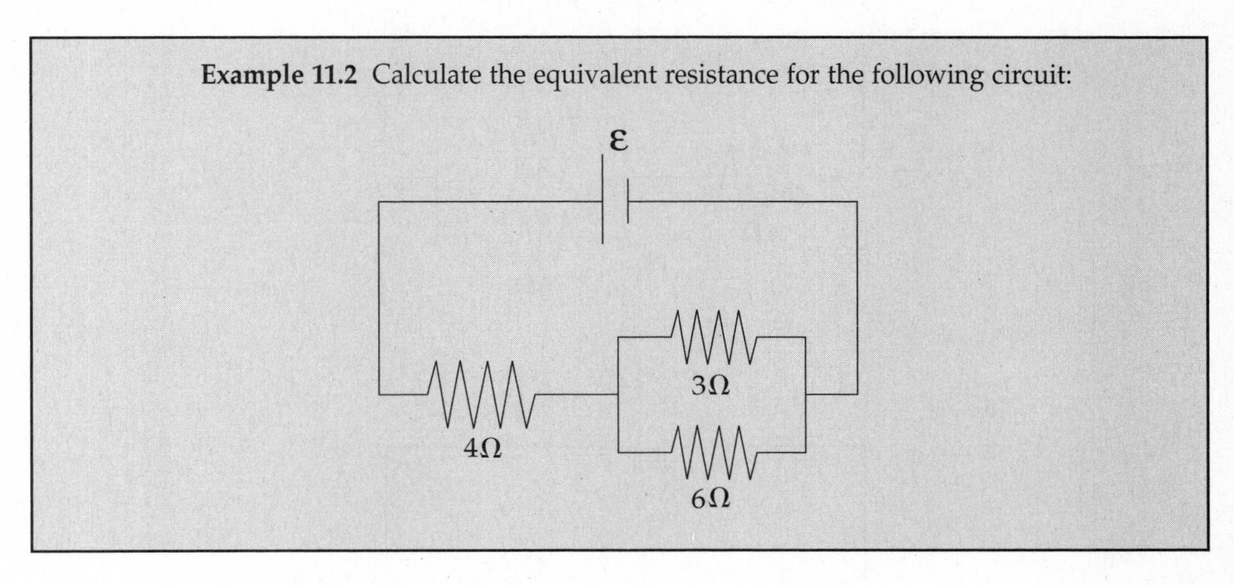

Solution. First find the equivalent resistance of the two parallel resistors:

$$\frac{1}{R_P} = \frac{1}{3\ \Omega} + \frac{1}{6\ \Omega} \quad \Rightarrow \quad \frac{1}{R_P} = \frac{1}{2\ \Omega} \quad \Rightarrow \quad R_P = 2\ \Omega$$

This resistance is in series with the 4 Ω resistor, so the overall equivalent resistance in the circuit is $R = 4\ \Omega + 2\ \Omega = 6\ \Omega$.

Example 11.3 Determine the current through each resistor, the voltage drop across each resistor, and the power given off (dissipated) as heat in each resistor of this circuit:

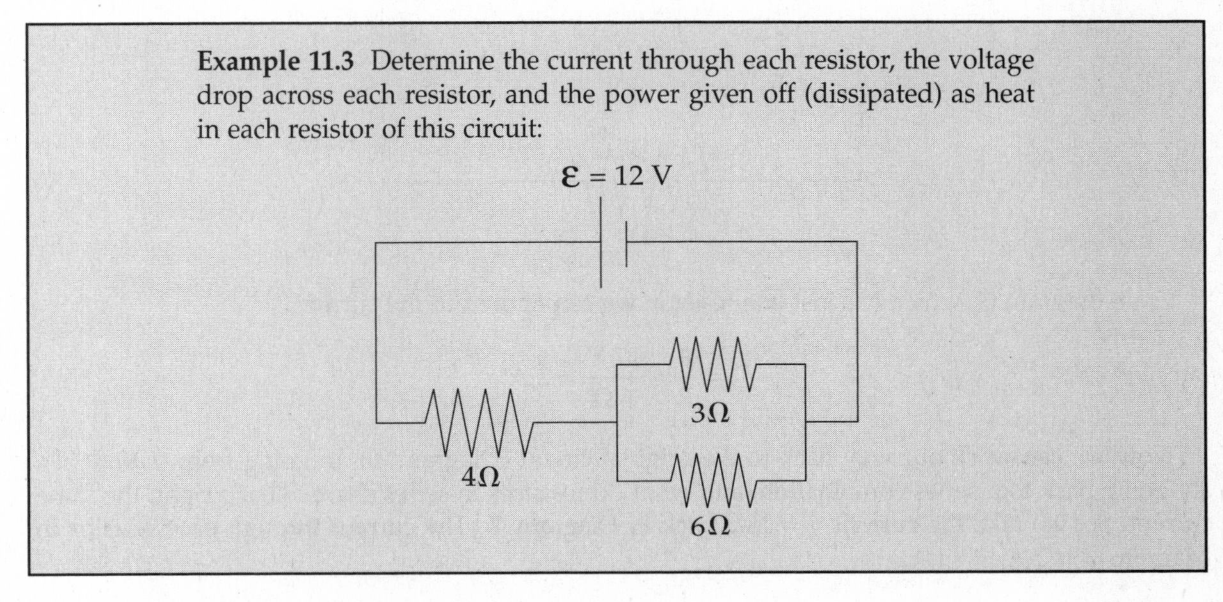

Solution. You might want to redraw the circuit each time we replace a combination of resistors by its equivalent resistance. From our work in the preceding example, we have

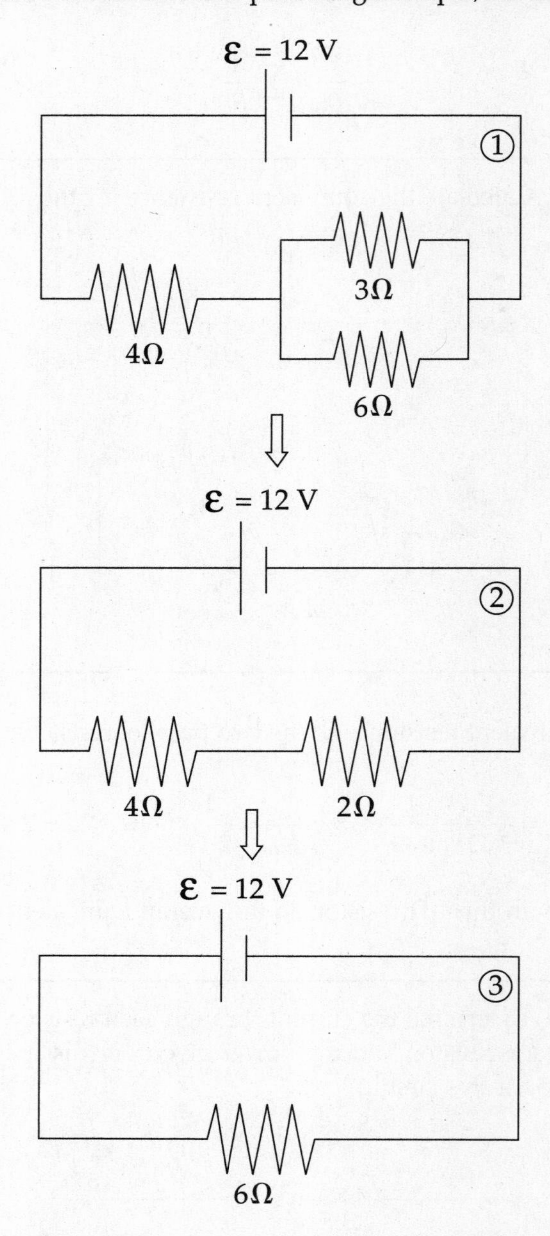

From diagram ③, which has just one resistor, we can figure out the current:

$$I = \frac{\mathcal{E}}{R} = \frac{12\text{ V}}{6\,\Omega} = 2\text{ A}$$

Now we can work our way back to the original circuit (Diagram ①). In going from ③ to ②, we are going back to a series combination, and what do resistors in series share? That's right, the same current. So, we take the current, I = 2 A, back to Diagram ②. The current through each resistor in Diagram ② is 2 A.

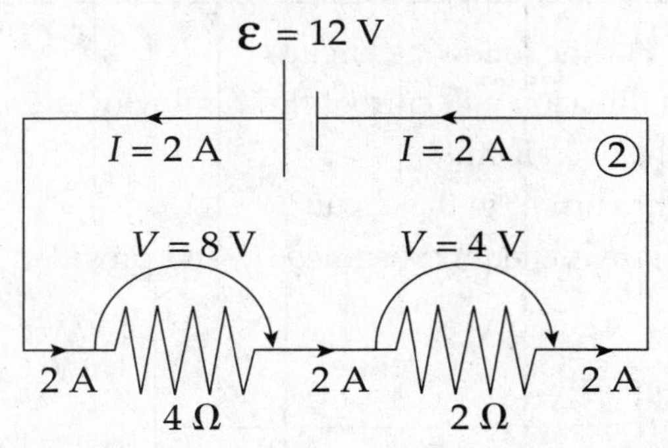

Since we know the current through each resistor, we can figure out the voltage drop across each resistor using the equation $V = IR$. The voltage drop across the 4 Ω resistor is (2 A)(4 Ω) = 8 V, and the voltage drop across the 2 Ω resistor is (2 A)(2 Ω) = 4 V. Notice that the total voltage drop across the two resistors is 8 V + 4 V = 12 V, which matches the emf of the battery, which is to be expected.

Now for the last step; going from diagram ② back to diagram ①. Nothing needs to be done with the 4 Ω resistor; nothing about it changes in going from diagram ② to ①, but the 2 Ω resistor in diagram ② goes back to the parallel combination. And what do resistors in parallel share? The same voltage drop. So we take the voltage drop, $V = 4$ V, back to diagram ①. The voltage drop across each of the two parallel resistors in diagram ① is 4 V.

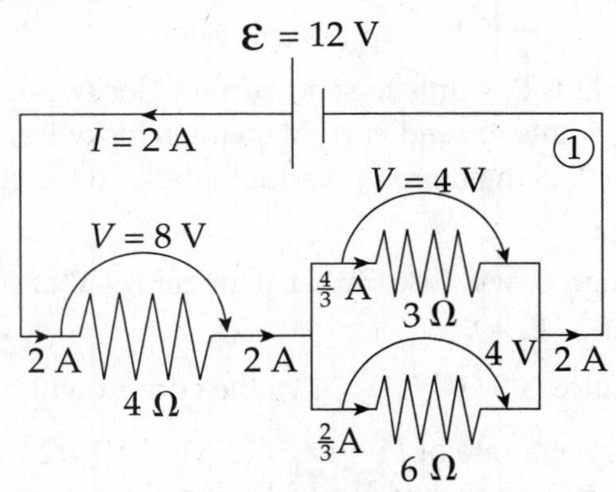

Since we know the voltage drop across each resistor, we can figure out the current through each resistor by using the equation $I = V/R$. The current through the 3 Ω resistor is (4 V)/(3 Ω) = $\frac{4}{3}$ A, and the current through the 6 Ω resistor is (4 V)/(6 Ω) = $\frac{2}{3}$ A. Note that the current entering the parallel combination (2 A) equals the total current passing through the individual resistors $\left(\frac{4}{3}\text{ A} + \frac{2}{3}\text{ A}\right)$. Again this was expected.

Finally, we will calculate the power dissipated as heat by each resistor. We can use any of the equivalent formulas: $P = IV$, $P = I^2R$, or $P = V^2/R$.

$$\text{For the 4 Ω resistor: } P = IV = (2\text{ A})(8\text{ V}) = 16\text{ W}$$

$$\text{For the 3 Ω resistor: } P = IV = (\tfrac{4}{3}\text{ A})(4\text{ V}) = \tfrac{16}{3}\text{ W}$$

$$\text{For the 6 Ω resistor: } P = IV = (\tfrac{2}{3}\text{ A})(4\text{ V}) = \tfrac{8}{3}\text{ W}$$

So, the resistors are dissipating a total of

$$16\text{ W} + \tfrac{16}{3}\text{ W} + \tfrac{8}{3}\text{ W} = 24\text{ W}$$

If the resistors are dissipating a total of 24 J every second, then they must be provided with that much power. This is easy to check: $P = IV = (2\text{ A})(12\text{ V}) = 24$ W.

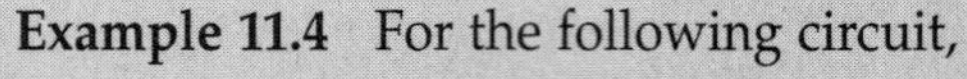

Example 11.4 For the following circuit,

(a) In which direction will current flow and why?

(b) What's the overall emf?

(c) What's the current in the circuit?

(d) At what rate is energy consumed by, and provided to, this circuit?

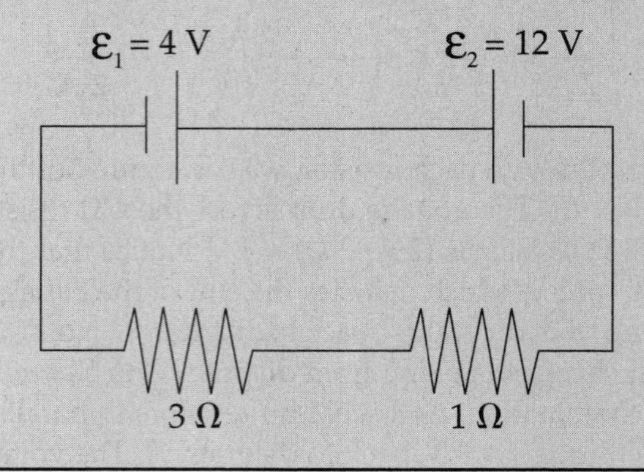

Solution.

(a) The battery whose emf is ε_1 wants to send current clockwise, while the battery whose emf is ε_2 wants to send current counterclockwise. Since $\varepsilon_2 > \varepsilon_1$, the battery whose emf is ε_2 is the more powerful battery, so the current will flow counterclockwise.

(b) Charges forced through ε_1 will lose, rather than gain, 4 V of potential, so the overall emf of this circuit is $\varepsilon_2 - \varepsilon_1 = 8$ V.

(c) Since the total resistance is 3 Ω + 1 Ω = 4 Ω, the current will be $I = (8 \text{ V})/(4\ \Omega) = 2$ A.

(d) ε_2 will provide energy at a rate of $P_2 = IV_2 = (2 \text{ A})(12 \text{ V}) = 24$ W, while ε_1 will absorb at a rate of $P_1 = IV_1 = (2 \text{ A})(4 \text{ V}) = 8$ W. Finally, energy will be dissipated in these resistors at a rate of $I^2R_1 + I^2R_2 = (2 \text{ A})^2(3\ \Omega) + (2 \text{ A})^2(1\ \Omega) = 16$ W. Once again, energy is conserved; the power delivered (24 W) equals the power taken (8 W + 16 W = 24 W).

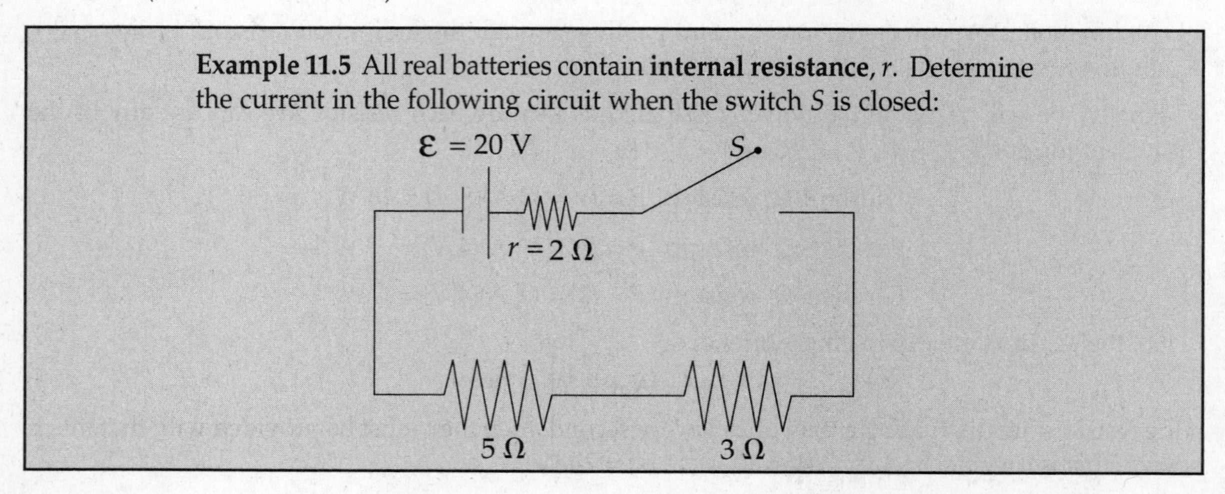

Example 11.5 All real batteries contain **internal resistance**, *r*. Determine the current in the following circuit when the switch *S* is closed:

Solution. Before the switch is closed, there is no complete conducting pathway from the positive terminal of the battery to the negative terminal, so no current flows through the resistors. However, once the switch is closed, the resistance of the circuit is 2 Ω + 3 Ω + 5 Ω = 10 Ω, so the current in the circuit is I = (20 V)/(10 Ω) = 2 A. Often the battery and its internal resistance are enclosed in a dashed box:

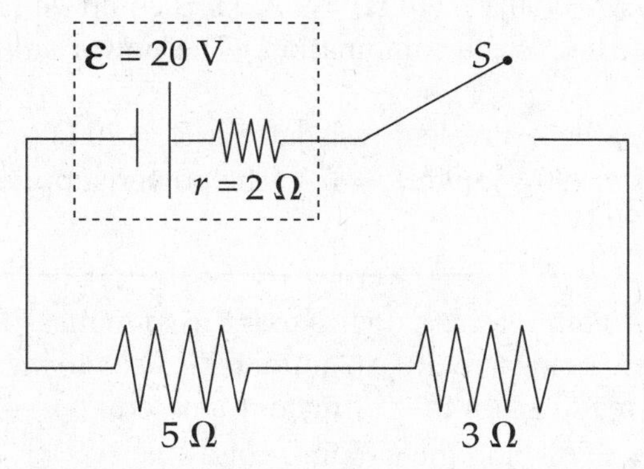

In this case, a distinction can be made between the emf of the battery and the actual voltage it provides once the current has begun. Since I = 2 A, the voltage drop across the internal resistance is Ir = (2 A)(2 Ω) = 4 V, so the effective voltage provided by the battery to the rest of the circuit—called the **terminal voltage**—is lower than the ideal emf. It is $V = \mathcal{E} - Ir = 20n - 4$ V = 16 V.

> **Example 11.6** A student has three 30 Ω resistors and an ideal 90 V battery. (A battery is *ideal* if it has a negligible internal resistance.) Compare the current drawn from—and the power supplied by—the battery when the resistors are arranged in parallel versus in series.

Solution. Resistors in series always provide an equivalent resistance that's greater than any of the individual resistances, and resistors in parallel always provide an equivalent resistance that's smaller than their individual resistances. So, hooking up the resistors in parallel will create the smallest resistance and draw the greatest total current:

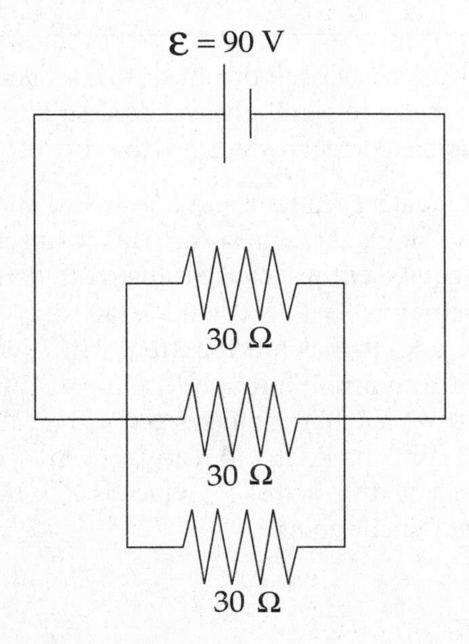

In this case, the equivalent resistance is

$$\frac{1}{R_P}=\frac{1}{30\ \Omega}+\frac{1}{30\ \Omega}+\frac{1}{30\ \Omega}\ \Rightarrow\ \frac{1}{R_P}=\frac{1}{10\ \Omega}\ \Rightarrow\ R_P=10\ \Omega$$

and the total current is $I = \mathcal{E}/R_P = (90\ \text{V})/(10\ \Omega) = 9$ A. (You could verify that 3 A of current would flow in each of the three branches of the combination.) The power supplied by the battery will be $P = IV = (9\ \text{A})(90\ \text{V}) = 810$ W.

If the resistors are in series, the equivalent resistance is $R_S = 30\ \Omega + 30\ \Omega + 30\ \Omega = 90\ \Omega$, and the current drawn is only $I = \mathcal{E}/R_S = (90\ \text{V})/(90\ \Omega) = 1$ A. The power supplied by the battery in this case is just $P = IV = (1\ \text{A})(90\ \text{V}) = 90$ W.

Example 11.7 A **voltmeter** is a device that's used to measure the voltage between two points in a circuit. An **ammeter** is used to measure current. Determine the readings on the voltmeter (denoted —Ⓥ—) and the ammeter (denoted —Ⓐ—) in the circuit below.

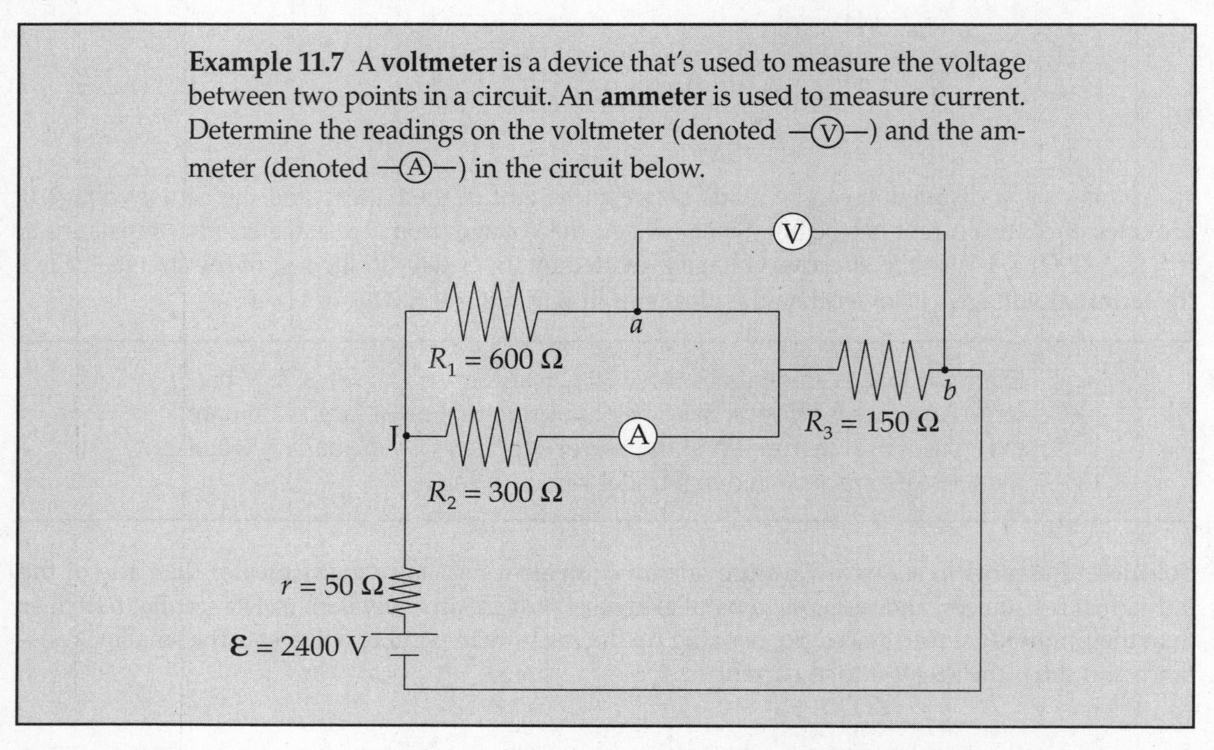

Solution. We consider the ammeter to be ideal; this means it has negligible resistance, so it doesn't alter the current that it's trying to measure. Similarly, we consider the voltmeter to have an extremely high resistance, so it draws negligible current away from the circuit.

Our first goal is to find the equivalent resistance in the circuit. The 600 Ω and 300 Ω resistors are in parallel; they're equivalent to a single 200 Ω resistor. This is in series with the battery's internal resistance, r, and R_3. The overall equivalent resistance is therefore $R = 50\ \Omega + 200\ \Omega + 150\ \Omega = 400\ \Omega$, so the current supplied by the battery is $I = \mathcal{E}/R\ (2400\ \text{V})/(400\ \Omega) = 6$ A. At the junction marked J, this current splits. Since R_1 is twice R_2, half as much current will flow through R_1 as through R_2; the current through R_1 is $I_1 = 2$ A, and the current through R_2 is $I_2 = 4$ A. The voltage drop across each of these resistors is $I_1R_1 = I_2R_2 = 1200$ V (matching voltages verify the values of currents I_1 and I_2). Since the ammeter is in the branch that contains R_2, it will read $I_2 = 4$ A.

The voltmeter will read the voltage drop across R_3, which is $V_3 = IR_3 = (6\ \text{A})(150\ \Omega) = 900$ V. So the potential at point b is 900 V lower than at point a.

Example 11.8 The diagram below shows a point a at potential $V = 20$ V connected by a combination of resistors to a point (denoted G) that is **grounded**. *The* ***ground*** *is considered to be at potential zero.* If the potential at point a is maintained at 20 V, what is the current through R_3?

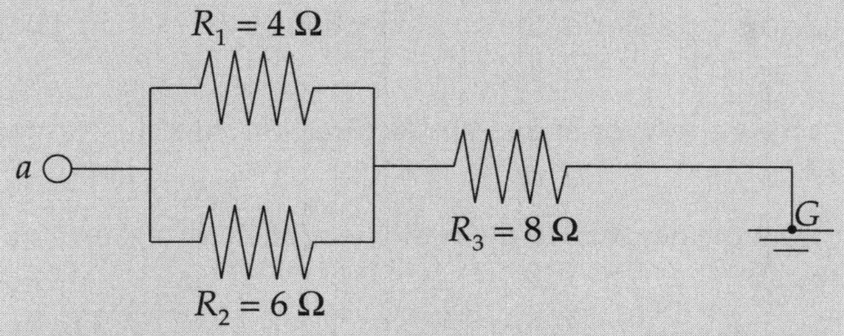

$R_1 = 4\ \Omega$

a

$R_3 = 8\ \Omega$

G

$R_2 = 6\ \Omega$

Solution. R_1 and R_2 are in parallel; their equivalent resistance is R_P, where

$$\frac{1}{R_P} = \frac{1}{4\ \Omega} + \frac{1}{6\ \Omega} \quad \Rightarrow \quad R_P = \frac{24}{10}\ \Omega = 2.4\ \Omega$$

R_P is in series with R_3, so the equivalent resistance is:

$$R = R_P + R_3 = (2.4\Omega) + (8\Omega) = 10.4\Omega$$

and the current that flow through R_3 is

$$I_3 = \frac{V}{R} = \frac{20\ \text{V}}{10.4\ \Omega} = 1.9\ \text{A}$$

Kirchhoff's Rules

When the resistors in a circuit cannot be classified as either in series or in parallel, we need another method for analyzing the circuit. The rules of Gustav Kirchhoff (pronounced "Keer koff") can be applied to any circuit:

The Loop Rule. The sum of the potential differences (positive and negative) that traverse any closed loop in a circuit must be zero.

The Junction Rule. The total current that enters a junction must equal the total current that leaves the junction. (This is also known as the **Node Rule**.)

The Loop Rule just says that, starting at any point, by the time we get back to that same point by following any closed loop, we have to be back to the same potential. Therefore, the total drop in potential must equal the total rise in potential. Put another way, the Loop Rule says that all the decreases in electrical potential energy (for example, caused by resistors in the direction of the current) must be balanced by all the increases in electrical potential energy (for example, caused by a source of emf from the negative to positive terminal). So the Loop Rule is basically a statement of the Law of Conservation of Energy.

Similarly, the Junction Rule simply says that the charge (per unit time) that goes into a junction must equal the charge (per unit time) that comes out. This is basically a re-statement of the Law of Conservation of Charge.

In practice, the Junction Rule is straightforward to apply. The most important things to remember about the Loop Rule can be summarized as follows:

- When going across a resistor in the *same* direction as the current, the potential *drops* by *IR*.
- When going across a resistor in the *opposite* direction from the current, the potential *increases* by *IR*.
- When going from the negative to the positive terminal of a source of emf, the potential *increases* by $\mathcal{E}$.
- When going from the positive to the negative terminal of a source of emf, the potentil *decreases* by $\mathcal{E}$.

Example 11.9 Use Kirchhoff's Rules to determine the current through R_2 in the following circuit:

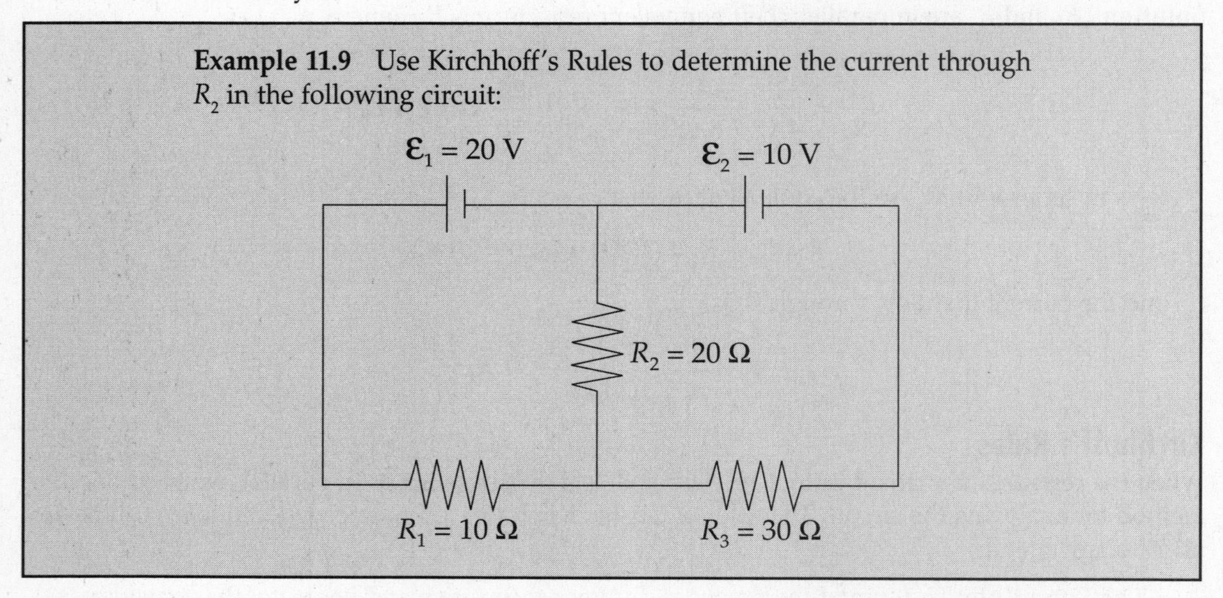

Solution. First let's label some points in the circuit.

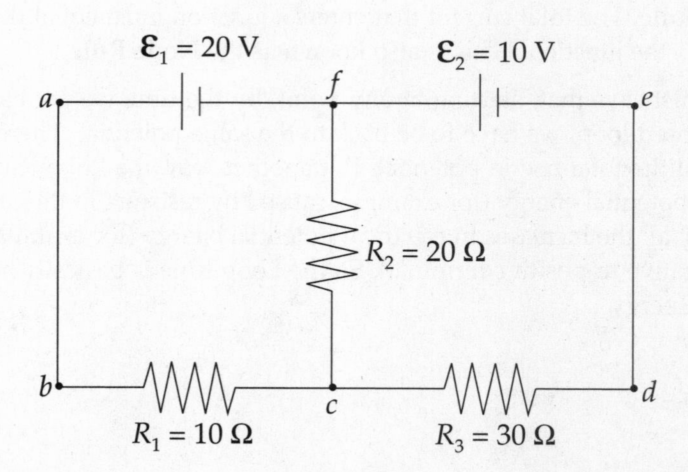

The points c and f are junctions (nodes). We have two nodes and three branches: one branch is *fabc*, another branch is *cdef*, and the third branch is *cf*. Each branch has one current throughout. If we label the current in *fabc* I_1 and the current in branch *cdef* I_2 (with the directions as shown in the diagram below), then the current in branch *cf* must be $I_1 - I_2$, by the Junction Rule: I_1 comes into c, and a total of $I_2 + (I_1 - I) = I_1$ comes out.

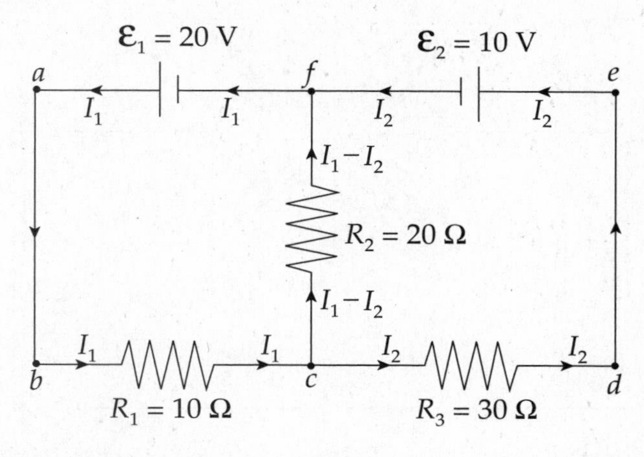

Now pick a loop; say, *abcfa*. Starting at a, we go to b, then across R_1 in the direction of the current, so the potential drops by I_1R_1. Then we move to c, then up through R_2 in the direction of the current, so the potential drops by $(I_1 - I_2)R_2$. Then we reach f, turn left and travel through ε_1 from the negative to the positive terminal, so the potential increases by ε_1. We now find ourselves back at a. By the Loop Rule, the total change in potential around this closed loop must be zero, and

$$-I_1R_1 - (I_1 - I_2)R_2 + \varepsilon_1 = 0 \quad (1)$$

Since we have two unknowns (I_1 and I_2), we need two equations, so now pick another loop; let's choose *cdefc*. From c to d, we travel across the resistor in the direction of the current, so the potential drops by I_2R_3. From e to f, we travel through ε_2 from the positive to the negative terminal, so the potential *drops* by ε_2. Heading down from f to c, we travel across R_2 but in the direction opposite to the current, so the potential *increases* by $(I_1 - I_2)R_2$. At c, our loop is completed, so

$$-I_2R_3 - \varepsilon_2 + (I_1 - I_2)R_2 = 0 \quad (2)$$

Substituting in the given numerical values for R_1, R_2, R_3, ε_1, and ε_2, and simplifying, these two equations become

$$3I_1 - 2I_2 = 2 \quad (1')$$

$$2I_1 - 5I_2 = 1 \quad (2')$$

Solving this pair of simultaneous equations, we get

$$I_1 = \tfrac{8}{11}\text{ A} = 0.73\text{ A} \quad \text{and} \quad I_2 = \tfrac{1}{11}\text{ A} = 0.09\text{ A}$$

So the current through R_2 is $I_1 - I_2 = \tfrac{7}{11}$ A = 0.64 A.

The choice of directions of the currents at the beginning of the solution was arbitrary. Don't worry about trying to guess the actual direction of the current in a particular branch. Just pick a direction, stick with it, and obey the Junction Rule. At the end, when you solve for the values of the branch current, a negative value will alert you that the direction of the current is actually opposite to the direction you originally chose for it in your diagram.

Example 11.10 In the circuit diagram below, Resistors ① and ③ have fixed resistances (R_1 and R_3, respectively), which are known. Resistor ② is a **variable** (or **adjustable**) resistor, and the resistance of Resistor ④ is unknown.

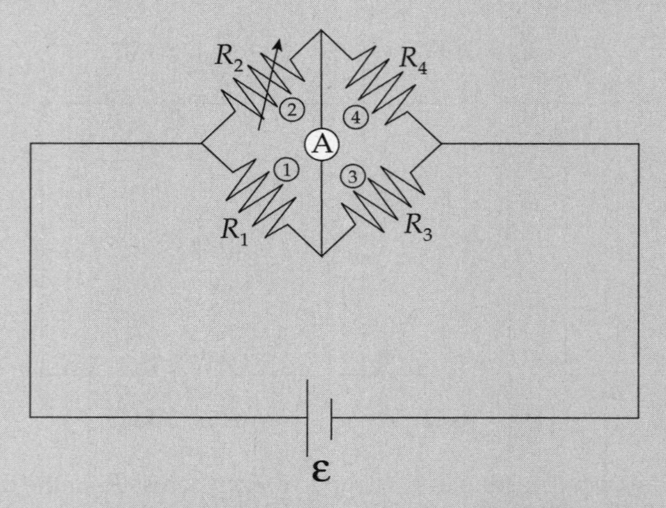

Show that if Resistor ② is adjusted until the ammeter shown registers no current through its branch, then

$$R_1 R_4 = R_2 R_3$$

(This circuit arrangement is known as a **Wheatstone bridge**.)

Solution. If no current flows through the ammeter, then Resistors ① and ③ are in series, and so are Resistors ② and ④. Therefore, the current flowing through ① and ③ is equal to $I_{1\text{-}3} = \mathcal{E}/(R_1 + R_3)$. Similarly, the current flowing through ② and ④ is $I_{2\text{-}4} = \mathcal{E}/(R_2 + R_4)$. Now consider the three points marked *a*, *b*, and *c* below:

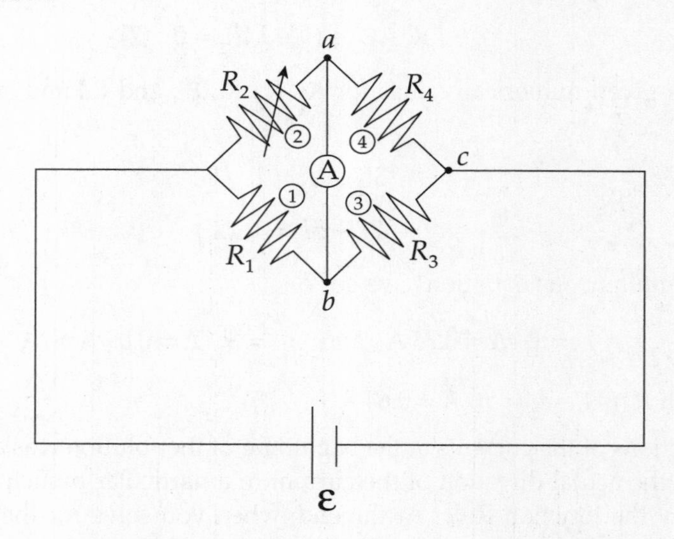

Since no current flows between Points *a* and *b*, there must be no potential difference between *a* and *b*; they're at the same potential. So the voltage drop from *a* to *c* must equal the voltage drop from *b* to *c*. The voltage drop from *a* to *c* is

$$V_{ac} = I_{2\text{-}4}R_4 = \frac{\varepsilon}{R_2 + R_4}R_4$$

and the voltage drop from *b* to *c* is

$$V_{bc} = I_{1\text{-}3}R_3 = \frac{\varepsilon}{R_1 + R_3}R_3$$

Setting these equal to each other, we get

$$\begin{aligned}
\frac{\varepsilon}{R_2 + R_4}R_4 &= \frac{\varepsilon}{R_1 + R_3}R_3 \\
\frac{R_4}{R_2 + R_4} &= \frac{R_3}{R_1 + R_3} \\
R_4(R_1 + R_3) &= R_3(R_2 + R_4) \\
R_1R_4 + R_3R_4 &= R_2R_3 + R_3R_4 \\
R_1R_4 &= R_2R_3
\end{aligned}$$

By knowing R_1, R_2, and R_3, we can solve for the unknown resistance: $R_4 = R_2R_3/R_1$.

RESISTANCE–CAPACITANCE (RC) CIRCUITS

Capacitors are typically charged by batteries. Once the switch in the diagram on the left is closed, electrons are attracted to the positive terminal of the battery and leave the top plate of the capacitor. Electrons also accumulate on the bottom plate of the capacitor, and this continues until the voltage across the capacitor plates matches the emf of the battery. When this condition is reached, the current stops and the capacitor is fully charged.

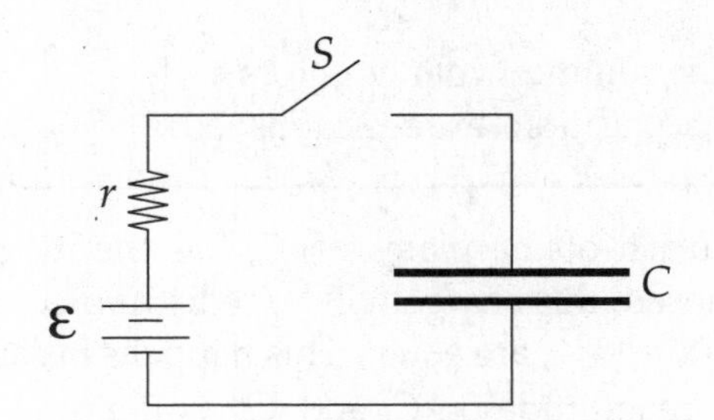

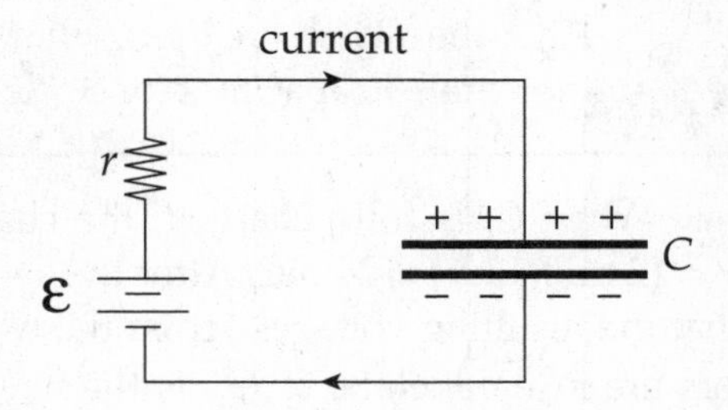

Example 11.11 Find the charge stored and the voltage across each capacitor in the following circuit, given that $\mathcal{E}$ = 180 V, C_1 = 30 μF, C_2 = 60 μF, and C_3 = 90 μF.

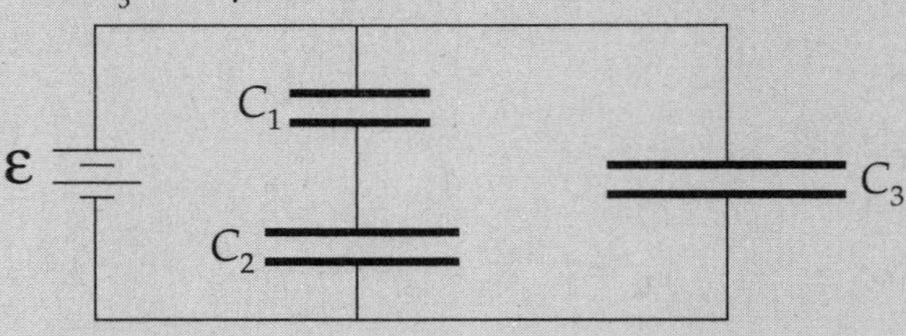

Solution. Once the charging currents stop, the voltage across C_3 is equal to the voltage across the battery, so V_3 = 180 V. This gives us $Q_3 = C_3V_3$ = (90 μF)(180 V) = 16.2 mC. Since C_1 and C_2 are in series, they must store identical amounts of charge, and, from the diagram, the sum of their voltages must equal the voltage of the battery. So, if we let Q be the charge on each of these two capacitors, then $Q = C_1V_1 = C_2V_2$, and $V_1 + V_2$ = 180 V. The equation $C_1V_1 = C_2V_2$ becomes (30 μF)V_1 = (60 μF)V_2, so $V_1 = 2V_2$. Substituting this into $V_1 + V_2$ = 180 V gives us V_1 = 120 V and V_2 = 60 V. The charge stored on each of these capacitors is

$$(30\ \mu\text{F})(120\ \text{V}) = C_1V_1 = C_2V_2 = (60\ \mu\text{F})(60\ \text{V}) = 3.6\ \text{mC}$$

Example 11.12 In the diagram below, C_1 = 2 mF and C_2 = 4 mF. When Switch S is open, a battery (which is not shown) is connected between points a and b and charges capacitor C_1 so that V_{ab} = 12 V. The battery is then disconnected.

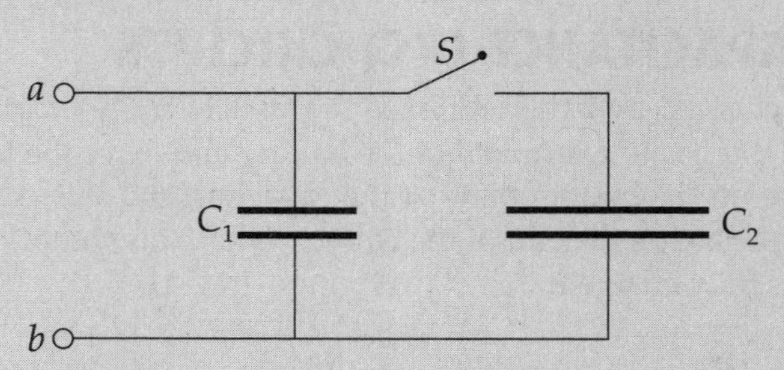

After the switch is closed, what will be the common voltage across each of the parallel capacitors (once electrostatic conditions are reestablished)?

Solution. When C_1 is fully charged, the charge on (each of the plates of) C_1 has the magnitude $Q = C_1V$ = (2 mF)(12 V) = 24 mC. After the switch is closed, this charge will be redistributed in such a way that the resulting voltages across the two capacitors, V', are equal. This happens because the capacitors are in parallel. So if Q_1' is the new charge magnitude on C_1 and Q_2' is the new charge magnitude on C_2, we have $Q_1' + Q_2' = Q$, so $C_1V' + C_2V' = Q$, which gives us:

$$V' = \frac{Q}{C_1 + C_2} = \frac{24\ \text{mC}}{2\ \text{mF} + 4\ \text{mF}} = 4\ \text{V}$$

CHARGING OR DISCHARGING A CAPACITOR IN AN RC CIRCUIT

When a battery is hooked up to an uncharged capacitor and current begins to flow, the current is not constant. So far, we have assumed that the currents in our circuits have been constant, but RC circuits are different.

Charging a Capacitor

Consider the RC circuit shown below, where the charge on the capacitor is zero:

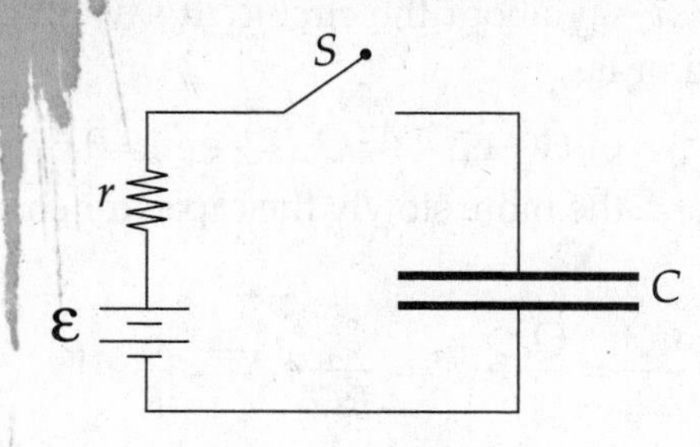

When the switch is closed (time $t = 0$), there is no charge on the capacitor, which means that there's no voltage across the capacitor, so Kirchhoff's Loop Rule gives us $\varepsilon - IR = 0$, and the initial current in the circuit is $I = \varepsilon/R$. But, as time passes, charge begins to accumulate on the capacitor, and a voltage is created that opposes the emf of the battery. At any time t after the switch is closed, Kirchhoff's Loop Rule gives us:

$$\varepsilon - I(t)R - V(t) = 0$$

where $I(t)$ is the current at time t and $V(t)$ is the voltage across the capacitor at time t. Since $V(t) = Q(t)/C$ and $I(t) = dQ(t)/dt$, the equation above becomes

$$\varepsilon - R\frac{dQ}{dt} - \frac{Q}{C} = 0 \quad \Rightarrow \quad \frac{dQ}{dt} = \frac{C\varepsilon - Q}{RC}$$

We will solve for $Q(t)$ using separation of variables.

$$\frac{dQ}{C\varepsilon - Q} = \frac{dt}{RC}$$

Now integrate both sides to get

$$\int_0^q \frac{dQ}{C\varepsilon - Q} = \int_0^t \frac{dt}{RC}$$

$$-\left[\ln(C\varepsilon - q) - \ln(C\varepsilon)\right] = \frac{t}{RC}$$

$$\left[\ln(\frac{C\varepsilon - q}{C\varepsilon})\right] = -\frac{t}{RC}$$

$$\frac{C\varepsilon - q}{C\varepsilon} = e^{-\frac{t}{RC}}$$

$$C\varepsilon - q = C\varepsilon e^{-\frac{t}{RC}}$$

$$q(t) = C\varepsilon(1 - e^{-\frac{t}{RC}})$$

This gives us an equation, $q(t)$, for how the charge on the capacitor changes over time.

The product $C\mathcal{E}$ is equal to the final charge on the capacitor (remember that $Q = CV$). The quantity RC that appears in the exponent is called the **time constant** for the circuit and is represented by τ:

$$\tau = RC$$

Therefore, letting Q_f replace the quantity $C\mathcal{E}$, the equation above becomes

$$Q(t) = Q_f(1 - e^{-t/\tau}) \quad (*)$$

What does the time constant, τ, say about the circuit? It says that, after a time interval of τ has elapsed, the charge on the capacitor is

$$Q(\tau) = Q_f(1 - e^{-\tau/\tau}) = Q_f(1 - e^{-1}) \approx 0.63Q_f$$

Therefore, the greater the value of τ, the more slowly the capacitor charges. By differentiating (*) with respect to t,

$$\frac{dQ}{dt} = \frac{Q_f}{\tau}e^{-t/\tau} = \frac{Q_f}{RC}e^{-t/\tau} = \frac{\mathcal{E}}{R}e^{-t/\tau}$$

we can get the current in the circuit as a function of time:

$$I(t) = \frac{\mathcal{E}}{R}e^{-t/\tau}$$

Therefore, the charge builds up gradually on the capacitor,

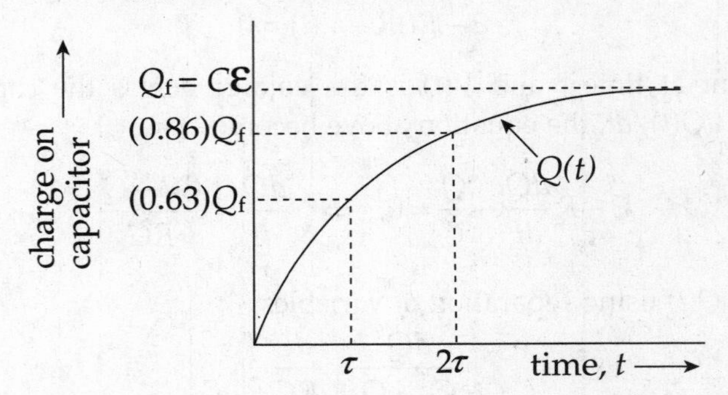

as the current in the circuit drops gradually to zero:

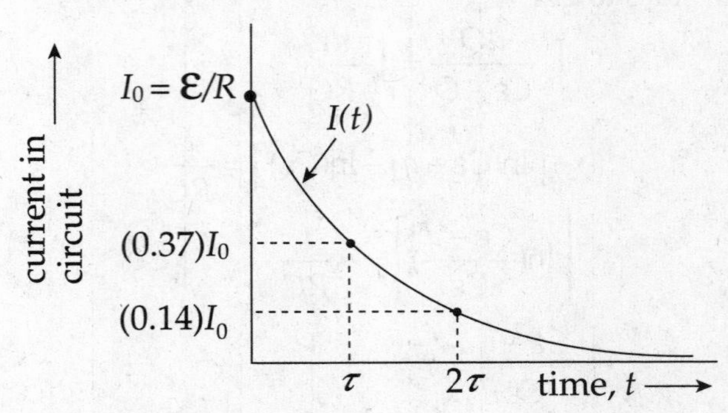

Discharging a Capacitor

Consider the RC circuit shown below, where the charge on the capacitor is Q_0:

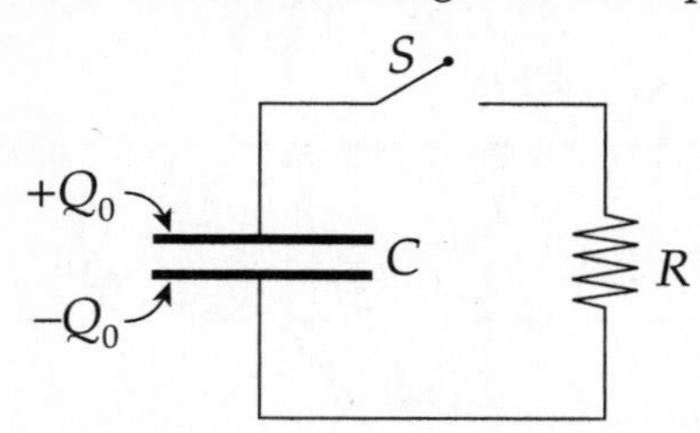

When the switch is closed (time $t = 0$), the voltage across the capacitor is $V_0 = Q_0/C$, so $V_0 - IR = 0$, and the initial current in the circuit is $I_0 = V_0/R$. But as time passes, charge leaks off the capacitor and its voltage decreases, which causes the current to decrease.

The equation for $Q(t)$ is

$$Q(t) = Q_0 e^{-t/RC} = Q_0 e^{-t/\tau}$$

The voltage across the plates as a function of time is

$$V(t) = \frac{Q(t)}{C} = \frac{Q_0}{C} e^{-t/\tau}$$

and, by using the equation $I(t) = -dQ/dt$, the current in the circuit is

$$I(t) = \frac{Q_0}{\tau} e^{-t/\tau} = \frac{Q_0}{RC} e^{-t/\tau} = \frac{V_0}{R} e^{-t/\tau}$$

Example 11.13 In the circuit below, $\mathcal{E} = 20$ V, $R = 1000\Omega$, and $C = 2$ mF. If the capacitor is initially uncharged, how long will it take (after the switch S is closed) for the capacitor to be 99% charged?

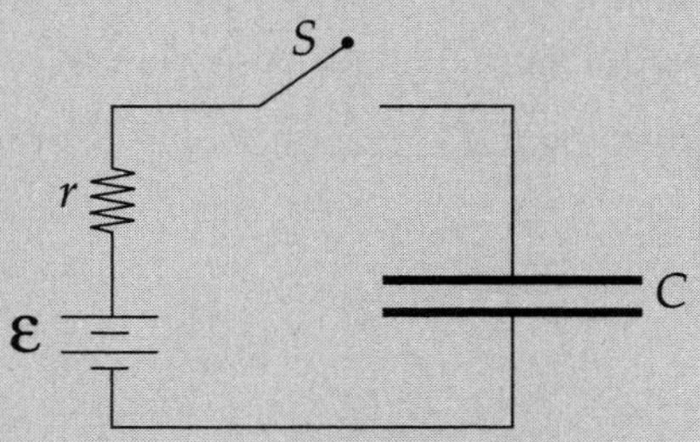

Solution. The equation $Q(t) = Q_f(1 - e^{-t/\tau})$, where $\tau = RC$, gives the charge on the capacitor as a function of time. If we want $Q(t)$ to equal $0.99Q_f$, then we have to wait until time t, when $e^{-t/\tau} = 0.01$. Solving this equation for t, we get

$$\begin{aligned} e^{-t/\tau} &= 0.01 \\ -\frac{t}{\tau} &= \ln(0.01) \\ t &= -\ln(0.01)\tau \\ &= (4.6)\tau \end{aligned}$$

Since the time constant is

$$\tau = RC = (1000\ \Omega)(2\ \text{mF}) = 2\ \text{s}$$

the time required is $(4.6)\tau = (4.6)(2\ \text{s}) = 9.2\ \text{s}$.

Example 11.14

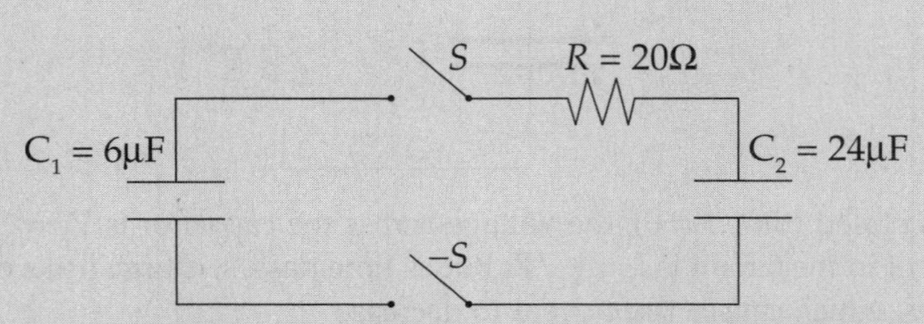

Capacitors 1 and 2, of capacitance $C_1 = 6\ \mu\text{F}$ and $C_2 = 24\ \mu\text{F}$, respectively, are connected in a circuit as shown above with a resistor of resistance $R = 20\ \Omega$ and two switches. Capacitor 1 is initially charged to a voltage $V_o = 30$ V, and capacitor 2 is initially uncharged. Both of the switches S are then closed simultaneously.

(a) What is the initial current in the circuit?

(b) What are the final charges on each of the capacitors 1 and 2 after equilibrium has been reached?

Solution.

(a) Use Ohm's Law to solve for the initial current. The voltage will decrease with time, but the initial voltage is 30 V.

$$V = IR \Rightarrow 30 = I(20) \Rightarrow I = 1.5\ \text{A}$$

(b) The initial charge stored on Capacitor 1 is $Q = CV = (6)(30) = 180\ \mu\text{C}$. This charge will redistribute until equilibrium is reached and both capacitors have the same potential difference across them. Rearrange $C = \frac{Q}{V}$, to solve for the potential difference across each capacitor: $V = \frac{Q}{C}$.

$$V = \frac{Q_1}{C_1} = \frac{Q_2}{C_2} \Rightarrow \frac{Q_1}{6} = \frac{180 - Q_1}{24}$$

$$Q_1 = 36\mu\text{C and } Q_2 = 144\mu\text{C}$$

CHAPTER 11 REVIEW QUESTIONS

SECTION I: MULTIPLE CHOICE

1. A wire made of brass and a wire made of silver have the same length, but the diameter of the brass wire is 4 times the diameter of the silver wire. The resistivity of brass is 5 times greater than the resistivity of silver. If R_B denotes the resistance of the brass wire and R_S denotes the resistance of the silver wire, which of the following is true?

   (A) $R_B = \frac{5}{16} R_S$
   (B) $R_B = \frac{4}{5} R_S$
   (C) $R_B = \frac{5}{4} R_S$
   (D) $R_B = \frac{5}{2} R_S$
   (E) $R_B = \frac{16}{5} R_S$

2. For an ohmic conductor, doubling the voltage without changing the resistance will cause the current to

   (A) decrease by a factor of 4
   (B) decrease by a factor of 2
   (C) remain unchanged
   (D) increase by a factor of 2
   (E) increase by a factor of 4

3. If a 60-watt light bulb operates at a voltage of 120 V, what is the resistance of the bulb?

   (A) 2Ω
   (B) 30Ω
   (C) 240Ω
   (D) 720Ω
   (E) 7200Ω

4. A battery whose emf is 40 V has an internal resistance of 5Ω. If this battery is connected to a 15 Ω resistor R, what will the voltage drop across R be?

   (A) 10 V
   (B) 30 V
   (C) 40 V
   (D) 50 V
   (E) 70 V

5. 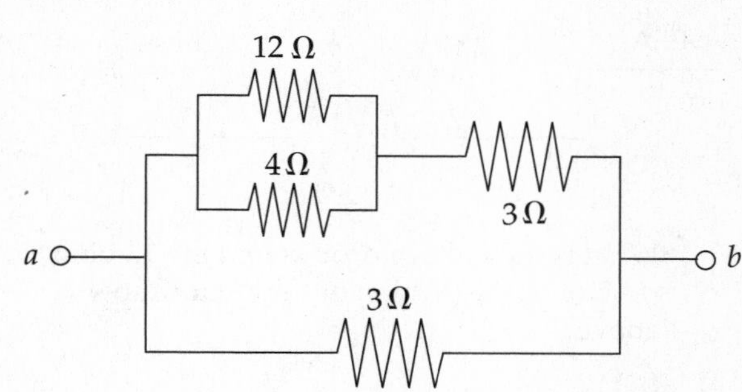

   Determine the equivalent resistance between points a and b.

   (A) 0.167Ω
   (B) 0.25Ω
   (C) 0.333Ω
   (D) 1.5Ω
   (E) 2Ω

6. 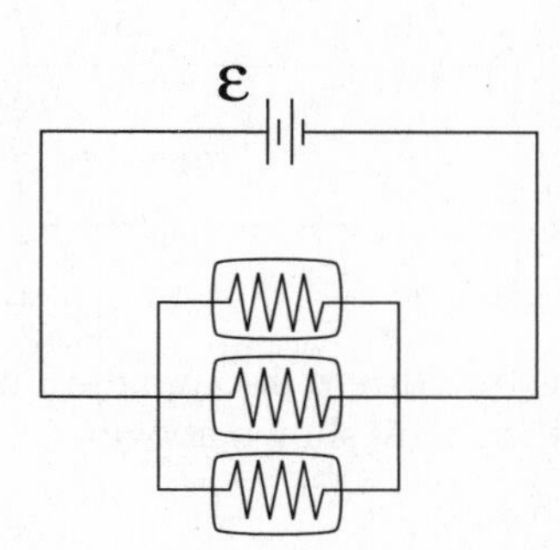

   Three identical light bulbs are connected to a source of emf, as shown in the diagram above. What will happen if the middle bulb burns out?

   (A) All the bulbs will go out.
   (B) The light intensity of the other two bulbs will decrease (but they won't go out).
   (C) The light intensity of the other two bulbs will increase.
   (D) The light intensity of the other two bulbs will remain the same.
   (E) More current will be drawn from the source of emf.

7.

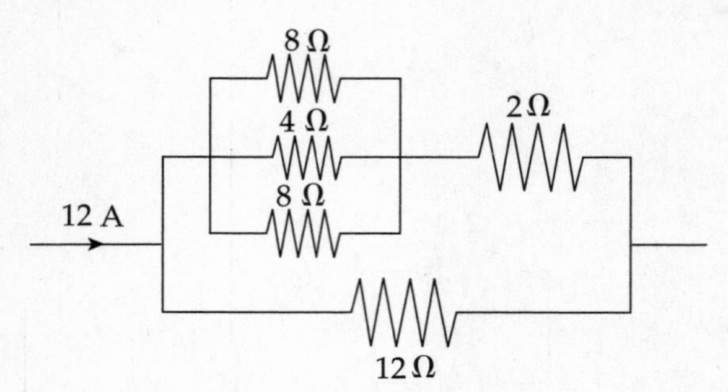

What is the voltage drop across the 12-ohm resistor in the portion of the circuit shown above?

(A) 24 V
(B) 36 V
(C) 48 V
(D 72 V
(E) 144 V

8.

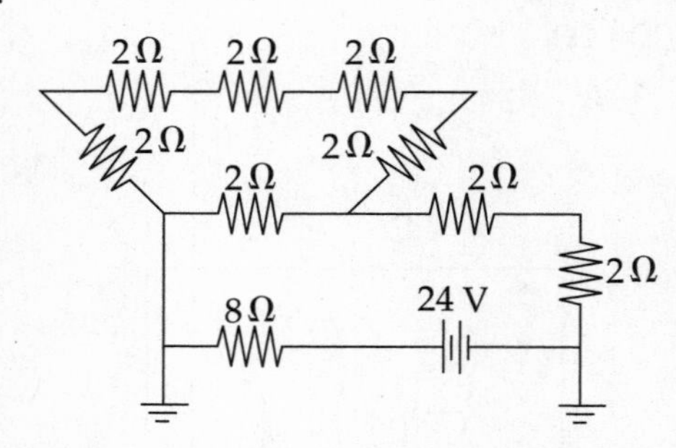

What is the current through the 8-ohm resistor in the circuit shown above?

(A) 0.5 A
(B) 1.0 A
(C) 1.25 A
(D) 1.5 A
(E) 3.0 A

9. How much energy is dissipated as heat in 20 s by a 100 Ω resistor that carries a current of 0.5 A?

(A) 50 J
(B) 100 J
(C) 250 J
(D) 500 J
(E) 1000 J

10.

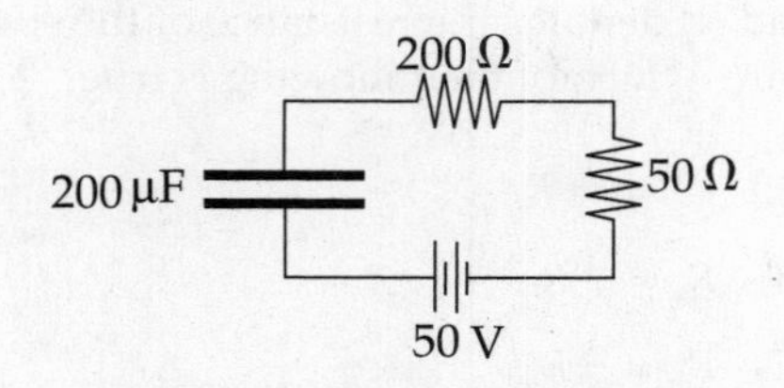

What is the time constant for the circuit above?

(A) 0.01 s
(B) 0.025 s
(C) 0.04 s
(D) 0.05 s
(E) 0.1 s

SECTION II: FREE RESPONSE

1. Consider the following circuit:

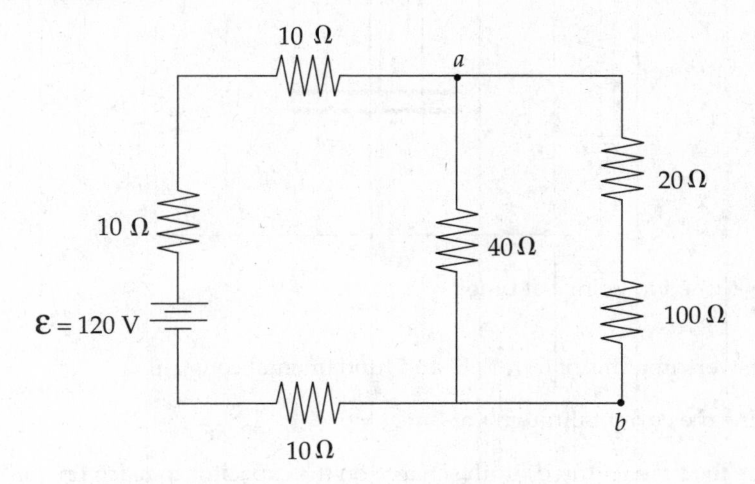

(a) At what rate does the battery deliver energy to the circuit?

(b) Find the current through the 20 Ω resistor.

(c) (i) Determine the potential difference between points *a* and *b*.

(ii) At which of these two points is the potential higher?

(d) Find the energy dissipated by the 100 Ω resistor in 10 s.

(e) Given that the 100 Ω resistor is a solid cylinder that's 4 cm long, composed of a material whose resistivity is 0.45 Ω·m, determine its radius.

2. The diagram below shows an uncharged capacitor, two resistors, and a battery whose emf is $\mathcal{E}$.

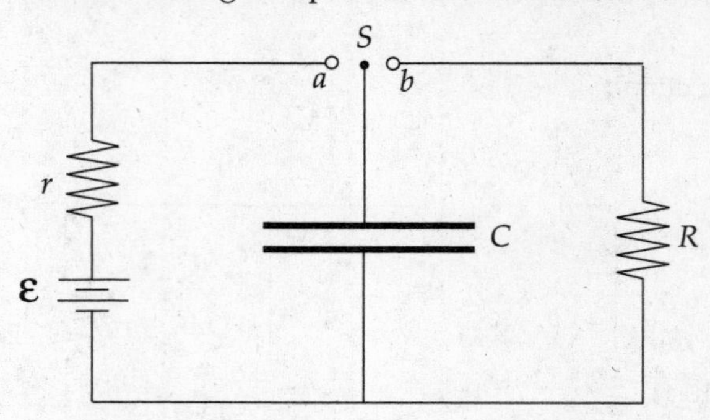

The switch S is turned to point a at time $t = 0$.

(Express all answers in terms of C, r, R, $\mathcal{E}$, and fundamental constants.)

(a) Determine the current through r at time $t = 0$.

(b) Compute the time required for the charge on the capacitor to reach one-half its final value.

(c) When the capacitor is fully charged, which plate is positively charged?

(d) Determine the electrical potential energy stored in the capacitor when the current through r is zero.

When the current through r is zero, the switch S is then moved to Point b; for the following parts, consider this event time $t = 0$.

(e) Determine the current through R as a function of time.

(f) Find the power dissipated in R as a function of time.

(g) Determine the total amount of energy dissipated as heat by R.

DIRECT CURRENT CIRCUITS SUMMARY

General Information

- **Current** is the flow rate of charge: $I = \frac{dQ}{dt}$. The units are amps (A).
- The resistance of length of material depends on the resistivity, length and cross-sectional area. $R = \frac{\rho L}{A}$. The units of resistance are ohms (Ω).
- Many electrical devices obey **Ohm's Law**: $\Delta V = IR$
- **Power** is the rate that electrical energy is transferred. $P = IV = I^2R = \frac{V^2}{R}$
- **Capacitance** is given by: $C = \frac{Q}{\Delta V}$

Circuits with Resistors

- The equivalent resistance of resistors added in series is the sum of the resistance of each device: $R_{eq} = R_1 + R_2 + \ldots = \sum_i R_i$
- The reciprocal of the equivalent resistance of resistors added in parallel is the sum of the reciprocals of each resistor: $\frac{1}{R_{eq}} = \frac{1}{R_1} + \frac{1}{R_2} + \ldots = \sum_i \frac{1}{R_i}$

Kirchhoff's Rules:

1. **The Loop Rule**. The sum of the potential differences around a closed loop in a circuit must be zero. $\sum_{loop} V = 0$
2. **The Junction (Node) Rule**. The total current that enters a junction must equal the total current that leaves the junction.

Circuits with Resistors and Capacitors (RC Circuits)

- When a battery is connected to a resistor and uncharged capacitors in a circuit, the charge builds up on the capacitor exponentially. This is due to the fact that as more charge is added to the capacitor it becomes more difficult for additional charge to be added. The equation for the charge build up is $q(t) = C\varepsilon(1 - e^{-\frac{t}{RC}})$.
- The quantity RC is called the time constant: $\tau = RC$

- During this time the current is decreasing exponentially because the capacitor is acting like an opposing battery as it charges up. The equation for the current is $I(t) = \frac{\varepsilon}{R}(e^{-\frac{t}{RC}})$.
- When a charged capacitor that has an initial voltage, V_0, and initial charge, Q_0, is connect in a circuit to a resistor both the charge on the plate and the current through the circuits decrease exponentially. The equations for the charge and the current are given by:

$$Q(t) = Q_0(e^{-\frac{t}{RC}}) \text{ and } I(t) = \frac{V_0}{R}(e^{-\frac{t}{RC}})$$

12

Magnetic Forces and Fields

INTRODUCTION

In Chapter 9, we learned that electric charges are the sources of electric fields and that other charges experience an electric force in those fields. The charges generating the field were assumed to be at rest, because if they weren't, then another force field would have been generated in addition to the electric field. Electric charges *that move* are the sources of **magnetic fields**, and other charges that move can experience a magnetic force in these fields.

THE MAGNETIC FORCE ON A MOVING CHARGE

If a particle with charge q moves with velocity **v** through a magnetic field **B**, it will experience a magnetic force, $\mathbf{F}_B$, with magnitude:

$$F_B = |q| vB \sin\theta \quad (1)$$

where θ is the angle between **v** and **B**. From this equation, we can see that if the charge is at rest, then $v = 0$ immediately gives us $F_B = 0$. This tells us that magnetic forces only act on moving charges. Also, if **v** is parallel (or antiparallel) to **B**, then $F_B = 0$ since, in either of these cases, $\sin\theta = 0$. So, only charges that cut across the magnetic field lines will experience a magnetic force. Furthermore, the magnetic force is maximized when **v** is perpendicular to **B**, since if $\theta = 90°$, then $\sin\theta$ is equal to 1, its maximum value.

The direction of $\mathbf{F}_B$ is always perpendicular to both **v** and **B** and depends on the sign of the charge q and the direction of $\mathbf{v} \times \mathbf{B}$ (which is given by the right-hand rule).

$$\text{direction of } \mathbf{F}_B = \begin{cases} \text{same as the direction of } \mathbf{v}\times\mathbf{B} \text{ if } q \text{ is positive} \\ \text{opposite to the direction of } \mathbf{v}\times\mathbf{B} \text{ if } q \text{ is negative} \end{cases} \quad (2)$$

Equations (1) and (2) can be summarized by a single equation:

$$\mathbf{F}_B = q(\mathbf{v} \times \mathbf{B})$$

Note that there are fundamental differences between the electric force and magnetic force on a charge. First, a magnetic force acts on a charge only if the charge is moving; the electric force acts on a charge whether it moves or not. Second, the direction of the magnetic force is always perpendicular to the magnetic field, while the electric force is always parallel (or antiparallel) to the electric field.

The SI unit for the magnetic field is the **tesla** (abbreviated **T**) which is one newton per ampere-meter. Another common unit for magnetic field strength is the **gauss** (abbreviated **G**); 1 G = 10^{-4} T.

Example 12.1 A charge $+q = +6 \times 10^{-6}$ C moves with speed $v = 4 \times 10^5$ m/s through a magnetic field of strength $B = 0.4$ T, as shown in the figure below. What is the magnetic force experienced by q?

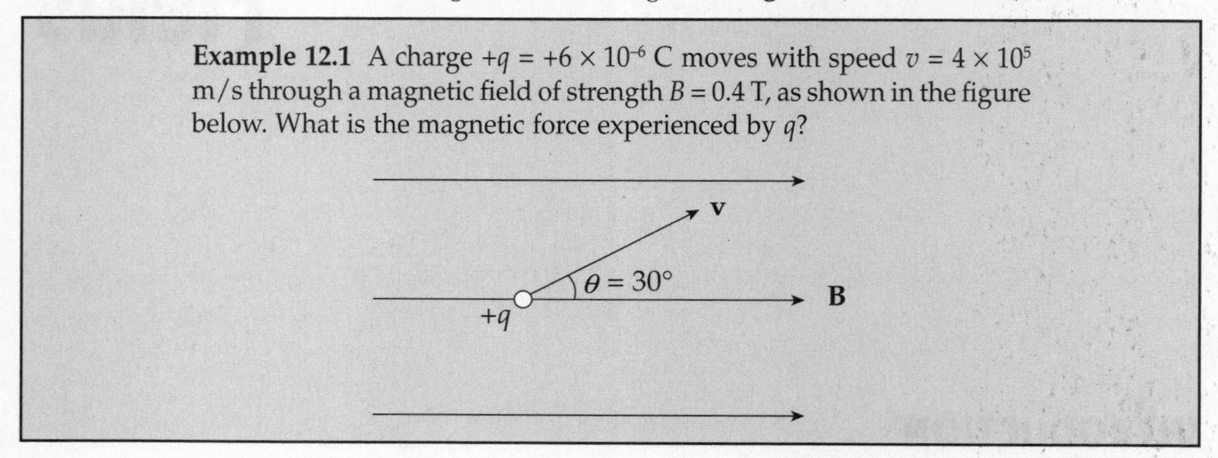

Solution. The magnitude of $\mathbf{F}_B$ is

$$F_B = qvB \sin\theta = (6 \times 10^{-6} \text{ C})(4 \times 10^5 \text{ m/s})(0.4 \text{ T}) \sin 30° = 0.48 \text{ N}$$

By the right-hand rule, the direction of $\mathbf{v} \times \mathbf{B}$ is into the plane of the page, which is symbolized by ⊗. Because q is a positive charge, the direction of $\mathbf{F}_B$ on q is also into the plane of the page.

Example 12.2 A particle of mass m and charge $+q$ is projected with velocity $\mathbf{v}$ (in the plane of the page) into a uniform magnetic field $\mathbf{B}$ that points into the page. How will the particle move

Solution. Since $\mathbf{v}$ is perpendicular to $\mathbf{B}$, the particle will feel a magnetic force of strength qvB, which will be directed perpendicular to $\mathbf{v}$ (and to $\mathbf{B}$) as shown

Since $\mathbf{F}_B$ is always perpendicular to $\mathbf{v}$, the particle will undergo uniform circular motion; $\mathbf{F}_B$ will provide the centripetal force. Notice that, because $\mathbf{F}_B$ is always perpendicular to $\mathbf{v}$, the magnitude of $\mathbf{v}$ will not change, just its direction. *Magnetic forces alone cannot change the speed of a charged particle, they can only change its direction of motion.* The radius of the particle's circular path is found from the equation $F_B = F_C$:

$$qvB = \frac{mv^2}{r} \quad \Rightarrow \quad r = \frac{mv}{qB}$$

Example 12.3 A particle of charge $-q$ is shot into a region that contains an electric field, $\mathbf{E}$, crossed with a perpendicular magnetic field, $\mathbf{B}$. If $E = 2 \times 10^4$ N/C and $B = 0.5$ T, what must be the speed of the particle if it is to cross this region without being deflected

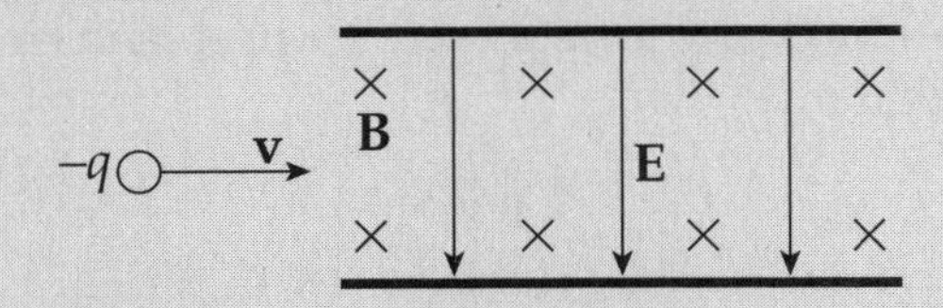

Solution. If the particle is to pass through undeflected, the electric force it feels has to be canceled by the magnetic force. In the diagram on the previous page, the electric force on the particle is directed upward (since the charge is negative and **E** is downward), and the magnetic force is directed downward (since the charge is negative and **v** × **B** is upward). So $\mathbf{F}_E$ and $\mathbf{F}_B$ point in opposite directions, and in order for their magnitudes to balance, qE must equal qvB, so v must equal E/B, which in this case gives

$$v = \frac{E}{B} = \frac{2\times10^4 \text{ N/C}}{0.5 \text{ T}} = 4\times10^4 \text{ m/s}$$

Example 12.4 A particle with charge $+q$, traveling with velocity **v**, enters a uniform magnetic field **B**, as shown below. Describe the particle's subsequent motion.

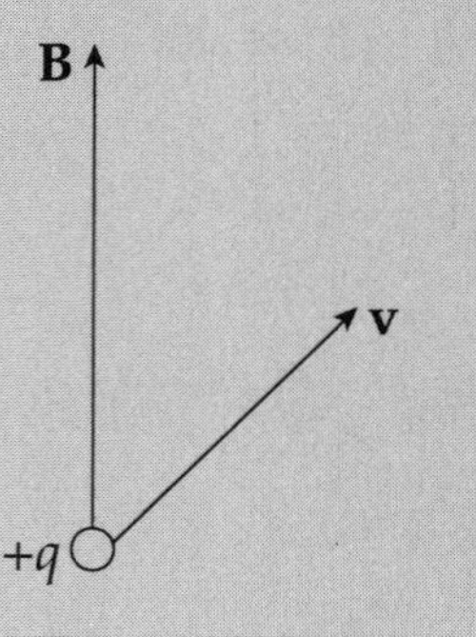

Solution. If the particle's velocity were parallel to **B**, then it would be unaffected by **B**. If **v** were perpendicular to **B**, then it would undergo uniform circular motion (as we saw in Example 12.2). In this case, **v** is neither purely parallel nor perpendicular to **B**. It has a component ($\mathbf{v}_1$) that's parallel to **B** and a component ($\mathbf{v}_2$) that's perpendicular to **B**.

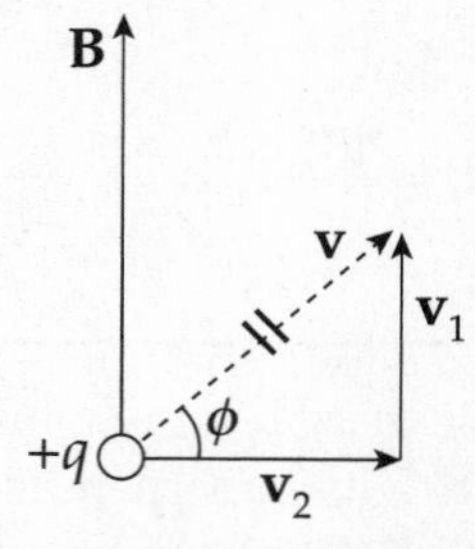

Component $\mathbf{v}_1$ will not be changed by **B**, so the particle will continue upward in the direction of **B**. However, the presence of $\mathbf{v}_2$ will create circular motion. The superposition of these two types of motion will cause the particle's trajectory to be a *helix*; it will spin in circular motion while traveling upward with the speed $v_1 = v \sin \phi$:

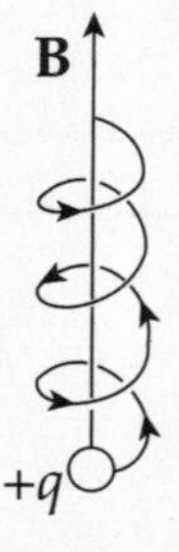

THE MAGNETIC FORCE ON A CURRENT-CARRYING WIRE

Since magnetic fields affect moving charges, they should also affect current-carrying wires. After all, a wire that contains a current contains charges that move.

Let a wire of length ℓ be immersed in magnetic field **B**. If the wire carries a current I, then the magnitude of the magnetic force it feels is

$$F_B = I\ell B \sin \theta$$

where is the angle between $\boldsymbol{\ell}$ and **B**. Here, the direction of $\boldsymbol{\ell}$ is the direction of the current, I. The direction of $\mathbf{F}_B$ is the same as the direction of $\boldsymbol{\ell} \times \mathbf{B}$, and these properties can be summarized in a single equation:

$$\mathbf{F}_B = I(\boldsymbol{\ell} \times \mathbf{B})$$

Example 12.5 A U-shaped wire of mass m is lowered into a magnetic field **B** that points out of the plane of the page. How much current I must pass through the wire in order to cause the net force on the wire to be zero?

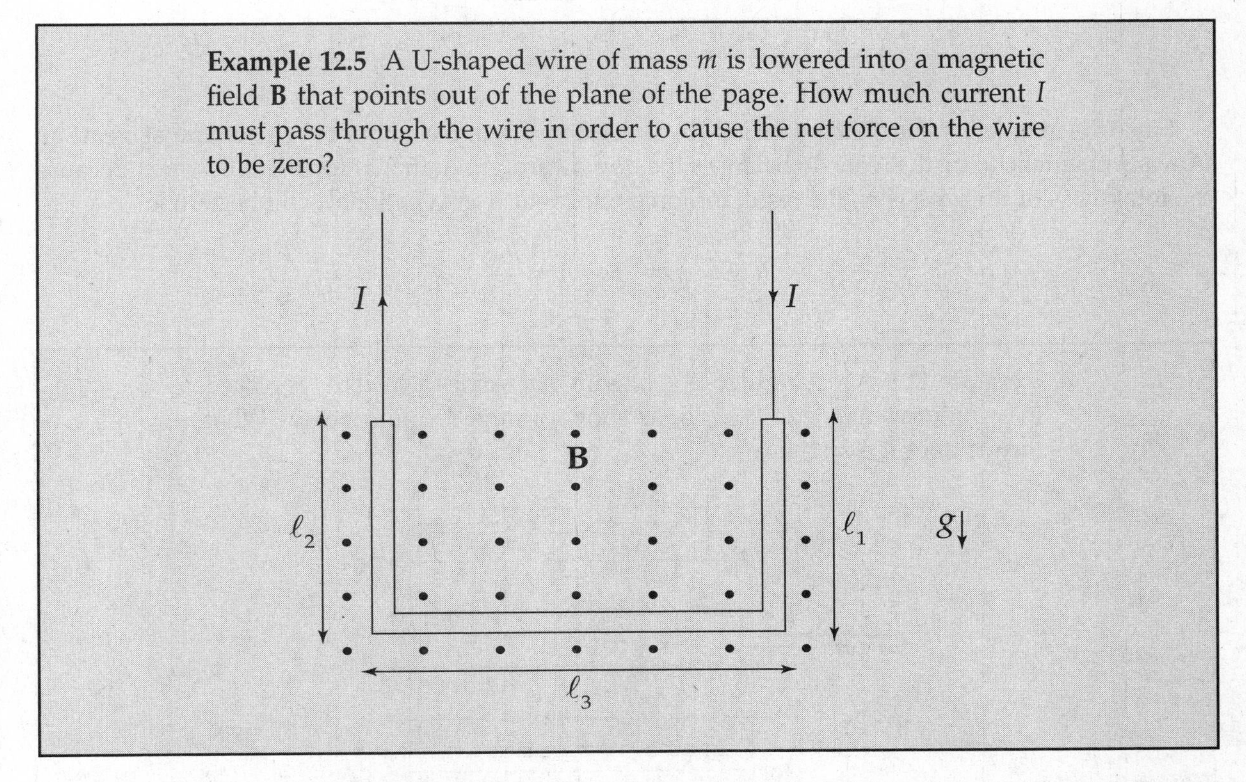

Solution. The total magnetic force on the wire is equal to the sum of the magnetic forces on each of the three sections of wire. The force on the first section (the right, vertical one), $\mathbf{F}_{B1}$, is directed to the left (applying the right-hand rule to $\boldsymbol{\ell}_1 \times \mathbf{B}$; $\boldsymbol{\ell}_1$ points downward), and the force on the third piece (the left, vertical one), $\mathbf{F}_{B3}$, is directed to the right. Since these pieces are the same length, these two oppositely-directed forces have the same magnitude, $I\ell_1 B = I\ell_3 B$, and they cancel. So the net magnetic force on the wire is the magnetic force on the middle piece. Since $\boldsymbol{\ell}_2$ points to the left and **B** is out of the page, $\boldsymbol{\ell}_2 \times \mathbf{B}$ [and, therefore, $\mathbf{F}_{B2} = I(\boldsymbol{\ell}_2 \times \mathbf{B})$] is directed upward.

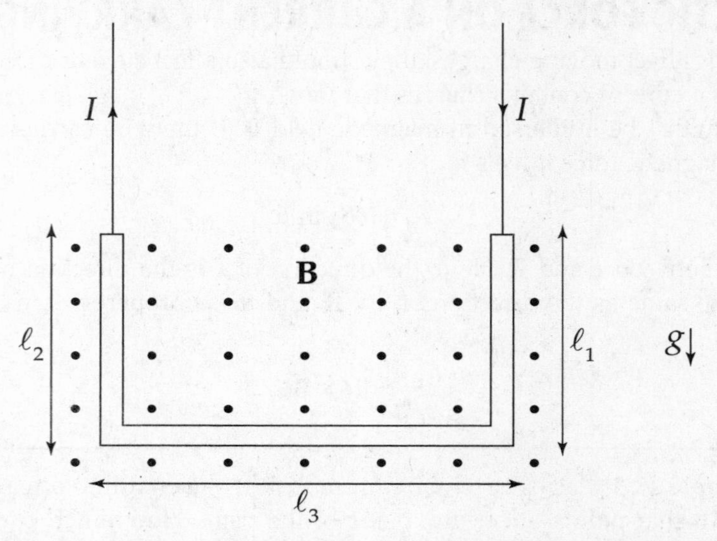

Since the magnetic force on the wire is $I\ell_2B$, directed upward, the amount of current must create an upward magnetic force that exactly balances the downward gravitational force on the wire. Because the total mass of the wire is m, the resultant force (magnetic + gravitational) will be zero if

$$I\ell_2 B = mg \quad \Rightarrow \quad I = \frac{mg}{\ell_2 B}$$

Example 12.6 A rectangular loop of wire that carries a current I is placed in a uniform magnetic field, **B**, as shown in the diagram below. What torque does it experience?

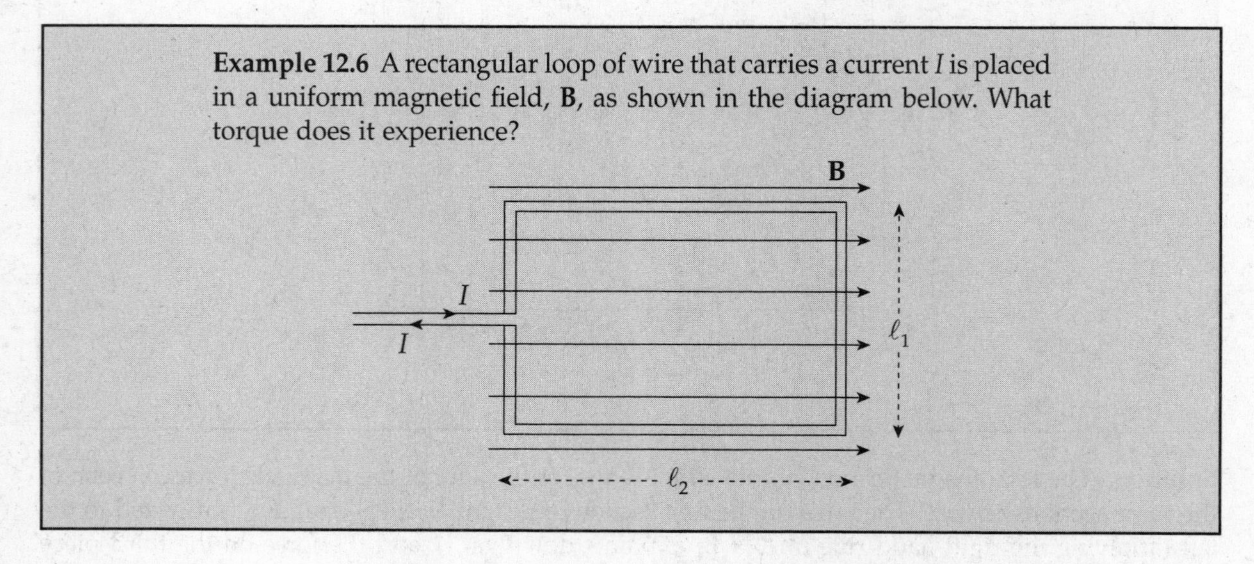

Solution. Ignoring the tiny gap in the vertical left-hand wire, we have two wires of length ℓ_1 and two of length ℓ_2. There is no magnetic force on either of the sides of the loop of length ℓ_2, because the current in the top side is parallel to **B** and the current in the bottom side is antiparallel to **B**. The magnetic force on the right-hand side points out of the plane of the page, while the magnetic force on the left-hand side points into the plane of the page.

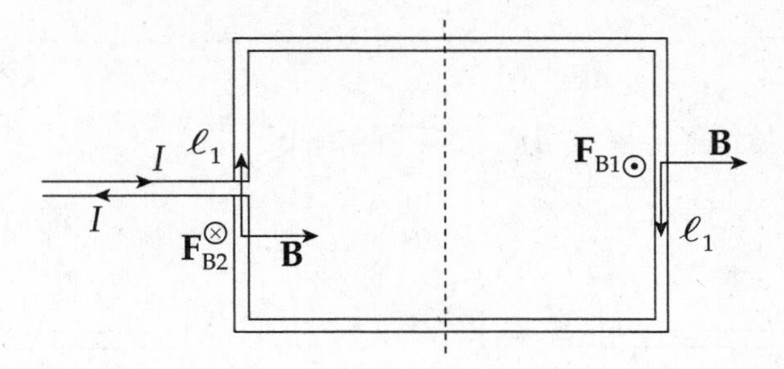

If the loop is free to rotate, then each of these two forces exerts a torque that tends to turn the loop in such a way that the right-hand side rises out of the plane of the page and the left-hand side rotates into the page. Relative to the axis shown above (which cuts the loop in half), the torque of $\mathbf{F}_{B1}$ is

$$\tau_1 = rF_{B1} \sin\theta = (\tfrac{1}{2}\ell_2)(I\ell_1 B)\sin 90° = \tfrac{1}{2}I\ell_1\ell_2 B$$

and the torque of $\mathbf{F}_{B2}$ is

$$\tau_2 = rF_{B2} \sin\theta = (\tfrac{1}{2}\ell_2)(I\ell_1 B)\sin 90° = \tfrac{1}{2}I\ell_1\ell_2 B$$

Since both these torques rotate the loop in the same direction, the net torque on the loop is

$$\tau_1 + \tau_2 = I\ell_1\ell_2 B$$

Example 12.7 The middle portion of the wire shown below is bent into the shape of a semicircle of radius r. The wire carries a current I. What's the total magnetic force that acts on the wire in the field, **B**?

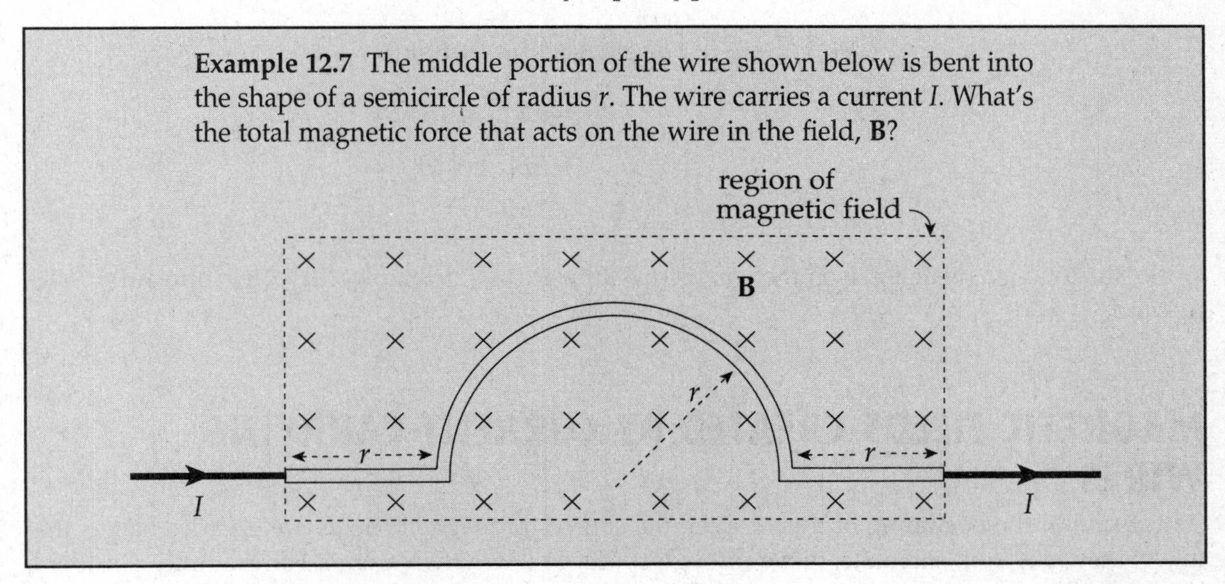

Solution. Let's first find the magnetic force that acts on the two straight sections. The magnitude of the magnetic force on each section is $I\ell B = IrB$, and since $\boldsymbol{\ell}$ points to the right in both sections, the magnetic force on each of them points upward.

On the semicircle, the direction of $\boldsymbol{\ell}$ is not constant, so we have to split this section into small pieces with lengths $d\ell = rd\theta$ and then use the equation $d\mathbf{F} = I(d\boldsymbol{\ell} \times \mathbf{B})$. In the diagram on the next page, the direction of $d\mathbf{F}$ is radially away from the center of the semicircle.

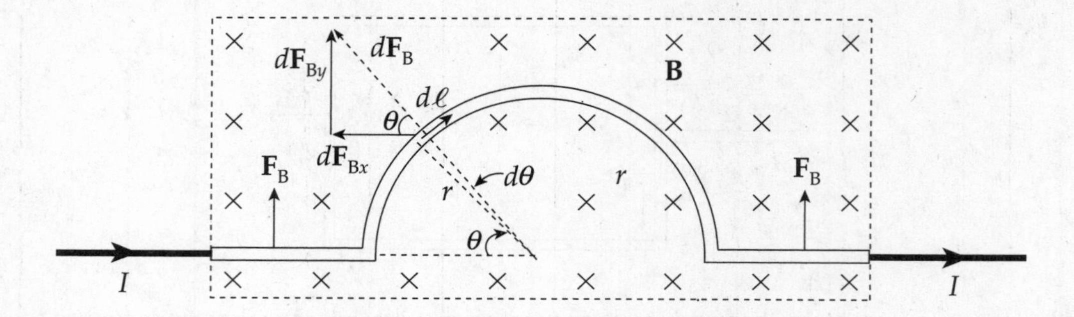

The total horizontal force on the semicircle, $F_{\text{B}x}$, can be found by integrating $dF_{\text{B}x} = dF \cos\theta$ from $\theta = 0$ to $\theta = \pi$:

$$\begin{aligned} F_{\text{B}x} = \int dF_{\text{B}x} = \int dF\cos\theta &= \int I(r\,d\theta)B\cos\theta \\ &= IrB\int_0^{\pi}\cos\theta\,d\theta \\ &= IrB[\sin\theta]_0^{\pi} \\ &= 0 \end{aligned}$$

The total vertical force on the semicircle, $F_{\text{B}y}$, can be found by integrating $dF_{\text{B}y} = dF \sin\theta$ from $\theta = 0$ to $\theta = \pi$:

$$\begin{aligned} F_{\text{B}y} = \int dF_{\text{B}y} = \int dF\sin\theta &= \int I(r\,d\theta)B\sin\theta \\ &= IrB\int_0^{\pi}\sin\theta\,d\theta \\ &= IrB[-\cos\theta]_0^{\pi} \\ &= 2IrB \end{aligned}$$

So, the total magnetic force on the wire is (IrB, upward) + ($2IrB$, upward) + (IrB, upward) = $4IrB$, upward.

MAGNETIC FIELDS CREATED BY CURRENT-CARRYING WIRES

As we said at the beginning of this chapter, the sources of magnetic fields are electric charges that move; they may spin, circulate, move through space, or flow down a wire. For example, consider a long, straight wire that carries a current I. The current generates a magnetic field in the surrounding space, of magnitude:

$$B = \frac{\mu_0}{2\pi}\frac{I}{r}$$

where r is the distance from the wire. The symbol μ_0 denotes a fundamental constant called the permeability of free space. Its value is:

$$\mu_0 = 4\pi \times 10^{-7}\,\text{N/A}^2 = 4\pi \times 10^{-7}\,\text{T}\cdot\text{m/A}$$

The magnetic field lines are actually circles whose centers are on the wire. The direction of these circles is determined by a variation of the right-hand rule. Imagine grabbing the wire in your right hand with your thumb pointing in the direction of the current. Then the direction in which your fingers curl around the wire gives the direction of the magnetic field line.

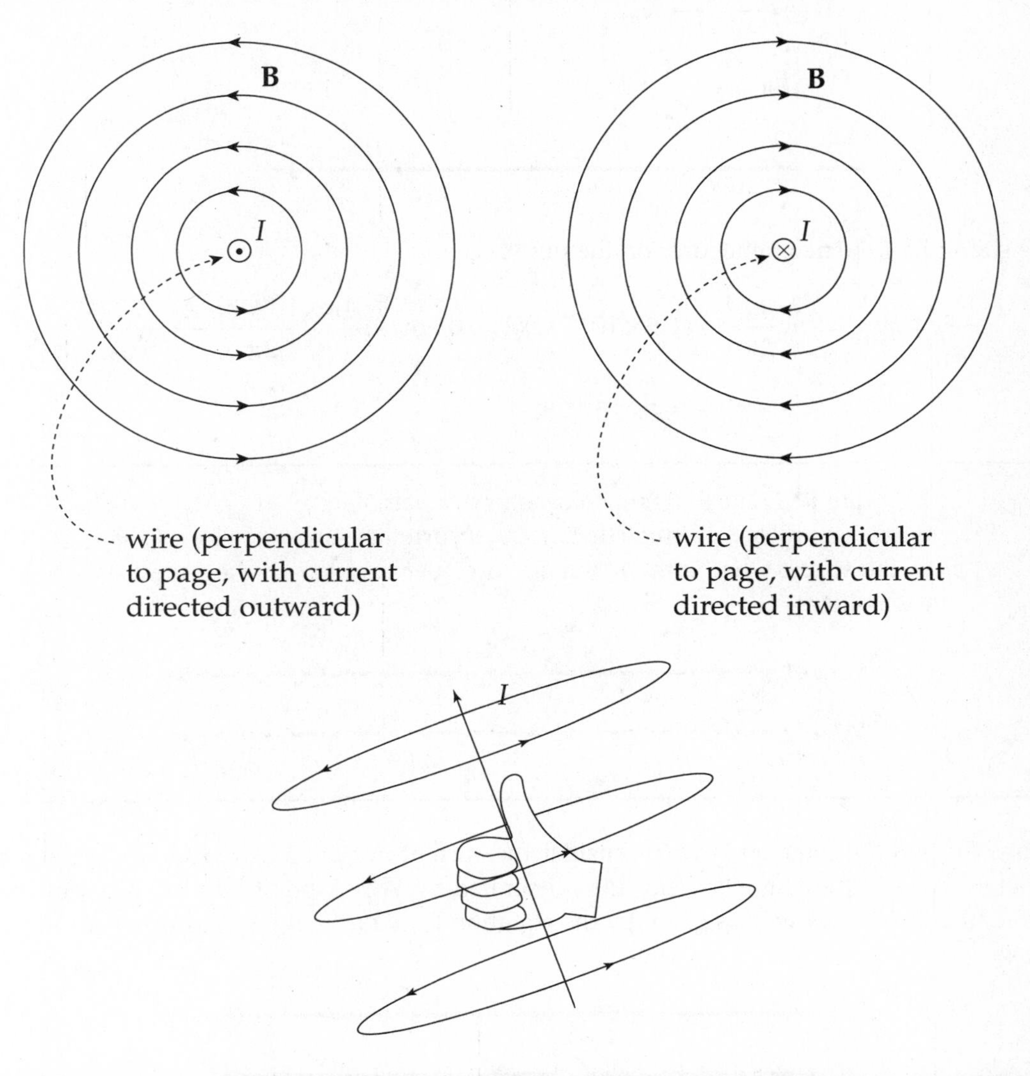

Example 12.8 The diagram below shows a proton moving with a speed of 2×10^5 m/s, initially parallel to, and 4 cm from, a long, straight wire. If the current in the wire is 20 A, what's the magnetic force on the proton?

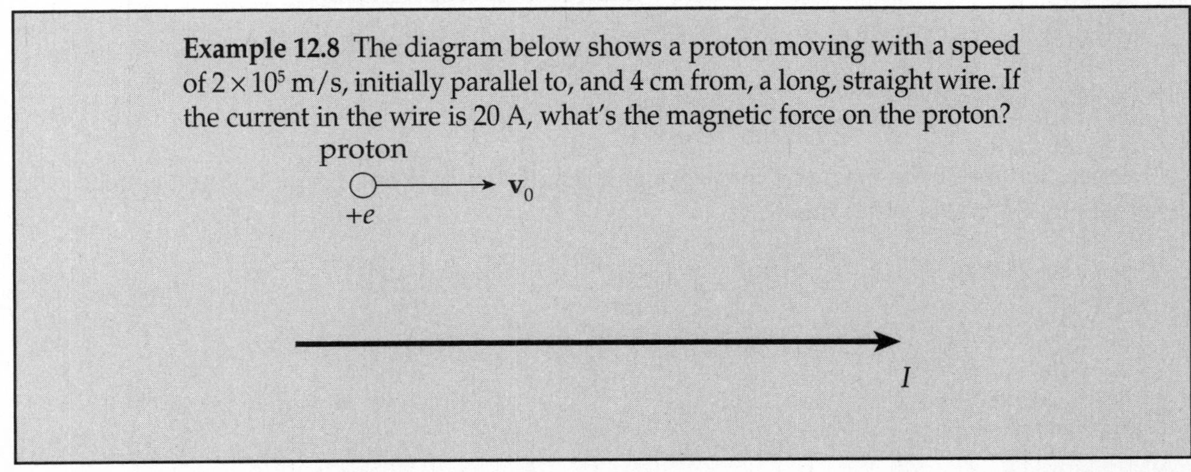

Solution. Above the wire (where the proton is), the magnetic field lines generated by the current-carrying wire point out of the plane of the page, so $\mathbf{v}_0 \times \mathbf{B}$ points downward. Since the proton's charge is positive, the magnetic force $\mathbf{F}_B = q(\mathbf{v}_0 \times \mathbf{B})$ is also directed down, toward the wire.

The strength of the magnetic force on the proton is

$$F_B = qv_0B = ev_0 \frac{\mu_0}{2\pi}\frac{I}{r} = (1.6\times10^{-19}\ \text{C})(2\times10^{5}\ \text{m/s})\frac{4\pi\times10^{-7}\ \text{N/A}^2}{2\pi}\frac{20\ \text{A}}{0.04\ \text{m}}$$

$$= 3.2\times10^{-18}\ \text{N}$$

Example 12.9 The diagram below shows a pair of long, straight, parallel wires, separated by a small distance, r. If currents I_1 and I_2 are established in the wires, what is the magnetic force per unit length they exert on each other?

Solution. To find the force on Wire 2, consider the current in Wire 1 as the source of the magnetic field. Below Wire 1, the magnetic field lines generated by Wire 1 point into the plane of the page. Therefore, the force on Wire 2, as given by the equation $\mathbf{F}_{B2} = I_2(\boldsymbol{\ell}_2 \times \mathbf{B}_1)$, points upward.

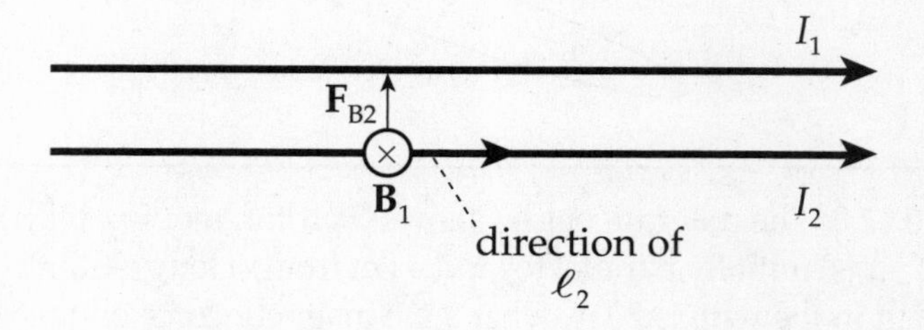

The magnitude of the magnetic force per unit length felt by Wire 2, due to the magnetic field generated by Wire 1, is found this way:

$$F_{B2} = I_2\ell_2B_1 = I_2\ell_2\frac{\mu_0}{2\pi}\frac{I_1}{r} \quad\Rightarrow\quad \frac{F_{B2}}{\ell_2} = \frac{\mu_0}{2\pi}\frac{I_1I_2}{r}$$

By Newton's Third Law, this is the same force that Wire 1 feels due to the magnetic field generated by Wire 2. The force is attractive because the currents point in the same direction; if one of the currents were reversed, then the force between the wires would be repulsive.

The Magnetic Field of a Solenoid

Imagine taking a long piece of wire and winding it in a helix around the length of a hollow tube; this is a **solenoid**. Let N be the total number of turns and L the length of the solenoid; then the number of turns per unit length is N/L, which we'll call n. If a current I is established in the wire, then a magnetic field is generated in the space that surrounds the solenoid. If the solenoid is much longer than it is wide, then the magnetic field inside will be parallel to the central axis and nearly uniform. The strength of the field inside is given by the equation

$$B = \mu_0 nI$$

So, the magnetic field inside a solenoid can be increased either by winding the wire more tightly (to increase n) or by increasing the current.

Example 12.10 A tightly-wound solenoid has a length of 30 cm, a diameter of 2 cm, and contains a total of 10,000 turns. If it carries a current of 5 A, what's the magnitude of the magnetic field inside the solenoid?

Solution. Since the solenoid is so much longer than it is wide, the equation $B = \mu_0 nI$ can be used to calculate the field inside:

$$B = \mu_0 nI = \mu_0 \frac{N}{L} I = (4\pi \times 10^{-7}\ \text{T} \cdot \text{m/A}) \frac{10{,}000}{0.30\ \text{m}} (5\ \text{A}) = 0.2\ \text{T}$$

THE BIOT-SAVART LAW

In this section, you'll learn one way to determine the magnetic field created by an electric current. Let's say we want to determine the magnitude and direction of the magnetic field, at Point P, created by the current I in the diagram below.

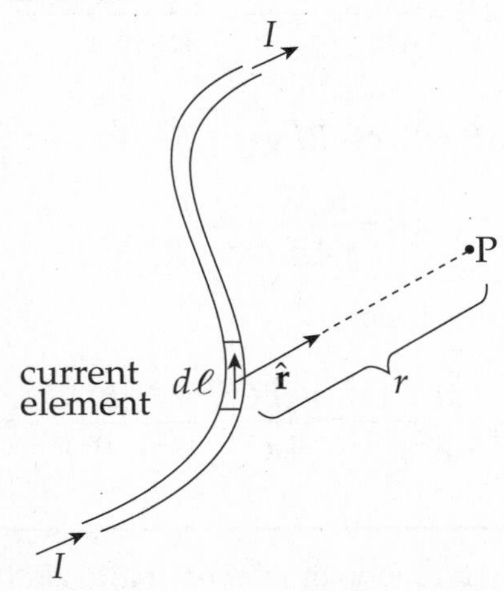

Consider a small section of the current, an infinitesimal section of length $d\ell$, and assign it the direction of the current so we obtain a **current element**, $d\boldsymbol{\ell}$. Let $\mathbf{r}$ be the position vector from the current element to the Point P, and form the unit vector $\hat{\mathbf{r}} = \mathbf{r}/r$. Then the magnitude and direction of the magnetic field at P, due to the current element is

$$d\mathbf{B} = \frac{\mu_0}{4\pi} \frac{I(d\boldsymbol{\ell} \times \hat{\mathbf{r}})}{r^2}$$

This is called the **Biot–Savart Law**. By adding all the contributions to the magnetic field, we can get the total magnetic field due to the entire current.

Example 12.11 Use the Biot–Savart Law to show that the magnetic field due to an infinitely-long straight wire carrying a current I is given by the equation $B = (\mu_0/2\pi)I/R$, where R is the distance from the wire. You should use the integration formula

$$\int_{-\infty}^{\infty} \frac{dx}{(x^2 + a^2)^{3/2}} = \frac{2}{a^2}$$

Solution. Refer to the diagram below:

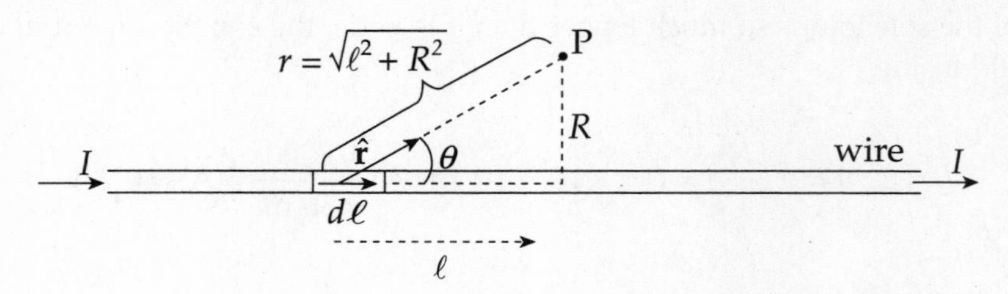

The direction of $d\boldsymbol{\ell} \times \hat{\mathbf{r}}$ is out of the plane of the page (because P is above the wire). The contribution from the current element to the total magnetic field at P is given by the Biot–Savart Law:

$$dB = \frac{\mu_0}{4\pi} \frac{I|d\boldsymbol{\ell} \times \hat{\mathbf{r}}|}{r^2} = \frac{\mu_0}{4\pi} \frac{I\,d\ell \sin\theta}{r^2}$$

From the diagram, we see that $\sin\theta = R/r = R/\sqrt{\ell^2 + R^2}$, so

$$dB = \frac{\mu_0 IR}{4\pi} \frac{d\ell}{(\ell^2 + R^2)^{3/2}}$$

and, integrating along the entire wire, we get:

$$B = \int dB = \int_{-\infty}^{\infty} \frac{\mu_0 IR}{4\pi} \frac{d\ell}{(\ell^2 + R^2)^{3/2}} = \frac{\mu_0 IR}{4\pi} \int_{-\infty}^{\infty} \frac{d\ell}{(\ell^2 + R^2)^{3/2}} = \frac{\mu_0 IR}{4\pi} \left[\frac{2}{R^2}\right] = \frac{\mu_0}{2\pi} \frac{I}{R}$$

Example 12.12 A circular loop of wire has radius R. If the loop carries a current I, what's the magnetic field created at the center of the loop?

Solution.

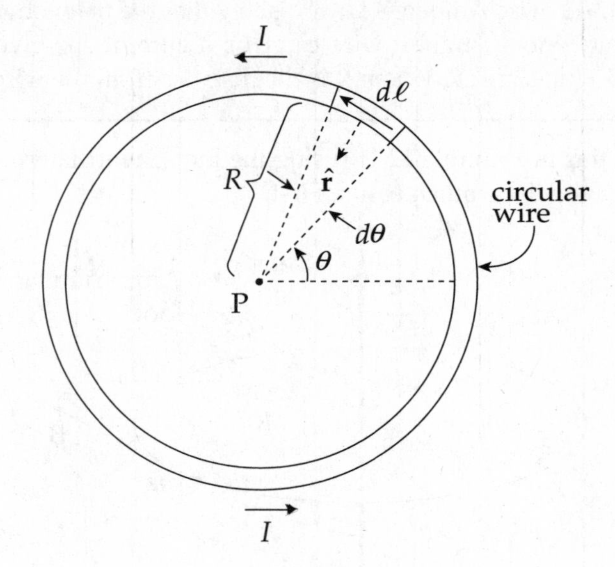

Every current element $d\ell$ is perpendicular to $\hat{\mathbf{r}}$, and $r = R$, so the Biot–Savart Law gives us:

$$dB = \frac{\mu_0}{4\pi}\frac{I\,d\ell\sin\theta}{r^2} = \frac{\mu_0}{4\pi}\frac{I\,d\ell\sin 90^\circ}{R^2} = \frac{\mu_0 I}{4\pi R^2}d\ell$$

Writing $d\ell = R\,d\theta$, we integrate around the circle to get

$$B = \int dB = \int_0^{2\pi}\frac{\mu_0 I}{4\pi R^2}R\,d\theta = \frac{\mu_0 I}{4\pi R}\int_0^{2\pi}d\theta = \frac{\mu_0 I}{2R}$$

If the current is counterclockwise (as shown above), then $d\ell \times \hat{\mathbf{r}}$, and the magnetic field, points out of the plane of the page, but if the current is clockwise, the magnetic field at the center points into the page.

AMPERE'S LAW

In Chapter 9, we learned about Gauss's Law, which can be used to calculate the electric field due to a configuration of electric charges. Gauss's Law says that the electric flux through a closed Gaussian surface (which is actually the integral of $\mathbf{E} \cdot d\mathbf{A}$ over the area of the surface) is equal to a constant $(1/\varepsilon_0)$ times the total electric charge enclosed by the surface.

There's an analogous law for calculating the magnetic field due to a configuration of electric currents. This law says that the integral of $\mathbf{B} \cdot d\mathbf{s}$ around a closed loop (called an **Amperian loop**) is equal to a constant (μ_0) times the total current that passes through the loop:

$$\oint_{\text{loop}} \mathbf{B} \cdot d\mathbf{s} = \mu_0 I_{\text{through loop}}$$

This is **Ampere's Law** and we'll use it to find the magnetic field at points on an Amperian loop surrounding the source current. (The circle on the integral sign in Ampere's Law emphasizes that the integral must always be taken around a *closed* path.)

Example 12.13 Use Ampere's Law to show that the magnetic field due to an infinitely-long straight wire carrying a current I is given by the equation $B = (\mu_0/2\pi)I/R$, where R is the distance from the wire.

Solution. When we did this in Example 12.11 (using the Biot–Savart Law), we ran into a messy integral. Using Ampere's Law will be much easier. First,

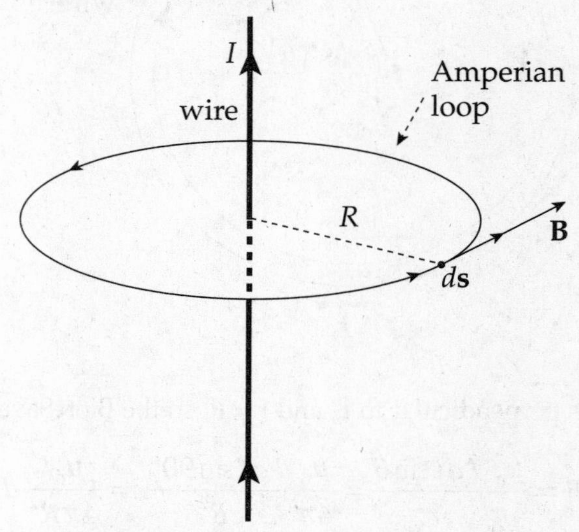

The magnetic field must have the same magnitude at every point on the Amperian circle, and the direction of the field must always be perpendicular to the circle (which means it points radially away from or toward the wire) or always tangent to the circle. We can rule out the first case in two ways: (1) Magnetic field lines *always* form closed curves that encircle the current that generates them; (2) If **B** were perpendicular to the Amperian circle, then $\mathbf{B} \cdot d\mathbf{s}$ would be zero everywhere along the circle, so Ampere's Law would give $0 = \mu_0 I$, a contradiction. So **B** must be tangent to the Amperian circle.

Since **B** has constant magnitude on the circle and is always parallel to $d\mathbf{s}$ (which is a small section of the Amperian loop),

$$\mathbf{B} \cdot d\mathbf{s} = B\, ds \cos\theta = B\, ds$$

The current that passes through the loop is I, so Ampere's Law gives us

$$\oint_{\text{loop}} B\,ds = \mu_0 I$$

$$B \oint_{\text{loop}} ds = \mu_0 I$$

$$B(2\pi R) = \mu_0 I$$

$$B = \frac{\mu_0 I}{2\pi R}$$

Example 12.14 The figure below is a cut-away view of a long coaxial cable that's composed of a solid cylindrical conductor (of radius R_1) surrounded by a thin conducting cylindrical shell (of radius R_2).

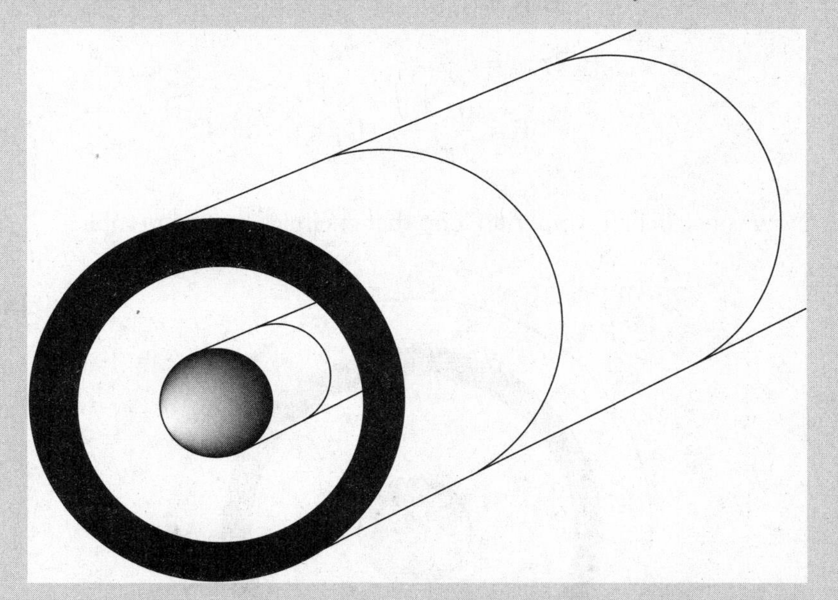

The inner cylinder carries a current of I_1 and the outer cylindrical shell carries a smaller current of I_2 in the opposite direction. Use Ampere's Law to find the magnitude of the magnetic field (a) in the space between the inner cylinder and the shell, and (b) outside the outer shell.

Solution.

(a) Construct an Amperian loop of radius r in the space between the solid cylinder and the outer shell. The magnetic field must be of constant magnitude along (and tangent to) this loop.

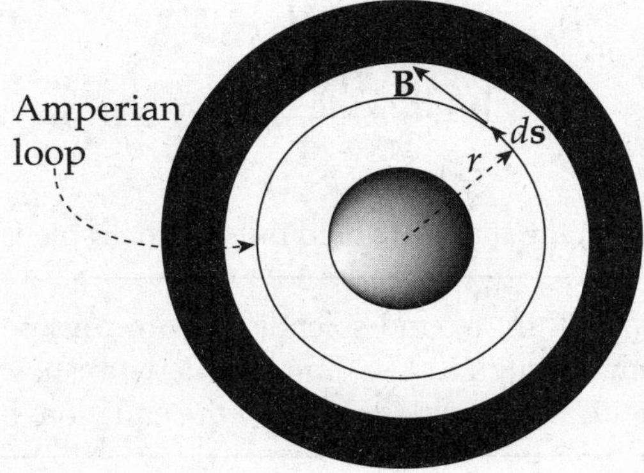

The only current that passes through the loop is I_1 (the current I_2 in the outer shell does not pierce the interior of the Amperian loop), so Ampere's Law gives us:

$$\oint_{\text{loop}} B\,ds = \mu_0 I_1$$

$$B(2\pi r) = \mu_0 I_1$$

$$B = \frac{\mu_0}{2\pi}\frac{I_1}{r} \quad (R_1 < r < R_2)$$

(b) Now construct an Amperian loop that encircles the entire cable.

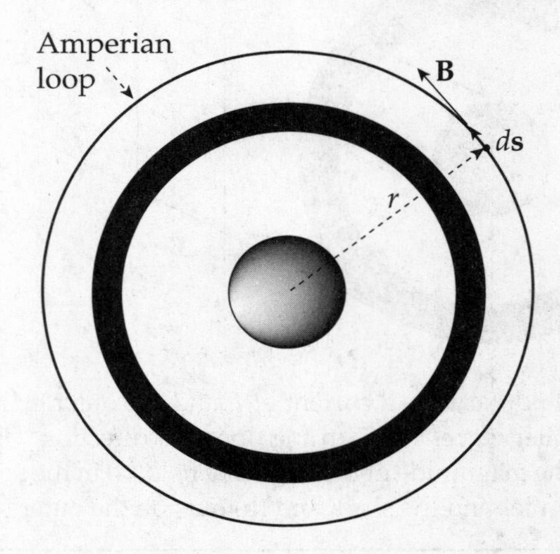

Unlike in the loop constructed in part (a), *both* currents pass through this loop. If they had the same direction, we'd add them to get the total current, but since the currents here have *opposite* directions, the net current through the loop is $I_1 - I_2$. Ampere's Law then gives us:

$$\oint_{\text{loop}} B\,ds = \mu_0 (I_1 - I_2)$$

$$B(2\pi r) = \mu_0 (I_1 - I_2)$$

$$B = \frac{\mu_0}{2\pi}\frac{(I_1 - I_2)}{r} \quad (r > R_2)$$

Note that if $I_1 = I_2$, then the magnetic field outside the cable is zero.

> **Example 12.15** A tightly wound solenoid with n coils per length has a current I running through it. Use Ampere's Law to show the magnetic field inside this ideal solenoid is given by the equation, $B = \mu_0 nI$.

Solution.

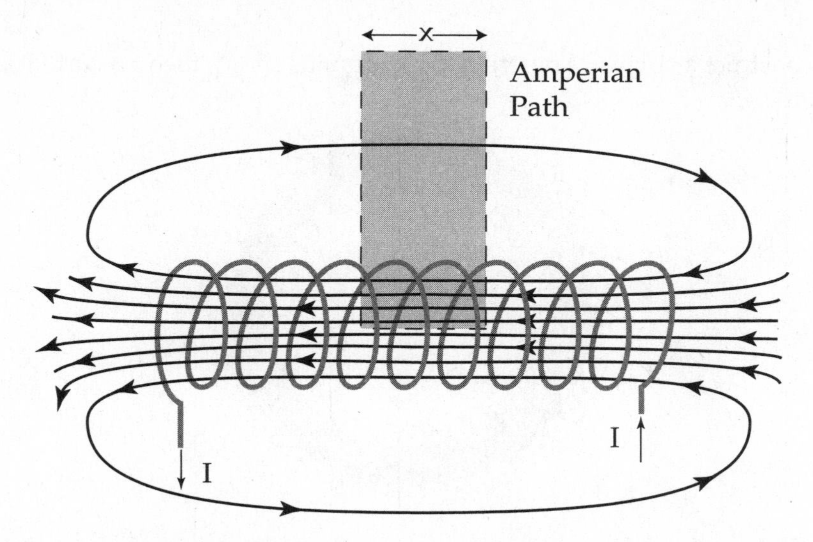

Consider the diagram above. The two vertical sections of the Amperian path do not matter because the path is perpendicular to the magnetic field (remember Ampere's Law is a dot product for **B** · **ds**). The top section of the path is approximated as zero for an ideal solenoid because the field is not weak outside a tightly-wound solenoid. Therefore, the only part of the path we will use is the length x inside of the solenoid. We will assume N total coils pass through the shaded region of our Amperian path.

$$\oint_{path} Bds = \mu_0 I$$

$$Bx = \mu_0 NI$$

$$B = \frac{\mu_0 NI}{x}$$

$$B = \mu_0 nI$$

Example 12.16 The figure below is a cross section of a *toroidal solenoid* (a solenoid bent into a doughnut shape) of inner radius R_1 and outer radius R_2. It consists of N windings, and the wire carries a current I. What's the magnetic field in each of the following regions?

(a) $r < R_1$
(b) $R_1 < r < R_2$
(c) $r > R_2$

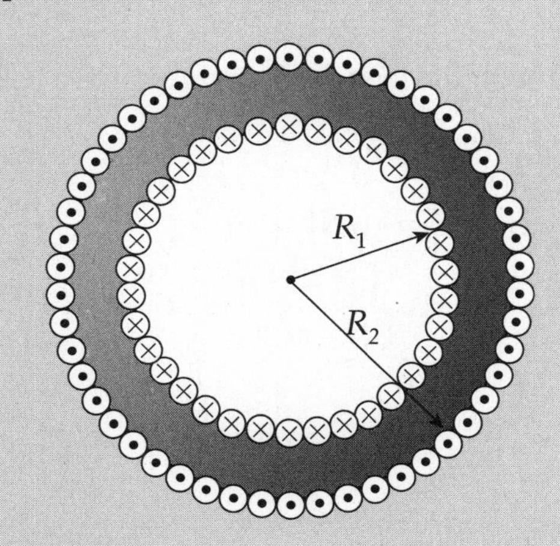

Solution.

(a) If we construct a circular Amperian loop of radius $r < R_1$, then no current will pass through it.

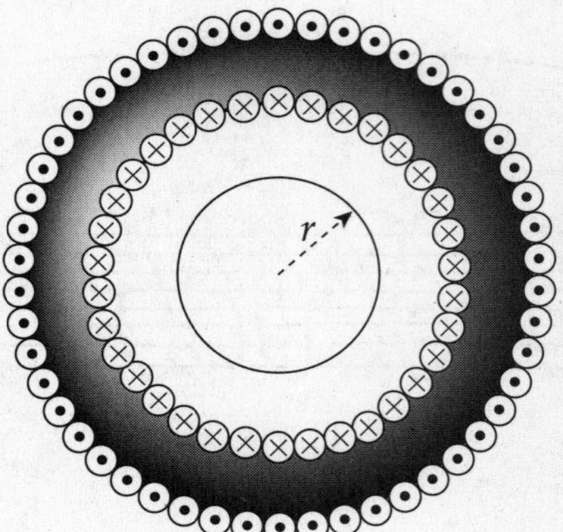

Since $\oint_{\text{loop}} \mathbf{B} \cdot d\mathbf{s} = \oint_{\text{loop}} B\,ds = B(2\pi r)$ and $I_{\text{through loop}} = 0$, Ampere's Law immediately tells us that $B = 0$ in this region (the hole of the doughnut).

(b) Construct a circular Amperian loop of radius r such that $R_1 < r < R_2$. Each of the N windings that brings current into the plane of the page passes through the loop, but the N windings that bring current out of the plane of the page do not pass through the loop.

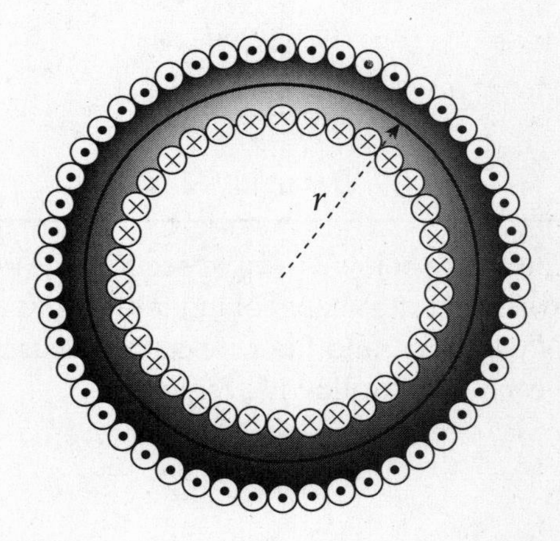

So, the net current through the loop is NI, and Ampere's Law gives us:

$$\oint_{\text{loop}} B\,ds = \mu_0 I_1$$

$$B(2\pi r) = \mu_0 NI$$

$$B = \frac{\mu_0}{2\pi}\frac{NI}{r} \quad (R_1 < r < R_2)$$

(c) Construct a circular Amperian loop of radius $r > R_2$.

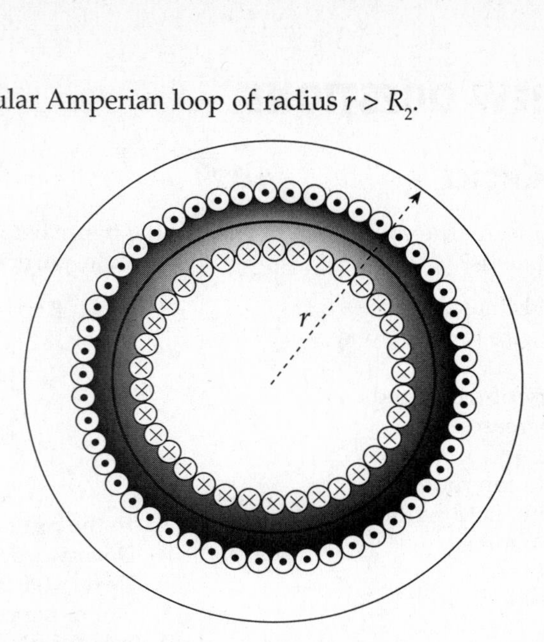

For every current contribution of I punching upward through the loop, there is an equal current punching downward through the loop, so the net current through the loop is zero, and $B = 0$ in this region.

Note that the results of this example show that for an ideal toroidal solenoid, the magnetic field is contained within the space enclosed by the windings.

CHAPTER 12 REVIEW QUESTIONS

SECTION I: MULTIPLE CHOICE

1. Which of the following is/are true concerning magnetic forces and fields?

    I. The magnetic field lines due to a current-carrying wire radiate away from the wire.
    II. The kinetic energy of a charged particle can be increased by a magnetic force.
    III. A charged particle can move through a magnetic field without feeling a magnetic force.

    (A) I only
    (B) II and III only
    (C) I and II only
    (D) III only
    (E) I and III only

2. The velocity of a particle of charge $+4.0 \times 10^{-9}$ C and mass 2×10^{-4} kg is perpendicular to a 0.1-tesla magnetic field. If the particle's speed is 3×10^{4} m/s, what is the acceleration of this particle due to the magnetic force?

    (A) 0.0006 m/s^2
    (B) 0.006 m/s^2
    (C) 0.06 m/s^2
    (D) 0.6 m/s^2
    (E) None of the above

3. In the figure below, what is the direction of the magnetic force $\mathbf{F}_B$?

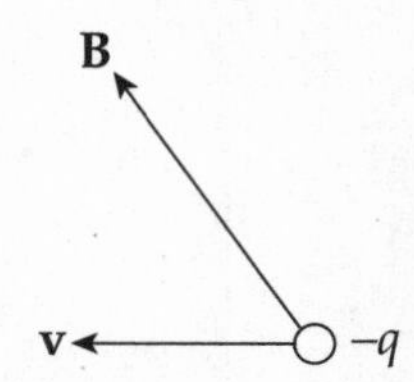

    (A) To the right
    (B) Downward, in the plane of the page
    (C) Upward, in the plane of the page
    (D) Out of the plane of the page
    (E) Into the plane of the page

4. In the figure below, what must be the direction of the particle's velocty, **v**?

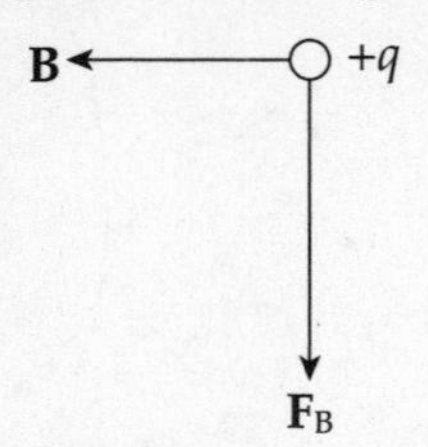

    (A) To the right
    (B) Downward, in the plane of the page
    (C) Upward, in the plane of the page
    (D) Out of the plane of the page
    (E) Into the plane of the page

5. Due to the magnetic force, a positively-charged particle executes uniform circular motion within a uniform magnetic field, **B**. If the charge is q and the radius of its path is r, which of the following expressions gives the magnitude of the particle's linear momentum?

    (A) qBr
    (B) qB/r
    (C) $q/(Br)$
    (D) $B/(qr)$
    (E) $r/(qB)$

6. A straight wire of length 2 m carries a 10-amp current. How strong is the magnetic field at a distance of 2 cm from the wire?

    (A) 1×10^{-6} T
    (B) 1×10^{-5} T
    (C) 2×10^{-5} T
    (D) 1×10^{-4} T
    (E) 2×10^{-4} T

7. Two long, straight wires are hanging parallel to each other and are 1 cm apart. The current in Wire 1 is 5 A, and the current in Wire 2 is 10 A, in the same direction. Which of the following best describes the magnetic force per unit length felt by the wires?

   (A) The force per unit length on Wire 1 is twice the force per unit length on Wire 2.
   (B) The force per unit length on Wire 2 is twice the force per unit length on Wire 1.
   (C) The force per unit length on Wire 1 is 0.0003 N/m, away from Wire 2.
   (D) The force per unit length on Wire 1 is 0.001 N/m, toward Wire 2.
   (E) The force per unit length on Wire 1 is 0.001 N/m, away from Wire 2.

8. In the figure below, what is the magnetic field at the Point P, which is midway between the two wires?

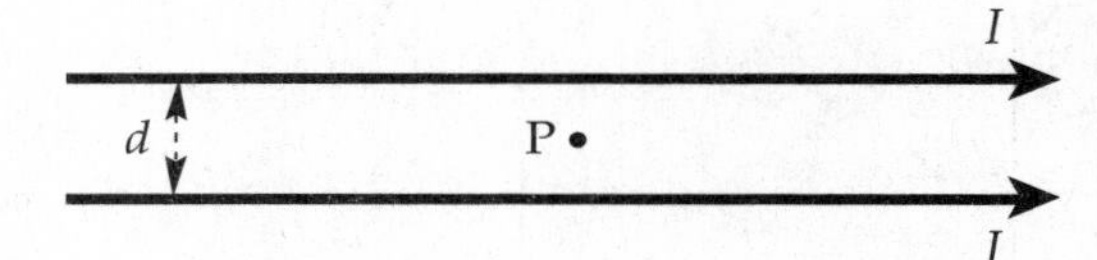

   (A) $2\mu_0 I/(\pi d)$, out of the plane of the page
   (B) $2\mu_0 I/(\pi d)$, into the plane of the page
   (C) $\mu_0 I/(2\pi d)$, out of the plane of the page
   (D) $\mu_0 I/(2\pi d)$, into the plane of the page
   (E) Zero

9. How many windings must a solenoid of length 80 cm have in order to establish a magnetic field of strength 0.2 T inside the solenoid, if it carries a current of 20 amps?

   (A) 1000
   (B) 6400
   (C) 10,000
   (D) 32,000
   (E) 64,000

10. The value of $\oint \mathbf{B} \cdot d\mathbf{s}$ along a closed path in a magnetic field **B** is 6.28×10^{-6} T · m. What is the total current that passes through this closed path?

   (A) 0.1 A
   (B) 0.5 A
   (C) 1 A
   (D) 4 A
   (E) 5 A

SECTION II: FREE RESPONSE

1. The diagram below shows a simple mass spectrograph. It consists of a source of ions (charged atoms) that are accelerated (essentially from rest) by the voltage V and enter a region containing a uniform magnetic field, **B**. The polarity of V may be reversed so that both positively-charged ions (cations) and negatively-charged ions (anions) can be accelerated. Once the ions enter the magnetic field, they follow a semicircular path and strike the front wall of the spectrograph, on which photographic plates are constructed to record the impact.

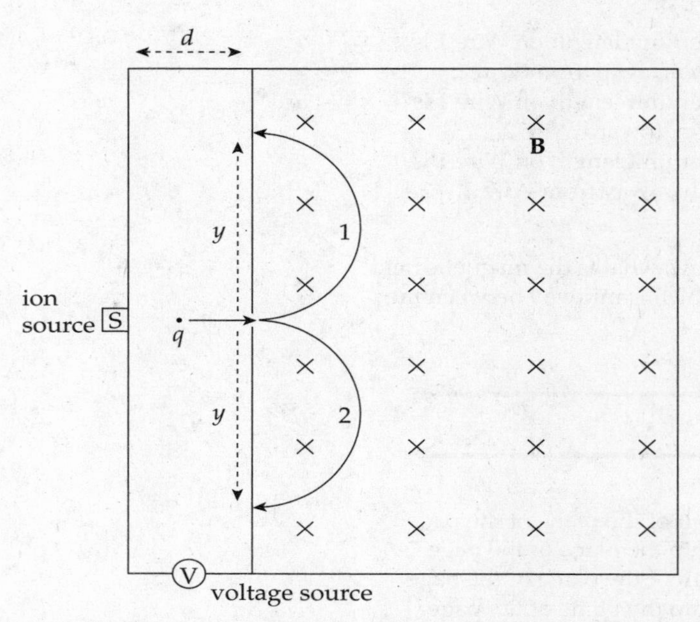

(a) What is the acceleration of an ion of charge q just before it enters the magnetic field?

(b) Find the speed with which an ion of charge q enters the magnetic field.

(c) (i) Which semicircular path, 1 or 2, would a cation follow?

(ii) Which semicircular path, 1 or 2, would an anion follow?

(d) Determine the mass of a cation entering the apparatus in terms of y, q, B, and V.

(e) Once a cation of charge q enters the magnetic field, how long does it take to strike the photographic plate?

(f) What is the work done by the magnetic force in the spectrograph on a cation of charge q?

2. A wire of diameter d and resistivity ρ is bent into a rectangular loop (of side lengths a and b) and fitted with a small battery that provides a voltage V. The loop is placed at a distance c from a very long, straight wire that carries a current I in the direction indicated in the diagram.

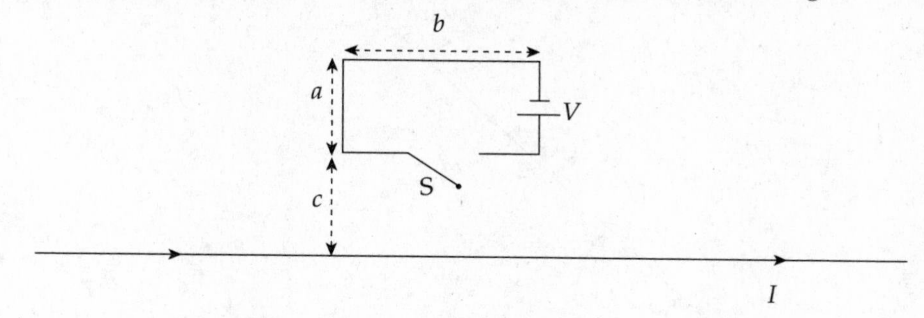

(Express all answers in terms of $a, b, c, d, \rho, V, I, m, B, x$, and fundamental constants.)

(a) When the switch S is closed, find the current in the rectangular loop.

(b) What is the magnetic force (magnitude and direction) exerted on the loop by the long, straight wire?

(c) The wire of the rectangular loop is then reshaped into a circle. What will be the radius of the circular loop?

(d) If the loop constructed in part (c) were then threaded around the long, straight wire (so that the straight wire passed through the center of the circular loop), what would be the magnetic force on the loop now?

(e) In the following diagram, two fixed L-shaped wires, separated by a distance x, are connected by a wire that's free to slide vertically.

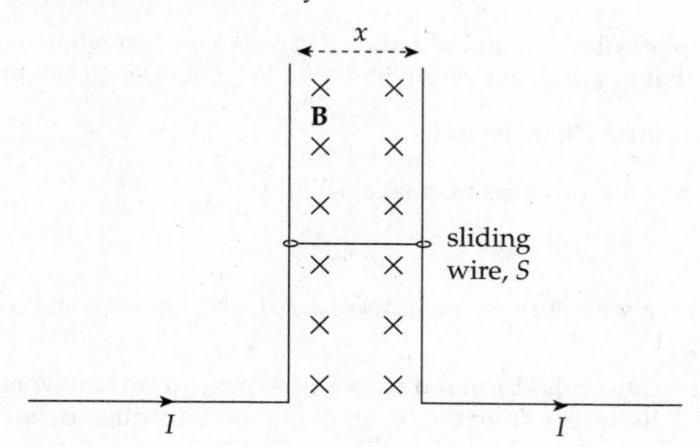

The mass of the sliding wire, S, is m. If the sliding wire S crosses a region that contains a uniform magnetic field **B**, how much current must be carried by the wire to keep S from sliding down (due to its weight)?

3. The figure below shows two long, straight wires connected by a circular arc of radius x that subtends a central angle ϕ. The current in the wire is I.

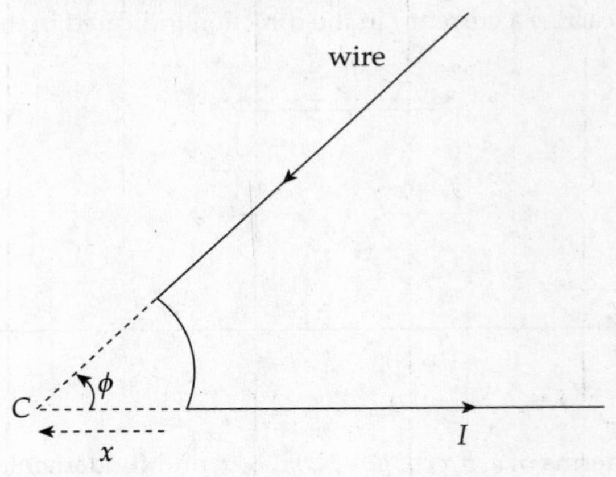

(a) Find the magnetic field (magnitude and direction) created at Point C. Write your answer in terms of x, ϕ, I, and fundamental constants.

(b) A particle of charge $+q$ is placed at Point C and released. Find the magnetic force on the particle.

(c) A second long, straight wire is set up perpendicular to the plane of the page through C, carrying the same current, I (directed out of the page), as the wire pictured in the diagram. Determine the magnetic force per unit length between the wires.

4. For a conducting rod that carries a current I, the **current density** is defined as the current per unit area: $J = I/A$.

Part 1. A homogeneous cylindrical rod of radius R carries a current whose current density, J, is uniform (constant); that is, J does not vary with the radial distance, r, from the center of the rod.

(a) Determine the current, I, in the rod.

(b) Calculate the magnitude of the magnetic field for

(i) $r < R$

(ii) $r > R$, writing your answers in terms of r, R, I, and fundamental constants.

Part 2. A nonhomogeneous cylindrical rod of radius R carries a current whose current density, J, varies with the radial distance, r, from the center of the rod according to the equation $J = \sigma r$, where σ is a constant.

(c) What are the units of σ?

(d) Determine the current, I, in the rod.

(e) Calculate the magnitude of the magnetic field for

(i) $r < R$

(ii) $r > R$, writing your answers in terms of r, R, and I.

MAGNETIC FORCES AND FIELDS SUMMARY

MAGNETIC FORCE

- When a charged particle moves through a magnetic field it will experience a force if the velocity is not parallel with the magnetic field. The force is given by the equation

$$\mathbf{F}_B = q(\mathbf{v} \times \mathbf{B})$$

- Notice that this is a cross product of the velocity and the magnetic field vectors, so the directions of the force can be obtained using the right-hand rule from chapter 1.
- Since the force will always be perpendicular to the velocity the force will not change the speed of the particle, it will only change its direction. This fact can be used to cause the particle to travel in a circle.
- A current-carrying wire will also experience a force when it is in a magnetic field because the wire has moving charge within it. The force is given by the equations

$$\mathbf{F}_B = I(\boldsymbol{\ell} \times \mathbf{B})$$

MAGNETIC FIELDS

- One can calculate the magnetic field created by a current using the **Biot-Savart Law** and **Ampere's Law**. Ampere's Law is easier to use, but the magnetic field created by the current needs to be more symmetric for it to be used.

- **The Biot-Savart Law** states $B = \frac{\mu_o I}{4\pi} \int \frac{d\boldsymbol{\ell} x \hat{\mathbf{r}}}{r^2}$

- **Ampere's Law** is $\oint_{loop} \boldsymbol{B} \bullet d\boldsymbol{s} = \mu_0 I_{\text{through loop}}$

- Use Ampere's Law to find the magnetic field due to the following current-carrying devices: long wires, coaxial cables, solenoids, and toroids.

13

Electromagnetic Induction

INTRODUCTION

In Chapter 12, we learned that electric currents generate magnetic fields, and we will now see how magnetism can generate electric currents.

MOTIONAL EMF

The figure below shows a conducting wire of length ℓ, moving with constant velocity **v** in the plane of the page through a uniform magnetic field **B** that's perpendicular to the page. The magnetic field exerts a force on the moving conduction electrons in the wire. With **B** pointing into the page, the direction of $\mathbf{v} \times \mathbf{B}$ is upward, so the magnetic force, $\mathbf{F}_B$, on these electrons (which are negatively-charged) is downward.

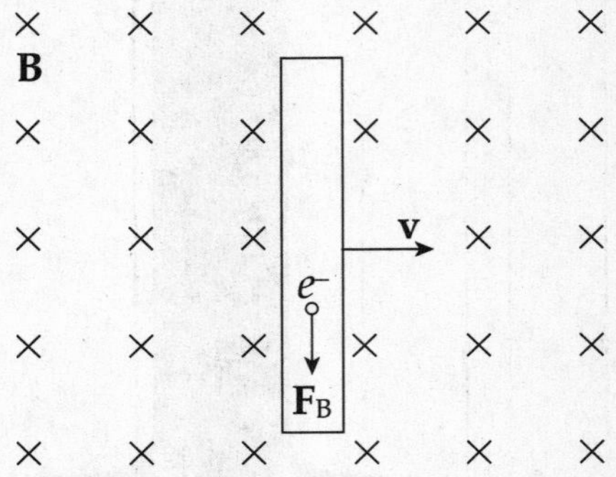

B
v
e^-
$\mathbf{F}_B$

As a result, electrons will be pushed to the lower end of the wire, which will leave an excess of positive charge at its upper end. This separation of charge creates a uniform electric field, **E**, within the wire (pointing downward).

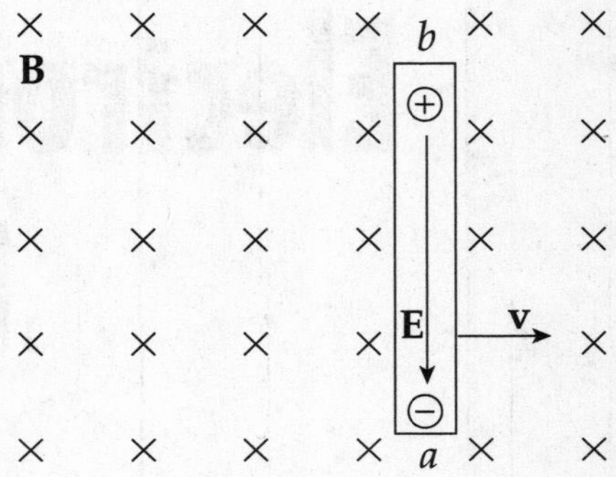

b
B
E
v
a

A charge q in the wire feels two forces: an electric force, $\mathbf{F}_E = q\mathbf{E}$, and a magnetic force,

$$\mathbf{F}_B = q(\mathbf{v} \times \mathbf{B})$$

If q is negative, $\mathbf{F}_E$ is upward and $\mathbf{F}_B$ is downward; if q is positive, $\mathbf{F}_E$ is downward and $\mathbf{F}_B$ is upward. So, in both cases, the forces act in opposite directions. Once the magnitude of $\mathbf{F}_E$ equals the magnitude of $\mathbf{F}_B$, the charges in the wire are in electromagnetic equilibrium. This occurs when $qE = qvB$; that is, when $E = vB$.

The presence of the electric field creates a potential difference between the ends of the rod. Since negative charge accumulates at the lower end (which we'll call point a) and positive charge accumulates at the upper end (point b), point b is at a higher electric potential.

The potential difference V_{ba} is equal to $E\ell$ and, since $E = vB$, the potential difference can be written as $vB\ell$.

Now, imagine that the rod is sliding along a pair of conducting rails connected at the left by a stationary bar. The sliding rod now completes a rectangular circuit, and the potential difference V_{ba} causes current to flow.

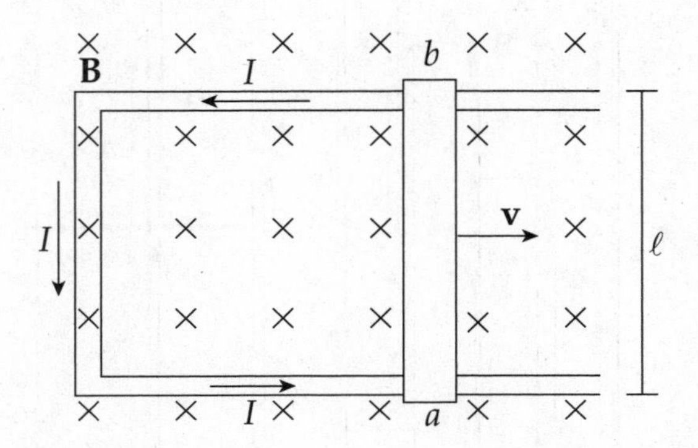

The motion of the sliding rod through the magnetic field creates an electromotive force, called **motional emf**:

$$\mathcal{E} = vB\ell$$

The existence of a current in the sliding rod causes the magnetic field to exert a force on it. Using the formula $\mathbf{F}_B = I(\boldsymbol{\ell} \times \mathbf{B})$, the fact that $\boldsymbol{\ell}$ points upward (in the direction of the current) and $\mathbf{B}$ is into the page, tells us that the direction of $\mathbf{F}_B$ on the rod is to the left. An external agent must provide this same amount of force to the right to maintain the rod's constant velocity and keep the current flowing. The power that the external agent must supply is $P = Fv = I\ell Bv$, and the electrical power delivered to the circuit is $P = IV_{ba} = I\mathcal{E} = IvB\ell$. Notice that these two expressions are identical. The energy provided by the external agent is transformed first into electrical energy and then thermal energy as the conductors making up the circuit dissipate heat.

FARADAY'S LAW OF ELECTROMAGNETIC INDUCTION

Electromotive force can be created by the motion of a conductor through a magnetic field, but there's another way to create an emf from a magnetic field.

Magnetic Flux

Think back to chapter 10 and electric flux. The electric flux through a surface of area A is equal to the product of A and the electric field that's perpendicular to it. That is, $\Phi_E = E_{\perp}A = \mathbf{E} \cdot \mathbf{A} = EA \cos \theta$. If $\mathbf{E}$ varies over the area A, then we write $\Phi_E = \int \mathbf{E} \cdot d\mathbf{A}$.

The idea of magnetic flux is exactly the same. The **magnetic flux**, Φ_B, through an area A is equal to the product of A and the magnetic field perpendicular to it: $\Phi_B = B_{\perp}A = \mathbf{B} \cdot \mathbf{A} = BA \cos \theta$. Again, if $\mathbf{B}$ varies over the area, then we must write $\Phi_B = \int \mathbf{B} \cdot d\mathbf{A}$.

Example 13.1 The figure below shows two views of a circular loop of radius 3 cm placed within a uniform magnetic field, **B** (magnitude 0.2 T).

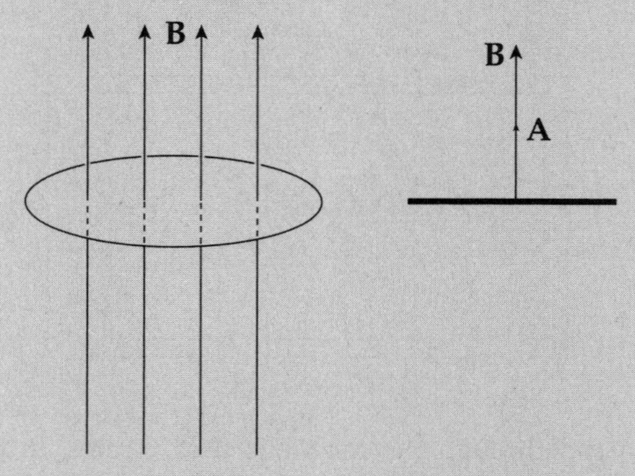

(a) What's the magnetic flux through the loop?

(b) What would be the magnetic flux through the loop if the loop were rotated 45°?

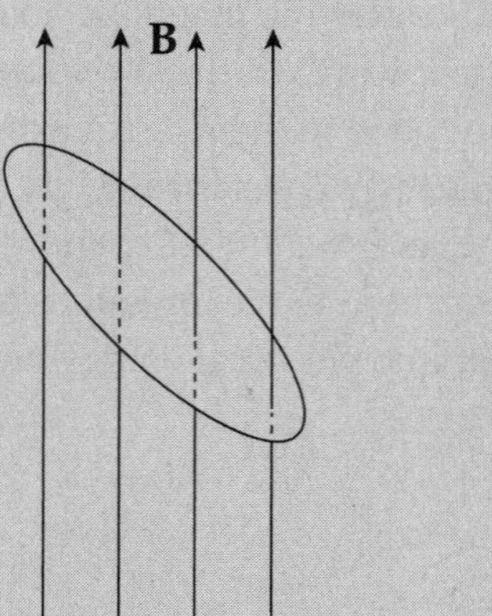

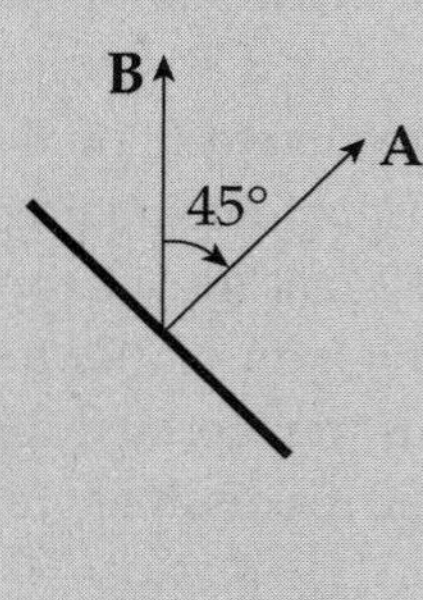

(c) What would be the magnetic flux through the loop if the loop were rotated 90°?

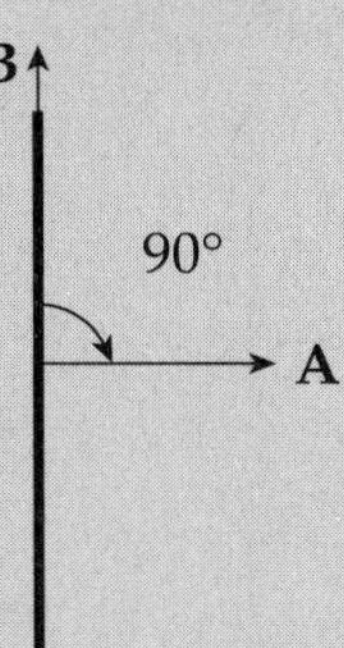

Solution.

(a) Since **B** is parallel to **A**, the magnetic flux is equal to BA:

$$\Phi_B = BA = B \cdot \pi r^2 = (0.2 \text{ T}) \cdot \pi(0.03 \text{ m})^2 = 5.7 \times 10^{-4} \text{ T·m}^2$$

The SI unit for magnetic flux, the tesla·meter², is called a **weber** (abbreviated **Wb**). So $\Phi_B = 5.7 \times 10^{-4}$ Wb.

(b) Since the angle between **B** and **A** is 45°, the magnetic flux through the loop is

$$\Phi_B = BA \cos 45° = B \cdot \pi r^2 \cos 45° = (0.2 \text{ T}) \cdot \pi(0.03 \text{ m})^2 \cos 45° = 4.0 \times 10^{-4} \text{ Wb}$$

(c) If the angle between **B** and **A** is 90°, the magnetic flux through the loop is zero, since $\cos 90° = 0$.

The concept of magnetic flux is crucial, because changes in magnetic flux induce emf. According to **Faraday's Law of Electromagnetic Induction**, the magnitude of the emf induced in a circuit is equal to the rate of change of the magnetic flux through the circuit. This can be written mathematically in the form

$$|\mathcal{E}_{avg}| = \left|\frac{\Delta\Phi_B}{\Delta t}\right|$$

or, if we let $\Delta t \to 0$, we get

$$|\mathcal{E}| = \left|\frac{d\Phi_B}{dt}\right|$$

This induced emf can produce a current, which will then create its own magnetic field. The direction of the induced current is determined by the polarity of the induced emf and is given by **Lenz's Law**: The induced current will always flow in the direction that opposes the change in magnetic flux that produced it. If this were not so, then the magnetic flux created by the induced current would magnify the change that produced it, and energy would not be conserved. Lenz's Law can be included mathematically with Faraday's Law by the introduction of a minus sign; this leads to a single equation that expresses both results:

$$\mathcal{E}_{avg} = -\frac{\Delta\Phi_B}{\Delta t} \quad \text{or} \quad \mathcal{E} = -\frac{d\Phi_B}{dt}$$

Example 13.2 The circular loop of Example 13.1 rotates at a constant angular speed through 45° in 0.5 s.

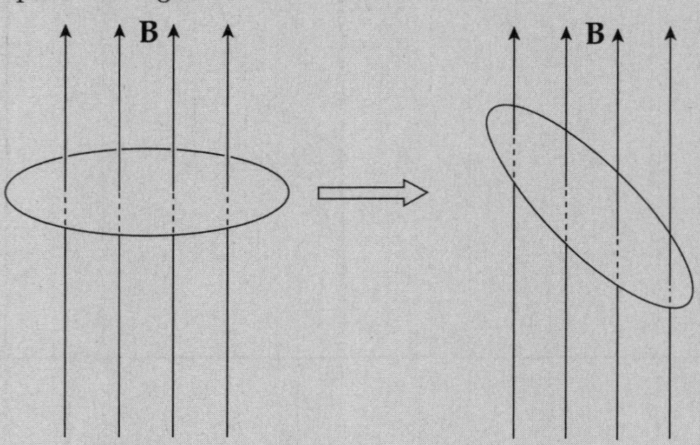

(a) What's the induced emf in the loop?

(b) In which direction will current be induced to flow?

Solution.

(a) As we found in Example 13.1, the magnetic flux through the loop changes when the loop rotates. Using the values we determined earlier, Faraday's Law gives

$$\mathcal{E}_{avg} = -\frac{\Delta\Phi_B}{\Delta t} = -\frac{(4.0\times10^{-4}\text{ Wb})-(5.7\times10^{-4}\text{ Wb})}{0.5\text{ s}} = 3.4\times10^{-4}\text{ V}$$

(b) The original magnetic flux was 5.7×10^{-4} Wb upward, and was decreased to 4.0×10^{-4} Wb. So the change in magnetic flux is -1.7×10^{-4} Wb upward, or, equivalently, $\Delta\Phi_B = 1.7 \times 10^{-4}$ Wb, downward. To oppose this change we would need to create some magnetic flux upward. The current would be induced in the counterclockwise direction (looking down on the loop), because the right-hand rule tells us that then the current would produce a magnetic field that would point up.

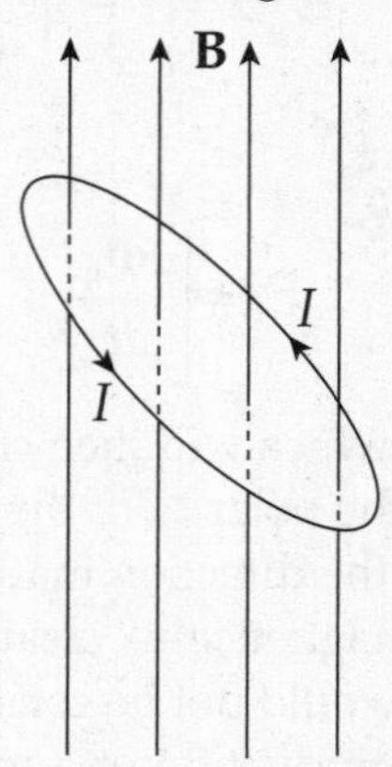

The current will flow only while the loop rotates, because emf is induced only when magnetic flux is changing. If the loop rotates 45° and then stops, the current will disappear.

Example 13.3 Again consider the conducting rod that's moving with constant velocity **v** along a pair of parallel conducting rails (separated by a distance ℓ), within a uniform magnetic field, **B**:

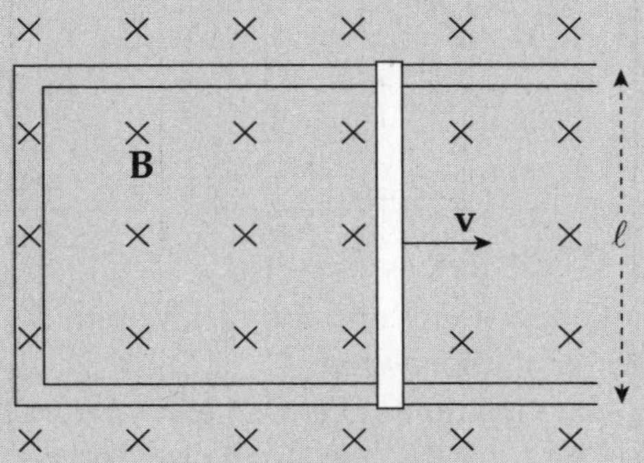

Find the induced emf and the direction of the induced current in the rectangular circuit.

Solution. The area of the rectangular loop is ℓx, where x is the distance from the left-hand bar to the moving rod:

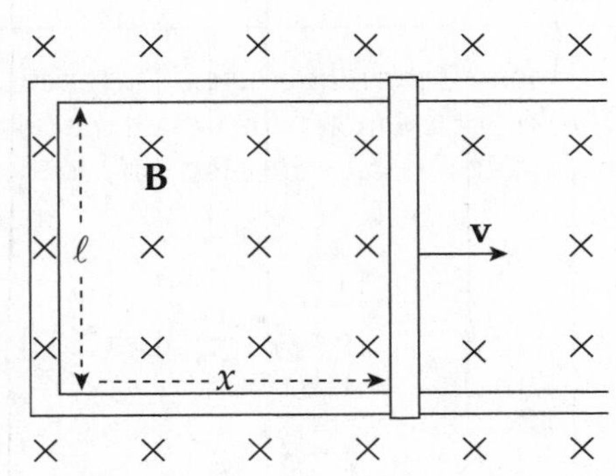

Because the area is changing, the magnetic flux through the loop is changing, which means that an emf will be induced in the loop. To calculate the induced emf, we first write $\Phi_B = BA = B\ell x$, then since $\Delta x/\Delta t = v$, we get

$$\mathcal{E}_{avg} = -\frac{\Delta\Phi_B}{\Delta t} = -\frac{\Delta(B\ell x)}{\Delta t} = -B\ell\frac{\Delta x}{\Delta t} = -B\ell v$$

We can figure out the direction of the induced current from Lenz's Law. As the rod slides to the right, the magnetic flux into the page increases. How do we oppose an increasing into-the-page flux? By producing out-of-the-page flux. In order for the induced current to generate a magnetic field that points out of the plane of the page, the current must be directed counterclockwise (according to the right-hand rule).

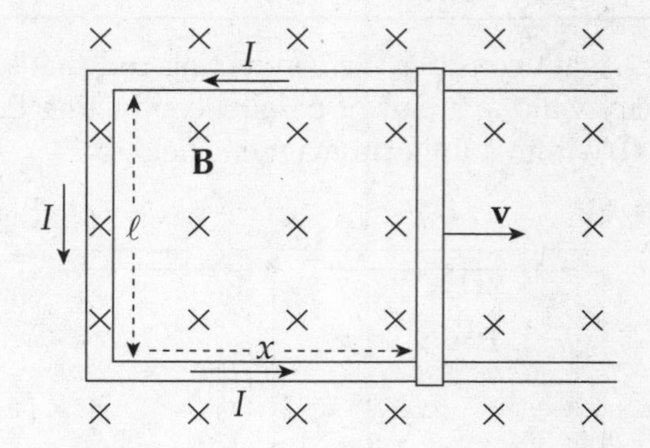

Note that the magnitude of the induced emf and the direction of the current agree with the results we derived earlier, in the section on motional emf.

This example also shows how a violation of Lenz's Law would lead directly to a violation of the Law of Conservation of Energy. The current in the sliding rod is directed upward, as given by Lenz's Law, so the conduction electrons are drifting downward. The force on these drifting electrons—and thus, the rod itself—is directed to the left, opposing the force that's pulling the rod to the right. If the current were directed downward, in violation of Lenz's Law, then the magnetic force on the rod would be to the right, causing the rod to accelerate to the right with ever-increasing speed and kinetic energy, without the input of an equal amount of energy from an external agent.

Example 13.4 A permanent magnet creates a magnetic field in the surrounding space. The end of the magnet at which the field lines emerge is designated the **north pole** (N), and the other end is the **south pole** (S):

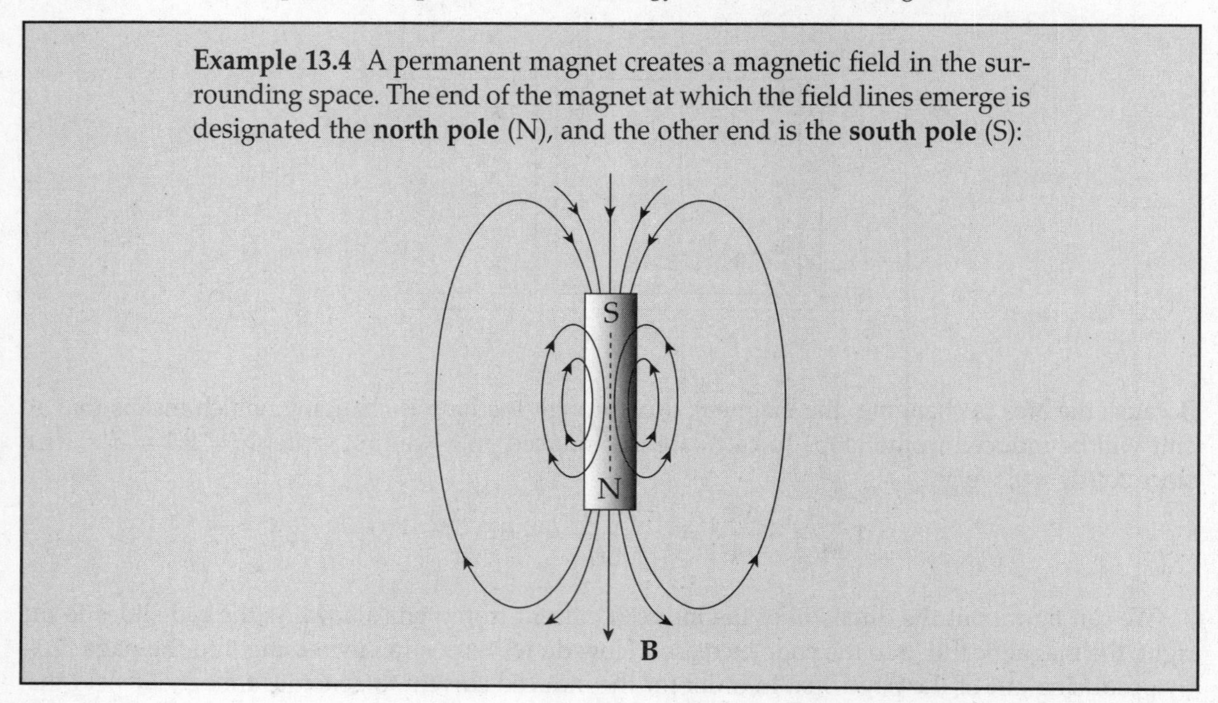

(a) The figure below shows a bar magnet moving down, through a circular loop of wire. What will be the direction of the induced current in the wire?

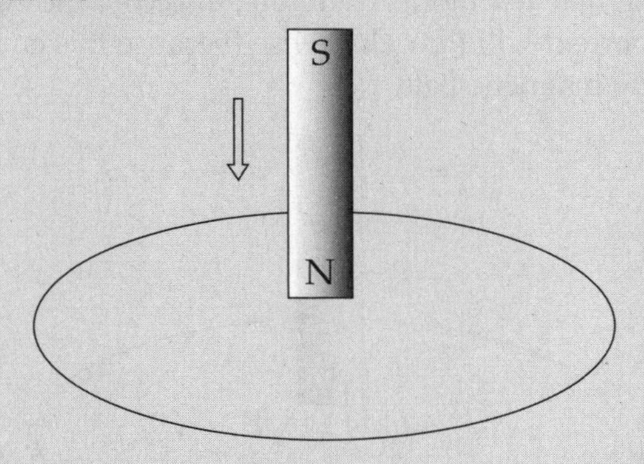

(b) What will be the direction of the induced current in the wire if the magnet is moved as shown in the following diagram?

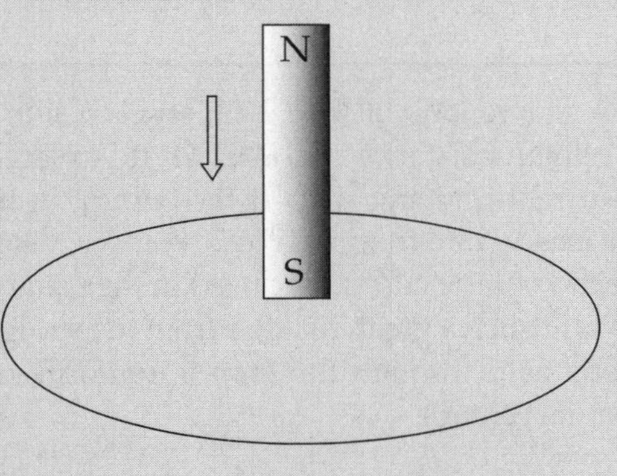

Solution.

(a) The magnetic flux down, through the loop, increases as the magnet is moved. By Lenz's Law, the induced emf will generate a current that opposes this change. How do we oppose a change of *more flux downward*? By creating flux *upward*. So, according to the right-hand rule, the induced current must flow counterclockwise (because this current will generate an upward-pointing magnetic field):

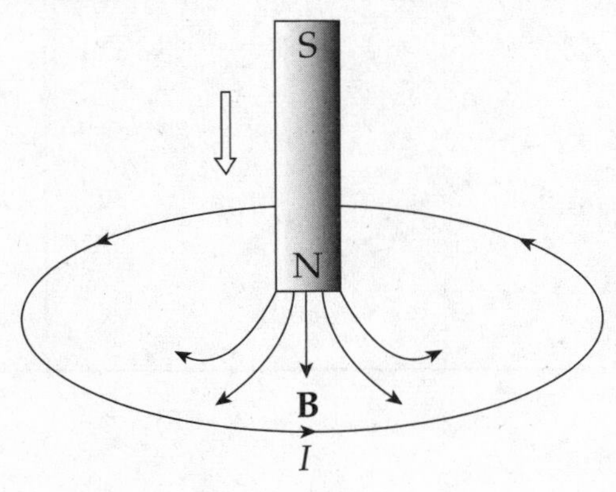

(b) In this case, the magnetic flux through the loop is upward and, as the south pole moves closer to the loop, the magnetic field strength increases so the magnetic flux through the loop increases upward. How do we oppose a change of *more flux upward*? By creating flux *downward*. Therefore, in accordance with the right-hand rule, the induced current will flow clockwise (because this current will generate a downward-pointing magnetic field):

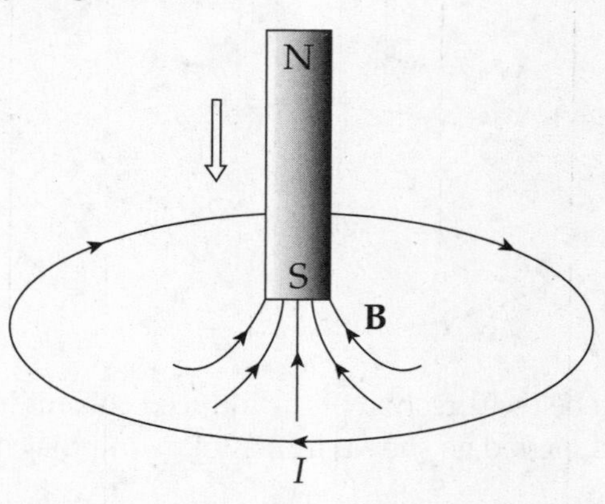

Example 13.5 A square loop of wire 2 cm on each side contains 5 tight turns and has a total resistance of 0.0002 Ω. It is placed 20 cm from a long, straight, current-carrying wire. If the current in the straight wire is increased at a steady rate from 20 A to 50 A in 2 s, determine the magnitude and direction of the current induced in the square loop. (Because the square loop is at such a great distance from the straight wire, assume that the magnetic field through the loop is uniform and equal to the magnetic field at its center.)

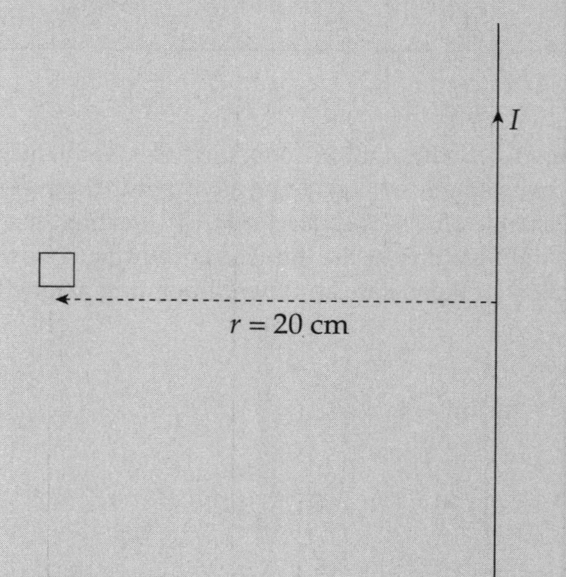

Solution. At the position of the square loop, the magnetic field due to the straight wire is directed out of the plane of the page and its strength is given by the equation $B = (\mu_0/2\pi)(I/r)$. As the current in the straight wire increases, the magnetic flux through the turns of the square loop changes, inducing an emf and current. There are $N = 5$ turns; Faraday's Law becomes $\mathcal{E}_{avg} = -N(\Delta\Phi_B/\Delta t)$, and

$$\mathcal{E}_{avg} = -N\frac{\Delta\Phi_B}{\Delta t} = -N\frac{\Delta(BA)}{\Delta t} = -NA\frac{\Delta B}{\Delta t} = -NA\frac{\mu_0}{2\pi r}\frac{\Delta I}{\Delta t}$$

Substituting the given numerical values, we get

$$\begin{aligned}\mathcal{E}_{avg} &= -NA\frac{\mu_0}{2\pi r}\frac{\Delta I}{\Delta t}\\ &= -(5)(0.02\text{ m})^2\frac{4\pi\times10^{-7}\text{ T}\cdot\text{m/A}}{2\pi(0.20\text{ m})}\frac{(50\text{ A}-20\text{ A})}{2\text{ s}}\\ &= 3\times10^{-8}\text{ V}\end{aligned}$$

The magnetic flux through the loop is out of the page and increases as the current in the straight wire increases. To oppose an increasing out-of-the-page flux, the direction of the induced current should be clockwise, thereby generating an into-the-page magnetic field (and flux).

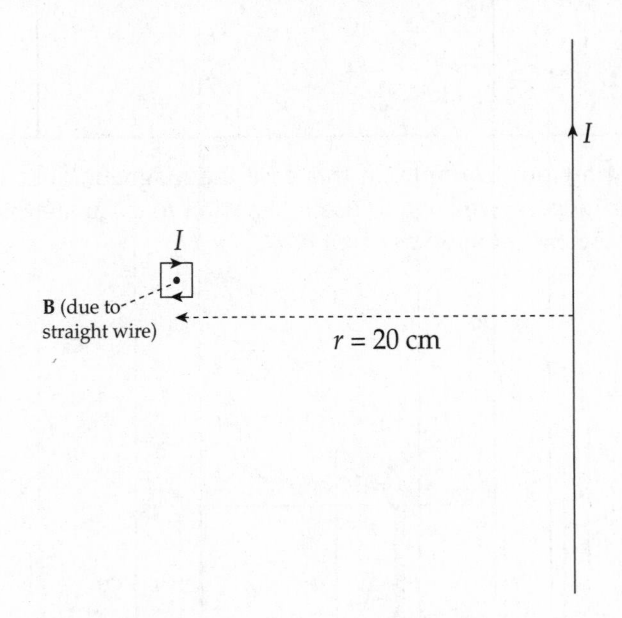

The value of the current in the loop will be

$$I = \frac{\mathcal{E}}{R} = \frac{3\times10^{-8}\text{ V}}{0.0002\ \Omega} = 1.5\times10^{-4}\text{ A}$$

Example 13.6 A rectangular loop of wire 10 cm by 4 cm has a total resistance of 0.005 Ω. It is placed 2 cm from a long, straight, current-carrying wire. If the current in the straight wire is increased at a steady rate from 20 A to 50 A in 2 s, determine the direction of the current induced in the rectangular loop.

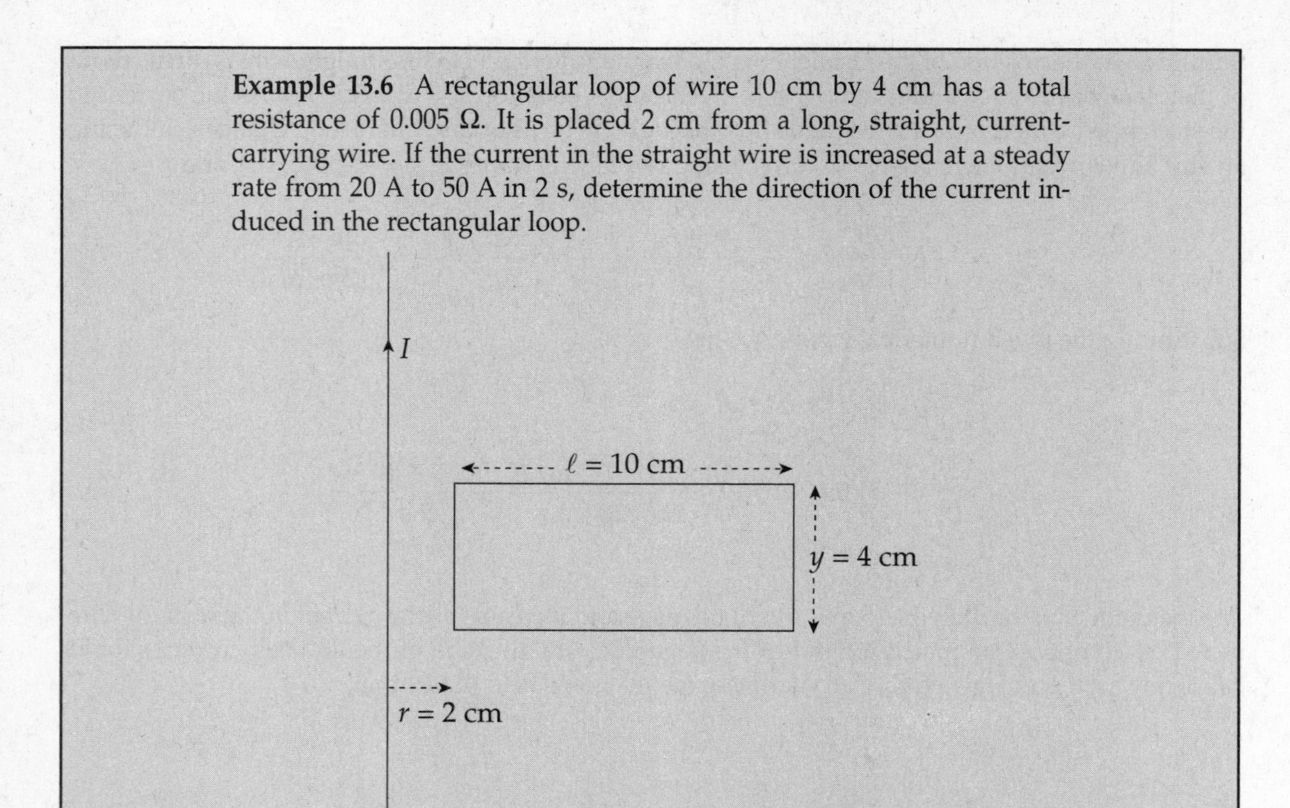

Solution. Unlike in the previous example, in this case the magnetic field varies greatly over the interior of the rectangular loop, so we have to use integration to calculate the magnetic flux. Take a narrow strip of width dx and height y, whose area is $dA = y\,dx$:

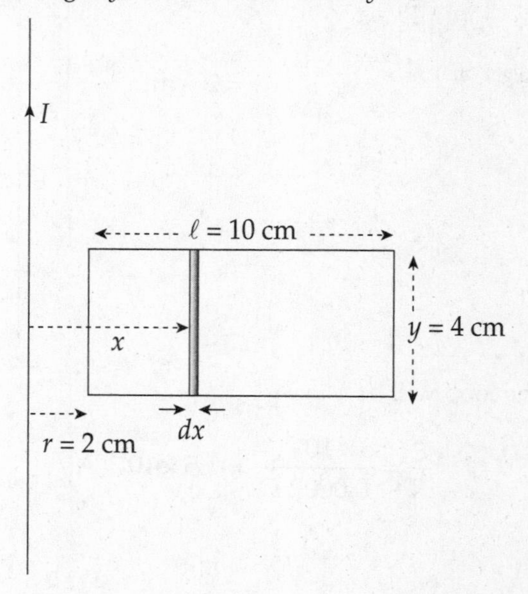

The magnetic field strength everywhere in this strip is $B = (\mu_0/2\pi)(I/x)$, so the magnetic flux through the strip is

$$d\Phi_B = B \cdot dA = B \cdot (y\,dx) = \frac{\mu_0}{2\pi}\frac{I}{x} \cdot y\,dx$$

Integrating from $x = r$ to $x = r + \ell$ gives the total magnetic flux through the loop:

$$\Phi_B = \int d\Phi_B = \int_{x=r}^{x=r+\ell} \frac{\mu_0}{2\pi}\frac{I}{x} \cdot y\,dx = \frac{\mu_0 Iy}{2\pi}\int_r^{r+\ell}\frac{dx}{x} = \frac{\mu_0 Iy}{2\pi}\left[\ln x\right]_r^{r+\ell} = \frac{\mu_0 Iy}{2\pi}\ln\frac{r+\ell}{r}$$

Faraday's Law then gives us:

$$\mathcal{E} = -\frac{d\Phi_B}{dt} = -\frac{d}{dt}\left(\frac{\mu_0 Iy}{2\pi}\ln\frac{r+\ell}{r}\right) = -\left(\frac{\mu_0 y}{2\pi}\ln\frac{r+\ell}{r}\right)\frac{dI}{dt}$$

Ignoring the minus sign (which only reminds us that Lenz's Law has to be obeyed), the value of the current induced in the loop is

$$\begin{aligned} I = \frac{\mathcal{E}}{R} &= \left(\frac{\mu_0 y}{2\pi R}\ln\frac{r+\ell}{r}\right)\frac{dI}{dt} \\ &= \frac{(4\pi\times10^{-7}\ \text{T}\cdot\text{m/A})(0.04\ \text{m})}{2\pi(0.005\ \Omega)} \cdot \ln\frac{2\ \text{cm}+10\ \text{cm}}{2\ \text{cm}} \cdot \frac{50\ \text{A}-20\ \text{A}}{2\ \text{s}} \\ &= 4.3\times10^{-5}\ \text{A} \end{aligned}$$

At the position of the rectangular loop, the magnetic field due to the straight wire is directed into the plane of the page. So, the magnetic flux through the loop is into the page and increases as the current in the straight wire increases. To oppose an increasing into-the-page flux, the direction of the induced current will be counterclockwise, generating an out-of-the-page magnetic field (and flux).

Induced Electric Fields

In Examples 13.4, 13.5, and 13.6, an electric current was induced in a stationary conducting loop by a changing magnetic flux through the loop, but what was the source of the force that pushed these charges around the loop? It was not the magnetic force, because the loop was stationary. It must therefore have been an electric force that was produced by an electric field. The changing magnetic field produced the electric field, which acted on the free charges in the conducting loop and caused the current.

Let's look at a specific situation. The figure below is a view down the central axis of an ideal solenoid (*ideal* means that the magnetic field is uniform and parallel to the axis and that no magnetic field exists outside). Surrounding the solenoid is a loop of wire.

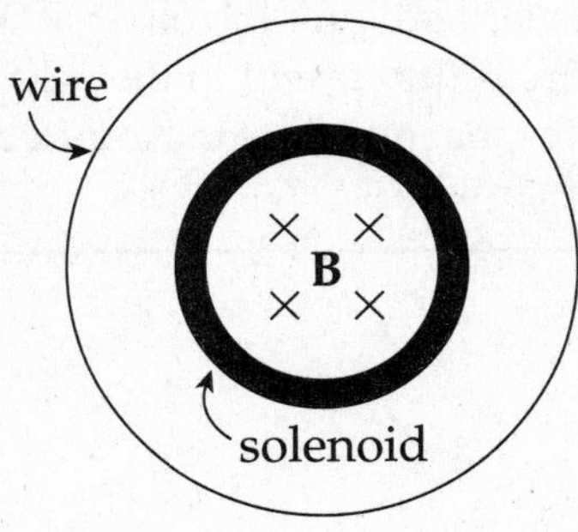

Now suppose that the current in the solenoid is increased, which increases the strength of its magnetic field. This changes the magnetic flux through the wire and induces a current (which is directed counterclockwise). The force that pushes the charges around this wire is $\mathbf{F}_E = q\mathbf{E}$; $\mathbf{E}$ is the induced electric field.

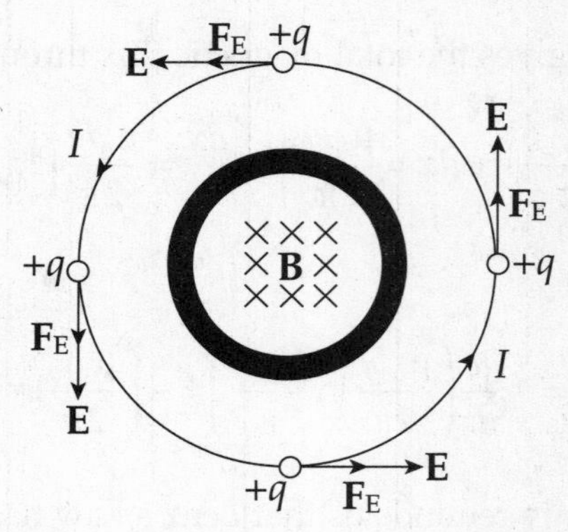

The work done on a charge q as it makes one revolution is F_E times the distance around the wire, $F_E \cdot 2\pi r = qE\cdot(2\pi r)$. So the work per unit charge, which is the definition of emf, is equal to $E(2\pi r)$. But Faraday's Law says that the emf is equal to $-d\Phi_B/dt$:

$$E(2\pi r) = -d\Phi_B/dt$$

This equation can be generalized by saying that the work done per unit charge by the induced electric field around a closed path is equal to $\oint \mathbf{E} \cdot d\boldsymbol{\ell}$. Therefore,

$$\oint \mathbf{E} \cdot d\boldsymbol{\ell} = -\frac{d\Phi_B}{dt}$$

This is a restatement of Faraday's Law that includes the electric field, **E**, induced by the changing magnetic flux. Notice that this electric field is different from the ones we studied in the previous chapters. Electric fields created by stationary source charges, called electrostatic fields, are conservative, meaning that the work done by them on charges moved along closed paths is always zero. However, as we've just seen in the example of the solenoid above, the electric field induced by a *changing magnetic flux* does not share this property. The work done by $\mathbf{E}_{\text{induced}}$ on a charge as it moves around a closed path is equal to $\oint \mathbf{E} \cdot d\boldsymbol{\ell}$, which is not zero because $d\Phi_B/dt$ is not zero. Because of this, the electric field induced by a changing magnetic flux is nonconservative.

Example 13.7 For the situation just described, assume that the solenoid has 15,000 turns per meter and a radius of r = 2 cm. The radius of the circular loop is R = 4 cm. If the current in the solenoid is increased at a rate of 10 A/s, what is the magnitude of the induced electric field at each position along the circular wire?

Solution. From Faraday's Law, we have

$$\oint \mathbf{E}\cdot d\boldsymbol{\ell} = -\frac{d\Phi_B}{dt}$$

$$E(2\pi R) = -\frac{d\Phi_B}{dt}$$

$$|E| = \frac{1}{2\pi R}\left|\frac{d\Phi_B}{dt}\right|$$

Assuming that the magnetic field exists only inside the solenoid, the magnetic flux through the circular wire is $BA = B(\pi r^2)$, where $B = \mu_0 nI$:

$$\Phi_B = (\mu_0 nI)\cdot(\pi r^2)$$

Therefore,

$$\begin{aligned}
|E| &= \frac{1}{2\pi R}\left|\frac{d\Phi_B}{dt}\right| \\
&= \frac{1}{2\pi R}\cdot\frac{d}{dt}(\mu_0 nI\cdot\pi r^2) \\
&= \frac{\mu_0 nr^2}{2R}\frac{dI}{dt} \\
&= \frac{(4\pi\times10^{-7}\ \text{T}\cdot\text{m/A})(15{,}000/\text{m})(0.02\ \text{m})^2}{2(0.04\ \text{m})}(10\ \text{A/s}) \\
&= 9.4\times10^{-4}\ \text{V/m}
\end{aligned}$$

INDUCTANCE

Placing a capacitor in series with a resistor and battery in an electric circuit causes the current to drop exponentially from $\mathcal{E}/R$ (when the switch is closed) to zero, as charge builds up on the capacitor and causes the voltage across the capacitor to oppose the emf of the battery. A similar thing happens when an inductor is placed in series with a resistor and a source of emf in an electric circuit.

An **inductor** is a circuit element that opposes changes in current. The prototypical inductor is a coil (a solenoid) that looks like this in diagrams:

How does this coil oppose changes in current in a circuit? Well, when current exists in the coil, a magnetic field is created, and magnetic flux passes through the loops. If the current changes, then the magnetic field changes proportionately, as does the magnetic flux through the loops. But, according to Faraday's Law of Induction, a changing magnetic flux induces an emf that opposes the change that produced it. In other words, the changing magnetic flux through the coil, due to the current in the coil itself, produces a **self-induced emf**. Since this self-induced emf opposes the change in the current that produced it, it's also called **back emf**.

Assume that the inductor contains N turns and that Φ_B is the magnetic flux through each turn. Then the total magnetic flux through the entire coil is $N\Phi_B$; this is proportional to the current, so $N\Phi_B = LI$ for some constant, L. The proportionality constant L is called the **inductance** (or **self-inductance**) of the coil:

$$L = \frac{N\Phi_B}{I}$$

The SI unit for inductance is the weber per ampere, which is renamed a **henry (H)**.

> **Example 13.8** An ideal solenoid of cross sectional area A and length ℓ contains n turns per unit length. What is its self-inductance?

Solution. The magnetic field inside an ideal solenoid is parallel to its central axis and has strength $B = \mu_0 nI$. Since the solenoid's length is ℓ, the total number of turns, N, is $n\ell$. Therefore,

$$L = \frac{N\Phi_B}{I} = \frac{N \cdot BA}{I} = \frac{n\ell \cdot BA}{I} = \frac{n\ell \cdot (\mu_0 nI)A}{I} = \mu_0 n^2 A\ell$$

Or, in terms of the total number of turns in the solenoid, $L = \mu_0 N^2 A/\ell$.

The self-induced emf in an inductor now follows directly from the definition of L and Faraday's Law. Since $N\,\Phi_B = LI$, we have $N(d\,\Phi_B/dt) = L(dI/dt)$. But by Faraday's Law, $\mathcal{E} = -N(d\,\Phi_B/dt)$. So,

$$\mathcal{E} = -L\frac{dI}{dt}$$

The self-induced emf is proportional to the rate of change of the current. The faster the current tries to change, the greater is the self-induced (back) emf. But if the current is steady, then $\mathcal{E} = 0$.

RL Circuits

Consider the following circuit, which contains a battery, an inductor, and a resistor:

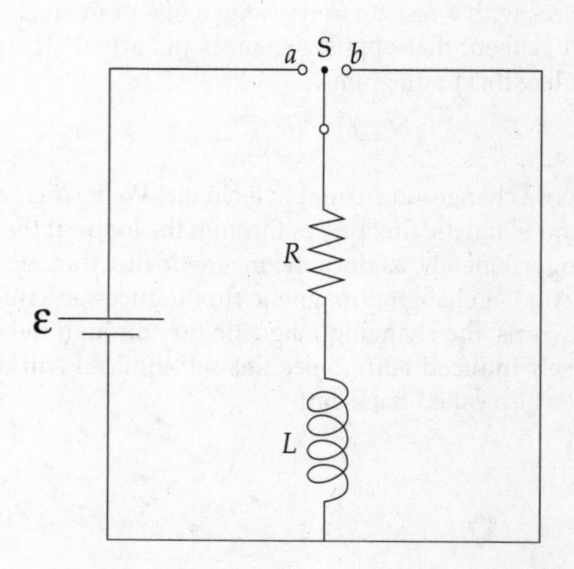

Move the switch S to point a. If the inductor were absent, the current would rise abruptly to $\mathcal{E}/R$, but the inductor opposes this rapid increase in current by producing a back emf. The result is a current that increases gradually in time, according to the equation

$$I(t) = \frac{\mathcal{E}}{R}\left(1 - e^{-t/\left(\frac{L}{R}\right)}\right)$$

The ratio of L to R (in the exponent) is called the **inductive time constant**: $\tau_L = L/R$, and the equation for the rising current in the **RL circuit** (a circuit that contains a resistor and an inductor) can be written as:

$$I(t) = \frac{\mathcal{E}}{R}\left(1 - e^{-t/\tau_L}\right)$$

which is similar to the equation for the gradually rising charge on the plates of a capacitor in an RC circuit. Notice that for an RL circuit the current increases over time and for an RC circuit the current decreases over time.

As current builds in the circuit, the source of emf provides electrical energy. Some of this energy is dissipated as heat by the resistor, and the remainder is stored in the magnetic field of the inductor. The amount of stored energy is

$$U_L = \frac{1}{2}LI^2$$

where L is the inductance of the inductor carrying a current I.

Example 13.9 In a particular circuit, assume that $\mathcal{E}$ = 12V, R = 40 Ω, and L = 5 mH. How much energy is stored in the inductor's magnetic field when the current reaches its maximum steady-state value?

Solution. When the current in the circuit reaches its maximum steady-state value, $I = \mathcal{E}/R$. At this point, the energy stored in the magnetic field of the inductor is

$$U_L = \frac{1}{2}LI^2 = \frac{1}{2}L \cdot \left(\frac{\mathcal{E}}{R}\right)^2 = \frac{L\mathcal{E}^2}{2R^2} = \frac{(5\times10^{-3}\text{ H})(12\text{ V})^2}{2(40\ \Omega)^2} = 2.3\times10^{-4}\text{J}$$

Once the current has reached its steady-state value of $\mathcal{E}/R$, imagine that you move the switch to point b. Without the inductor, the current would drop abruptly to zero (because the source of emf has been taken out of the circuit). But the inductor would oppose this abrupt decrease in the current by producing a self-induced emf. The presence of this emf would cause the current to die out gradually, according to the equation

$$I(t) = \frac{\mathcal{E}}{R}e^{-t/\tau_L}$$

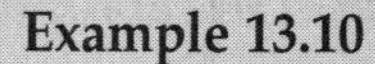

Example 13.10

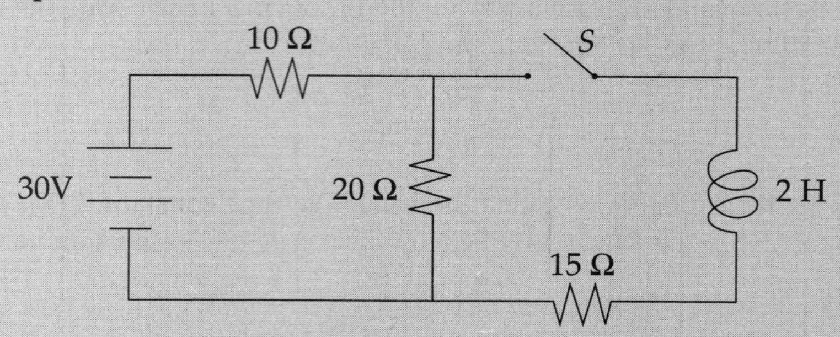

A circuit is connected as shown above.

(a) Determine the current through the 10 Ω resistor when the switch is open.

(b) Determine the current through the 15 Ω resistor when the switch is first closed.

(c) Determine the current through the 10 Ω resistor when the switch has been closed a long time.

Solution.

(a) When the switch is open the circuit is a basic series circuit with a total resistance of 30 Ω. Use Ohm's Law to solve for the current:

$$V = IR \quad \Rightarrow \quad 30 = I(30) \quad \Rightarrow \quad I = 1\text{ A}$$

(b) When the switch is first closed the inductor will not let the current go through that part of the branch because it opposes changes in current. Therefore the current through the 15 Ω resistor is zero.

(c) After the switch has been closed a long time the inductor acts like a wire because the current is constant. Now we solve for the resistance of the parallel part of the circuit. $\frac{1}{R} = \frac{1}{15} + \frac{1}{20} \Rightarrow R = 8.57\ \Omega$

Then the total resistance is that value added to 10 Ω. $R_{tot} = 18.57\ \Omega$

The current leaving the battery will be the current through the 10 Ω resistor. Solve for that current using Ohm's Law.

$$V = IR \quad \Rightarrow \quad 30 = I(18.57) \quad \Rightarrow \quad I = 1.62\text{ A}$$

LC Circuits

The RC circuit we looked at in Chapter 11 and the RL circuit above are similar in that they both experience an exponential increase or decrease of current. However, an **LC circuit**, which is a circuit that contains both an inductor and a capacitor, behaves quite differently.

The following figure shows a capacitor and an inductor in a simple series circuit.

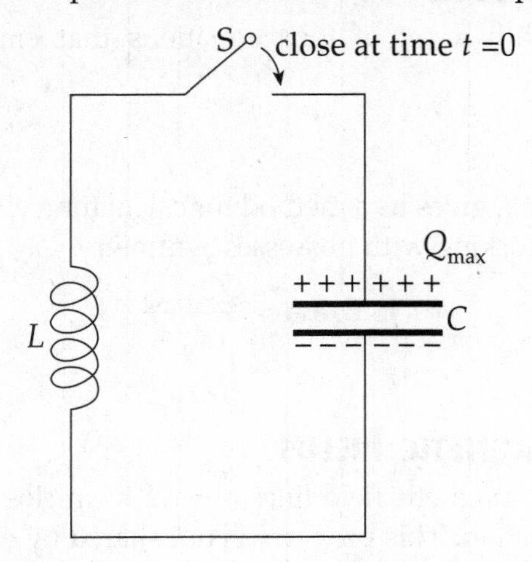

Let's assume that initially there is no current and that the capacitor is charged; at time $t = 0$, the switch S is closed and the circuit is complete. The presence of the inductor prevents the capacitor from discharging abruptly, so the current rises gradually and the energy in the electric field of the capacitor is transferred to energy in the magnetic field of the inductor. Once the capacitor has discharged, all the energy is in the inductor's magnetic field. The current in the inductor (now at its maximum), delivers charge to the capacitor, but in the opposite direction from its original charge configuration. The current gradually returns to zero as the magnetic-field energy in the inductor is transformed back to electric-field energy in the capacitor.

The capacitor starts to discharge again, but this time it sends current in the opposite direction. The current rises gradually, reaching a maximum value as the charge on the capacitor reaches zero. The inductor continues to deliver charge to the capacitor as the current gradually returns to zero, and it finds itself right back where it started. The circuit oscillates, and this defines one cycle of oscillation.

The frequency of oscillation is related to inductance and capacitance by the equation

$$f = \frac{1}{2\pi\sqrt{LC}}$$

The charge on the capacitor, as a function of time (assuming that it possessed its maximum charge, Q_{max}, at time $t = 0$) is

$$Q(t) = Q_{max} \cos (\omega t)$$

where $\omega = 2\pi f$.

Example 13.11 Find an equation for the current in the LC circuit as a function of time. How long does it take for the circuit to complete a full oscillation?

Solution. By definition, the current is the time derivative of the charge. So,

$$I(t) = \frac{dQ}{dt} = \frac{d}{dt}\left[Q_{max} \cos(\omega t)\right] = -\omega Q_{max} \sin(\omega t)$$

The time required to complete a cycle (the period) would be the reciprocal of the frequency:

$$T = \frac{1}{f} = 2\pi\sqrt{LC}$$

MAXWELL'S EQUATIONS

We'll finish up this chapter with a set of four equations that embody the subject of electromagnetism.

1. Gauss's Law

This law, first studied in Chapter 9, gives us a method for calculating electric fields and is particularly useful when the system we're working with possesses symmetry.

$$\int_{\substack{\text{closed}\\ \text{surface}}} \mathbf{E} \cdot d\mathbf{A} = \frac{Q_{\text{enclosed}}}{\varepsilon_0}$$

2. Gauss's Law for Magnetic Fields

We mentioned (Chapter 12) that magnetic field lines always form closed loops that encircle the current that generates them. (Remember: This property is not shared by electrostatic fields, which radiate away from positive source charges and toward negative ones.) The fact that magnetic field lines always close upon themselves tells us that the magnetic flux through any closed surface must be zero; as much magnetic flux will enter the closed surface as will exit it. Mathematically, this says:

$$\int_{\substack{\text{closed}\\ \text{surface}}} \mathbf{B} \cdot d\mathbf{A} = 0$$

Another statement of this law is that there are no magnetic monopoles. If there were, a single isolated magnetic north pole would generate a magnetic field that radiates away, and an isolated magnetic south pole would generate a magnetic field that radiates inward, toward the pole. A closed surface surrounding such a magnetic charge (if one existed) would have a nonzero flux through it. No magnetic monopoles like this have ever been observed.

3. Faraday's Law

As we discussed in this chapter, a changing magnetic flux induces an emf; this also means that a changing magnetic field produces an electric field. The equation is

$$\varepsilon = \oint \mathbf{E} \cdot d\ell = -\frac{d\Phi_B}{dt}$$

This equation is typically used to find the emf and the electric field induced by a changing magnetic flux.

4. The Ampere-Maxwell Law

The final equation in this set begins with Ampere's Law, which provides us with a method for calculating magnetic fields and is particularly useful when systems are symmetrical. It reads

$$\oint \mathbf{B} \cdot d\mathbf{s} = \mu_0 I_{\text{enclosed}}$$

You may notice that the first two equations listed in this section possess a sort of symmetry; there's a Gauss's Law for E-fields and one for B-fields. The third equation says that a changing B-field produces an E-field. One question we haven't asked yet, which is of interest here is, *Does a changing E-field produce a B-field?* The answer is "yes," and the equation can be amended in the following way

[Maxwell added the missing piece, which is proportional to the rate of change of the electric flux, $\mu_0\varepsilon_0(d\,\Phi_E/dt)$]:

$$\oint \mathbf{B}\cdot d\mathbf{s} = \mu_0 I + \mu_0\varepsilon_0 \frac{d\Phi_E}{dt}$$

This shows that a changing electric field, which appears on the right-hand side of the equation [in the term $\mu_0\varepsilon_0(d\,\Phi_E/dt)$], will produce a magnetic field, which appears on the left-hand side of the equation.

The quantity $\varepsilon_0(d\,\Phi_E/dt)$ has units of current and is called displacement current, I_D. It differs from the current I, which is conduction current (I_C), because while I_C is composed of moving electric charge, I_D is not. The Ampere–Maxwell equation can be written in terms of conduction current and displacement current as follows:

$$\oint \mathbf{B}\cdot d\mathbf{s} = \mu_0(I_C + I_D)$$

CHAPTER 13 REVIEW QUESTIONS

SECTION I: MULTIPLE CHOICE

1. A metal rod of length L is pulled upward with constant velocity **v** through a uniform magnetic field **B** that points out of the plane of the page.

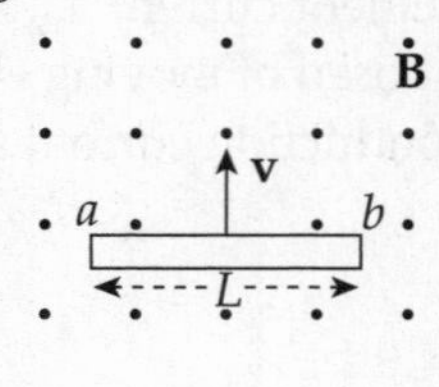

What is the potential difference between points a and b?

(A) 0

(B) $\frac{1}{2}vBL$, with point a at the higher potential

(C) $\frac{1}{2}vBL$, with point b at the higher potential

(D) vBL, with point a at the higher potential

(E) vBL, with point b at the higher potential

2. A circular disk of radius a is rotating at a constant angular speed ω in a uniform magnetic field, **B**, which is directed out of the plane of the page.

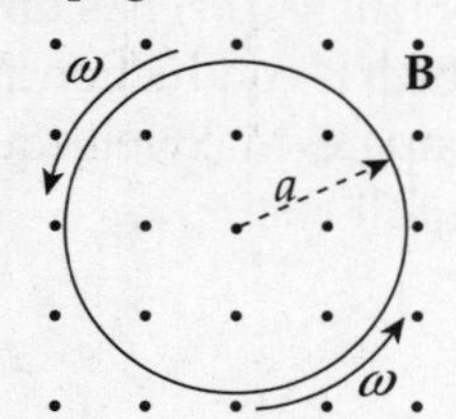

Determine the induced emf between the center of the disk and the rim.

(A) $\frac{1}{2}\omega \mathbf{B}a$

(B) $\frac{1}{2}\mathbf{B}a$

(C) $\frac{1}{2}\omega \mathbf{B}a^2$

(D) $\omega \mathbf{B}a^2$

(E) $2\pi\omega \mathbf{B}a^2$

3. A conducting rod of length 0.2 m and resistance 10 ohms between its endpoints slides without friction along a U-shaped conductor in a uniform magnetic field B of magnitude 0.5 T perpendicular to the plane of the conductor, as shown in the diagram below.

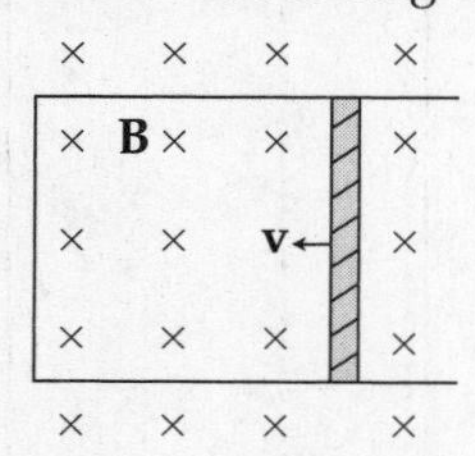

If the rod is moving with velocity **v** = 3 m/s to the left, what is the magnitude and direction of the current induced in the rod?

	Current	Direction
(A)	0.03 A	down
(B)	0.03 A	up
(C)	0.3 A	down
(D)	0.3 A	up
(E)	3 A	down

4. In the figure below, a small, circular loop of wire (radius r) is placed on an insulating stand inside a hollow solenoid of radius R. The solenoid has n turns per unit length and carries a current I. If the current in the solenoid is decreased at a steady rate of a amps/s, determine the induced emf, $\mathcal{E}$, and the direction of the induced current in the loop.

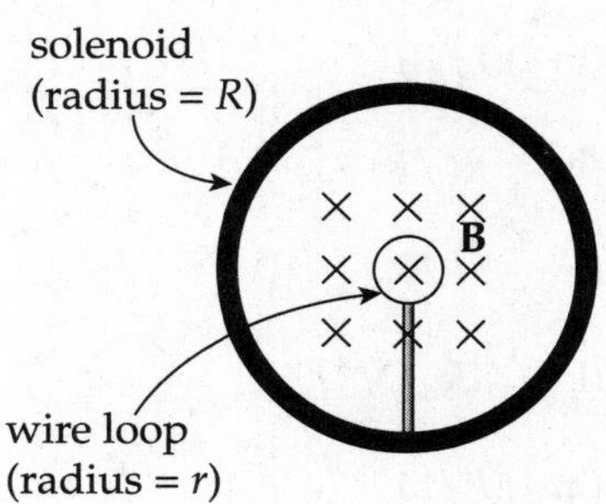

(A) $\mathcal{E} = \mu_0 \pi n r^2 a$; induced current is clockwise

(B) $\mathcal{E} = \mu_0 \pi n r^2 a$; induced current is couterclockwise

(C) $\mathcal{E} = \mu_0 \pi n R^2 a$; induced current is clockwise

(D) $\mathcal{E} = \mu_0 \pi n R^2 a$; induced current is couterclockwise

(E) $\mathcal{E} = \mu_0 \pi I n R^2 a$; induced current is counterclockwise

5. In the figure below, a permanent bar magnet is pulled upward with a constant velocity through a loop of wire.

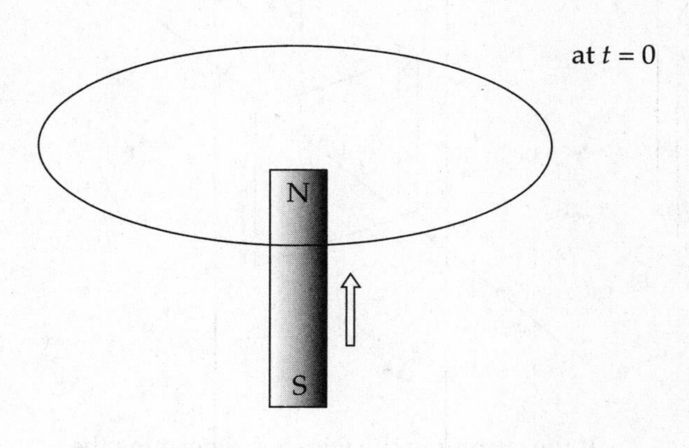

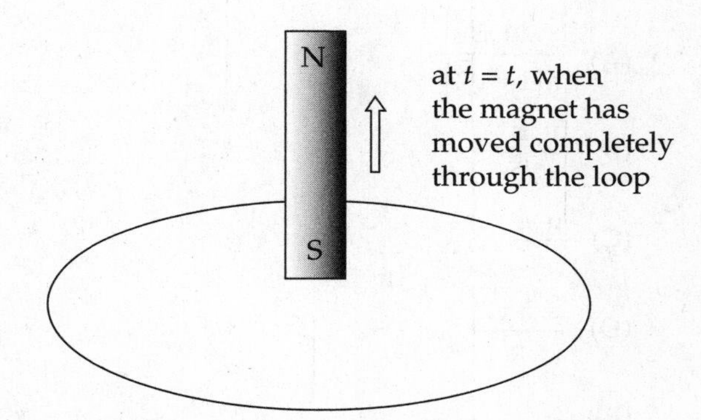

Which of the following best describes the direction(s) of the current induced in the loop (looking down on the loop from above)?

(A) Always clockwise
(B) Always counterclockwise
(C) First clockwise, then counterclockwise
(D) First counterclockwise, then clockwise
(E) No current will be induced in the loop

6. A square loop of wire (side length = s) surrounds a long, straight wire such that the wire passes through the center of the square.

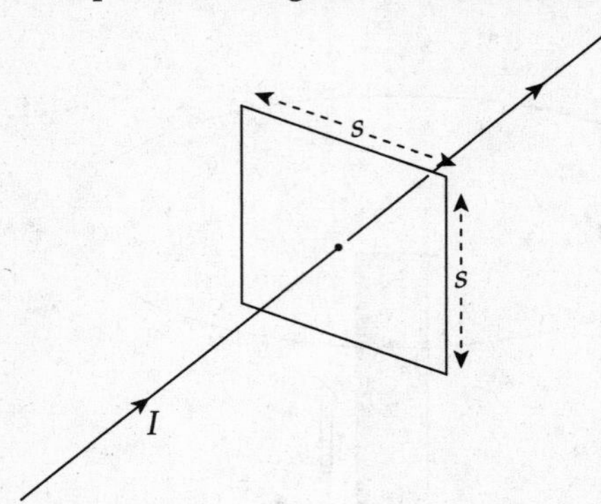

If the current in the wire is I, determine the current induced in the square loop.

(A) $\dfrac{2\mu_0 Is}{\pi(1+\sqrt{2})}$

(B) $\dfrac{\mu_0 Is}{\pi\sqrt{2}}$

(C) $\dfrac{\mu_0 Is}{\pi}$

(D) $\dfrac{\mu_0 Is\sqrt{2}}{\pi}$

(E) 0

Questions 7–9

A circuit contains a solenoid of inductance L in series with a resistor of resistance R and a battery with terminal voltage ε. At time $t = 0$, a switch is closed and the circuit is completed.

7. How long does it take for the current to reach 3/4 of its maximum (steady-state) value?

(A) $(\ln 4)(L/R)$

(B) $(\ln \frac{3}{4})(L/R)$

(C) $(\ln \frac{4}{3})(L/R)$

(D) $(\ln \frac{4}{3})(R/L)$

(E) $(\ln 4)(R/L)$

8. When the current reaches its maximum value, how much energy is stored in the magnetic field of th solenoid?

(A) $L^2\varepsilon^2/(4R^2)$
(B) $L^2\varepsilon^2/(2R^2)$
(C) $L\varepsilon^2/(4R^2)$
(D) $L\varepsilon^2/(2R^2)$
(E) 0

9. When the current reaches its maximum value, what is the total magnetic flux through the solenoid?

(A) $L\varepsilon$
(B) $L\varepsilon/R$
(C) $\varepsilon/(RL)$
(D) RL/ε
(E) 0

10. Which one of Maxwell's equations states that a changing electric field produces a magnetic field?

(A) Gauss's Law
(B) Gauss's Law for Magnetism
(C) Biot-Savart Law
(D) Ampere-Maxwell Law
(E) Faraday's Law

SECTION II: FREE RESPONSE

1. The diagram below shows two views of a metal rod of length ℓ rotating with constant angular speed ω about an axis that is in the plane of the page. The rotation takes place in a uniform magnetic field **B** whose direction is parallel to the angular velocity ω.

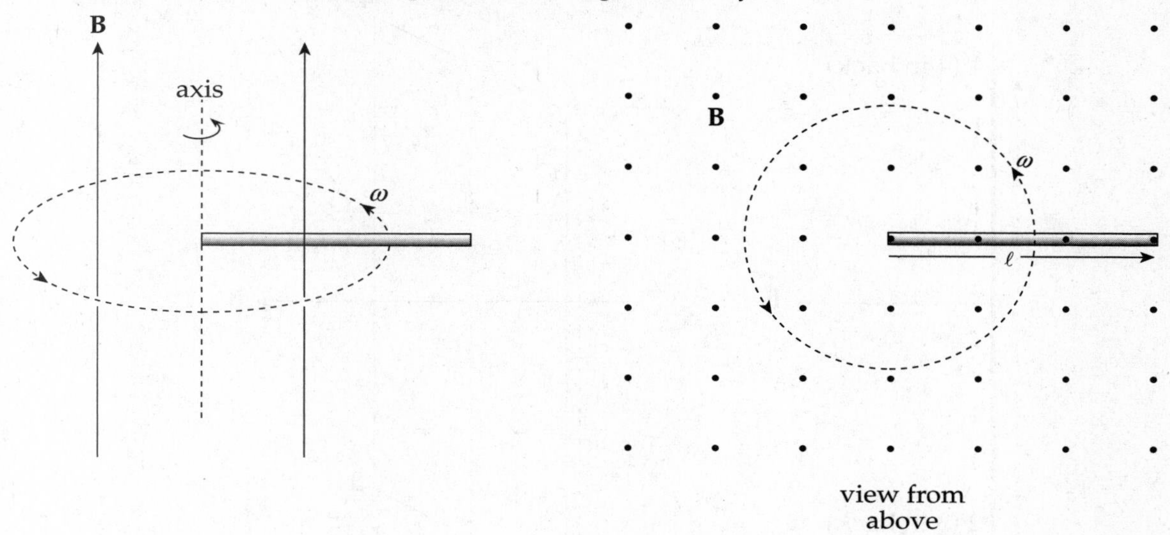

(a) What is the emf induced between the ends of the rod?

(b) What is the polarity (+ or –) of the rotating end?

In the following diagram, a metal rod of length ℓ moves with constant velocity **v** parallel to a long, straight wire carrying a steady current I. The lower end of the rod maintains a distance of a from the straight wire.

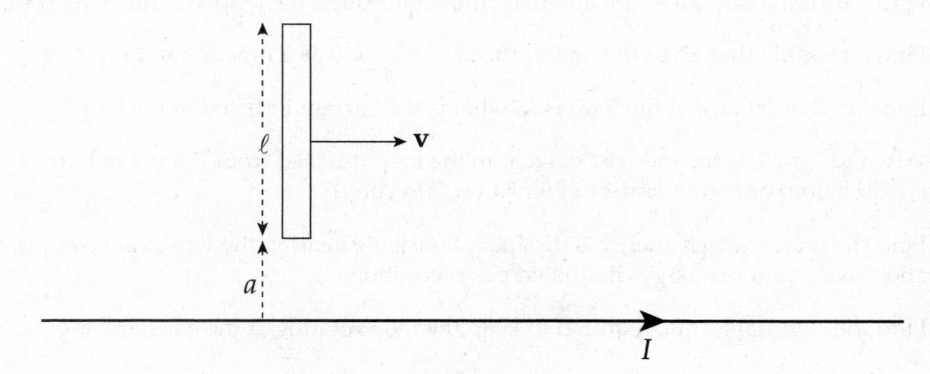

(c) What is the emf induced between the ends of the rod?

(d) What is the polarity (+ or –) of the end that is farther from the straight wire?

2. A rectangular loop of wire (side lengths a and b) rotates with constant angular speed ω in a uniform magnetic field **B**. At time $t = 0$, the plane of the loop is perpendicular to **B**, as shown in the figure on the left. The magnetic field **B** is directed to the right (in the $+x$ direction), and the rotation axis is the y axis (with ω in the $+y$ direction), and the four corners of the loop are labeled 1, 2, 3, and 4. (Express answers in terms of a, b, ω, B, and fundamental constants.)

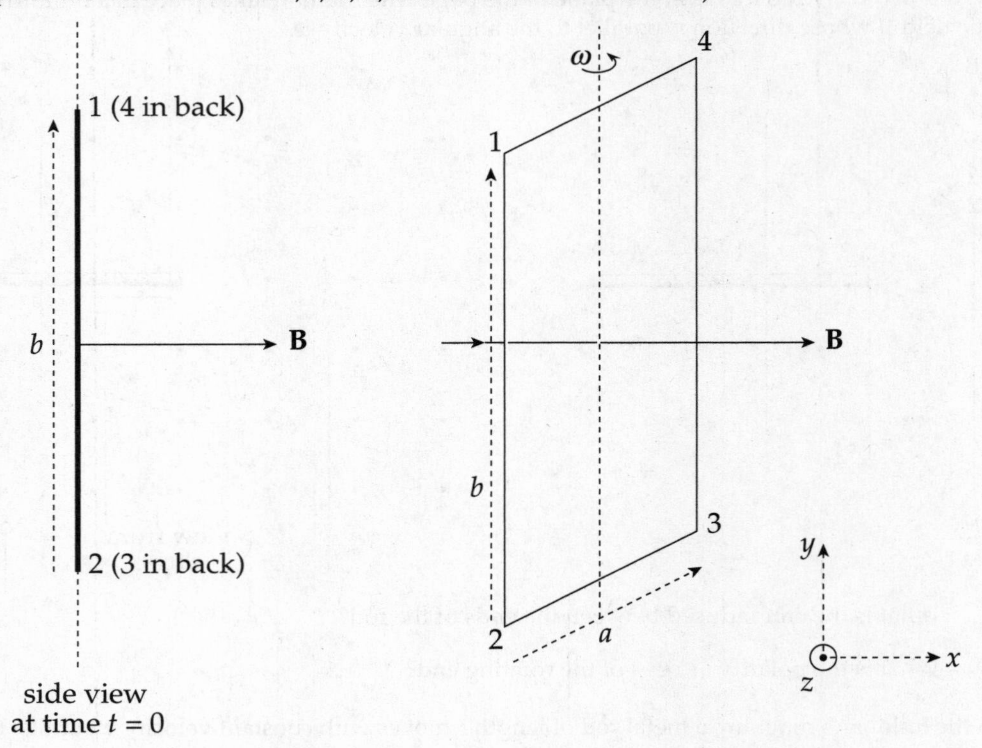

(a) Find a formula that gives the magnetic flux Φ_B through the loop as a function of time, t.

(b) Find a formula that gives the emf induced in the loop as a function of time, t.

(c) If the total resistance of the loop is R, what is the current induced in the loop?

(d) When $\omega t = \pi/2$, is the induced current in the loop directed from Point 1 to Point 2 ($-y$ direction) or from Point 2 to Point 1 ($+y$ direction)?

(e) Find the rate at which energy is dissipated (as joule heat) in the wires that comprise the loop, and the amount of energy dissipated per revolution.

(f) Find the external torque required to keep the loop rotating at the constant angular speed ω.

3. The figure below shows a toroidal solenoid of mean radius R and N total windings. The cross-sections of the toroid are circles of radius a (which is much smaller than R, so variations in the magnetic field strength within the space enclosed by the windings may be neglected).

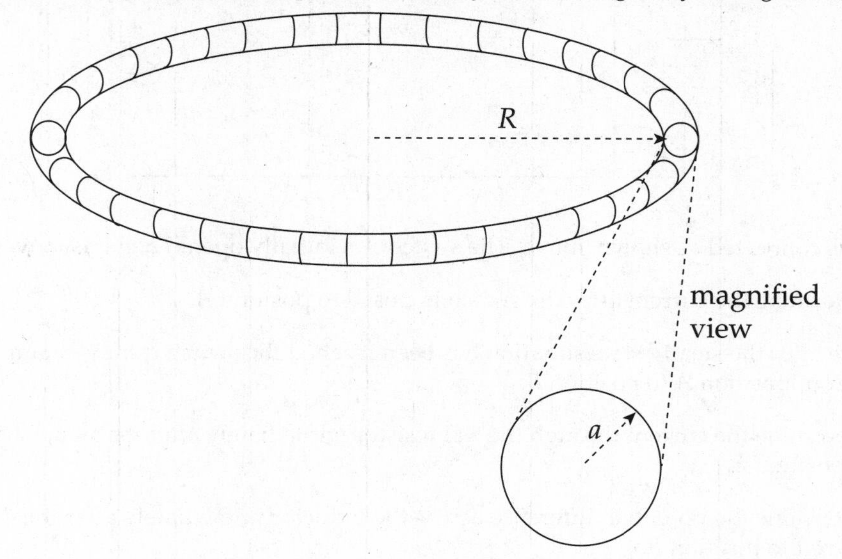

(a) Use Ampere's Law to find the magnetic field strength within the toroid. Write your answer in terms of N, I, R, and fundamental constants.

A circular loop of wire of radius $2a$ is placed around the toroid as shown:

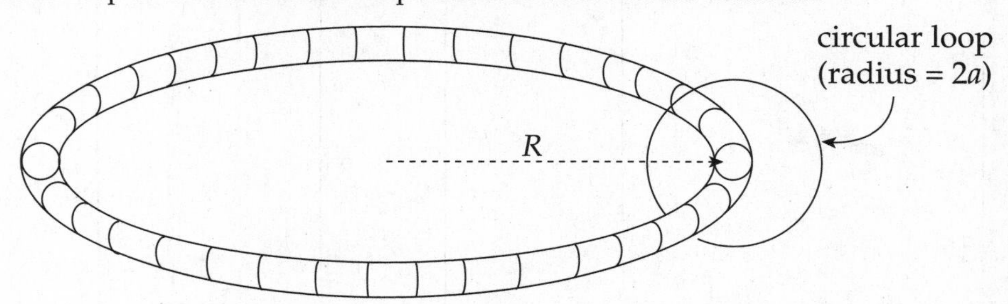

Assume that the current in the toroid is varied sinusoidally according to the equation $I(t) = I_0 \sin \omega t$, where I_0 and ω are fixed constants.

(b) Determine the emf induced in the circular wire loop.

(c) Determine the electric field induced at the position of the circular wire loop.

(d) What is the self-inductance of the toroidal solenoid?

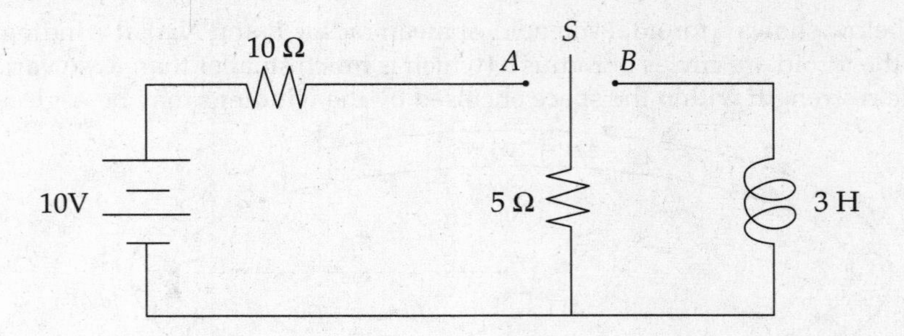

4. A circuit is connected as shown above. The switch S is initially open. Then it is moved to position A.

(a) Determine the current after the switch is closed to position A.

Some time after the steady state situation has been reached the switch is moved almost instantaneously from position A to position B.

(b) Determine the current through the 5 Ω resistor immediately after the switch has been moved to position B.

(c) Determine the potential difference across the inductor immediately after the switch has been moved to position B.

ELECTROMAGNETIC INDUCTION SUMMARY

MOTIONAL EMF

- A conducting bar with a component of its velocity perpendicular to the magnetic field will have an induced emf across its length. This is due to the fact that each charge in the bar will have a force on it and the charges will separate until the magnetic force and the electric force caused by the attraction of the separate charges are equal. The **motional emf** is given by the equation, $\varepsilon = B\ell v$.

MAGNETIC FLUX, FARADAY'S LAW, AND LENZ'S LAW

- **Magnetic flux** is determined by the number of magnetic field lines that pass through a surface. It is given by the dot product of the magnetic field and area vector: $\Phi_B = \mathbf{B} \bullet \mathbf{A} = \int \mathbf{B} \bullet d\mathbf{A}$
- When the magnetic flux is changing, an emf is induced in a conductor. This is known as **Faraday's Law** of Electromagnetic Induction. The general form of Faraday's Law is:
  $\varepsilon = -\frac{d\Phi_B}{dt} = -\int \mathbf{E} \bullet d\boldsymbol{\ell}$. The first part of the equation is used to calculate the induced emf from a changing magnetic flux and the last part can be used to calculate the induced electric field created in the conductor. The last part of the equation comes from $\Delta V = -\int \mathbf{E} \bullet d\boldsymbol{\ell}$, where the potential difference is replaced with the emf.
- **Lenz's Law** is used to determine the direction of the induced current and explains the negative sign in Faraday's Law. The induced current will flow in the direction that opposes the change in the magnetic flux that produced it.

INDUCTANCE

- An **inductor** is an electric device that has a **self-induced emf** that opposes the change in the current in the circuit. The inductance, L, is given by the equation $L = \frac{N\Phi_B}{I}$.
- The emf of an inductor is $\varepsilon = -L\frac{dI}{dt}$, where the negative sign indicates the emf is opposing the change in the current through the circuit.
- The energy stored in the magnetic field on an inductor is $U_L = \frac{1}{2}LI^2$.

Resistor-Inductor (RL) Circuits

- When a switch is closed in a typical RL circuit the current increases exponentially over time. This is due to the fact that the inductor opposes the change in the current so will not let it increase from zero to a maximum instantaneously. The equation for the current is $I(t) = \frac{\varepsilon}{R}\left(1 - e^{-t/\left(\frac{L}{R}\right)}\right)$, where the time constant is now given by, $\tau_L = \frac{L}{R}$.

Inductor-Capacitor (LC) Circuits

- When a switch is closed in a typical LC circuit the energy oscillates between being stored in the electric field of the capacitor and the magnetic field of the inductor. The frequency of the oscillation is $f = \frac{1}{2\pi\sqrt{LC}}$ and the charge of the plate is given by $Q(t) = Q_{\max}\cos(\omega t)$.

14

Solutions to the Chapter Review Questions

CHAPTER 2 REVIEW QUESTIONS

SECTION I: MULTIPLE CHOICE

1. **A** Traveling once around a circular path means that the final position coincides with the initial position. Therefore, the displacement is zero. The average speed, which is *total* distance traveled divided by elapsed time, cannot be zero. Since the velocity changed (because its direction changed), there was a nonzero acceleration. Therefore, only Statement I is true.

2. **C** By definition $\bar{\mathbf{a}} = \Delta\mathbf{v} / \Delta t$. We determine $\Delta\mathbf{v} = \mathbf{v}_2 - \mathbf{v}_1 = \mathbf{v}_2 + (-\mathbf{v}_1)$ geometrically as follows:

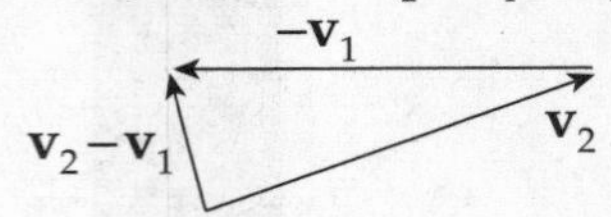

Since Δt is a positive scalar, the direction of $\bar{\mathbf{a}}$ is the same as the direction of $\Delta\mathbf{v}$, which is displayed above; choice (C) is best.

3. **C** Statement I is false since a projectile experiencing only the constant acceleration due to gravity can travel in a parabolic trajectory. Statement II is true: Zero acceleration means no change in speed (or direction). Statement III is false: An object whose speed remains constant but whose velocity vector is changing direction is accelerating.

4. **C** The baseball is still under the influence of Earth's gravity. Its acceleration throughout the *entire* flight is constant, equal to g downward.

5. **A** Use Big Five #3 with $v_0 = 0$:

$$\Delta s = v_0 t + \tfrac{1}{2}at^2 = \tfrac{1}{2}at^2 \quad \Rightarrow \quad t = \sqrt{\frac{2\Delta s}{a}} = \sqrt{\frac{2(200\text{ m})}{5\text{ m/s}^2}} = 9\text{ s}$$

6. **D** Use Big Five #5 with $v_0 = 0$ (calling *down* the positive direction):

$$v^2 = v_0^2 + 2a\Delta s = 2a\Delta s \quad \Rightarrow \quad \Delta s = \frac{v^2}{2a} = \frac{v^2}{2g} = \frac{(30\text{ m/s})^2}{2(10\text{ m/s}^2)} = 45\text{ m}$$

7. **C** Apply Big Five #3 to the vertical motion, calling *down* the positive direction:

$$\Delta y = v_{0y}t + \tfrac{1}{2}a_y t^2 = \tfrac{1}{2}a_y t^2 = \tfrac{1}{2}gt^2 \quad \Rightarrow \quad t = \sqrt{\frac{2\Delta y}{g}} = \sqrt{\frac{2(80\text{ m})}{10\text{ m/s}^2}} = 4\text{ s}$$

Note that the stone's initial horizontal speed ($v_{0x} = 10$ m/s) is irrelevant.

8. **B** First we determine the time required for the ball to reach the top of its parabolic trajectory (which is the time required for the vertical velocity to drop to zero).

$$v_y \overset{\text{set}}{=} 0 \quad \Rightarrow \quad v_{0y} - gt = 0 \quad \Rightarrow \quad t = \frac{v_{0y}}{g}$$

The total flight time is equal to twice this value:

$$T = 2t = 2\frac{v_{0y}}{g} = 2\frac{v_0 \sin\theta_0}{g} = \frac{2(10\ \text{m/s})\sin 30°}{10\ \text{m/s}^2} = 1\ \text{s}$$

9. **C** After 4 seconds, the stone's vertical speed has changed by $\Delta v_y = a_y t = (10\ \text{m/s}^2)(4\ \text{s}) = 40\ \text{m/s}$. Since $v_{0y} = 0$, the value of v_y at $t = 4$ is 40 m/s. The horizontal speed does not change. Therefore, when the rock hits the water, its velocity has a horizontal component of 30 m/s and a vertical component of 40 m/s.

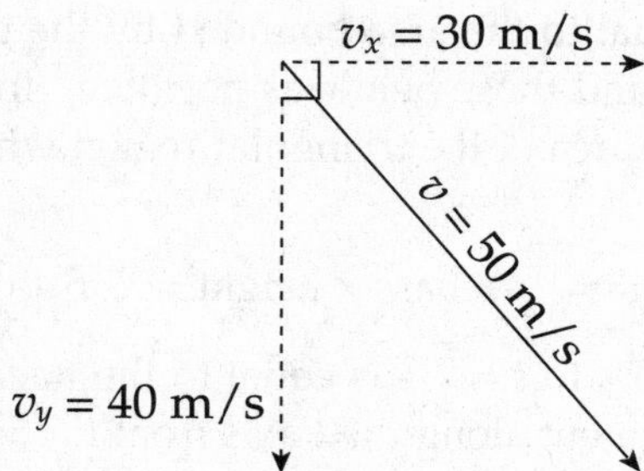

By the Pythagorean theorem, the magnitude of the total velocity, v, is 50 m/s.

10. **E** Since the acceleration of the projectile is always downward (because it's gravitational acceleration), the vertical speed decreases as the projectile rises and increases as the projectile falls. Statements A, B, C, and D are all false.

11. **D** The velocity is the derivative of the position function. Take the derivative of $x(t)$ and plug $t = 4$ s, to determine the velocity.

$$v = \frac{dx}{dt} = 10 - 18t^2$$

$$v(4) = 10 - 18(4)^2 = -278\,\text{m}$$

12. **D** The displacement of the object is the area between the velocity function and the t-axis. Divide the area up to 3 seconds into one rectangle created for the first second and one triangle for the next two seconds.

$$Area = (4)(1) + \frac{1}{2}(2)(4) = 8$$

Remember that the area is the displacement, not the position. Since the object started at $x = 3$ m, when $t = 0$ s, it is now 8 meters away from that point. Therefore its position is 11 m.

SECTION II: FREE RESPONSE

1. (a) At time $t = 1$ s, the car's velocity starts to decrease as the acceleration (which is the slope of the given v vs. t graph) changes from positive to negative.

(b) The average velocity between $t = 0$ and $t = 1$ s is $\frac{1}{2}(v_{t=0} + v_{t=1}) = \frac{1}{2}(0 + 20 \text{ m/s}) = 10 \text{ m/s}$, and the average velocity between $t = 1$ and $t = 5$ is $\frac{1}{2}(v_{t=1} + v_{t=5}) = \frac{1}{2}(20 \text{ m/s} + 0) = 10 \text{ m/s}$. The two average velocities are the same.

(c) The displacement is equal to the area bounded by the graph and the t axis, taking areas above the t axis as positive and those below as negative. In this case, the displacement from $t = 0$ to $t = 5$ s is equal to the area of the triangular region whose base is the segment along the t axis from $t = 0$ to $t = 5$ s:

$$\Delta s\ (t = 0 \text{ to } t = 5 \text{ s}) = \tfrac{1}{2} \times \text{base} \times \text{height} = \tfrac{1}{2}(5 \text{ s})(20 \text{ m/s}) = 50 \text{ m}$$

The displacement from $t = 5$ s to $t = 7$ s is equal to the negative of the area of the triangular region whose base is the segment along the t axis from $t = 5$ s to $t = 7$ s:

$$\Delta s\ (t = 5 \text{ s to } t = 7 \text{ s}) = -\tfrac{1}{2} \times \text{base} \times \text{height} = -\tfrac{1}{2}(2 \text{ s})(10 \text{ m/s}) = -10 \text{ m}$$

Therefore, the displacement from $t = 0$ to $t = 7$ s is

$$\Delta s\ (t = 0 \text{ to } t = 5 \text{ s}) + \Delta s\ (t = 5 \text{ s to } t = 7 \text{ s}) = 50 \text{ m} + (-10 \text{ m}) = 40 \text{ m}$$

(d) The acceleration is the slope of the v vs. t graph. The segment of the graph from $t = 0$ to $t = 1$ s has a slope of $a = \Delta v/\Delta t = (20 \text{ m/s} - 0)/(1 \text{ s} - 0) = 20 \text{ m/s}^2$, and the segment of the graph from $t = 1$ s to $t = 7$ s has a slope of $a = \Delta v/\Delta t = (-10 \text{ m/s} - 20 \text{ s})/(7 \text{ s} - 1 \text{ s}) = -5 \text{ m/s}^2$. Therefore, the graph of a vs. t is

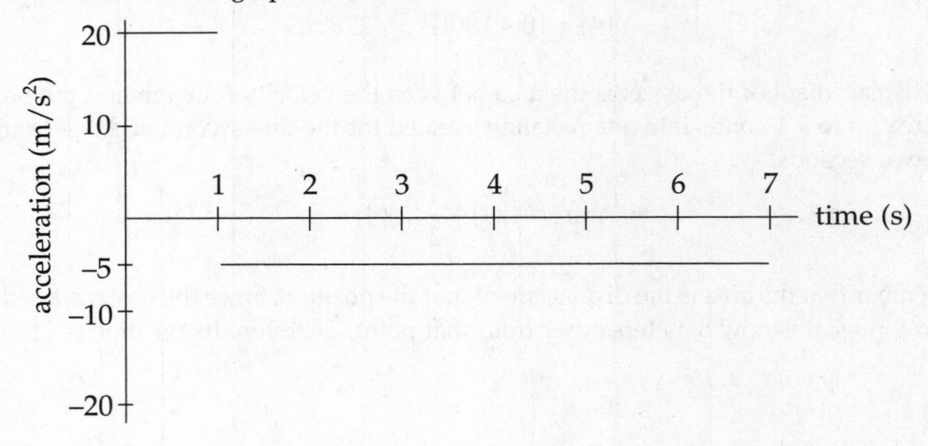

(e) One way to determine the displacement is to determine equations for $v(t)$ and integrate to get $s(t)$. The segment from $t = 0$ to $t = 1$ s connects the points (0, 0) and (1, 20); the slope is 20, and we get $v = 20t$. The segment from $t = 1$ to $t = 7$ connects the points (1, 20) and (7, –10); the slope is –5, and we get $v = -5t + 25$. In summary,

$$v(t) = \begin{cases} 20t & 0 \le t \le 1 \\ -5t + 25 & 1 \le t \le 7 \end{cases}$$

Therefore, since $s(t) = \int v(t)\,dt$, we find that

$$s(t) = \begin{cases} 10t^2 & 0 \le t \le 1 \\ -\frac{5}{2}t^2 + 25t - \frac{25}{2} & 1 \le t \le 7 \end{cases}$$

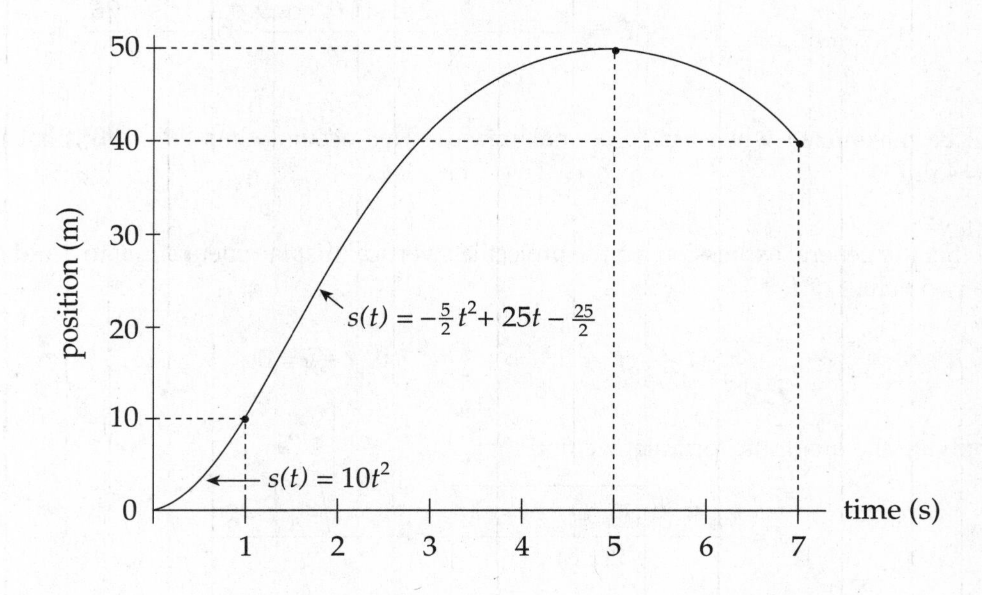

2. (a) The maximum height of the projectile occurs at the time at which its vertical velocity drops to zero:

$$v_y \overset{\text{set}}{=} 0 \;\Rightarrow\; v_{0y} - gt = 0 \;\Rightarrow\; t = \frac{v_{0y}}{g}$$

The vertical displacement of the projectile at this time is computed as follows:

$$\Delta y = v_{0y}t - \tfrac{1}{2}gt^2 \;\Rightarrow\; H = v_{0y}\frac{v_{0y}}{g} - \tfrac{1}{2}g\left(\frac{v_{0y}}{g}\right)^2 = \frac{v_{0y}^2}{2g} = \frac{v_0^2 \sin^2\theta_0}{2g}$$

(b) The total flight time is equal to twice the time computed in part (a):

$$T = 2t = 2\frac{v_{0y}}{g}$$

The horizontal displacement at this time gives the projectile's range:

$$\Delta x = v_{0x}t \quad \Rightarrow \quad R = v_{0x}T = \frac{v_{0x} \cdot 2v_{0y}}{g} = \frac{2v_0^2 \sin\theta_0 \cos\theta_0}{g} \text{ or } \frac{v_0^2 \sin 2\theta_0}{g}$$

(c) For any given value of v_0, the range,

$$\Delta x = v_{0x}t \quad \Rightarrow \quad R = v_{0x}T = \frac{v_{0x} \cdot 2v_{0y}}{g} = \frac{2v_0^2 \sin\theta_0 \cos\theta_0}{g} \text{ or } \frac{v_0^2 \sin 2\theta_0}{g}$$

will be maximized when $\sin 2\theta_0$ is maximized. This occurs when $2\theta_0 = 90°$, that is, when $\theta_0 = 45°$.

(d) Set the general expression for the projectile's vertical displacement equal to h and solve for the two values of t:

$$v_{0y}t - \tfrac{1}{2}gt^2 \overset{\text{set}}{=} h \quad \Rightarrow \quad \tfrac{1}{2}gt^2 - v_{0y}t + h = 0$$

Applying the quadratic formula, we find that

$$t = \frac{v_{0y} \pm \sqrt{(-v_{0y})^2 - 4(\tfrac{1}{2}g)(h)}}{2(\tfrac{1}{2}g)} = \frac{v_{0y} \pm \sqrt{v_{0y}^2 - 2gh}}{g}$$

Therefore, the two times at which the projectile crosses the horizontal line at height h are

$$t_1 = \frac{v_{0y} - \sqrt{v_{0y}^2 - 2gh}}{g} \quad \text{and} \quad t_2 = \frac{v_{0y} + \sqrt{v_{0y}^2 - 2gh}}{g}$$

so the amount of time that elapses between these events is

$$\Delta t = t_2 - t_1 = \frac{2\sqrt{v_{0y}^2 - 2gh}}{g}$$

3. (a) If we need to find how long the cannonball takes to reach the plane of the wall, we are dealing with the horizontal direction. The only equation we need is $x = v_x t$.

$220 = (50 \cos 40°)t \Rightarrow t = 5.74$ s

(b) To determine whether the cannonball hits the wall, we need to know the vertical displacement of the ball when it reaches the plane of the wall.

$$\Delta y = v_{0y}t + \frac{1}{2}gt^2$$

$$\Delta y = (50 \sin 40°)(5.74) + \frac{1}{2}(-9.8)(5.74)^2$$

$$\Delta y = 22.8 \text{ m}$$

Since the wall is 30 m tall, the cannonball strikes the wall 7.2 m below the top of the wall.

4. (a) Integrating $a(t)$ with respect to time gives the velocity, $v(t)$:

$$v(t) = \int a(t)\,dt = \int 6t\,dt = 3t^2 + v_0 \quad \Rightarrow \quad v(t) = 3t^2 + 2$$

Setting this equal to 14, we solve for t:

$$v(t) = 14 \quad \Rightarrow \quad 3t^2 + 2 = 14 \quad \Rightarrow \quad t = 2 \text{ s}$$

(Note that we discarded the solution $t = -2$.)

(b) Integrating $v(t)$ with respect to time gives the position, $x(t)$:

$$x(t) = \int v(t)\,dt = \int (3t^2 + 2)\,dt = t^3 + 2t + x_0 \quad \Rightarrow \quad x(t) = t^3 + 2t + 4$$

Therefore, the particle's position at $t = 3$ s is

$$x(3) = \left[t^3 + 2t + 4\right]_{t=3} = 37 \text{ m}$$

CHAPTER 3 REVIEW QUESTIONS

SECTION I: MULTIPLE CHOICE

1. **B** Because the person is not accelerating, the net force he feels must be zero. Therefore, the magnitude of the upward normal force from the floor must balance that of the downward gravitational force. Although these two forces have equal magnitudes, they do not form an action/reaction pair because they both act on the same object (namely, the person). The forces in an action/reaction pair always act on different objects.

2. **D** First draw a free-body diagram:

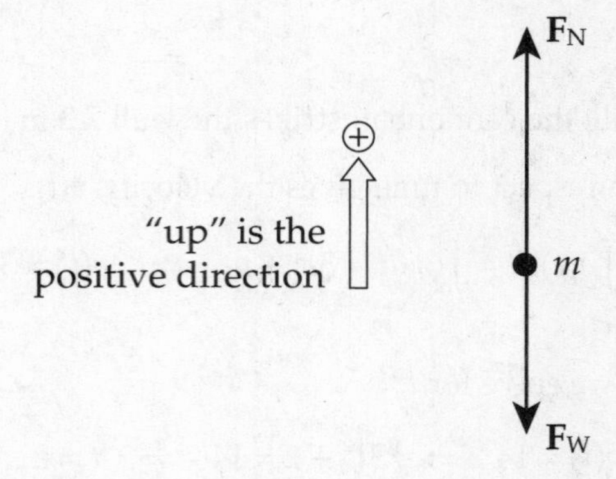

The person exerts a downward force on the scale, and the scale pushes up on the person with an equal (but opposite) force, $\mathbf{F}_N$. Thus, the scale reading is F_N, the magnitude of the normal force. Since $F_N - F_w = ma$, we have $F_N = F_w + ma = (800 \text{ N}) + [800 \text{ N}/(10 \text{ m/s}^2)](5 \text{ m/s}^2) = 1200 \text{ N}$.

3. **A** The net force that the object feels on the inclined plane is $mg \sin \theta$, the component of the gravitational force that is parallel to the ramp. Since $\sin \theta = (5 \text{ m})/(20 \text{ m}) = 1/4$, we have $F_{net} = (2 \text{ kg})(10 \text{ N/kg})(1/4) = 5 \text{ N}$.

4. **C** The net force on the block is $F - F_f = F - \mu_k F_N = F - \mu_k F_w = (18 \text{ N}) - (0.4)(20 \text{ N}) = 10 \text{ N}$. Since $F_{net} = ma = (F_w/g)a$, we find that $10 \text{ N} = [(20 \text{ N})/(10 \text{ m/s}^2)]a$, which gives $a = 5 \text{ m/s}^2$.

5. **A** The force pulling the block down the ramp is $mg \sin \theta$, and the maximum force of static friction is $\mu_s F_N = \mu_s mg \cos \theta$. If $mg \sin \theta$ is greater than $\mu_s mg \cos \theta$, then there is a net force down the ramp, and the block will accelerate down. So, the question becomes, "Is $\sin \theta$ greater than $\mu_s \cos \theta$?" Since $\theta = 30°$ and $\mu_s = 0.5$, the answer is "yes."

6. **E** One way to attack this question is to notice that if the two masses happen to be equal, that is, if $M = m$, then the blocks won't accelerate (because their weights balance). The only expression given that becomes zero when $M = m$ is the one given in choice (E). If we draw a free-body diagram,

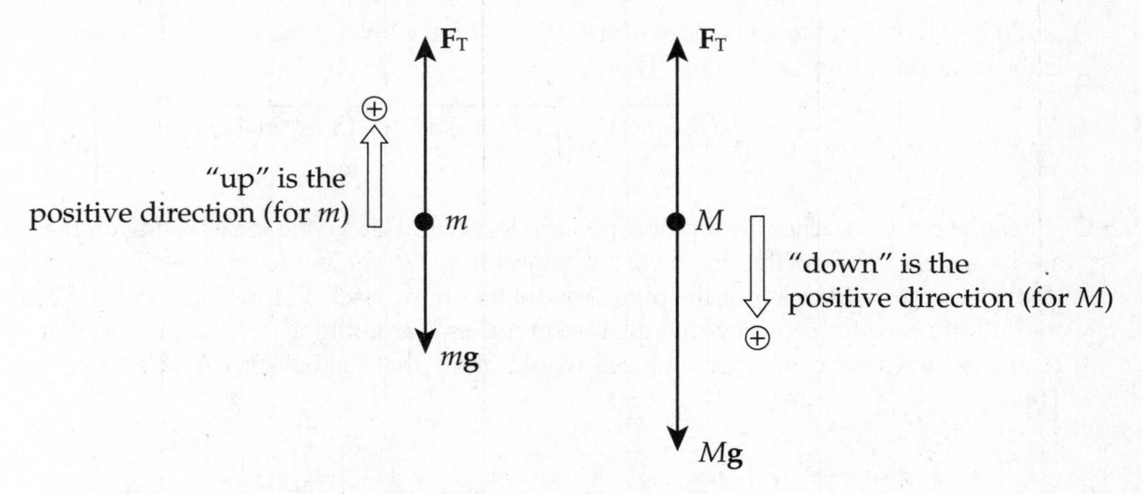

Newton's Second Law gives us the following two equations:

$$F_T - mg = ma \qquad (1)$$

$$Mg - F_T = Ma \qquad (2)$$

Adding these equations yields $Mg - mg = ma + Ma = (M + m)a$, so

$$a = \frac{Mg - mg}{M + m} = \frac{M - m}{M + m}g$$

7. **E** If $\mathbf{F}_{net} = \mathbf{0}$, then $\mathbf{a} = \mathbf{0}$. No acceleration means constant speed (possibly, but not necessarily, zero) with no change in direction. Therefore, statements B, C, and D are false, and statement A is not necessarily true.

8. **D** The horizontal motion across the frictionless tables is unaffected by (vertical) gravitational acceleration. It would take as much force to accelerate the block across the table on Earth as it would on the Moon. (If friction *were* taken into account, then the smaller weight of the block on the Moon would imply a smaller normal force by the table and hence a smaller frictional force. Less force would be needed on the Moon in this case.)

9. **D** The maximum force which static friction can exert on the crate is $\mu_s F_N = \mu_s F_w = \mu_s mg$ = (0.4)(100 kg)(10 N/kg) = 392 N. Since the force applied to the crate is only 344 N, static friction is able to apply that same magnitude of force on the crate, keeping it stationary. [Choice (B) is incorrect because the static friction force is *not* the reaction force to **F**; both **F** and $\mathbf{F}_{f\,(static)}$ act on the same object (the crate) and therefore cannot form an action/reaction pair.]

10. **E** Neither the velocity nor the acceleration is constant because the direction of each of these vectors is always changing as the object moves along its circular path. And the net force on the object is not zero, because a centripetal force must be acting to provide the necessary centripetal acceleration to maintain the object's circular motion.

11. **B** When the bucket is at the lowest point in its vertical circle, it feels a tension force $\mathbf{F}_T$ upward and the gravitational force $\mathbf{F}_w$ downward. The net force toward the center of the circle, which is the centripetal force, is $F_T - F_w$. Thus,

$$F_T - F_w = m\frac{v^2}{r} \quad \Rightarrow \quad v = \sqrt{\frac{r(F_T - mg)}{m}} = \sqrt{\frac{(0.60\text{ m})[50\text{ N} - (3\text{ kg})(10\text{ N/kg})]}{3\text{ kg}}} = 2\text{ m/s}$$

12. **C** When the bucket reaches the topmost point in its vertical circle, the forces acting on the bucket are its weight, $\mathbf{F}_w$, and the downward tension force, $\mathbf{F}_T$. The net force, $\mathbf{F}_w + \mathbf{F}_T$, provides the centripetal force. In order for the rope to avoid becoming slack, $\mathbf{F}_T$ must not vanish. Therefore, the cut-off speed for ensuring that the bucket makes it around the circle is the speed at which $\mathbf{F}_T$ just becomes zero; any greater speed would imply that the bucket would make it around. Thus,

$$F_w + F_T = m\frac{v^2}{r} \quad \Rightarrow \quad F_w + 0 = m\frac{v^2_{\text{cut-off}}}{r} \quad \Rightarrow \quad v_{\text{cut-off}} = \sqrt{\frac{rF_w}{m}} = \sqrt{gr}$$

$$= \sqrt{(10\text{ m/s}^2)(0.60\text{ m})}$$

$$= 2.4\text{ m/s}$$

13. **D** Centripetal acceleration is given by the equation $a_c = v^2/r$. Since the object covers a distance of $2\pi r$ in 1 revolution,

$$a_c = \frac{v^2}{r} = \frac{(2\pi r)^2}{r} = 4\pi^2 r$$

Section II: Free Response

1. (a) The forces acting on the crate are $\mathbf{F}_T$ (the tension in the rope), $\mathbf{F}_w$ (the weight of the block), $\mathbf{F}_N$ (the normal force exerted by the floor), and $\mathbf{F}_f$ (the force of kinetic friction):

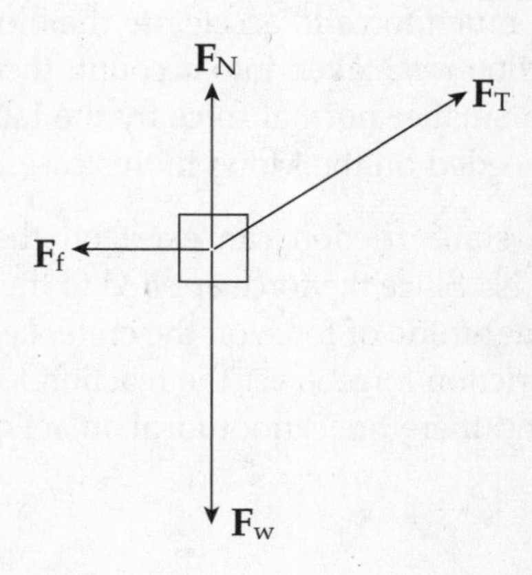

(b) First break $\mathbf{F}_T$ into its horizontal and vertical components:

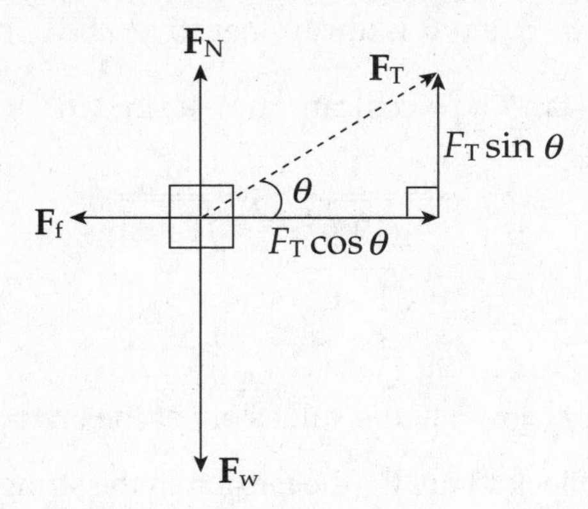

Since the net vertical force on the crate is zero, we have $F_N + F_T \sin\theta = F_w$, so $F_N = F_w - F_T \sin\theta = mg - F_T \sin\theta$.

(c) From part (b), we see that the net horizontal force acting on the crate is

$$F_T \cos\theta - F_f = F_T \cos\theta - \mu F_N = F_T \cos\theta - \mu(mg - F_T \sin\theta)$$

so the crate's horizontal acceleration across the floor is

$$a = \frac{F_{net}}{m} = \frac{F_T \cos\theta - \mu(mg - F_T \sin\theta)}{m}$$

(d) In order to maximize the crate's acceleration, we want to maximize the net horizontal force, $F_T \cos\theta - \mu(mg - F_T \sin\theta) = F_T(\cos\theta + \mu \sin\theta) - \mu mg$. Since F_T, μ, m, and g are all fixed, maximizing the net horizontal force depends on maximizing the expression

$$\cos\theta + \mu \sin\theta$$

Call this $f(\theta)$. To determine the value of θ at which $f(\theta) = \cos\theta + \mu\sin\theta$ attains an extreme value, we take the derivative of f and set it equal to zero:

$$\begin{aligned} f'(\theta) = -\sin\theta + \mu\cos\theta &\overset{\text{set}}{=} 0 \\ \mu\cos\theta &= \sin 0 \\ \mu &= \tan\theta \\ \theta &= \tan^{-1}\mu \end{aligned}$$

[That this value of θ does indeed maximize f can be verified by noticing that for θ in the interval $0 \le \theta \le \frac{1}{2}\pi$, $f''(\theta) = -\cos\theta - \mu\sin,\theta$ is always negative, and

$$\begin{aligned} f\left(\tan^{-1}\mu\right) &= \cos\left(\tan^{-1}\mu\right) + \mu\sin\left(\tan^{-1}\mu\right) \\ &= \frac{1}{\sqrt{1+\mu^2}} + \mu\frac{\mu}{\sqrt{1+\mu^2}} \\ &= \sqrt{1+\mu^2} \end{aligned}$$

is greater than $f(0) = 1$ or $f(\frac{1}{2}\pi) = \mu$, the values of f at the endpoints of the interval.]

2. (a) The forces acting on Block #1 are $\mathbf{F}_T$ (the tension in the string connecting it to Block #2), $\mathbf{F}_{w1}$ (the weight of the block), and $\mathbf{F}_{N1}$ (the normal force exerted by the tabletop):

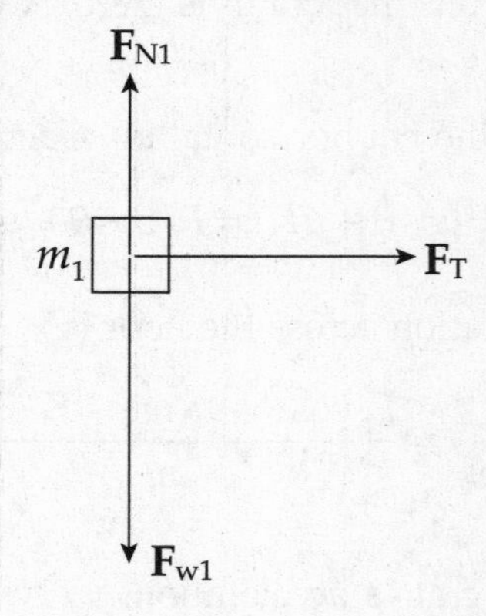

(b) The forces acting on Block #2 are $\mathbf{F}$ (the pulling force), $\mathbf{F}_T$ (the tension in the string connecting it to Block #1), $\mathbf{F}_{w2}$ (the weight of the block), and $\mathbf{F}_{N2}$ (the normal force exerted by the tabletop):

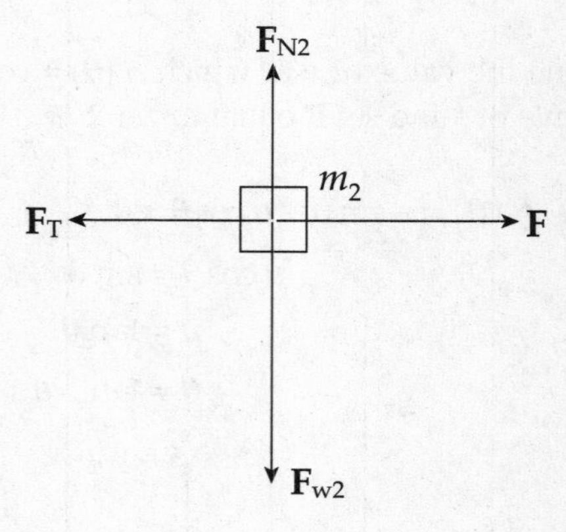

(c) Newton's Second Law applied to Block #2 yields $F - F_T = m_2a$, and applied to Block #1 yields $F_T = m_1a$. Adding these equations, we find that $F = (m_1 + m_2)a$, so

$$a = \frac{F}{m_1 + m_2}$$

(d) Substituting the result of part (c) into the equation $F_T = m_1 a$, we get

$$F_T = m_1 a = \frac{m_1}{m_1 + m_2} F$$

(e) (i) Since the force **F** must accelerate all three masses—m_1, m, and m_2—the common acceleration of all parts of the system is

$$a = \frac{F}{m_1 + m + m_2}$$

(e) (ii) Let $\mathbf{F}_{T1}$ denote the tension force in the connecting string acting on Block #1, and let $\mathbf{F}_{T2}$ denote the tension force in the connecting string acting on Block #2. Then, Newton's Second Law applied to Block #1 yields $F_{T1} = m_1 a$ and applied to Block #2 yields $F - F_{T2} = m_2 a$. Therefore, using the value for a computed above, we get

$$\begin{aligned} F_{T2} - F_{T1} &= (F - m_2 a) - m_1 a \\ &= F - (m_1 + m_2)a \\ &= F - (m_1 + m_2)\frac{F}{m_1 + m + m_2} \\ &= F\left(1 - \frac{m_1 + m_2}{m_1 + m + m_2}\right) \\ &= F\frac{m}{m_1 + m + m_2} \end{aligned}$$

3. (a) First draw free-body diagrams for the two boxes:

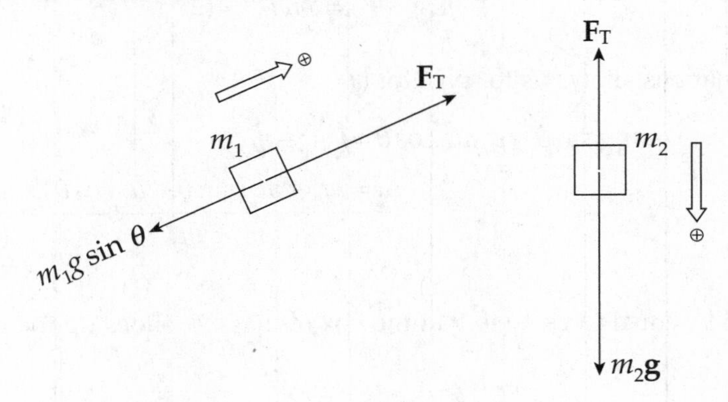

Applying Newton's Second Law to the boxes yields the following two equations:

$$F_T - m_1 g \sin\theta = m_1 a \qquad (1)$$

$$m_2 g - F_T = m_2 a \qquad (2)$$

Adding the equations allows us to solve for a:

$$m_2g - m_1g\sin\theta = (m_1 + m_2)a$$

$$a = \frac{m_2 - m_1\sin\theta}{m_1 + m_2}g$$

(i) For a to be positive, we must have $m_2 - m_1 \sin\theta > 0$, which implies that $\sin\theta < m_2/m_1$, or, equivalently, $\theta < \sin^{-1}(m_2/m_1)$.

(ii) For a to be zero, we must have $m_2 - m_1 \sin\theta = 0$, which implies that $\sin\theta = m_2/m_1$, or, equivalently, $\theta = \sin^{-1}(m_2/m_1)$.

(b) Including the force of kinetic friction, the force diagram for m_1 is

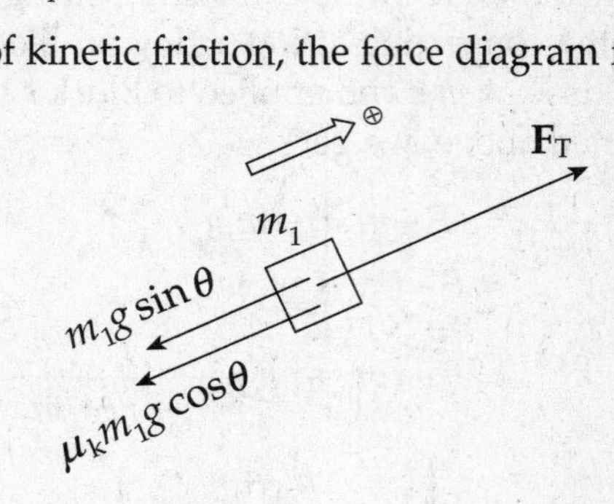

Since $F_f = \mu_k F_N = \mu_k m_1 g \cos\theta$, applying Newton's Second Law to the boxes yields these two equations:

$$F_T - m_1g\sin\theta - \mu_k mg\cos\theta = m_1a \qquad (1)$$

$$m_2g - F_T = m_2a \qquad (2)$$

Adding the equations allows us to solve for a:

$$m_2g - m_1g\sin\theta - \mu_k mg\cos\theta = (m_1 + m_2)a$$

$$\frac{a = m_2 - m_1(\sin\theta + \mu_k\cos\theta)}{m_1 + m_2}g$$

If we want a to be equal to zero (so that the box of mass m_1 slides up the ramp with constant velocity), then

$$m_2 - m_1(\sin\theta + \mu_k\cos\theta) = 0$$

$$\sin\theta + \mu_k\cos\theta = \frac{m_2}{m_1}$$

4. (a) The forces acting on the sky diver are $\mathbf{F}_r$, the force of air resistance (upward), and $\mathbf{F}_w$, the weight of the sky diver (downward):

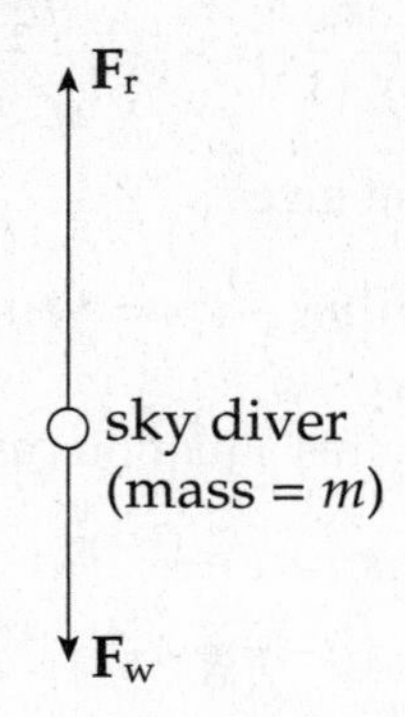

(b) Since $F_{net} = F_w - F_r = mg - kv$, the sky diver's acceleration is

$$a = \frac{F_{net}}{m} = \frac{mg - kv}{m}$$

(c) Terminal speed occurs when the sky diver's acceleration becomes zero, because then the descent velocity becomes constant. Setting the expression derived in part (b) equal to 0, we find the speed $v = v_t$ at which this occurs:

$$v = v_t \text{ when } a = 0 \quad \Rightarrow \quad \frac{mg - kv_t}{m} = 0 \quad \Rightarrow \quad v_t = \frac{mg}{k}$$

(d) The sky diver's descent speed is initially v_0 and the acceleration is (close to) g. However, once the parachute opens, the force of air resistance provides a large (speed-dependent) upward acceleration, causing her descent velocity to decrease. The slope of the v vs. t graph (the acceleration) is not constant but instead decreases to zero as her descent speed decreases from v_0 to v_t. Therefore, the graph is not linear.

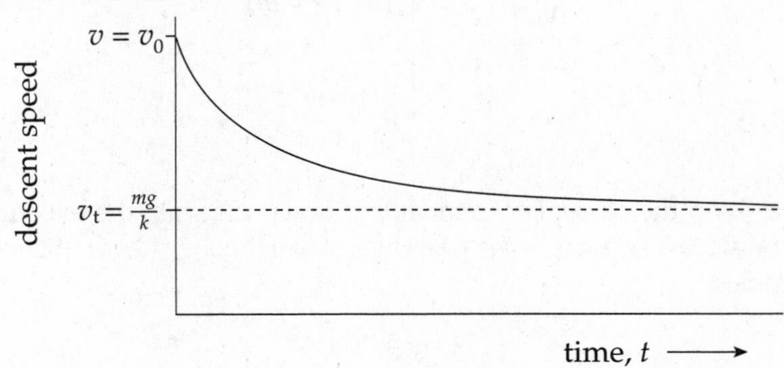

(e) Since $a = dv/dt$ and, from part (b), $a = (mg - kv)/m$, we have

$$\frac{dv}{dt} = \frac{mg - kv}{m} \quad \Rightarrow \quad \frac{dv}{mg - kv} = \frac{dt}{m}$$

Integrating both sides of this equation gives

$$-\tfrac{1}{k}\ln|mg - kv| = \tfrac{1}{m}t + c$$

where c is a constant of integration. This equation can be rewritten in the form:

$$mg - kv = Ce^{(-k/m)t}$$
$$kv = mg - Ce^{(-k/m)t}$$

Since $v = v_0$ at $t = 0$, we can determine the constant C:

$$kv_0 = mg - C \quad \Rightarrow \quad C = mg - kv_0$$

Therefore, the equation for the sky diver's descent speed as a function of time is

$$v(t) = \tfrac{1}{k}\left[mg - (mg - kv_0)\,e^{(-k/m)t}\right]$$

5. (a) The forces acting on a person standing against the cylinder wall are gravity ($\mathbf{F}_w$, downward), the normal force from the wall ($\mathbf{F}_N$, directed toward the center of the cylinder), and the force of static friction ($\mathbf{F}_f$, directed upward):

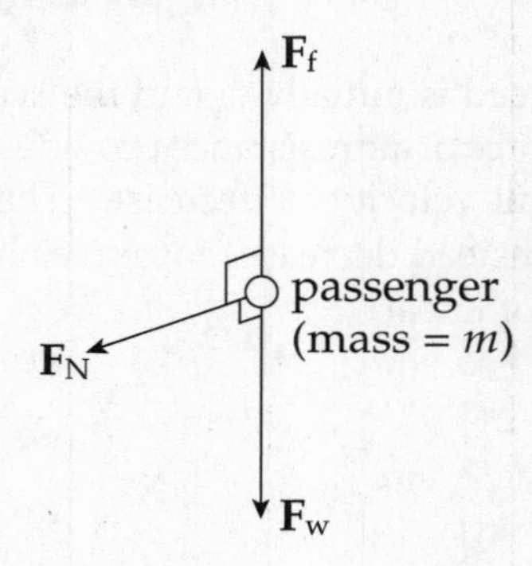

(b) In order to keep the passenger from sliding down the wall, the maximum force of static friction must be at least as great as the passenger's weight: $F_{f\,(max)} \geq mg$. Since $F_{f\,(max)} = \mu_s F_N$, this condition becomes

$$\mu_s F_N \geq mg$$

Now, consider the circular motion of the passenger. Neither $\mathbf{F}_f$ nor $\mathbf{F}_w$ has a component toward the center of the path, so the centripetal force is provided entirely by the normal force:

$$F_N = \frac{mv^2}{r}$$

Substituting this expression for F_N into the previous equation, we get

$$\mu_s \frac{mv^2}{r} \geq mg$$

$$\mu_s \geq \frac{gr}{v^2}$$

Therefore, the coefficient of static friction between the passenger and the wall of the cylinder must satisfy this condition in order to keep the passenger from sliding down.

(c) Since the mass m canceled out in deriving the expression for μ_s, the conditions are independent of mass. Thus, the inequality $\mu_s \geq gr/v^2$ holds for both the adult passenger of mass m and the child of mass $m/2$.

6. (a) The forces acting on the car are gravity ($\mathbf{F}_w$, downward), the normal force from the road ($\mathbf{F}_N$, upward), and the force of static friction ($\mathbf{F}_f$, directed toward the center of curvature of the road):

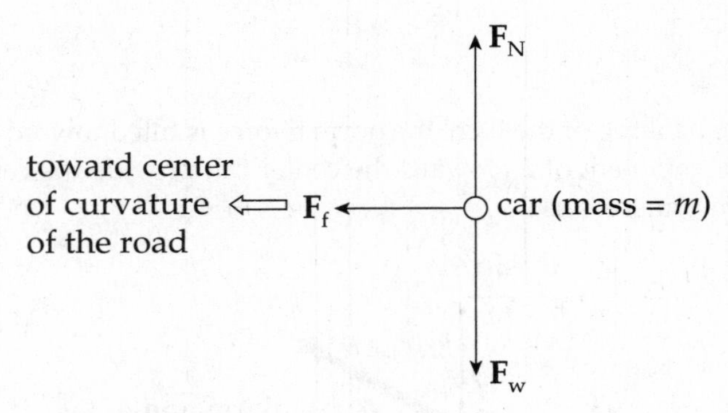

(b) The force of static friction [we assume static friction because we *don't* want the car to slide (that is, skid)] provides the necessary centripetal force:

$$F_f = \frac{mv^2}{r}$$

Therefore, to find the maximum speed at which static friction can continue to provide the necessary force, we write

$$F_{f(\max)} = \frac{mv_{\max}^2}{r}$$

$$\mu_s F_N = \frac{mv_{\max}^2}{r}$$

$$\mu_s mg = \frac{mv_{\max}^2}{r}$$

$$v_{\max} = \sqrt{\mu_s gr}$$

(c) Ignoring friction, the forces acting on the car are gravity ($\mathbf{F}_w$, downward) and the normal force from the road ($\mathbf{F}_N$, which is now tilted toward the center of curvature of the road):

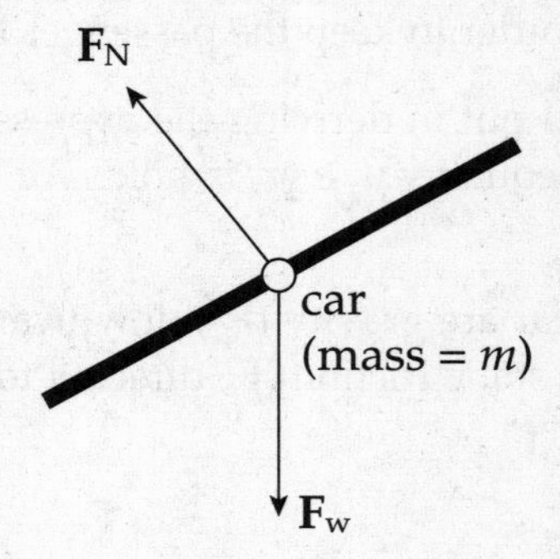

(d) Because of the banking of the turn, the normal force is tilted toward the center of curvature of the road. The component of $\mathbf{F}_N$ toward the center can provide the centripetal force, making reliance on friction unnecessary.

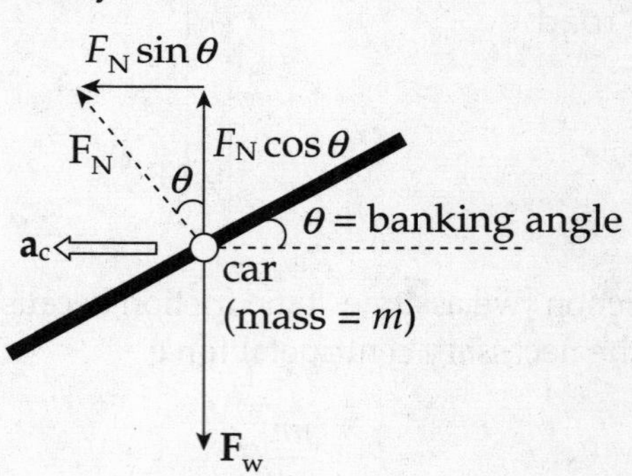

There's no vertical acceleration, so $F_N \cos\theta = F_w = mg$, so $F_N = mg/\cos\theta$. The component of F_N toward the center of curvature of the turn, $F_N \sin\theta$, provides the centripetal force:

$$F_N \sin\theta = \frac{mv^2}{r}$$
$$\frac{mg}{\cos\theta}\sin\theta = \frac{mv^2}{r}$$
$$g\tan\theta = \frac{v^2}{r}$$
$$\theta = \tan^{-1}\frac{v^2}{gr}$$

CHAPTER 4 REVIEW QUESTIONS

SECTION I: MULTIPLE CHOICE

1. **A** Since the force **F** is perpendicular to the displacement, the work it does is zero.

2. **B** By the work–energy theorem,

$$W = \Delta K = \tfrac{1}{2}m(v^2 - v_0^2) = \tfrac{1}{2}(4\text{ kg})[(6\text{ m/s})^2 - (3\text{ m/s})^2] = 54\text{ J}$$

3. **B** Since the box (mass m) falls through a vertical distance of h, its gravitational potential energy decreases by mgh. The length of the ramp is irrelevant here.

4. **C** Since the centripetal force always points along a radius toward the center of the circle, and the velocity of the object is always tangent to the circle (and thus perpendicular to the radius), the work done by the centripetal force is zero. Alternatively, since the object's speed remains constant, the work–energy theorem tells us that no work is being performed.

5. **A** The gravitational force points downward while the book's displacement is upward. Therefore, the work done by gravity is $-mgh = -(2\text{ kg})(10\text{ N/kg})(1.5\text{ m}) = -30\text{ J}$.

6. **D** The work done by gravity as the block slides down the inclined plane is equal to the potential energy at the top (mgh).

$$mgh = W = \Delta K = \tfrac{1}{2}m(v^2 - v_0^2) = \tfrac{1}{2}mv^2 \quad \Rightarrow \quad v = \sqrt{2gh} = \sqrt{2(10)(6.4)\sin 30^\circ} = 8\text{ m/s}$$

7. **D** Since a nonconservative force (namely, friction) is acting during the motion, we use the modified Conservation of Mechanical Energy equation.

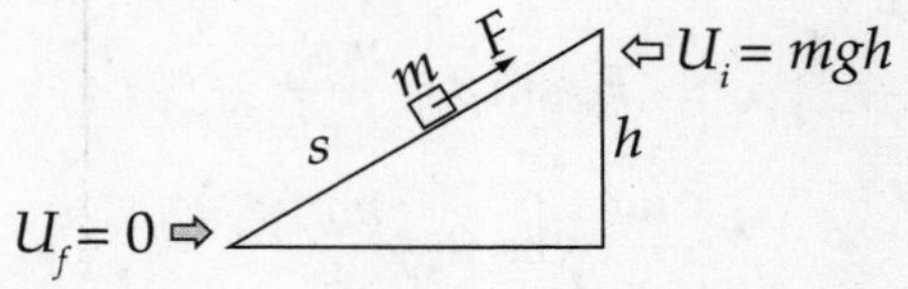

$$K_i + U_i + W_{\text{friction}} = K_i + U_f$$
$$0 + mgh - Fs = K_f + 0$$
$$mgh - Fs = K_f$$

8. **E** Apply Conservation of Mechanical Energy (including the negative work done by $\mathbf{F}_r$, the force of air resistance):

$$K_i + U_i + W_r = K_f + U_f$$
$$0 + mgh - F_r h = \tfrac{1}{2}mv^2 + 0$$
$$v = \sqrt{\frac{2h(mg - F_r)}{m}}$$
$$= \sqrt{\frac{2(40\text{ m})[(4\text{ kg})(10\text{ N/kg}) - 20\text{ N}]}{4\text{ kg}}}$$
$$= 20\text{ m/s}$$

9. **E** Because the rock has lost half of its gravitational potential energy, its kinetic energy at the halfway point is half of its kinetic energy at impact. Since K is proportional to v^2, if $K_{\text{at halfway point}}$ is equal to $\frac{1}{2}K_{\text{at impact}}$, then the rock's speed at the halfway point is $\sqrt{1/2} = 1/\sqrt{2}$ its speed at impact.

10. **D** Using the equation $P = Fv$, we find that $P = (200\text{ N})(2\text{ m/s}) = 400\text{ W}$.

SECTION II: FREE RESPONSE

1. (a) Applying Conservation of Energy,

$$K_A + U_A = K_{\text{at }H/2} + U_{\text{at }H/2}$$
$$0 + mgH = \tfrac{1}{2}mv^2 + mg(\tfrac{1}{2}H)$$
$$\tfrac{1}{2}mgH = \tfrac{1}{2}mv^2$$
$$v = \sqrt{gH}$$

(b) Applying Conservation of Energy again,

$$K_A + U_A = K_B + U_B$$
$$0 + mgH = \tfrac{1}{2}mv_B^2 + 0$$
$$v_B = \sqrt{2gH}$$

(c) By the work–energy theorem, we want the work done by friction to be equal (but opposite) to the kinetic energy of the box at Point B:

$$W = \Delta K = \tfrac{1}{2}m(v_C^2 - v_B^2) = -\tfrac{1}{2}mv_B^2 = -\tfrac{1}{2}m(\sqrt{2gH})^2 = -mgH$$

Therefore,

$$W = -mgH \quad \Rightarrow \quad -F_f x = -mgH \quad \Rightarrow \quad -\mu_k mgx = -mgH \quad \Rightarrow \quad \mu_k = H/x$$

(d) Apply Conservation of Energy (including the negative work done by friction as the box slides up the ramp from B to C):

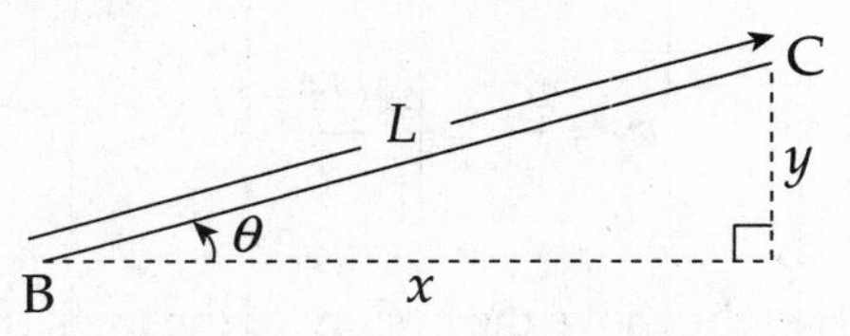

$$K_B + U_B + W_f = K_C + U_C$$
$$\tfrac{1}{2}m(\sqrt{2gH})^2 + 0 - F_f L = 0 + mgy$$
$$mgH + 0 - F_f L = 0 + mgy$$
$$mg(H - y) - (\mu_k mg \cos\theta)(L) = 0$$
$$\mu_k = \frac{H - y}{L\cos\theta} = \frac{H - y}{x}$$

(e) The result of part (b) reads $v_B = \sqrt{2gH}$. Therefore, by Conservation of Mechanical Energy (with the work done by the frictional force on the slide included), we get

$$K_A + U_A + W_f = K'_B + U_B$$
$$0 + mgH + W_f = \tfrac{1}{2}m\left(\tfrac{1}{2}v_B\right)^2 + 0$$
$$mgH + W_f = \tfrac{1}{2}m\left(\tfrac{1}{2}\sqrt{2gH}\right)^2$$
$$mgH + W_f = \tfrac{1}{4}mgH$$
$$W_f = -\tfrac{3}{4}mgH$$

2. (a) The centripetal acceleration of the car at Point C is given by the equation $a = v_C^2/r$, where v_C is the speed of the car at C. To find v_C^2, we apply Conservation of Energy:

$$\begin{aligned} K_A + U_A &= K_C + U_C \\ 0 + mgH &= \tfrac{1}{2}mv_C^2 + mgr \\ mg(H-r) &= \tfrac{1}{2}mv_C^2 \\ v_C^2 &= 2g(H-r) \end{aligned}$$

Therefore,

$$a_c = \frac{v_C^2}{r} = \frac{2g(H-r)}{r}$$

(b) In terms of θ, the car's height above the bottom of the track (Point B) is given by the equation $h = r + (-r\cos\theta)$,

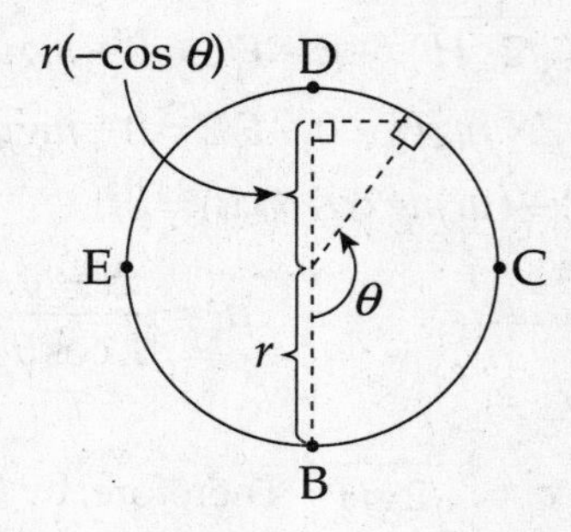

so we get

$$\begin{aligned} K_A + U_A &= K + U \\ 0 + mgH &= \tfrac{1}{2}mv^2 + mg(r - r\cos\theta) \\ mg[H - r(1-\cos\theta)] &= \tfrac{1}{2}mv^2 \\ v &= \sqrt{2g[H - r(1-\cos\theta)]} \end{aligned}$$

(c) When the car reaches Point D, the forces acting on the car are its weight, $\mathbf{F}_w$, and the downward normal force, $\mathbf{F}_N$, from the track. Thus, the net force, $\mathbf{F}_w + \mathbf{F}_N$, provides the centripetal force. In order for the car to maintain contact with the track, $\mathbf{F}_N$ must not vanish. Therefore, the cut-off speed for ensuring that the car makes it safely around the track is the speed at which $\mathbf{F}_N$ just becomes zero; any greater speed would imply that the car would make it around. Thus,

$$F_w + F_N = m\frac{v^2}{r} \quad\Rightarrow\quad F_w + 0 = m\frac{v_{\text{cut-off}}^2}{r} \quad\Rightarrow\quad v_{\text{cut-off}} = \sqrt{\frac{rF_w}{m}} = \sqrt{gr}$$

(d) Using the cut-off speed calculated in part (c), we now apply Conservation of Mechanical Energy:

$$\begin{aligned} K_A + U_A &= K_D + U_D \\ 0 + mgH &= \tfrac{1}{2}mv^2_{\text{cut-off}} + mg(2r) \\ mgH &= \tfrac{1}{2}m(gr) + mg(2r) \\ &= \tfrac{5}{2}mgr \\ H &= \tfrac{5}{2}r \end{aligned}$$

(e) First, we calculate the car's kinetic energy at Point B; then, we determine the distance x the car must travel from B to F for the work done by friction to eliminate this kinetic energy. So, applying Conservation of Mechanical Energy, we find

$$\begin{aligned} K_A + U_A &= K_B + U_B \\ 0 + mg(6r) &= \tfrac{1}{2}mv_B^2 + 0 \\ 6mgr &= \tfrac{1}{2}mv_B^2 \end{aligned}$$

Now, by the work–energy theorem,

$$W = \Delta K = \tfrac{1}{2}mv_F^2 - \tfrac{1}{2}mv_B^2 = -\tfrac{1}{2}mv_B^2 \quad \Rightarrow \quad \begin{aligned} -F_f x &= -\tfrac{1}{2}mv_B^2 \\ -\mu mgx &= -6mgr \\ x &= \frac{6r}{\mu} = \frac{6r}{0.5} = 12r \end{aligned}$$

3. (a) The point where $x = x_1$ is a local minimum of the function $U(x)$, and the point where $x = x_3$ is a local maximum. Therefore, at each of these locations, the derivative of $U(x)$ must be equal to 0.

$$\frac{dU}{dx} = 0 \;\Rightarrow\; 3 - 3(x-3)^2 = 0 \;\Rightarrow\; (x-3)^2 = 1 \;\Rightarrow\; x = 2 \text{ or } 4$$

Therefore, $x_1 = 2$ m and $x_3 = 4$ m.

(b) If the particle's total energy is $E_2 = \frac{1}{2}(E_1 + E_3) = \frac{1}{2}[U(x_1) + U(x_3)] = \frac{1}{2}[U(2) + U(4)] = \frac{1}{2}(4\text{ J} + 8\text{ J}) = 6$ J, then the particle can oscillate between the points marked a and b in the figure on the next page.

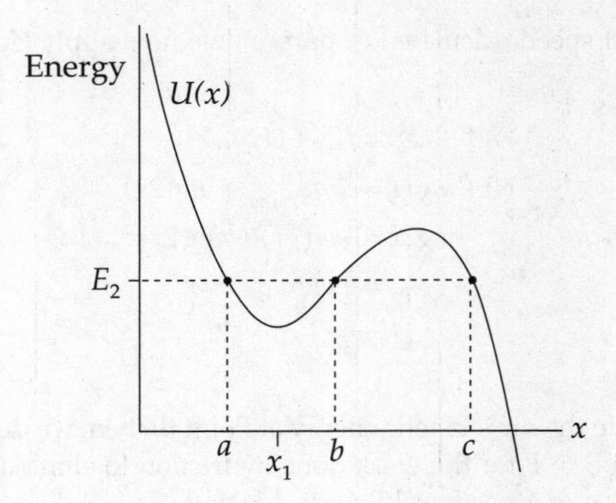

At points a and b the object has no kinetic energy (since $U = E_2$ at these points), so these are the turning points at which the object momentarily comes to rest before being accelerated back toward $x = x_1$ (where the potential energy is minimized). The particle cannot be found at $x < a$ or within the interval $b < x < c$, since $U > E_2$ in these intervals (and this would imply a negative K, which is impossible). If $x > c$, then since U decreases, K increases: the particle would move off with increasing speed.

(c) Since $K + U = E$, we have $K = E - U$. Therefore,

$$\begin{aligned} K(x_1) &= E - U(x_1) \\ &= E - \left[3(x-1) - (x-3)^3\right]_{x=2} \\ &= 58\text{ J} - [(3-(-1)]\text{ J} \\ &= 54\text{ J} \end{aligned}$$

Therefore,

$$K = \tfrac{1}{2}mv_1^2 \quad \Rightarrow \quad v_1 = \sqrt{\frac{2K}{m}} = \sqrt{\frac{2(54\text{ J})}{3\text{ kg}}} = 6\text{ m/s}$$

(d) Since $F(x) = -dU/dx = -[3 - 3(x - 3)^2] = 3(x - 3)^2 - 3$, dividing by m gives a:

$$a(x) = \frac{F(x)}{m} = \frac{[3(x-3)^2 - 3]\text{ J}}{3\text{ kg}} = (x-3)^2 - 1 \ \text{(in m/s}^2\text{)}$$

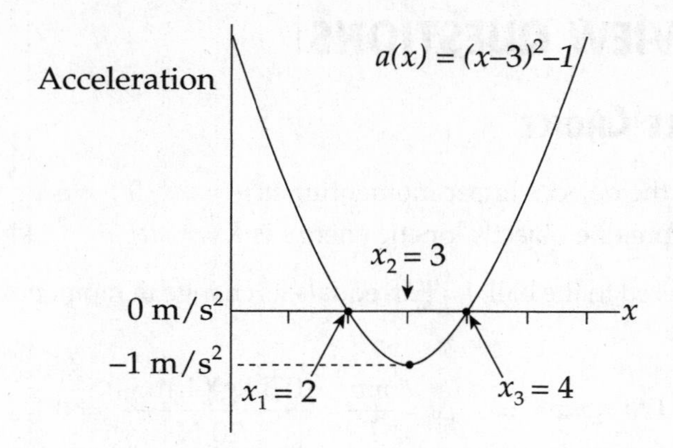

(e) At $x = \frac{1}{2}x_1$, the particle's energy is $E_3 = U(\frac{1}{2}x_1) = U(1) = \left[3(x-1)-(x-3)^3\right]_{x=1} = 8\text{ J}$. This is the particle's total energy because it has no initial kinetic energy (it is released from rest). As the particle passes through $x = x_1$, its potential energy decreases to

$$U(x_1) = U(2) = \left[3(x-1)-(x-3)^3\right]_{x=2} = 4\text{ J}$$

Therefore, since its total energy is 8 J, the particle's kinetic energy at $x = x_1$ must be $K = E - U = 8\text{ J} - 4\text{ J} = 4\text{ J}$. This implies that its speed is

$$K = \tfrac{1}{2}mv^2 \quad \Rightarrow \quad v = \sqrt{\frac{2K}{m}} = \sqrt{\frac{2(4\text{ J})}{3\text{ kg}}} = 1.6\text{ m/s}$$

4. (a) Work equals the integral of the dot product of the force and displacement.

$$W = \int F \bullet dx = \int_0^4 (3x+d)dx = \left(\frac{3}{2}x^2 + 5x\right)\Bigg|_0^4$$

$$W = 44\text{ J}$$

(b) The work-energy theorem states the work done on an object equals the change in kinetic energy of the object.

$$W = \Delta K$$

$$44 = \frac{1}{2}m\left(v_2^{\,2} - v_1^{\,2}\right)$$

$$44 = \frac{1}{2}(6)\left(v_2^{\,2} - 2^2\right)$$

$$v_2 = 4.3\text{m/s}$$

CHAPTER 5 REVIEW QUESTIONS

Section I: Multiple Choice

1. **C** The magnitude of the object's linear momentum is $p = mv$. If $p = 6\ \text{kg}\cdot\text{m/s}$ and $m = 2$ kg, then $v = 3$ m/s. Therefore, the object's kinetic energy is $K = \frac{1}{2}mv^2 = \frac{1}{2}(2\ \text{kg})(3\ \text{m/s})^2 = 9$ J.

2. **C** The impulse delivered to the ball, $J = F\Delta t$, equals its change in momentum. Since the ball started from rest, we have

$$F\Delta t = mv \quad\Rightarrow\quad \Delta t = \frac{mv}{F} = \frac{(0.5\ \text{kg})(4\ \text{m/s})}{20\ \text{N}} = 0.1\ \text{s}$$

3. **E** The impulse delivered to the box, $J = \bar{F}\Delta t$, equals its change in momentum. Thus,

$$\bar{F}\Delta t = \Delta p = p_{\text{f}} - p_{\text{i}} = m(v_{\text{f}} - v_{\text{i}}) \quad\Rightarrow\quad \bar{F} = \frac{m(v_{\text{f}} - v_{\text{i}})}{\Delta t} = \frac{(2\ \text{kg})(8\ \text{m/s} - 4\ \text{m/s})}{0.5\ \text{s}} = 16\ \text{N}$$

4. **D** The impulse delivered to the ball is equal to its change in momentum. The momentum of the ball was $m\mathbf{v}$ before hitting the wall and $m(-\mathbf{v})$ after. Therefore, the change in momentum is $m(-\mathbf{v}) - m\mathbf{v} = -2m\mathbf{v}$, so the magnitude of the momentum change (and the impulse) is $2mv$.

5. **B** By definition of *perfectly inelastic*, the objects move off together with one common velocity, $\mathbf{v}'$, after the collision. By Conservation of Linear Momentum,

$$\begin{aligned} m_1\mathbf{v}_1 + m_2\mathbf{v}_2 &= (m_1 + m_2)\mathbf{v}' \\ \mathbf{v}' &= \frac{m_1v_1 + m_2v_2}{m_1 + m_2} \\ &= \frac{(3\ \text{kg})(2\ \text{m/s}) + (5\ \text{kg})(-2\ \text{m/s})}{3\ \text{kg} + 5\ \text{kg}} \\ &= -0.5\ \text{m/s} \end{aligned}$$

6. **D** First, apply Conservation of Linear Momentum to calculate the speed of the combined object after the (perfectly inelastic) collision:

$$\begin{aligned} m_1v_1 + m_2v_2 &= (m_1 + m_2)v' \\ v' &= \frac{m_1v_1 + m_2v_2}{m_1 + m_2} \\ &= \frac{m_1v_1 + (2m_1)(0)}{m_1 + 2m_1} \\ &= \tfrac{1}{3}v_1 \end{aligned}$$

Therefore, the ratio of the kinetic energy after the collision to the kinetic energy before the collision is

$$\frac{K'}{K} = \frac{\frac{1}{2}m'v'^2}{\frac{1}{2}m_1v_1^2} = \frac{\frac{1}{2}(m_1 + 2m_1)(\frac{1}{3}v_1)^2}{\frac{1}{2}m_1v_1^2} = \frac{1}{3}$$

7. **C** Total linear momentum is conserved in a collision during which the net external force is zero. If kinetic energy is lost, then by definition, the collision is not elastic.

8. **B** First replace each rod by concentrating its mass at its center of mass position:

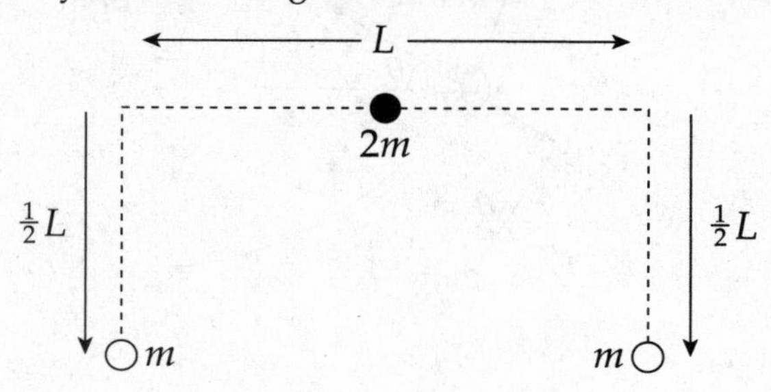

The center of mass of the two m's is at their midpoint, at a distance of $\frac{1}{2}L$ below the center of mass of the rod of mass $2m$:

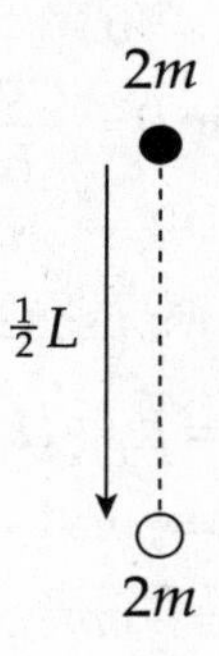

Now, applying the equation for locating the center of mass (letting $y = 0$ denote the position of the center of mass of the top horizontal rod), we find

$$y_{cm} = \frac{(2m)(0) + (2m)(\frac{1}{2}L)}{2m + 2m} = \frac{1}{4}L$$

9. **D** The linear momentum of the bullet must have the same magnitude as the linear momentum of the block in order for their combined momentum after impact to be zero. The block has momentum MV to the left, so the bullet must have momentum MV to the right. Since the bullet's mass is m, its speed must be $v = MV/m$.

10. **C** In a perfectly inelastic collision, kinetic energy is never conserved; some of the initial kinetic energy is always lost to heat and some is converted to potential energy in the deformed shapes of the objects as they lock together.

SECTION II: FREE RESPONSE

1. (a) First draw a free-body diagram:

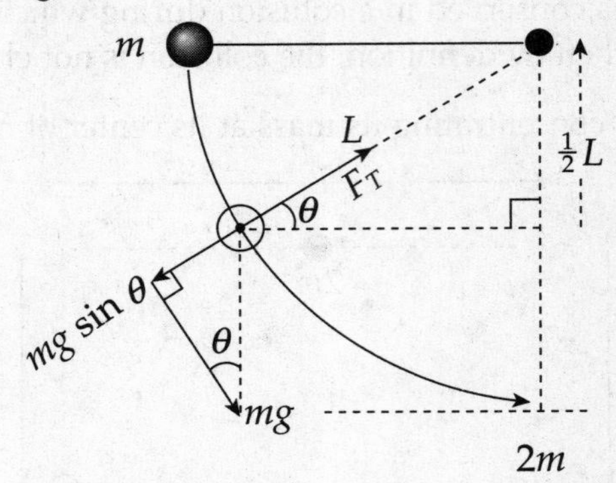

The net force toward the center of the steel ball's circular path provides the centripetal force. From the geometry of the diagram, we have

$$F_T - mg\sin\theta = \frac{mv^2}{L} \qquad (*)$$

In order to determine the value of mv^2, we use Conservation of Mechanical Energy:

$$\begin{aligned} K_i + U_i &= K_f + U_f \\ 0 + mgL &= \tfrac{1}{2}mv^2 + mg(\tfrac{1}{2}L) \\ \tfrac{1}{2}mgL &= \tfrac{1}{2}mv^2 \\ mgL &= mv^2 \end{aligned}$$

Substituting this result into Equation (*), we get

$$F_T - mg\sin\theta = \frac{mgL}{L}$$

$$F_T = mg(1+\sin\theta)$$

Now, from the free-body diagram we see that $\sin\theta = \tfrac{1}{2}L/L = \tfrac{1}{2}$, so

$$F_T = mg(1+\tfrac{1}{2}) = \tfrac{3}{2}mg$$

(b) Applying Conservation of Energy, we find the speed of the ball just before impact:

$$K_i + U_i = K_f + U_f$$
$$0 + mgL = \tfrac{1}{2}mv^2 + 0$$
$$v = \sqrt{2gL}$$

We can now use Conservation of Linear Momentum and the fact that kinetic energy is conserved to derive expressions for the speeds of the ball and block immediately after their collision. Since the collision is elastic, head-on, and the target object is at rest, this was done in Example 5.8. The velocity of the block after the collision is

$$v_2' = \frac{2m_1}{m_1 + m_2}v_1 = \frac{2m}{m + 4m}\sqrt{2gL} = \tfrac{2}{5}\sqrt{2gL}$$

(c) We can quote the result of Example 5.8 to find the velocity of the ball immediately after the collision:

$$v_1' = \frac{m_1 - m_2}{m_1 + m_2}v_1 = \frac{m - 4m}{m + 4m}\sqrt{2gL} = -\tfrac{3}{5}\sqrt{2gL}$$

Now, applying Conservation of Mechanical Energy, we find

$$K_i + U_i = K_f + U_f$$
$$\tfrac{1}{2}mv_1'^2 + 0 = 0 + mgh$$
$$h = \frac{v_1'^2}{2g} = \frac{\left(\tfrac{3}{5}\sqrt{2gL}\right)^2}{2g} = \tfrac{9}{25}L$$

2. (a) By Conservation of Linear Momentum, $mv = (m + M)v'$, so $v' = \dfrac{mv}{m + M}$

Now, by Conservation of Mechanical Energy,

$$K_i + U_i = K_f + U_f$$
$$\tfrac{1}{2}(m + M)v'^2 + 0 = 0 + (m + M)gy$$
$$\tfrac{1}{2}v'^2 = gy$$
$$\tfrac{1}{2}\left(\frac{mv}{m + M}\right)^2 = gy$$
$$v = \frac{m + M}{m}\sqrt{2gy}$$

(b) Use the result derived in part (a) to compute the kinetic energy of the block and bullet immediately after the collision:

$$K' = \tfrac{1}{2}(m+M)v'^2 = \tfrac{1}{2}(m+M)\left(\frac{mv}{m+M}\right)^2 = \tfrac{1}{2}\frac{m^2v^2}{m+M}$$

Since $K = \tfrac{1}{2}mv^2$, the difference is

$$\begin{aligned}\Delta K = K' - K &= \tfrac{1}{2}\frac{m^2v^2}{m+M} - \tfrac{1}{2}mv^2 \\ &= \tfrac{1}{2}mv^2\left(\frac{m}{m+M} - 1\right) \\ &= K\left(\frac{-M}{m+M}\right)\end{aligned}$$

Therefore, the fraction of the bullet's original kinetic energy that was lost is $M/(m + M)$. This energy is manifested as heat (the bullet and block are warmer after the collision than before), and some was used to break the intermolecular bonds within the wooden block to allow the bullet to penetrate.

(c) From the geometry of the diagram,

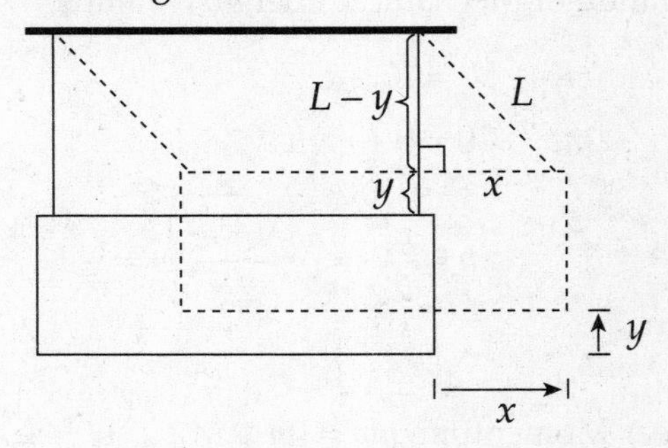

the Pythagorean theorem implies that $(L-y)^2 + x^2 = L^2$. Therefore,

$$L^2 - 2Ly + y^2 + x^2 = L^2 \quad \Rightarrow \quad y = \frac{x^2}{2L}$$

(where we have used the fact that y^2 is small enough to be neglected). Substituting this into the result of part (a), we derive the following equation for the speed of the bullet in terms of x and L instead of y:

$$v = \frac{m+M}{m}\sqrt{2gy} = \frac{m+M}{m}\sqrt{2g\frac{x^2}{2L}} = \frac{m+M}{m}x\sqrt{\frac{g}{L}}$$

(d) No; momentum is conserved only when the net external force on the system is zero (or at least negligible). In this case, the block and bullet feel a net nonzero force that causes it to slow down as it swings upward. Since its speed is decreasing as it swings upward, its linear momentum cannot remain constant.

3. (a) To apply Conservation of Linear Momentum to the collision, we recognize that momentum is a vector quantity and, therefore, linear momentum must be conserved separately in the horizontal (x) and vertical (y) directions. Before the collision, the linear momentum was horizontal only. Therefore, in the x direction:

$$mv = mv_1' \cos\theta_1 + mv_2' \cos\theta_2 \quad \Rightarrow \quad v = v_1' \cos\theta_1 + v_2' \cos\theta_2 \qquad (1)$$

in the y direction:

$$0 = mv_1' \sin\theta_1 - mv_2' \sin\theta_2 \quad \Rightarrow \quad 0 = v_1' \sin\theta_1 - v_2' \sin\theta_2 \qquad (2)$$

Since the collision is elastic, kinetic energy is also conserved; thus,

$$K = K_1' + K_2'$$
$$\tfrac{1}{2}mv^2 = \tfrac{1}{2}mv_1'^2 + \tfrac{1}{2}mv_2'^2$$
$$v^2 = v_1'^2 + v_2'^2 \qquad (3)$$

In order to find $K_1' = \frac{1}{2}mv_1'^2$ in terms of K, and θ_1 only, we need to manipulate the equations to eliminate v_2' and θ_2. The following approach will do this. Rewrite Equations (1) and (2) in the following forms:

$$v_2' \cos\theta_2 = v - v_1' \cos\theta_1 \qquad (1')$$
$$v_2' \sin\theta_2 = v_1' \sin\theta_1 \qquad (2')$$

Square both equations and add, exploiting the trig identity $\cos^2\theta + \sin^2\theta = 1$:

$$v_2'^2 \cos^2\theta_2 = v^2 - 2vv_1' \cos\theta_1 + v_1'^2 \cos^2\theta_1$$
$$v_2'^2 \sin^2\theta_2 = v_1'^2 \sin^2\theta_1$$
$$v_2'^2 = v^2 - 2vv_1' \cos\theta_1 + v_1'^2 \qquad (4)$$

This eliminates θ_2. Now, to eliminate v_2', substitute the value of $v_2'^2$ from Equation (3) into Equation (4):

$$v^2 - v_1'^2 = v^2 - 2vv_1' \cos\theta_1 + v_1'^2$$
$$-2v_1'^2 = -2vv_1' \cos\theta_1$$
$$v_1' = v \cos\theta_1$$

Therefore, we can write

$$K_1' = \tfrac{1}{2}mv_1'^2 = \tfrac{1}{2}m(v\cos\theta_1)^2 = (\cos^2\theta_1)\cdot\ \tfrac{1}{2}mv^2 = (\cos^2\theta_1)K_1$$

(b) Since, by Equation (3), $v_2'^2 = v^2 - v_1'^2$, the result of part (a) gives

$$\begin{aligned} K_2' = \tfrac{1}{2}mv_2'^2 = \tfrac{1}{2}m(v^2 - v_1'^2) &= \tfrac{1}{2}mv^2 - \tfrac{1}{2}mv_1'^2 \\ &= K_1 - K_1' \\ &= K_1 - (\cos^2\theta_1)K_1 \\ &= (1-\cos^2\theta_1)K_1 \\ &= (\sin^2\theta_1)K_1 \end{aligned}$$

(c) Square Equation (1) and Equation (2) from part (a) and add:

$$v^2 = v_1'^2\cos^2\theta_1 + 2v_1'v_2'\cos\theta_1\cos\theta_2 + v_2'^2\cos^2\theta_2$$
$$0 = v_1'^2\sin^2\theta_1 - 2v_1'v_2'\sin\theta_1\sin\theta_2 + v_2'^2\sin^2\theta_2$$

Adding gives:

$$v^2 = v_1'^2 + 2v_1'v_2'(\cos\theta_1\cos\theta_2 - \sin\theta_1\sin\theta_2) + v_2'^2$$
$$v^2 = v_1'^2 + 2v_1'v_2'\cos(\theta_1+\theta_2) + v_2'^2 \qquad (5)$$

Now, since by Equation (3), $v^2 = v_1'^2 + v_2'^2$, combining this with equation (5) gives

$$2v_1'v_2'\cos(\theta_1+\theta_2) = 0$$
$$\cos(\theta_1+\theta_2) = 0$$
$$\theta_1+\theta_2 = 90°$$

Thus, the objects' post-collision velocity vectors are perpendicular to each other.

CHAPTER 6 REVIEW QUESTIONS

SECTION I: MULTIPLE CHOICE

1. **B** We use the equation $v = r\omega$:

$$v = r\omega = 0.06 \text{ m} \times \left(\frac{5 \text{ rev}}{\text{s}} \times \frac{2\pi \text{ rad}}{\text{rev}} \right) = (0.6)\pi \text{ m/s} = 1.9 \text{ m/s}$$

2. **E** Use the equation *distance* = *rate* × *time* with $v = r\omega$:

$$s = vt = r\omega t = 0.06 \text{ m} \times \left(\frac{5 \text{ rev}}{\text{s}} \times \frac{2\pi \text{ rad}}{\text{rev}} \right) \times \left(40 \text{ min} \times \frac{60 \text{ s}}{\text{min}} \right) = 4500 \text{ m}$$

3. **C** By combining the equation for centripetal acceleration, $a_c = v^2/r$, with $v = r\omega$, we find

$$a_c = \frac{v^2}{r} = \frac{(r\omega)^2}{r} = \omega^2 r \quad \Rightarrow \quad \omega = \sqrt{\frac{a_c}{r}} = \sqrt{\frac{9 \text{ m/s}^2}{0.25 \text{ m}}} = 6 \text{ s}^{-1}$$

4. **C** Use Big Five #3 for rotational motion:

$$\Delta\theta = \omega_0 t + \tfrac{1}{2}\alpha t^2 = \tfrac{1}{2}\alpha t^2 \quad \Rightarrow \quad \alpha = \frac{2\Delta\theta}{t^2} = \frac{2(60 \text{ rad})}{(10 \text{ s})^2} = 1.2 \text{ rad/s}^2$$

5. **D** The torque is $\tau = rF = (0.20 \text{ m})(20 \text{ N}) = 4 \text{ N} \cdot \text{m}$.

6. **D** From the diagram,

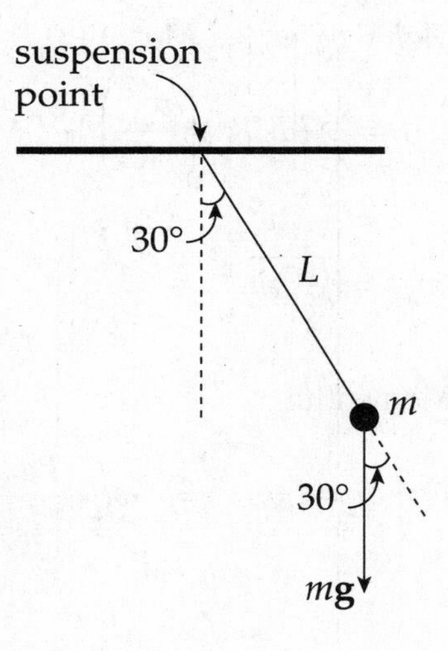

we calculate that

$$\begin{aligned} \tau = rF\sin\theta &= Lmg\sin\theta \\ &= (0.80 \text{ m})(0.50 \text{ kg})(10 \text{ N/kg})(\sin 30^\circ) \\ &= 2.0 \text{ N} \cdot \text{m} \end{aligned}$$

7. **B** The stick will remain at rest in the horizontal position if the torques about the suspension point are balanced:

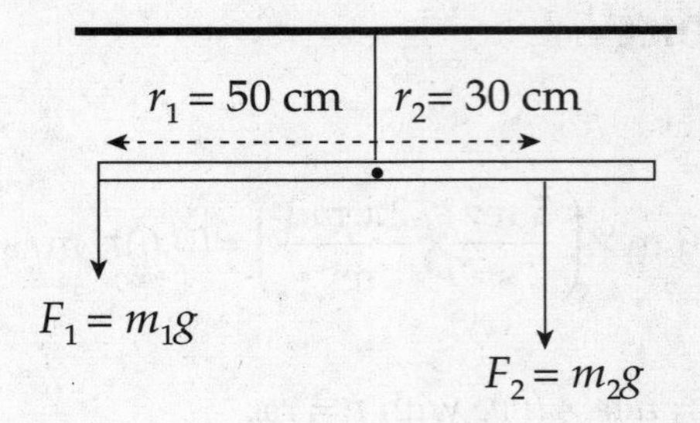

$$\tau_{CCW} = \tau_{CW}$$
$$r_1 F_1 = r_2 F_2$$
$$r_1 m_1 g = r_2 m_2 g$$
$$m_2 = \frac{r_1 m_1}{r_2} = \frac{(50\text{ cm})(3\text{ kg})}{30\text{ cm}} = 5\text{ kg}$$

8. **A** Each of the four masses is at distance L from the rotation axis, so

$$I = \sum m_i r_i^2 = mL^2 + mL^2 + mL^2 + mL^2 = 4mL^2$$

Note that the length of the rectangular frame, $\frac{8}{3}L$, is irrelevant here.

9. **A** By Conservation of Mechanical Energy,

$$K_i + U_i = K_f + U_f$$
$$0 + Mgh = (\tfrac{1}{2}I\omega^2 + \tfrac{1}{2}Mv_{cm}^2) + 0$$
$$= \tfrac{1}{2}(\tfrac{2}{5}MR^2)\left(\frac{v_{cm}}{R}\right)^2 + \tfrac{1}{2}Mv_{cm}^2$$
$$= \tfrac{7}{10}Mv_{cm}^2$$
$$v_{cm} = \sqrt{\tfrac{10}{7}gh}$$

10. **B** By Conservation of Angular Momentum,

$$L_i = L_f \quad \Rightarrow \quad I_i\omega_i = I_f\omega_f \quad \Rightarrow \quad \omega_f = \frac{I_i\omega_i}{I_f} = \frac{I_i\omega}{2I_i} = \frac{\omega}{2}$$

SECTION II: FREE RESPONSE

1. (a) Apply Conservation of Mechanical Energy, remembering to take into account the translational kinetic energy of the falling block (m_2), the rising block (m_1), and the rotating pulley:

$$K_i + U_i = K_f + U_f$$

$$0 + m_2gh = \left(\tfrac{1}{2}m_1v^2 + \tfrac{1}{2}m_2v^2 + \tfrac{1}{2}I\omega^2\right) + m_1gh$$

$$= \left[\tfrac{1}{2}m_1v^2 + \tfrac{1}{2}m_2v^2 + \tfrac{1}{2}\left(\tfrac{1}{2}MR^2\right)\left(\frac{v}{R}\right)^2\right] + m_1gh$$

$$(m_2 - m_1)gh = \tfrac{1}{2}v^2\left(m_1 + m_2 + \tfrac{1}{2}M\right)$$

$$v = \sqrt{\frac{2(m_2 - m_1)gh}{m_1 + m_2 + \frac{1}{2}M}}$$

(b) Using the equation $v = R\omega$, we find

$$\omega = \frac{v}{R} = \frac{1}{R}\sqrt{\frac{2(m_2 - m_1)gh}{m_1 + m_2 + \frac{1}{2}M}}$$

(c) Since Block 2 (and Block 1) moved a distance h during the motion, a point on the rim of the pulley must also move through a distance h. Therefore, $\Delta\theta = h/R$.

(d) Use Big Five #1 (for translational motion):

$$\Delta t = \frac{\Delta s}{\bar{v}} = \frac{\Delta s}{\frac{1}{2}v} = \frac{2\Delta s}{v} = \frac{2h}{v} = 2h\sqrt{\frac{m_1 + m_2 + \frac{1}{2}M}{2(m_2 - m_1)gh}} = h\sqrt{\frac{2m_1 + 2m_2 + M}{(m_2 - m_1)gh}}$$

2. (a) Consider the following diagram of the disk rolling down the ramp:

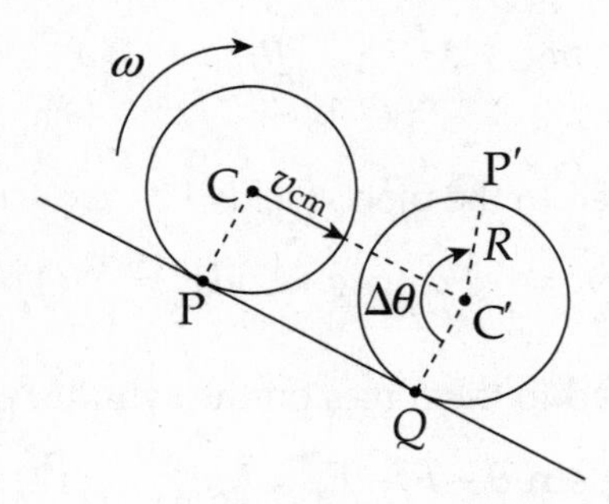

In a time interval Δt, the center of mass moves from C to C′, a distance equal to $v_{cm}\Delta t$. While this is occurring, Point P on the rim moves to the new position P′. In order for there to be no slipping, the arc length P′Q, which is $R\Delta\theta$, must equal the straight line distance PQ. But PQ = CC′, so we have

$$v_{cm}\Delta t = R\Delta\theta \;\Rightarrow\; v_{cm}\Delta t = R\omega\Delta t \;\Rightarrow\; v_{cm} = R\omega$$

(b) One way to derive the desired result is to notice that the velocity of T relative to P is equal to the velocity of T relative to C plus the velocity of C relative to P ($\mathbf{v}_{TP} = \mathbf{v}_{TC} + \mathbf{v}_{CP}$).

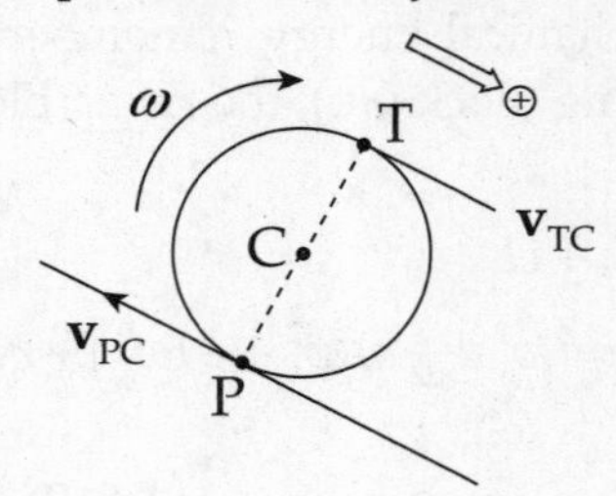

Since the velocity of C relative to P is the opposite of the velocity of P relative to C ($\mathbf{v}_{CP} = -\mathbf{v}_{PC}$), we have $\mathbf{v}_{TP} = \mathbf{v}_{TC} - \mathbf{v}_{PC}$. Relative to C, point T is moving forward with speed $R\omega$ and point P is moving backward with speed $R\omega$. Therefore, $\mathbf{v}_{TP} = (+R\omega) - (-R\omega) = +2R\omega = 2\mathbf{v}_{cm}$.

Another method is to think of the disk as executing pure rotation instantaneously about the contact Point P. From this point of view (with P as pivot), the distance to C is R, and the distance to T is $2R$. Therefore, the linear speed of C is $R\omega$ and that of T is $2R\omega$, which implies that the linear speed of point T is twice that of point C, as desired.

(c) The speed with which the block rises, v_b, is equal to the speed at which the thread wraps around the cylinder, which is $2v_{cm}$, as shown in part (b). Therefore, differentiating the equation $v_b = 2v_{cm}$ with respect to time, we get $a_b = 2a_{cm}$. That is, the acceleration of the block is equal to twice that of the (center of mass of the) cylinder.

(d) First draw free-body diagrams for the block and the cylinder:

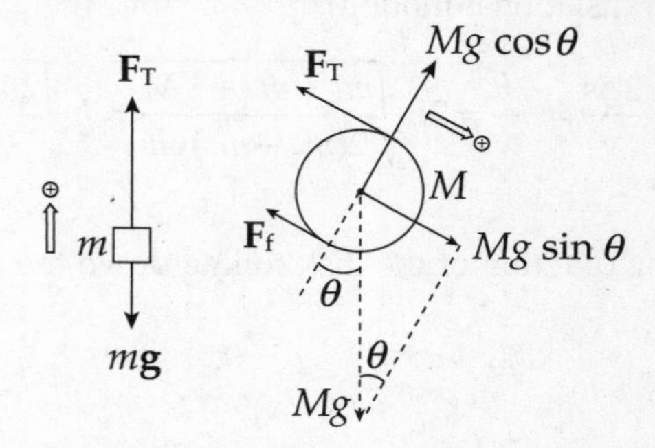

Newton's Second Law applied to the block yields $F_T - mg = ma_b$. Since $a_b = 2a_{cm}$, we write

$$F_T - mg = 2ma_{cm} \qquad (1)$$

Newton's Second Law applied to the forces on the cylinder gives us

$$Mg\sin\theta - F_f - F_T = Ma_{cm} \qquad (2)$$

The tension force and friction force each exert a torque about the center of mass of the cylinder. The torque of the friction force is counterclockwise (positive), and the torque of the tension force is negative. Therefore, $\tau_{net} = I\alpha$ becomes

$$RF_f - RF_T = \tfrac{1}{2}MR^2 \cdot \frac{a_{cm}}{R}$$
$$F_f - F_T = \tfrac{1}{2}Ma_{cm} \qquad (3)$$

Now for the algebra. Adding Equations (2) and (3) gives $Mg\sin\theta - 2F_T = \tfrac{3}{2}Ma_{cm}$. Multiplying both sides of Equation (1) by 2 gives $2F_T - 2mg = 4ma_{cm}$. Adding these last two equations allows us to find a_{cm}, the linear acceleration of the cylinder:

$$Mg\sin\theta - 2mg = \left(\tfrac{3}{2}M + 4m\right)a_{cm} \quad \Rightarrow \quad a_{cm} = \frac{M\sin\theta - 2m}{\tfrac{3}{2}M + 4m}g$$

(e) Since $a_b = 2a_{cm}$, the acceleration of the block is

$$a_b = \frac{2(M\sin\theta - 2m)}{\tfrac{3}{2}M + 4m}g$$

3. (a) Consider a rod of length $2L$ with the pivot at its center. Since the rod is uniform, its linear mass density is $M/(2L)$, so the mass of a segment of length dx is $M\,dx/(2L)$.

Applying the definition of I, we find

$$I = \int r^2\,dm = \int_{-L}^{L} x^2 \cdot \tfrac{M}{2L}\,dx = \tfrac{M}{2L}\left[\tfrac{1}{3}x^3\right]_{-L}^{L} = \tfrac{M}{2L}\left[\tfrac{2}{3}L^3\right] = \tfrac{1}{3}ML^2$$

(b) We apply Conservation of Angular Momentum to the collision of the bullet and assembly. Since the rods were at rest prior to the bullet's impact, only the bullet had angular momentum. With respect to the pivot point, its angular momentum was $Lm_bv_\perp$, where $v_\perp$ is the component of **v** that is perpendicular to the radius vector. From the definition of θ in the figure given with the question, $v_\perp = v\cos\theta$. After the collision, the angular momentum is equal to the angular velocity of the assembly times the rotational inertia of the four clay balls, the two rods, and the embedded bullet. We get:

$$Lm_bv\cos\theta = \left[4mL^2 + 2\left(\tfrac{1}{3}ML^2\right) + m_bL^2\right]\omega_f$$

$$\omega_f = \frac{m_bv\cos\theta}{L\left(4m + \tfrac{2}{3}M + m_b\right)}$$

(c) Using the equation $v = R\omega$, we find that

$$v_f = L\omega_f = \frac{m_b v \cos\theta}{4m + \frac{2}{3}M + m_b}$$

(d) After the collision, the assembly has rotational kinetic energy. The ratio of this energy to the (translational) kinetic energy of the bullet before impact is

$$\frac{K'}{K} = \frac{\frac{1}{2}I\omega_f^2}{\frac{1}{2}m_b v^2} = \frac{\left(4mL^2 + \frac{2}{3}ML^2 + m_b L^2\right)\left[\frac{m_b v \cos\theta}{L\left(4m + \frac{2}{3}M + m_b\right)}\right]^2}{m_b v^2}$$

$$= \frac{m_b \cos^2\theta}{4m + \frac{2}{3}M + m_b}$$

CHAPTER 7 REVIEW QUESTIONS

SECTION I: MULTIPLE CHOICE

1. **A** Gravitational force obeys an inverse-square law: $F_{grav} \propto 1/r^2$. Therefore, if r increases by a factor of 2, then F_{grav} decreases by a factor of $2^2 = 4$.

2. **E** Mass is an intrinsic property of an object and does not change with location. This eliminates choices A and C. If an object's height above the surface of the earth is equal to $2R_E$, then its distance from the center of the earth is $3R_E$. Thus, the object's distance from the earth's center increases by a factor of 3, so its weight decreases by a factor of $3^2 = 9$.

3. **C** The gravitational force that the moon exerts on the planet is equal in magnitude to the gravitational force that the planet exerts on the moon (Newton's Third Law).

4. **D** The gravitational acceleration at the surface of a planet of mass M and radius R is given by the equation $g = GM/R^2$. Therefore,

$$g_{Pluto} = G\frac{M_{Pluto}}{R_{Pluto}^2} = G\frac{\frac{1}{500}M_{Earth}}{\left(\frac{1}{15}R_{Earth}\right)^2} = \frac{15^2}{500}\cdot G\frac{M_{Earth}}{R_{Earth}^2} = \frac{225}{500}(10\ {}^{m/s^2}) = \frac{225}{50}\ m/s^2$$

5. **B** The gravitational pull by the earth provides the centripetal force on the satellite, so $GMm/R^2 = mv^2/R$. This gives $\frac{1}{2}mv^2 = GMm/2R$, so the kinetic energy K of the satellite is inversely proportional to R. Therefore if R increases by a factor of 2, then K decreases by a factor of 2.

6. **E** The gravitational pull by Jupiter provides the centripetal force on its moon:

$$G\frac{Mm}{R^2}=\frac{mv^2}{R}$$
$$G\frac{M}{R}=v^2$$
$$=\left(\frac{2\pi R}{T}\right)^2$$
$$=\frac{4\pi^2R^2}{T^2}$$
$$M=\frac{4\pi^2R^3}{GT^2}$$

7. **E** Let the object's distance from Body A be x; then its distance from Body B is $R - x$. In order for the object to feel no net gravitational force, the gravitational pull by A must balance the gravitational pull by B. Therefore, if we let M denote the mass of the object, then

$$G\frac{m_A M}{x^2}=G\frac{m_B M}{(R-x)^2}$$
$$\frac{m}{x^2}=\frac{4m}{(R-x)^2}$$
$$\frac{(R-x)^2}{x^2}=\frac{4m}{m}$$
$$\left(\frac{R-x}{x}\right)^2=4$$
$$\left(\frac{R}{x}-1\right)^2=4$$
$$\frac{R}{x}-1=2$$
$$x=R/3$$

8. **B** Kepler's Third Law says that $T^2 \propto R^3$ for a planet with a circular orbit of radius R. Since $T \propto R^{3/2}$, if R increases by a factor of 9, then T increases by a factor of $9^{3/2} = (3^2)^{3/2} = 3^3 = 27$.

9. **D** Apply Conservation of Mechanical Energy:

$$K_i+U_i=K_f+U_f$$
$$0-G\frac{Mm}{3R}=\tfrac{1}{2}mv_f^2-G\frac{Mm}{R}$$
$$\tfrac{1}{2}mv_f^2=G\frac{2Mm}{3R}$$
$$v_f=\sqrt{\frac{\tfrac{4}{3}GM}{R}}$$

10. **E** The force of gravity is the centripetal force, so we can solve for the velocity of each satellite and compare them.

$$F_g = F_C$$

$$\frac{Gm_A m_p}{R_A^{\,2}} = \frac{m_A v_A^{\,2}}{R_A}$$

$$v_A = \sqrt{\frac{Gm_p}{R_A}} \;\&\; v_B = \sqrt{\frac{Gm_p}{R_B}}$$

Now since $R_B = 3R_A$

$$v_A = \sqrt{\frac{Gm_p}{R_A}} \;\&\; v_B = \sqrt{\frac{Gm_p}{3R_A}} = \sqrt{\frac{1}{3}}\left(\sqrt{\frac{Gm_p}{R_A}}\right)$$

$$v_B = \sqrt{\frac{1}{3}}v_A = \frac{v_A}{\sqrt{3}}$$

SECTION II: FREE RESPONSE

1. (a) Combining Newton's Second Law with the Law of Gravitation, we find

$$a_1 = \frac{F_{2\text{-on-}1}}{m_1} = \frac{G\dfrac{m_1 m_2}{(\frac{1}{2}R)^2}}{m_1} = \frac{4Gm_2}{R^2}$$

The direction of $\mathbf{a}_1$ is toward Sphere 2.

(b) Combining Newton's Second Law with the Law of Gravitation, we find

$$a_2 = \frac{F_{1\text{-on-}2}}{m_2} = \frac{G\dfrac{m_1 m_2}{(\frac{1}{2}R)^2}}{m_2} = \frac{4Gm_1}{R^2}$$

The direction of $\mathbf{a}_2$ is toward Sphere 1.

(c) Since the two spheres start from rest, the total linear momentum of the system is clearly zero. Since no external forces act (we're in "deep space"), the total momentum must remain zero during their motion toward each other. Therefore, the magnitude of Sphere 1's linear momentum, m_1v_1, must always equal the magnitude of Sphere 2's linear momentum, m_2v_2. Thus, $v_2 = (m_1/m_2)v_1$. With this result, we can apply Conservation of Mechanical Energy:

$$
\begin{aligned}
K_i + U_i &= K_f + U_f \\
0 - \frac{Gm_1m_2}{R} &= \tfrac{1}{2}m_1v_1^2 + \tfrac{1}{2}m_2v_2^2 - \frac{Gm_1m_2}{R/2} \\
\frac{Gm_1m_2}{R} &= \tfrac{1}{2}m_1v_1^2 + \tfrac{1}{2}m_2\left(\tfrac{m_1}{m_2}v_1\right)^2 \\
\frac{Gm_2}{R} &= \tfrac{1}{2}v_1^2\left(1 + \tfrac{m_1}{m_2}\right) \\
v_1 &= \sqrt{\frac{2Gm_2}{R\left(1 + \frac{m_1}{m_2}\right)}} \\
&= m_2\sqrt{\frac{2G}{R(m_1 + m_2)}}
\end{aligned}
$$

(d) Using the result of part (c) and the relationship $v_2 = (m_1/m_2)v_1$, we get

$$
v_2 = \tfrac{m_1}{m_2}v_1 = \tfrac{m_1}{m_2} \cdot m_2\sqrt{\frac{2G}{R(m_1 + m_2)}} = m_1\sqrt{\frac{2G}{R(m_1 + m_2)}}
$$

(e) The centripetal force on each sphere is provided by the gravitational pull by the other sphere.

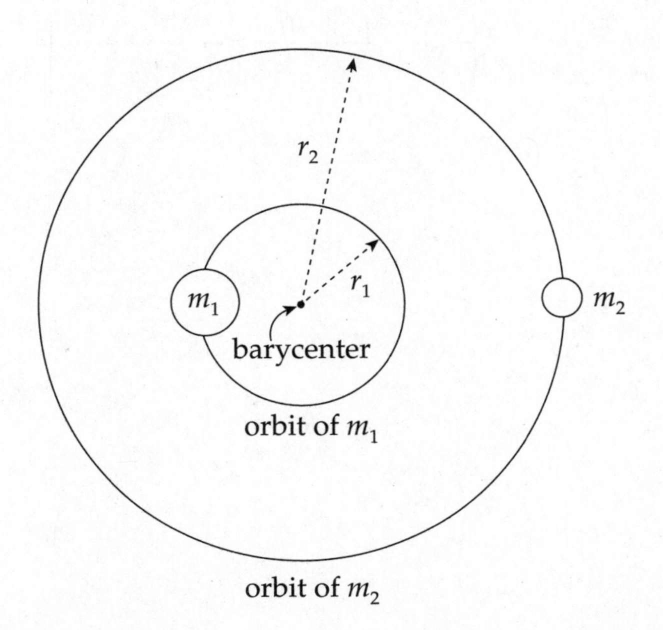

Therefore,

$$\frac{m_1 v_1^2}{r_1} = \frac{G m_1 m_2}{(r_1 + r_2)^2} = \frac{m_2 v_2^2}{r_2}$$

Since both spheres have the same orbit period, T, we get

$$\frac{m_1 v_1^2}{r_1} = \frac{m_2 v_2^2}{r_2} \quad \Rightarrow \quad \frac{m_1}{r_1}\left(\frac{2\pi r_1}{T}\right)^2 = \frac{m_2}{r_2}\left(\frac{2\pi r_2}{T}\right)^2 \quad \Rightarrow \quad m_1 r_1 = m_2 r_2 \qquad (1)$$

We can also derive (as we did in Example 7.9), that

$$\frac{T^2}{(r_1 + r_2)^3} = \frac{4\pi^2}{G(m_1 + m_2)} \quad \Rightarrow \quad r_1 + r_2 = \sqrt[3]{\frac{G(m_1 + m_2)T^2}{4\pi^2}} \qquad (2)$$

Substituting the result of Equation (1), $r_2 = (m_1/m_2)r_1$, into Equation (2), we can solve for r_1:

$$r_1 + r_2 = \sqrt[3]{\frac{G(m_1 + m_2)T^2}{4\pi^2}}$$

$$r_1 + \tfrac{m_1}{m_2} r_1 =$$

$$r_1\left(1 + \tfrac{m_1}{m_2}\right) =$$

$$r_1 = \frac{m_2}{m_1 + m_2} \sqrt[3]{\frac{G(m_1 + m_2)T^2}{4\pi^2}}$$

and then for r_2:

$$r_2 = \tfrac{m_1}{m_2} r_1 = \frac{m_1}{m_2} \cdot \frac{m_2}{m_1 + m_2} \sqrt[3]{\frac{G(m_1 + m_2)T^2}{4\pi^2}} = \frac{m_1}{m_1 + m_2} \sqrt[3]{\frac{G(m_1 + m_2)T^2}{4\pi^2}}$$

2. (a) The total energy of a satellite of mass m in an elliptical orbit with semimajor axis a around a planet of mass M is given by the equation $E = -GmM/2a$. From the figure, we see that $2a = r_1 + r_2$, so the satellite's kinetic energy at perigee is

$$K_1 = E - U_1 = -\frac{GmM}{r_1 + r_2} - \left(-\frac{GmM}{r_1}\right) = GmM\left(\frac{1}{r_1} - \frac{1}{r_1 + r_2}\right) = GmM\frac{r_2}{r_1(r_1 + r_2)}$$

Its speed at this point is then calculated as follows:

$$\tfrac{1}{2}mv_1^2 = GmM\frac{r_2}{r_1(r_1 + r_2)} \quad \Rightarrow \quad v_1 = \sqrt{\frac{2GMr_2}{r_1(r_1 + r_2)}}$$

(b) Employing the same method as used in part (a), we find

$$K_2 = E - U_2 = -\frac{GmM}{r_1 + r_2} - \left(-\frac{GmM}{r_2}\right) = GmM\left(\frac{1}{r_2} - \frac{1}{r_1 + r_2}\right) = GmM\frac{r_1}{r_2(r_1 + r_2)}$$

so

$$\tfrac{1}{2}mv_2^2 = GmM\frac{r_1}{r_2(r_1 + r_2)} \quad \Rightarrow \quad v_2 = \sqrt{\frac{2GMr_1}{r_2(r_1 + r_2)}}$$

(c) The ratio of v_1 to v_2 is

$$\frac{v_1}{v_2} = \frac{\sqrt{\frac{2GMr_2}{r_1(r_1 + r_2)}}}{\sqrt{\frac{2GMr_1}{r_2(r_1 + r_2)}}} = \frac{\sqrt{\frac{r_2}{r_1}}}{\sqrt{\frac{r_1}{r_2}}} = \sqrt{\frac{r_2}{r_1}} \cdot \sqrt{\frac{r_2}{r_1}} = \frac{r_2}{r_1}$$

(d) Since $\mathbf{v}_2$ is perpendicular to $\mathbf{r}_2$, the satellite's angular momentum is

$$L_2 = r_2 m v_2 = r_2 m\sqrt{\frac{2GMr_1}{r_2(r_1 + r_2)}} = m\sqrt{\frac{2GMr_1r_2}{r_1 + r_2}}$$

(e) The distance from X to the center of the Earth is equal to the semimajor axis, $a = \frac{1}{2}(r_1 + r_2)$. (This is because the sum of the distances from *any* point on the ellipse to the two foci must equal $2a = r_1 + r_2$, and X is equidistant from the two foci, one of which is at the center of the earth.) Therefore, the kinetic energy of the satellite when it's at X is

$$K_X = E - U_X = -\frac{GmM}{r_1 + r_2} - \left[-\frac{GmM}{\frac{1}{2}(r_1 + r_2)}\right] = \frac{GmM}{r_1 + r_2}$$

so its speed at this point can be determined:

$$\frac{1}{2}mv_X^2 = \frac{GmM}{r_1 + r_2} \quad \Rightarrow \quad v_X = \sqrt{\frac{2GM}{r_1 + r_2}}$$

(f) [See Example 7.8(c)] Use Kepler's Third Law for elliptical orbits:

$$\frac{T^2}{a^3} = \frac{4\pi^2}{GM} \quad \Rightarrow \quad T = \sqrt{\frac{4\pi^2 a^3}{GM}} = \sqrt{\frac{4\pi^2[\frac{1}{2}(r_1 + r_2)]^3}{GM}} = \sqrt{\frac{\pi^2(r_1 + r_2)^3}{2GM}}$$

CHAPTER 8 REVIEW QUESTIONS

SECTION I: MULTIPLE CHOICE

1. **D** The acceleration of a simple harmonic oscillator is not constant, since the restoring force—and, consequently, the acceleration—depends on position. Therefore Statement I is false. However, both Statements II and III are fundamental, defining characteristics of simple harmonic motion.

2. **C** The acceleration of the block has its maximum magnitude at the points where its displacement from equilibrium has the maximum magnitude (since $a = F/m = kx/m$). At the endpoints of the oscillation region, the potential energy is maximized and the kinetic energy (and hence the speed) is zero.

3. **E** By Conservation of Mechanical Energy, $K + U_S$ is a constant for the motion of the block. At the endpoints of the oscillation region, the block's displacement, x, is equal to $\pm A$. Since $K = 0$ here, all the energy is in the form of potential energy of the spring, $\frac{1}{2}kA^2$. Because $\frac{1}{2}kA^2$ gives the total energy at these positions, it also gives the total energy at any other position.

   Using the equation $U_S(x) = \frac{1}{2}kx^2$, we find that, at $x = \frac{1}{2}A$,

$$K + U_S = \frac{1}{2}kA^2$$
$$K + \frac{1}{2}k(\frac{1}{2}A)^2 = \frac{1}{2}kA^2$$
$$K = \frac{3}{8}kA^2$$

   Therefore,

$$K/E = \frac{3}{8}kA^2 / \frac{1}{2}kA^2 = \frac{3}{4}$$

4. **C** As we derived in Example 8.2, the maximum speed of the block is given by the equation $v_{max} = A\sqrt{k/m}$. Therefore, v_{max} is inversely proportional to $\sqrt{m}$. If m is increased by a factor of 2, then v_{max} will decrease by a factor of $\sqrt{2}$.

5. **D** The period of a spring–block simple harmonic oscillator is independent of the value of g. (Recall that $T = 2\pi\sqrt{m/k}$.) Therefore, the period will remain the same.

6. **D** The frequency of a spring–block simple harmonic oscillator is given by the equation $f = (1/2\pi)\sqrt{k/m}$. Squaring both sides of this equation, we get $f^2 = (k/4\pi^2)(1/m)$. Therefore, if f^2 is plotted vs. $(1/m)$, then the graph will be a straight line with slope $k/4\pi^2$. (Note: The slope of the line whose equation is $y = ax$ is a.)

7. **A** First we find the angular frequency:

$$\omega = 2\pi f = \sqrt{\frac{k}{m}} = \sqrt{\frac{400 \text{ N/m}}{4 \text{ kg}}} = 10 \text{ s}^{-1}$$

Therefore, the equation that gives the block's position, $x = A \sin(\omega t + \phi_0)$, becomes $x = 6 \sin(10t + \phi_0)$. Because the block is at $x = 6$ cm at time $t = 0$, the initial phase angle ϕ_0 must be $\frac{1}{2}\pi$.

8. **A** Starting with the equation $x = A\sin(\omega t + \phi_0)$, we differentiate with respect to t to find the velocity function: $v = \dot{x} = A\omega\cos(\omega t + \phi_0)$. Since the maximum value of $\cos(\omega t + \phi_0)$ is 1, the maximum value of v is $v_{max} = A\omega$.

9. **C** For small angular displacements, the period of a simple pendulum is essentially independent of amplitude.

10. **B** First draw a free-body diagram:

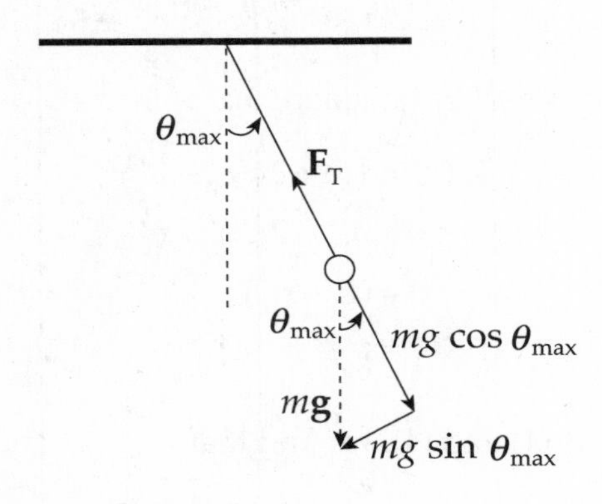

The net force toward the center of the bob's circular path is $F_T - mg \cos \theta_{max}$. This must provide the centripetal force, mv^2/L. But since the speed of the bob at this moment is zero ($v = 0$), we get $F_T = mg \cos \theta_{max}$. [The acceleration is purely tangential here, equal to $(mg \sin \theta_{max})/m = g \sin \theta_{max}$.]

SECTION II: FREE RESPONSE

1. (a) Since the spring is compressed to its $\frac{3}{4}$ natural length, the block's position relative to equilibrium is $x = -\frac{1}{4}L$. Therefore, from $F_S = -kx$, we find

$$a = \frac{F_S}{m} = \frac{-k(-\frac{1}{4}L)}{m} = \frac{kL}{4m}$$

(b) Let v_1 denote the velocity of Block 1 just before impact, and let v_1' and v_2' denote, respectively, the velocities of Block 1 and Block 2 immediately after impact. By conservation of linear momentum, we write $mv_1 = mv_1' + mv_2'$, or

$$v_1 = v_1' + v_2' \qquad (1)$$

The initial kinetic energy of Block 1 is $\frac{1}{2}mv_1^2$. If half is lost to heat, then $\frac{1}{4}mv_1^2$ is left to be shared by Block 1 and Block 2 after impact: $\frac{1}{4}mv_1^2 = \frac{1}{2}mv_1'^2 + \frac{1}{2}mv_2'^2$, or

$$v_1^2 = 2v_1'^2 + 2v_2'^2 \qquad (2)$$

Square Equation (1) and multiply by 2 to give

$$2v_1^2 = 2v_1'^2 + 4v_1'v_2' + 2v_2'^2 \qquad (1')$$

then subtract Equation (2) from Equation (1′):

$$v_1^2 = 4v_1'v_2' \qquad (3)$$

Square Equation (1) again,

$$v_1^2 = v_1'^2 + 2v_1'v_2' + v_2'^2$$

and substitute into this the result of Equation (3):

$$4v_1'v_2' = v_1'^2 + 2v_1'v_2' + v_2'^2$$
$$0 = v_1'^2 - 2v_1'v_2' + v_2'^2$$
$$0 = (v_1' - v_2')^2$$
$$v_1' = v_2' \qquad (4)$$

Thus, combining Equations (1) and (4), we find that

$$v_1' = v_2' = \tfrac{1}{2}v_1$$

(c) When Block 1 reaches its new amplitude position, A', all of its kinetic energy is converted to elastic potential energy of the spring. That is,

$$K_1' \to U_S' \quad \Rightarrow \quad \tfrac{1}{2}mv_1'^2 = \tfrac{1}{2}kA'^2$$

$$A'^2 = \frac{m}{k}v_1'^2$$

$$A'^2 = \frac{m}{k}(\tfrac{1}{2}v_1)^2$$

$$A'^2 = \frac{mv_1^2}{4k} \qquad (1)$$

But the original potential energy of the spring, $U_S = \frac{1}{2}k(-\frac{1}{4}L)^2$, gave K_1:

$$U_S \to K_1 \quad \Rightarrow \quad \tfrac{1}{2}k(-\tfrac{1}{4}L)^2 = \tfrac{1}{2}mv_1^2 \quad \Rightarrow \quad mv_1^2 = \tfrac{1}{16}kL^2 \qquad (2)$$

Substituting this result into Equation (1) gives

$$A'^2 = \frac{\frac{1}{16}kL^2}{4k} = \frac{L^2}{64} \quad \Rightarrow \quad A' = \tfrac{1}{8}L$$

(d) The period of a spring–block simple harmonic oscillator depends only on the spring constant k and the mass of the block. Since neither of these changes, the period will remain the same; that is, $T' = T_0$.

(e) As we showed in part (b), Block 2's velocity as it slides off the table is $\frac{1}{2}v_1$ (horizontally). The time required to drop the vertical distance H is found as follows (calling *down* the positive direction):

$$\Delta y = v_{0y} + \tfrac{1}{2}gt^2 \quad \Rightarrow \quad H = \tfrac{1}{2}gt^2 \quad \Rightarrow \quad t = \sqrt{\frac{2H}{g}}$$

Therefore,

$$R = (\tfrac{1}{2}v_1)t = \tfrac{1}{2}v_1\sqrt{\frac{2H}{g}}$$

Now, from Equation (2) of part (c), $v_1 = \sqrt{kL^2/16m}$, so

$$R = \tfrac{1}{2}\sqrt{\frac{kL^2}{16m}}\sqrt{\frac{2H}{g}} = \frac{L}{8}\sqrt{\frac{2kH}{mg}}$$

2. (a) By conservation of linear momentum,

$$mv = (m+M)v' \quad \Rightarrow \quad v' = \frac{mv}{m+M}$$

(b) When the block is at its amplitude position (maximum compression of spring), the kinetic energy it (and the embedded bullet) had just after impact will become potential energy of the spring:

$$K' \rightarrow U_S$$

$$\tfrac{1}{2}(m+M)\left(\frac{mv}{m+M}\right)^2 = \tfrac{1}{2}kA^2$$

$$A = \frac{mv}{\sqrt{k(m+M)}}$$

(c) Since the mass on the spring is $m + M$, $f = \dfrac{1}{2\pi}\sqrt{\dfrac{k}{m+M}}$.

(d) The position of the block is given by the equation $x = A\sin(\omega t + \phi_0)$, where $\omega = 2\pi f$ and A is the amplitude. Since $x = 0$ at time $t = 0$, the initial phase, ϕ_0, is 0. From the results of parts (b) and (c), we have

$$x = A\sin(2\pi f t) = \frac{mv}{\sqrt{k(m+M)}}\sin\left(t\sqrt{\frac{k}{m+M}}\right)$$

3. (a) By Conservation of Mechanical Energy, $K + U = E$, so

$$\tfrac{1}{2}Mv^2 + \tfrac{1}{2}k(\tfrac{1}{2}A)^2 = \tfrac{1}{2}kA^2$$

$$\tfrac{1}{2}Mv^2 = \tfrac{3}{8}kA^2$$

$$v = A\sqrt{\frac{3k}{4M}}$$

(b) Since the clay ball delivers no horizontal linear momentum to the block, horizontal linear momentum is conserved. Thus,

$$Mv = (M+m)v'$$

$$v' = \frac{Mv}{M+m} = \frac{MA}{M+m}\sqrt{\frac{3k}{4M}} = \frac{A}{M+m}\sqrt{\frac{3kM}{4}}$$

(c) Applying the general equation for the period of a spring–block simple harmonic oscillator,

$$T = 2\pi\sqrt{\frac{M+m}{k}}$$

(d) The total energy of the oscillator after the clay hits is $\frac{1}{2}kA'^2$, where A' is the new amplitude. Just after the clay hits the block, the total energy is

$$K' + U_S = \tfrac{1}{2}(M+m)v'^2 + \tfrac{1}{2}k(\tfrac{1}{2}A)^2$$

Substitute for v' from part (b), set the resulting sum equal to $\frac{1}{2}kA'^2$, and solve for A':

$$\tfrac{1}{2}(M+m)\left(\frac{A}{M+m}\sqrt{\frac{3kM}{4}}\right)^2 + \tfrac{1}{2}k(\tfrac{1}{2}A)^2 = \tfrac{1}{2}kA'^2$$

$$\frac{A^2 \cdot 3kM}{8(M+m)} + \tfrac{1}{8}kA^2 = \tfrac{1}{2}kA'^2$$

$$A^2\left(\frac{3M}{M+m}+1\right) = 4A'^2$$

$$A' = \tfrac{1}{2}A\sqrt{\frac{3M}{M+m}+1}$$

(e) No, since the period depends only on the mass and the spring constant k.

(f) Yes. For example, if the clay had landed when the block was at $x = A$, the speed of the block would have been zero immediately before the collision and immediately after. No change in the block's speed would have meant no change in K, so no change in E, so no change in $A = \sqrt{2E/k}$.

4. (a) Since the weight provides a clockwise torque (which is negative), we have $\tau = -dMg \sin \theta$.

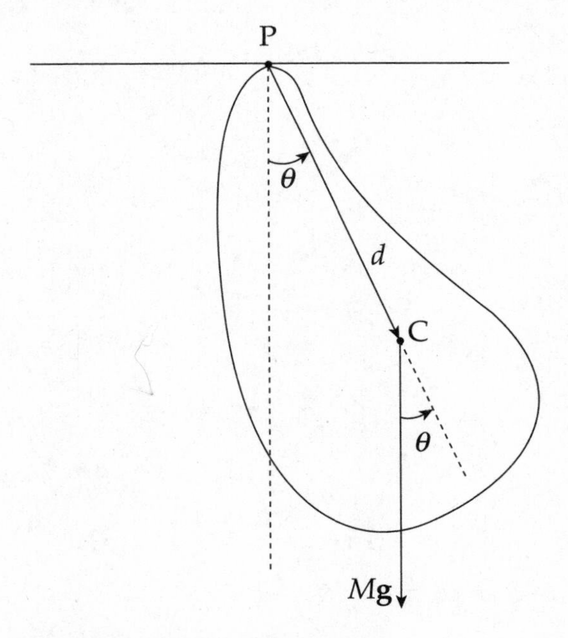

(b) If $\sin \theta \approx \theta$, then the equation in part (a) becomes

$$\tau = -dMg\theta \qquad (1)$$

(c) Since $\tau = I\alpha$, Equation (1) implies $I\alpha = -dMg\theta$. But by definition, $\alpha = d^2\theta/dt^2$, so we get

$$\frac{d^2\theta}{dt^2} = -\frac{dMg}{I}\theta \qquad (2)$$

Equation (2) is identical to Equation (*) given in the statement of the question with $z = \theta$ and $b = dMg/I$. Therefore,

$$T = \frac{2\pi}{\sqrt{b}} = \frac{2\pi}{\sqrt{dMg/I}} = 2\pi\sqrt{\frac{I}{dMg}}$$

(d) Using the result of part (c) with $I = \frac{1}{3}ML^2$ and $d = \frac{1}{2}L$, we get

$$T = 2\pi\sqrt{\frac{\frac{1}{3}ML^2}{\frac{1}{2}L \cdot Mg}} = 2\pi\sqrt{\frac{2L}{3g}}$$

CHAPTER 9 REVIEW QUESTIONS

Section I: Multiple Choice

1. **D** Electrostatic force obeys an inverse-square law: $F_E \propto 1/r^2$. Therefore, if r increases by a factor of 3, then F_E decreases by a factor of $3^2 = 9$.

2. **C** The strength of the electric force is given by kq^2/r^2, and the strength of the gravitational force is Gm^2/r^2. Since both of these quantities have r^2 in the denominator, we simply need to compare the numerical values of kq^2 and Gm^2. There's no contest: Since

$$kq^2 = (9 \times 10^9 \text{ N·m}^2/\text{C}^2)(1 \text{ C})^2 = 9 \times 10^9 \text{ N·m}^2$$

and

$$Gm^2 = (6.7 \times 10^{-11} \text{ N·m}^2/\text{kg}^2)(1 \text{ kg})^2 = 6.7 \times 10^{-11} \text{ N·m}^2$$

we see that $kq^2 >> Gm^2$, so F_E is much stronger than F_G.

3. **C** If the net electric force on the center charge is zero, the electrical repulsion by the $+2q$ charge must balance the electrical repulsion by the $+3q$ charge:

$$\frac{1}{4\pi\varepsilon_0}\frac{(2q)(q)}{x^2} = \frac{1}{4\pi\varepsilon_0}\frac{(3q)(q)}{y^2} \Rightarrow \frac{2}{x^2} = \frac{3}{y^2} \Rightarrow \frac{y^2}{x^2} = \frac{3}{2} \Rightarrow \frac{y}{x} = \sqrt{\frac{3}{2}}$$

4. **E** Since P is equidistant from the two charges, and the magnitudes of the charges are identical, the strength of the electric field at P due to $+Q$ is the same as the strength of the electric field at P due to $-Q$. The electric field vector at P due to $+Q$ points away from $+Q$, and the electric field vector at P due to $-Q$ points toward $-Q$. Since these vectors point in the same direction, the net electric field at P is (E to the right) + (E to the right) = ($2E$ to the right).

5. **D** The acceleration of the small sphere is

$$a = \frac{F_E}{m} = \frac{1}{4\pi\varepsilon_0}\frac{Qq}{mr^2}$$

As r increases (that is, as the small sphere is pushed away), a decreases. However, since a is always positive, the small sphere's speed, v, is always increasing.

6. **B** Since $\mathbf{F}_E$ (on q) = $q\mathbf{E}$, it must be true that $\mathbf{F}_E$ (on $-2q$) = $-2q\mathbf{E} = -2\mathbf{F}_E$.

7. **D** All excess electric charge on a conductor resides on the outer surface.

8. **B** The net electric flux through the Gaussian surface, Φ_E, is equal to $(1/\varepsilon_0)$ times the net charge *enclosed by the surface*, $Q_2 + Q_3$. However, the net electric field vector at P is equal to the sum of the electric field vectors due to each of the four charges individually.

9. **A** Since the charge is distributed uniformly throughout the sphere, the magnitude of its volume charge density is $\rho = Q/(\frac{4}{3}\pi R^3)$. Therefore, if we construct a spherical Gaussian surface of radius $r < R$, Gauss's Law gives

$$\Phi_E = \frac{Q_{\text{enclosed}}}{\varepsilon_0} \Rightarrow E(4\pi r^2) = \frac{\rho \cdot \frac{4}{3}\pi r^3}{\varepsilon_0} \Rightarrow E = \rho\frac{r}{3\varepsilon_0}$$

$$E = \frac{Q}{\frac{4}{3}\pi R^3}\frac{r}{3\varepsilon_0}$$

$$E = \frac{1}{4\pi\varepsilon_0}\frac{Q}{R^3}r$$

(Since the charge on the sphere is negative, the direction of **E** at any point is directed radially inward.)

10. **A** By definition, an electric dipole consists of two equal but opposite charges. Therefore a Gaussian surface enclosing both charges would enclose zero net charge. By Gauss's Law, the electric flux is zero as well (since $\Phi_E = Q_{\text{enclosed}}/\varepsilon_0$).

SECTION II: FREE RESPONSE

1. (a) From the figure below, we have $F_{1\text{-}2} = F_1/\cos 45°$.

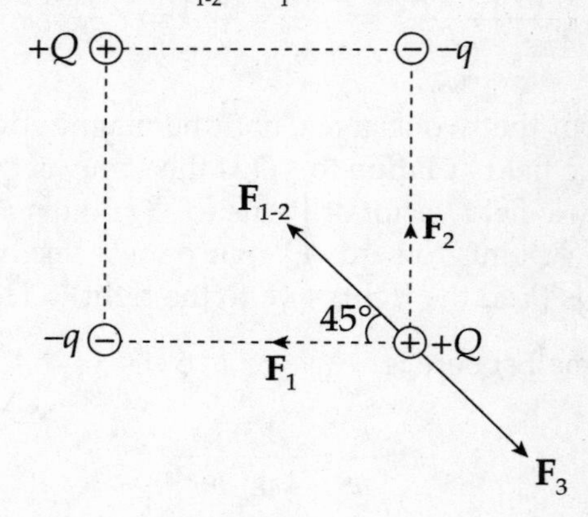

Since the net force on $+Q$ is zero, we want $F_{1\text{-}2} = F_3$. If s is the length of each side of the square, then:

$$F_{1-2} = F_3 \Rightarrow \frac{F_1}{\cos 45°} = F_3 \Rightarrow \frac{1}{\cos 45°}\frac{1}{4\pi\varepsilon_0}\frac{Qq}{s^2} = \frac{1}{4\pi\varepsilon_0}\frac{Q^2}{(s\sqrt{2})^2}$$

$$\sqrt{2}\cdot q = \frac{Q}{2}$$

$$q = \frac{Q}{2\sqrt{2}}$$

(b) No. If $q = Q/2\sqrt{2}$, as found in part (a), then the net force on $-q$ is not zero.

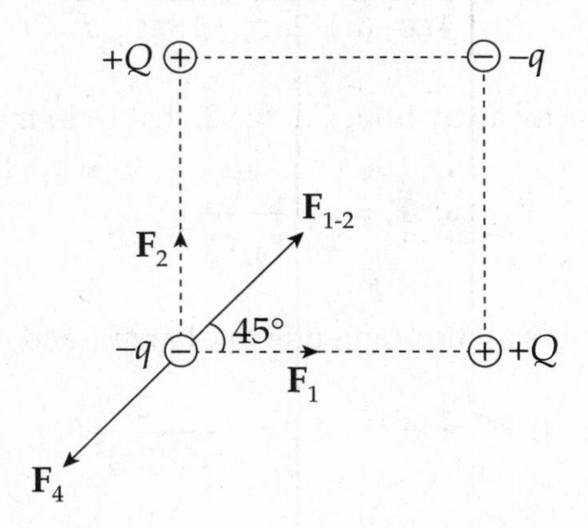

This is because $F_{1\text{-}2} \neq F_4$, as the following calculations show:

$$F_{1\text{-}2} = \frac{F_1}{\cos 45^\circ} = \sqrt{2}\frac{1}{4\pi\varepsilon_0}\frac{Qq}{s^2} = \sqrt{2}\frac{1}{4\pi\varepsilon_0}\frac{Q\frac{Q}{2\sqrt{2}}}{s^2} = \frac{1}{8\pi\varepsilon_0}\frac{Q^2}{s^2}$$

but

$$F_4 = \frac{1}{4\pi\varepsilon_0}\frac{q^2}{(s\sqrt{2})^2} = \frac{1}{4\pi\varepsilon_0}\frac{\left(\frac{Q}{2\sqrt{2}}\right)^2}{(s\sqrt{2})^2} = \frac{1}{64\pi\varepsilon_0}\frac{Q^2}{s^2}$$

(c) By symmetry, $E_1 = E_2$ and $E_3 = E_4$, so the net electric field at the center of the square is zero:

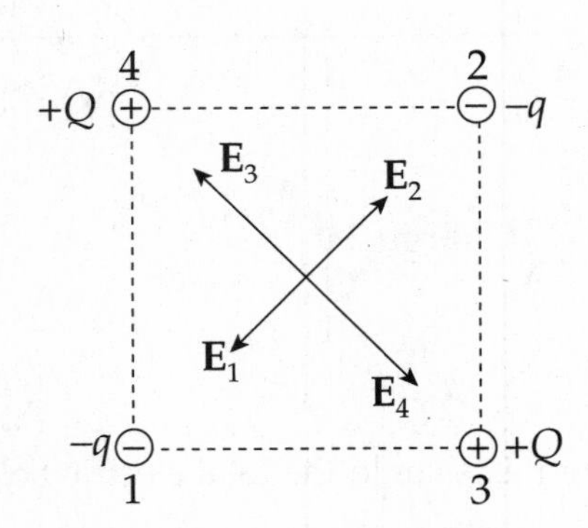

2. (a) The magnitude of the electric force on Charge 1 is

$$F_1 = \frac{1}{4\pi\varepsilon_0}\frac{(Q)(2Q)}{(a+2a)^2} = \frac{1}{18\pi\varepsilon_0}\frac{Q^2}{a^2}$$

The direction of $\mathbf{F}_1$ is directly away from Charge 2; that is, in the $+y$ direction, so

$$\mathbf{F}_1 = \frac{1}{18\pi\varepsilon_0}\frac{Q^2}{a^2}\hat{\mathbf{j}}$$

(b) The electric field vectors at the origin due to Charge 1 and due to Charge 2 are

$$\mathbf{E}_1 = \frac{1}{4\pi\varepsilon_0}\frac{Q}{a^2}(-\hat{\mathbf{j}}) \quad \text{and} \quad \mathbf{E}_2 = \frac{1}{4\pi\varepsilon_0}\frac{2Q}{(2a)^2}(+\hat{\mathbf{j}}) = \frac{1}{8\pi\varepsilon_0}\frac{Q}{a^2}(+\hat{\mathbf{j}})$$

Therefore, the net electric field at the origin is

$$\mathbf{E} = \mathbf{E}_1 + \mathbf{E}_2 = \frac{1}{4\pi\varepsilon_0}\frac{Q}{a^2}(-\hat{\mathbf{j}}) + \frac{1}{8\pi\varepsilon_0}\frac{Q}{a^2}(+\hat{\mathbf{j}}) = \frac{1}{8\pi\varepsilon_0}\frac{Q}{a^2}(-\hat{\mathbf{j}})$$

(c) No. At no point on the x axis will the individual electric field vectors due to each of the two charges point in exactly opposite directions.

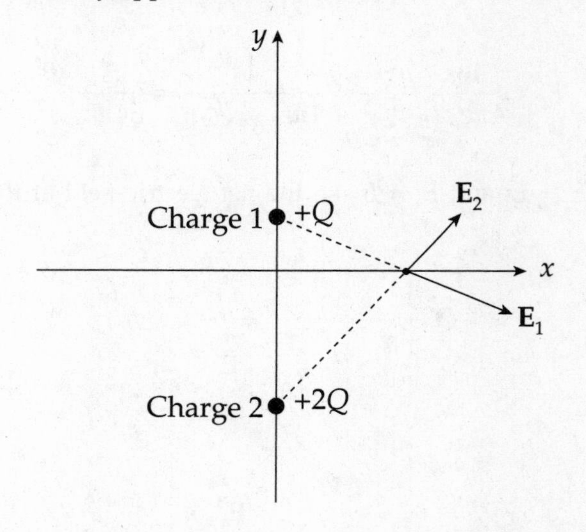

Therefore, at no point on the x axis could the total electric field be zero.

(d) Yes. There will be a Point P on the y axis between the two charges,

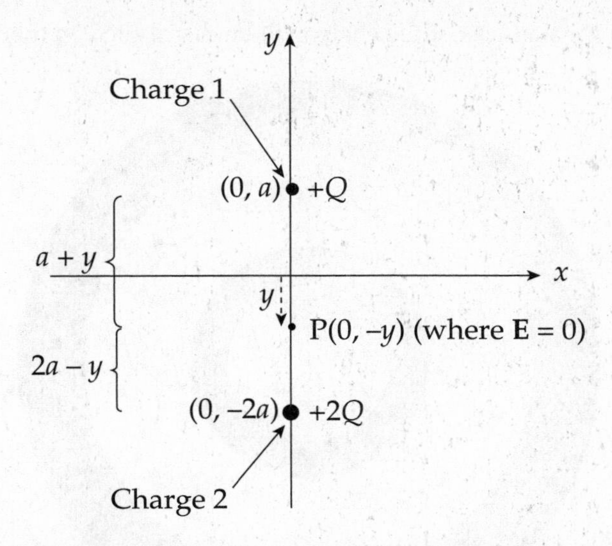

where the electric fields due to the individual charges will cancel each other out.

$$E_1 = E_2$$

$$\frac{1}{4\pi\varepsilon_0}\frac{Q}{(a+y)^2} = \frac{1}{4\pi\varepsilon_0}\frac{2Q}{(2a-y)^2}$$

$$\frac{1}{(a+y)^2} = \frac{2}{(2a-y)^2}$$

$$(2a-y)^2 = 2(a+y)^2$$

$$y^2 + 8ay - 2a^2 = 0$$

$$y = \frac{-8a \pm \sqrt{(8a)^2 - 4(-2a^2)}}{2}$$

$$= (-4 \pm 3\sqrt{2})a$$

Disregarding the value $y = (-4 - 3\sqrt{2})a$ (because it would place the point P below Charge 2 on the y axis, where the electric field vectors do not point in opposite directions), we have that $\mathbf{E} = \mathbf{0}$ at the point $P = (0, -y) = (0, (4 - 3\sqrt{2})a)$.

(e) Use the result of part (b) with Newton's Second Law:

$$\mathbf{a} = \frac{\mathbf{F}}{m} = \frac{-q\mathbf{E}}{m} = \frac{-q}{m}\left[\frac{1}{8\pi\varepsilon_0}\frac{Q}{a^2}(-\hat{\mathbf{j}})\right] = \frac{1}{8\pi\varepsilon_0}\frac{qQ}{ma^2}(+\hat{\mathbf{j}})$$

3. (a) The charges on the surfaces of the shells will be distributed as follows:

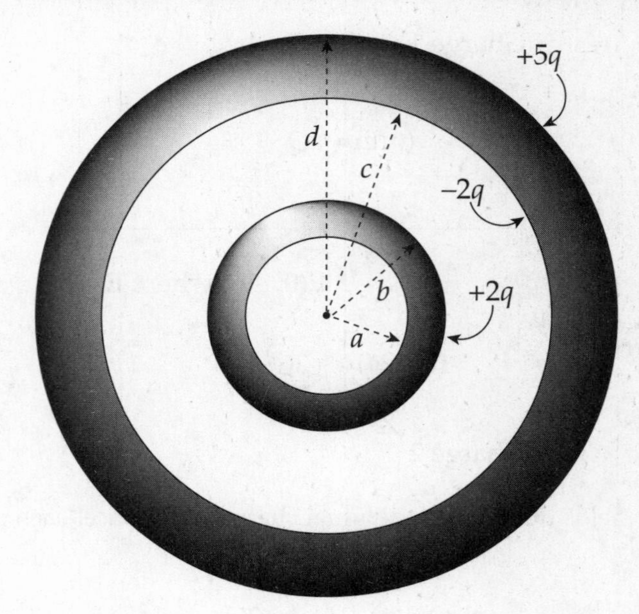

Therefore, by Gauss's Law, the electric field in the various regions are as follows:

(i) $r < a$ $E = 0$ (no charge inside)

(ii) $a < r < b$ $E = 0$ (within metallic shell)

(iii) $b < r < c$ $E = (1/4\pi\varepsilon_0)(2q/r^2)$

(iv) $c < r < d$ $E = 0$ (within metallic shell)

(v) $r > d$ $E = (1/4\pi\varepsilon_0)(5q/r^2)$

(b) Within each metallic shell, there is no excess charge. The inner surface of the inner shell also has no excess charge. See the figure above for part (a) for the charges residing on the other surfaces of the shells.

4. (a) Very close to the rod, the rod behaves as if it were "infinitely long." Construct a cylindrical Gaussian surface of radius x_1 and length L centered on the rod. By symmetry the electric field at P_1 must point radially away from the rod, so the field at P_1 must point in the positive x direction.

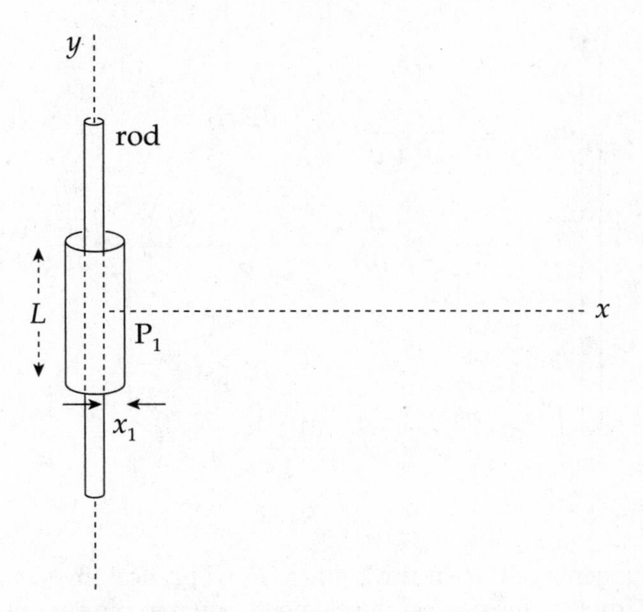

There is no electric flux through the ends of the cylinder (since **E** is parallel to the lids of the cylinder); the electric flux is perpendicular to the lateral surface area of the cylinder, so $\Phi_E = 0 + EA + 0 = E(2\pi x_1 L)$. Since the charge enclosed by the surface is λL, Gauss's Law gives

$$E(2\pi x_1 L) = \frac{\lambda L}{\varepsilon_0} \quad \Rightarrow \quad E = \frac{\lambda}{2\pi\varepsilon_0 x_1}$$

(b) The total charge on the rod is equal to its linear charge density times its total length: $Q = \lambda \ell$.

(c) If x_2 is not small compared to ℓ, then we must calculate the electric field at P_2 by integration. Consider two symmetrically-located line elements along the rod, each of length dy, at a distance y above and below the origin.

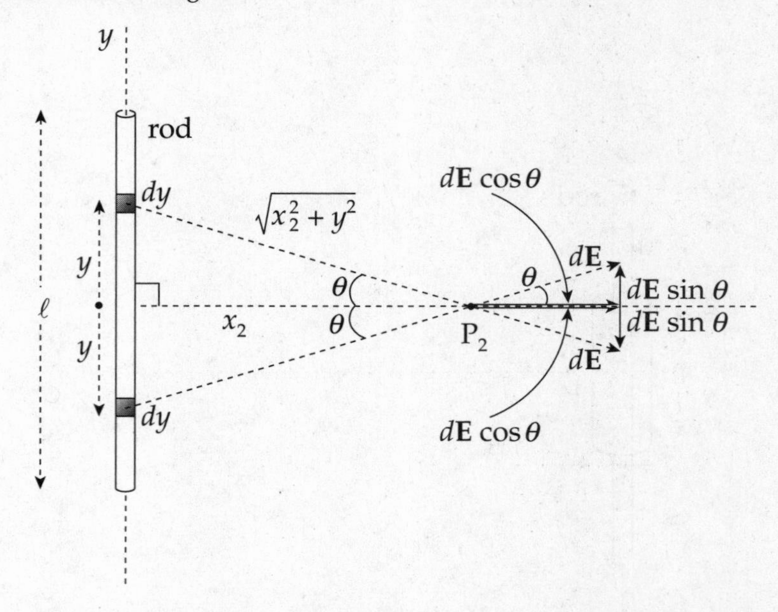

The vertical components of the electric fields at P_2 will cancel (by symmetry), leaving the net electric field at P_2 due to this pair of line elements equal to the sum of the horizontal components. Therefore,

$$dE_{\text{total}} = 2\, dE \cos\theta$$

$$E_{\text{total}} = \int_{y=0}^{y=\ell/2} 2\left[\frac{1}{4\pi\varepsilon_0}\frac{\lambda\, dy}{x_2^2 + y^2}\right]\frac{x_2}{\sqrt{x_2^2 + y^2}}$$

$$= \frac{\lambda x_2}{2\pi\varepsilon_0}\int_{y=0}^{y=\ell/2}\frac{dy}{(x_2^2 + y^2)^{3/2}}$$

$$= \frac{\lambda x_2}{2\pi\varepsilon_0}\left[\frac{y}{x_2^2\sqrt{x_2^2 + y^2}}\right]_{y=0}^{y=\ell/2}$$

$$= \frac{\lambda}{2\pi\varepsilon_0 x_2}\left[\frac{\ell/2}{\sqrt{x_2^2 + (\ell/2)^2}}\right]$$

If $x_2 = \ell$, then

$$E_{\text{total}} = \frac{\lambda}{2\pi\varepsilon_0 \ell}\left[\frac{\ell/2}{\sqrt{\ell^2 + (\ell/2)^2}}\right] = \frac{\lambda}{4\pi\varepsilon_0}\frac{1}{\sqrt{5\ell^2/4}} = \frac{\lambda}{2\pi\varepsilon_0 \ell\sqrt{5}} = \frac{Q/\ell}{2\pi\varepsilon_0 \ell\sqrt{5}} = \frac{Q}{2\pi\varepsilon_0 \ell^2\sqrt{5}}$$

5. (a) Volume charge density has units of charge per unit volume, that is, $[\rho] = \text{C/m}^3$. Since r/a has no units, ρ_s must also have units of C/m^3.

   (b) Since ρ varies with the distance from the center, consider a thin spherical shell of radius R and thickness dR. Its volume is $dV = 4\pi R^2\, dR$, so its charge is

$$dQ = \rho\, dV = \rho_s \frac{R}{a} \cdot 4\pi R^2\, dR = \frac{4\pi\rho_s}{a} R^3\, dR$$

   The charge on the entire sphere is found by integrating dQ from $R = 0$ to $R = a$:

$$\begin{aligned} Q = \int dQ &= \int_{R=0}^{R=a} \frac{4\pi\rho_s}{a} R^3\, dR \qquad (1) \\ &= \frac{4\pi\rho_s}{a}\frac{a^4}{4} \\ &= \pi\rho_s a^3 \end{aligned}$$

   (c) (i) Replace "$R = a$" with "$R = r$" in Equation (1) above to find the charge enclosed by a spherical Gaussian surface of radius $r < a$:

$$Q_{\text{within } r} = \int_{R=0}^{R=r} \frac{4\pi\rho_s}{a} R^3\, dR = \frac{4\pi\rho_s}{a}\frac{r^4}{4}$$

   Using the result of part (b), we can write this expression for $Q_{\text{within } r}$ in terms of Q, the total charge on the full sphere:

$$Q_{\text{within } r} = \frac{4\pi\rho_s}{a}\frac{r^4}{4} = \frac{4\frac{Q}{a^3}}{a}\frac{r^4}{4} = \frac{Q}{a^4} r^4$$

   So, by Gauss's Law,

$$\begin{aligned} \Phi_E &= \frac{1}{\varepsilon_0} \cdot Q_{\text{enclosed}} \\ E(4\pi r^2) &= \frac{1}{\varepsilon_0} \cdot \frac{Q}{a^4} r^4 \\ E &= \frac{1}{4\pi\varepsilon_0}\frac{Q}{a^4} r^2 \quad (r < a) \end{aligned}$$

(c) (ii) Outside the sphere, $Q_{enclosed} = Q$, so

$$E = \frac{1}{4\pi\varepsilon_0}\frac{Q}{r^2} \quad (r \geq a)$$

(d)

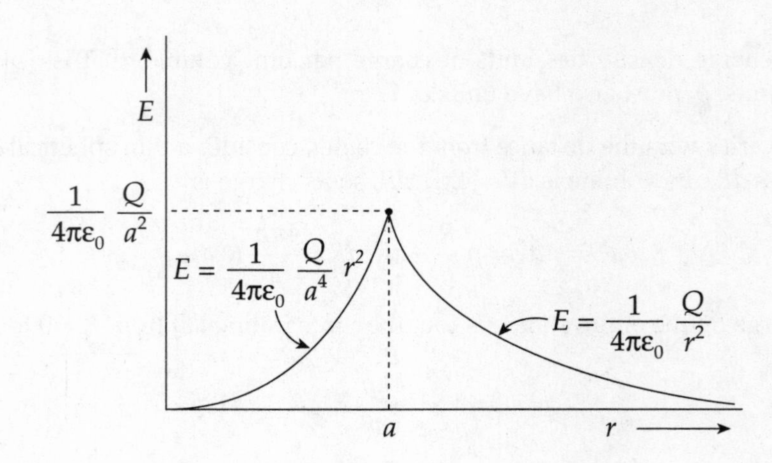

CHAPTER 10 REVIEW QUESTIONS

SECTION I: MULTIPLE CHOICE

1. **E** A counterexample for Statement I is provided by two equal positive charges; at the point midway between the charges, the electric field is zero, but the potential is not. A counterexample for Statement II is provided by an electric dipole (a pair of equal but opposite charges); at the point midway between the charges, the electric potential is zero, but the electric field is not. As for Statement III, consider a single positive point charge $+Q$. Then at a distance r from this source charge, the electric field strength is $E = kQ/r^2$ and the potential is $V = kQ/r$. Thus, $V = rE$, so V is not inversely proportional to E.

2. **C** By definition, $\Delta U_E = -W_E$, so if W_E is negative, then ΔU_E is positive. This implies that the potential energy, U_E, increases.

3. **D** You may be tempted to solve for the equivalent capacitance and then use that to determine the total charge stored on the capacitors, but we can already tell the potential difference across each capacitor is 9 V. The energy stored in a capacitor is given by the equation

$$U_c = \frac{1}{2}CV^2 = \frac{1}{2}(6\text{x}10^{-6})(9)^2$$
$$U_c = 2.43\text{x}10^{-4}\text{ J}$$

4. **B** Use the definition $\Delta V = -W_E/q$. If an electric field accelerates a negative charge doing positive work on it, then $W_E > 0$. If $q < 0$, then $-W_E/q$ is positive. Therefore, ΔV is positive, which implies that V increases.

5. **E** By definition,

$$V_{A\to B} = \Delta U_E/q, \text{ so } V_B - V_A = \Delta U_E/q$$

6. **C** Because **E** is uniform, the potential varies linearly with distance from either plate ($\Delta V = Ed$). Since Points 2 and 4 are at the same distance from the plates, they lie on the same equipotential. (The equipotentials in this case are planes parallel to the capacitor plates.)

7. **D** Once the spheres are connected by a conducting wire, they quickly form a single equipotential surface. Since the potential on a sphere of radius r carrying charge q is given by the equation $(1/4\pi\varepsilon_0)(q/r)$, we have

$$V_{\text{Sphere \#1}} = V_{\text{Sphere \#2}} \Rightarrow \frac{1}{4\pi\varepsilon_0}\frac{q_1}{a} = \frac{1}{4\pi\varepsilon_0}\frac{q_2}{4a} \Rightarrow q_1 = \frac{q_2}{4}$$

Now, since $q_1 + q_2 = -Q + -Q = -2Q$, the fact that q_1 must equal $q_2/4$ means that $(q_2/4) + q_2 = -2Q$, which gives $q_2 = (-8/5)Q$ and $q_1 = (-2/5)Q$.

8. **A** Since Q cannot change and C is increased (because of the dielectric), $\Delta V = Q/C$ must decrease. Also, since $U_E = Q^2/(2C)$, an increase in C with no change in Q implies a decrease in U_E.

9. **B** By definition, $W_E = -q\Delta V$, which gives

$$W_E = -q(V_B - V_A) = -(-0.05 \text{ C})(100 \text{ V} - 200 \text{ V}) = -5 \text{ J}$$

Note that neither the length of the segment AB nor that of the curved path from A to B is relevant.

10. **D**

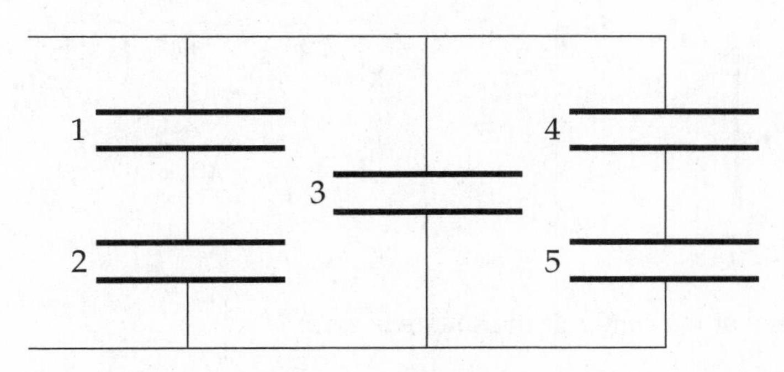

Capacitors 1 and 2 are in series, so their equivalent capacitance is $C_{1\text{-}2} = C/2$. (This is obtained from the equation $1/C_{1\text{-}2} = 1/C_1 + 1/C_2 = 1/C + 1/C = 2/C$.) Capacitors 4 and 5 are also in series, so their equivalent capacitance is $C_{4\text{-}5} = C/2$. The capacitances $C_{1\text{-}2}$, C_3, and $C_{4\text{-}5}$ are in parallel, so the overall equivalent capacitance is $(C/2) + C + (C/2) = 2C$.

11. **C** Points A and E are on the same equipotential line, so no work is done by the field to move the electron from A to E. Electric fields go in the direction of decreasing potential, so the field associated with these equipotential lines is generally going to the left. The field lines indicate the force on a positive test charge, so the force on an electron will be in the opposite direction. Therefore, the work done by the field on the electron is negative for the movement from E to C because the force on the electron is to the right and the movement is to the left.

SECTION II: FREE RESPONSE

1. (a) Labeling the four charges as given in the diagram, we get

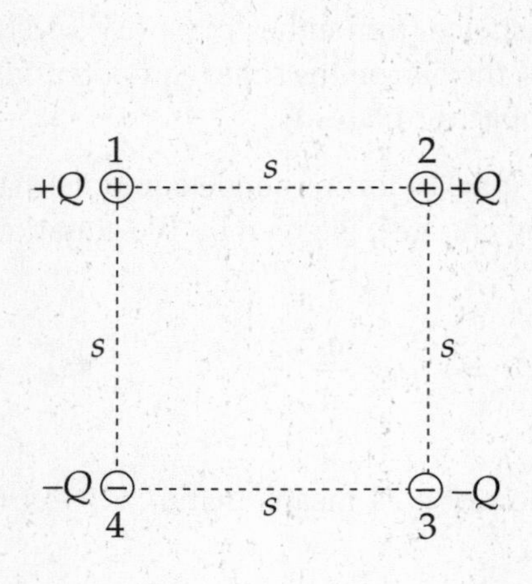

$$\begin{aligned} U_E &= \frac{1}{4\pi\varepsilon_0}\sum_{i<j}\frac{q_i q_j}{r_{ij}} \\ &= \frac{1}{4\pi\varepsilon_0}\left(\frac{q_1q_2}{r_{12}}+\frac{q_1q_3}{r_{13}}+\frac{q_1q_4}{r_{14}}+\frac{q_2q_3}{r_{23}}+\frac{q_2q_4}{r_{24}}+\frac{q_3q_4}{r_{34}}\right) \\ &= \frac{1}{4\pi\varepsilon_0}\left(\frac{Q^2}{s}+\frac{-Q^2}{s\sqrt{2}}+\frac{-Q^2}{s}+\frac{-Q^2}{s}+\frac{-Q^2}{s\sqrt{2}}+\frac{Q^2}{s}\right) \\ &= \frac{1}{4\pi\varepsilon_0}\frac{-2Q^2}{s\sqrt{2}} \\ &= -\frac{\sqrt{2}}{4\pi\varepsilon_0}\frac{Q^2}{s} \end{aligned}$$

(b) Let $\mathbf{E}_i$ denote the electric field at the center of the square due to charge i. Then by symmetry, $\mathbf{E}_1 = \mathbf{E}_3$, $\mathbf{E}_2 = \mathbf{E}_4$, and $E_1 = E_2 = E_3 = E_4$. The horizontal components of the four individual field vectors cancel, leaving only a downward-pointing electric field of magnitude $E_{total} = 4E_1 \cos 45°$:

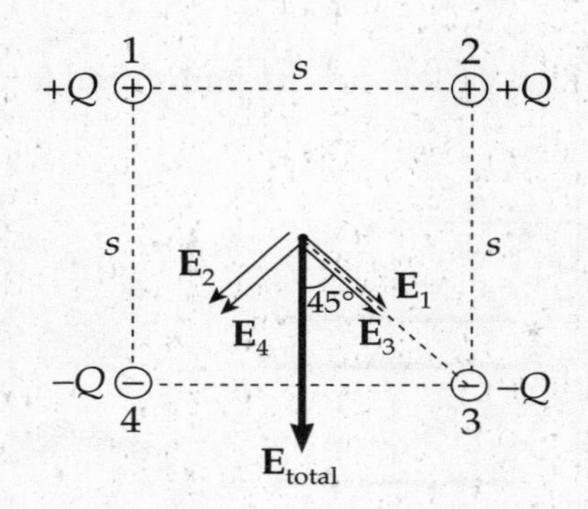

$$\begin{aligned} E_{total} &= 4E_1 \cos 45° \\ &= 4\frac{1}{4\pi\varepsilon_0}\frac{Q}{\left(\frac{1}{2}s\sqrt{2}\right)^2}\cdot\frac{\sqrt{2}}{2} \\ &= \frac{\sqrt{2}}{\pi\varepsilon_0}\frac{Q}{s^2} \end{aligned}$$

(c) The potential at the center of the square is zero:

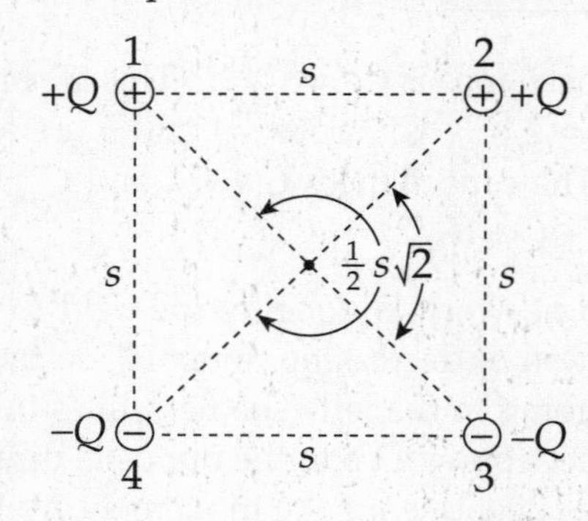

$$\begin{aligned} V &= \frac{1}{4\pi\varepsilon_0}\sum_i \frac{q_i}{r_i} \\ &= \frac{1}{4\pi\varepsilon_0}\left(\frac{q_1}{r_1}+\frac{q_2}{r_2}+\frac{q_3}{r_3}+\frac{q_4}{r_4}\right) \\ &= \frac{1}{4\pi\varepsilon_0}\left(\frac{Q}{\frac{1}{2}s\sqrt{2}}+\frac{Q}{\frac{1}{2}s\sqrt{2}}+\frac{-Q}{\frac{1}{2}s\sqrt{2}}+\frac{-Q}{\frac{1}{2}s\sqrt{2}}\right) \\ &= 0 \end{aligned}$$

(d) At every point on the center horizontal line shown, $r_1 = r_4$ and $r_2 = r_3$, so V will equal zero (just as it does at the center of the square):

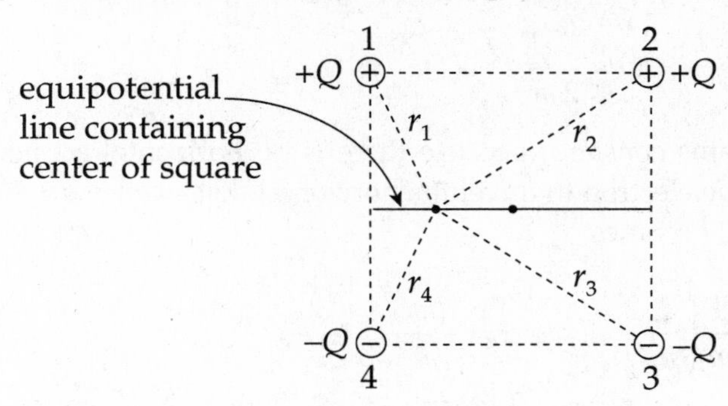

$$V = \frac{1}{4\pi\varepsilon_0}\sum_i \frac{q_i}{r_i}$$

$$= \frac{1}{4\pi\varepsilon_0}\left(\frac{q_1}{r_1}+\frac{q_2}{r_2}+\frac{q_3}{r_3}+\frac{q_4}{r_4}\right)$$

$$= \frac{1}{4\pi\varepsilon_0}\left(\frac{Q}{r_1}+\frac{Q}{r_2}+\frac{-Q}{r_3}+\frac{-Q}{r_4}\right)$$

$$= 0$$

(e) The work done by the electric force as q is displaced from Point A to Point B is given by the equation $W_E = -q\Delta V_{A\to B} = -q(V_B - V_A) = -qV_B$ (since $V_A = 0$).

$$W_E = -qV_B = -q\left[\frac{1}{4\pi\varepsilon_0}\sum_i \frac{q_i}{r_{i\to B}}\right]$$

$$= \frac{-q}{4\pi\varepsilon_0}\left(\frac{q_1}{r_{1\to B}}+\frac{q_2}{r_{2\to B}}+\frac{q_3}{r_{3\to B}}+\frac{q_4}{r_{4\to B}}\right)$$

$$= \frac{-q}{4\pi\varepsilon_0}\left(\frac{Q}{\frac{1}{2}s}+\frac{Q}{\frac{1}{2}s}+\frac{-Q}{\sqrt{s^2+(\frac{1}{2}s)^2}}+\frac{-Q}{\sqrt{s^2+(\frac{1}{2}s)^2}}\right)$$

$$= \frac{-q}{4\pi\varepsilon_0}\left(\frac{4Q}{s}+\frac{-2Q}{\frac{\sqrt{5}}{2}s}\right)$$

$$= \frac{-qQ}{\pi\varepsilon_0 s}\left(1-\frac{1}{\sqrt{5}}\right)$$

2. (a) The capacitance is $C = \varepsilon_0 A/d$. Since the plates are rectangular, the area A is equal to Lw, so $C = \varepsilon_0 Lw/d$.

(b) and (c) Since the electron is attracted upward, the top plate must be the positive plate:

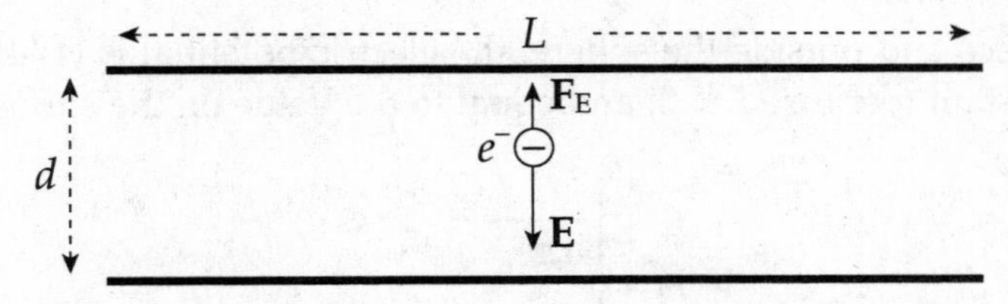

(d) The acceleration of the electron is $a = F_E/m = qE/m = eE/m$, vertically upward. Therefore, applying Big Five #3 for vertical motion, $\Delta y = v_{0y}t + \frac{1}{2}a_y t^2$, we get

$$\Delta y = \frac{1}{2}\frac{eE}{m}t^2 \quad (1)$$

To find t, notice that $v_{0x} = v_0$ remains constant (because there is no horizontal acceleration). Therefore, the time necessary for the electron to travel the horizontal distance L is $t = L/v_0$. In this time, Δy is $d/2$, so Equation (1) becomes

$$\frac{d}{2} = \frac{1}{2}\frac{eE}{m}\left(\frac{L}{v_0}\right)^2 \quad \Rightarrow \quad E = \frac{dmv_0^2}{eL^2}$$

(e) Substituting the result of part (d) into the equation $\Delta V = Ed$ gives

$$\Delta V = \frac{d^2mv_0^2}{eL^2}$$

Since $Q = C\Delta V$ (by definition), the result of part (a) now gives

$$Q = \frac{\varepsilon_0 Lw}{d} \cdot \frac{d^2mv_0^2}{eL^2} = \frac{\varepsilon_0 wdmv_0^2}{eL}$$

(f) Applying the equation $U_E = \frac{1}{2}C(\Delta V)^2$, we get

$$U_E = \frac{1}{2} \cdot \frac{\varepsilon_0 Lw}{d} \cdot \left(\frac{d^2mv_0^2}{eL^2}\right)^2 = \frac{\varepsilon_0 wd^3m^2v_0^4}{2e^2L^3}$$

3. (a) Outside the sphere, the sphere behaves as if all the charge were concentrated at the center. Inside the sphere, the electrostatic field is zero:

$$E(r) = \begin{cases} 0 & (r < a) \\ \dfrac{1}{4\pi\varepsilon_0}\dfrac{Q}{r^2} & (r > a) \end{cases}$$

(b) On the surface and outside the sphere, the electric potential is $(1/4\pi\varepsilon_0)(Q/r)$. Within the sphere, V is constant (because $E = 0$) and equal to the value on the surface. Therefore,

$$V(r) = \begin{cases} \dfrac{1}{4\pi\varepsilon_0}\dfrac{Q}{a} & (r \le a) \\ \dfrac{1}{4\pi\varepsilon_0}\dfrac{Q}{r} & (r > a) \end{cases}$$

(c) See diagrams:

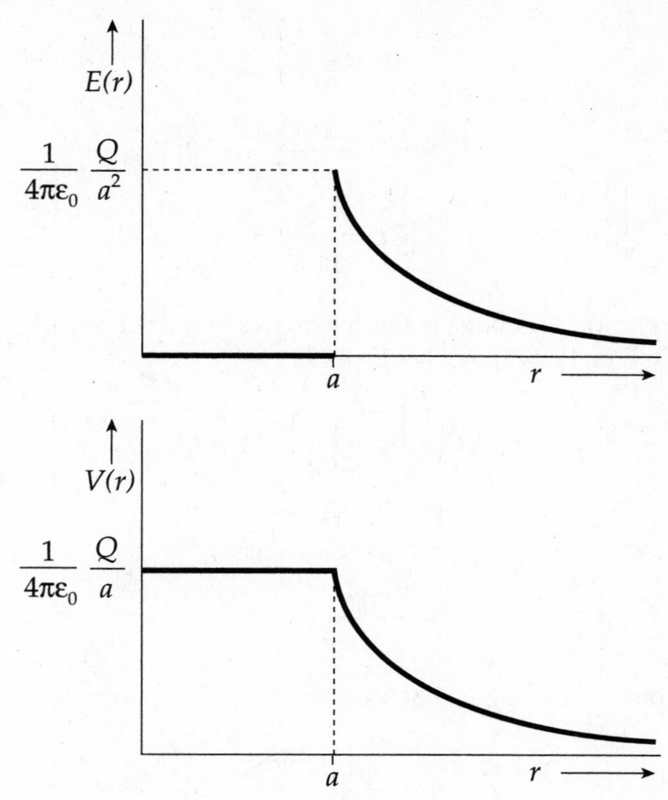

4. (a) To find the electric field inside the sphere, construct a spherical Gaussian surface of radius $R < a$. We need to determine the charge enclosed by this sphere. Therefore, consider a thin spherical shell of radius r and thickness dr. The charge within this shell is $dQ = \rho\, dV = \rho\,(4\pi r^2\, dr)$, so the total charge within a sphere of radius R is

$$\begin{aligned} Q_{\text{within } R} &= \int dQ \\ &= \int_{r=0}^{r=R} \rho_0 \left(\frac{r}{a}\right)^3 \cdot 4\pi r^2\, dr \qquad (1) \\ &= \frac{4\pi\rho_0}{a^3}\int_0^R r^5\, dr \\ &= \frac{2\pi\rho_0 R^6}{3a^3} \end{aligned}$$

For the Gaussian surface, Gauss's Law then gives

$$\Phi_E = \frac{1}{\varepsilon_0} Q_{\text{enclosed}}$$

$$E(4\pi R^2) = \frac{1}{\varepsilon_0} \frac{2\pi\rho_0 R^6}{3a^3}$$

$$E = \frac{\rho_0}{6\varepsilon_0 a^3} R^4$$

Now, to find the electric field outside the sphere, we first determine the sphere's total charge. Replacing $r = R$ in Equation (1) by $r = a$ gives

$$\text{total } Q = \int_{r=0}^{r=a} \rho_0 \left(\frac{r}{a}\right)^3 \cdot 4\pi r^2 \, dr$$

$$= \frac{4\pi\rho_0}{a^3} \int_0^a r^5 \, dr$$

$$= \frac{2\pi\rho_0 a^3}{3}$$

Thus, for points outside the sphere, the electric field is

$$E = \frac{1}{4\pi\varepsilon_0} \frac{Q}{r^2} = \frac{1}{4\pi\varepsilon_0} \frac{2\pi\rho_0 a^3}{3r^2} = \frac{\rho_0 a^3}{6\varepsilon_0 r^2}$$

In summary, then,

$$E(r) = \begin{cases} \dfrac{\rho_0}{6\varepsilon_0 a^3} r^4 & (r < a) \\ \dfrac{\rho_0 a^3}{6\varepsilon_0 r^2} & (r \geq a) \end{cases}$$

(b) We follow the same procedure used in Example 10.14: The potential at some distance, let's call it r_0, from the center is equal to the negative of the work done by the electric field as a charge q is brought to r_0 from infinity, divided by q. (By definition, $V = U_E/q = -W_E/q$, where we take $V = 0$ at infinity.) So, our first step (the big one) is to compute the work done by the electric field in bringing a charge q in from infinity to a point inside the sphere at a distance of $r_0 < a$ from the center:

$$
\begin{aligned}
W_{\rm E} &= W_{{\rm E},\,\infty\,{\rm to}\,a} + W_{{\rm E},\,a\,{\rm to}\,r_0} \\
&= \int_{\infty}^{a} q\cdot\frac{\rho_0 a^3}{6\varepsilon_0 r^2}\,dr + \int_{a}^{r_0} q\cdot\frac{\rho_0}{6\varepsilon_0 a^3}\,r^4\,dr \\
&= \frac{q\rho_0 a^3}{6\varepsilon_0}\int_{\infty}^{a}\frac{1}{r^2}\,dr + \frac{q\rho_0}{6\varepsilon_0 a^3}\int_{a}^{r_0} r^4\,dr \\
&= \frac{q\rho_0 a^3}{6\varepsilon_0}\left[-\frac{1}{r}\right]_{\infty}^{a} + \frac{q\rho_0}{6\varepsilon_0 a^3}\left[\frac{r^5}{5}\right]_{a}^{r_0} \\
&= -\frac{q\rho_0 a^2}{6\varepsilon_0} + \frac{q\rho_0}{30\varepsilon_0 a^3}(r_0^5 - a^5) \\
&= \frac{q\rho_0}{30\varepsilon_0}\left(\frac{r_0^5}{a^3} - 6a^2\right)
\end{aligned}
$$

Taking the negative of this result, dividing by q, and replacing r_0 by r gives our answer:

$$
V = \frac{-W_{\rm E}}{q} = \frac{\rho_0}{30\varepsilon_0}\left(6a^2 - \frac{r^5}{a^3}\right) \qquad (r < a)
$$

On the surface and outside the sphere, the potential is

$$
V = \frac{1}{4\pi\varepsilon_0}\frac{Q_{\rm total}}{r} = \frac{1}{4\pi\varepsilon_0}\frac{2\pi\rho_0 a^3}{3r} = \frac{\rho_0 a^3}{6\varepsilon_0 r} \qquad (r \ge a)
$$

In summary,

$$
V(r) = \begin{cases} \dfrac{\rho_0}{30\varepsilon_0}\left(6a^2 - \dfrac{r^5}{a^3}\right) & (r < a) \\[2ex] \dfrac{\rho_0 a^3}{6\varepsilon_0 r} & (r \ge a) \end{cases}
$$

As a check on these formulas, we note first that the function $V(r)$ is continuous (as it must always be); the only question about continuity is at $r = a$. But continuity here is assured because substituting $r = a$ into either of the expressions for $V(r)$ gives the same result:

$$
\left.\frac{\rho_0}{30\varepsilon_0}\left(6a^2 - \frac{r^5}{a^3}\right)\right|_{r=a} = \frac{\rho_0}{30\varepsilon_0}(6a^2 - a^2) = \frac{\rho_0 a^2}{6\varepsilon_0}
$$

$$
\left.\frac{\rho_0 a^3}{6\varepsilon_0 r}\right|_{r=a} = \frac{\rho_0 a^2}{6\varepsilon_0}
$$

Second, we can verify that $E = -dV/dr$:

$$-\frac{dV}{dr} = \begin{cases} -\dfrac{d}{dr}\left[\dfrac{\rho_0}{30\varepsilon_0}\left(6a^2 - \dfrac{r^5}{a^3}\right)\right] & (r < a) \\ -\dfrac{d}{dr}\left[\dfrac{\rho_0 a^3}{6\varepsilon_0 r}\right] & (r \geq a) \end{cases}$$

$$= \begin{cases} \dfrac{\rho_0}{6\varepsilon_0 a^3} r^4 & (r < a) \\ \dfrac{\rho_0 a^3}{6\varepsilon_0 r^2} & (r \geq a) \end{cases}$$

$$= E(r)$$

(c) See the diagrams:

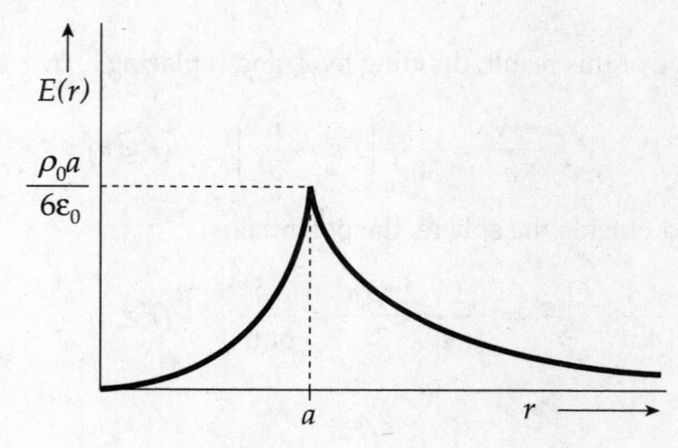

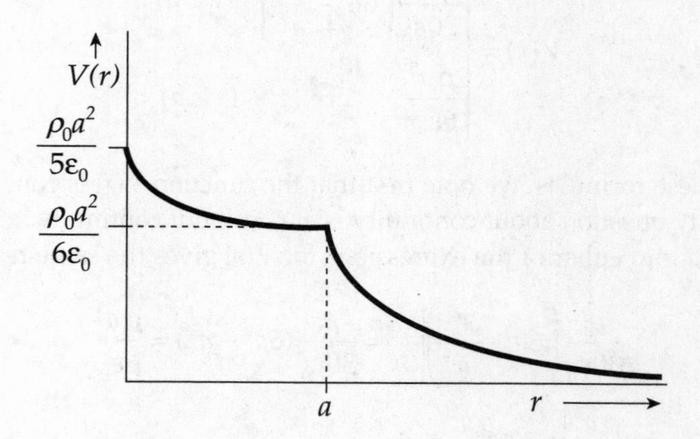

CHAPTER 11 REVIEW QUESTIONS

SECTION I: MULTIPLE CHOICE

1. **A** Let ρ_S denote the resistivity of silver and let A_S denote the cross-sectional area of the silver wire. Then

$$R_B = \frac{\rho_B L}{A_B} = \frac{(5\rho_S)L}{4^2 A_S} = \frac{5}{16}\frac{\rho_S L}{A_S} = \frac{5}{16}R_S$$

2. **D** The equation $I = V/R$ implies that increasing V by a factor of 2 will cause I to increase by a factor of 2.

3. **C** Use the equation $P = V^2/R$:

$$P = \frac{V^2}{R} \quad \Rightarrow \quad R = \frac{V^2}{P} = \frac{(120 \text{ V})^2}{60 \text{ W}} = 240 \ \Omega$$

4. **B** The current through the circuit is

$$I = \frac{\mathcal{E}}{r+R} = \frac{40 \text{ V}}{(5 \ \Omega) + (15 \ \Omega)} = 2 \text{ A}$$

Therefore, the voltage drop across R is $V = IR = (2 \text{ A})(15 \ \Omega) = 30 \text{ V}$.

5. **E** The 12 Ω and 4 Ω resistors are in parallel and are equivalent to a single 3 Ω resistor, because 1/(12 Ω) + 1/(4 Ω) = 1/(3 Ω). This 3 Ω resistor is in series with the top 3 Ω resistor, giving an equivalent resistance in the top branch of 3 + 3 = 6 Ω. Finally, this 6 Ω resistor is in parallel with the bottom 3 Ω resistor, giving an overall equivalent resistance of 2 Ω, because 1/(6 Ω) + 1/(3 Ω) = 1/(2 Ω).

6. **D** If each of the identical bulbs has resistance R, then the current through each bulb is ε/R. This is unchanged if the middle branch is taken out of the parallel circuit. (What *will* change is the total amount of current provided by the battery.)

7. **B** The three parallel resistors are equivalent to a single 2 Ω resistor, because 1/(8 Ω) + 1/(4 Ω) + 1/(8 Ω) = 1/(2 Ω). This 2 Ω resistance is in series with the given 2 Ω resistor, so their equivalent resistance is 2 + 2 = 4 Ω. Therefore, three times as much current will flow through this equivalent 4 Ω resistance in the top branch as through the parallel 12 Ω resistor in the bottom branch, which implies that the current through the bottom branch is 3 A, and the current through the top branch is 9 A. The voltage drop across the 12 Ω resistor is therefore $V = IR = (3 \text{ A})(12 \ \Omega) = 36 \text{ V}$.

8. **E** Since points a and b are grounded, they're at the same potential (call it zero).

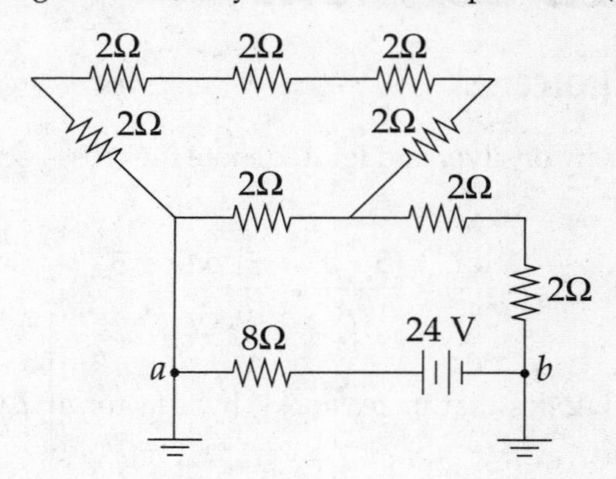

Traveling from b to a across the battery, the potential increases by 24 V, so it must decrease by 24 V across the 8 Ω resistor as we reach point a. Thus, $I = V/R = (24\text{ V})/(8\ \Omega) = 3\text{ A}$.

9. **D** The equation $P = I^2R$ gives

$$P = (0.5\text{ A})^2(100\ \Omega) = 25\text{ W} = 25\text{ J/s}$$

Therefore, in 20 s, the energy dissipated as heat is

$$E = Pt = (25\text{ J/s})(20\text{ s}) = 500\text{ J}$$

10. **D** The equivalent resistance is $R = 50 + 200 = 250\ \Omega$, so the time constant is

$$\tau = RC = (250\ \Omega)(200\ \mu\text{F}) = 0.05\text{ s}$$

SECTION II: FREE RESPONSE

1. (a) The two parallel branches, the one containing the 40 Ω resistor and the other a total of 120 Ω, is equivalent to a single 30 Ω resistance. This 30 Ω resistance is in series with the three 10 Ω resistors, giving an overall equivalent circuit resistance of 10 + 10 + 30 + 10 = 60 Ω. Therefore, the current supplied by the battery is $I = V/R = (120\text{ V})/(60\ \Omega) = 2\text{ A}$, so it must supply energy at a rate of $P = IV = (2\text{ A})(120\text{ V}) = 240\text{ W}$.

(b) Since three times as much current will flow through the 20 Ω resistor as through the branch containing 120 Ω of resistance, the current through the 20 Ω resistor must be 0.5 A.

(c) (i) $V_a - V_b = IR_{20} + IR_{100} = (0.5\text{ A})(20\ \Omega) + (0.5\text{ A})(100\ \Omega) = 60\text{ V}$.

(ii) Point a is at the higher potential (current flows from high to low potential).

(d) Because energy is equal to power multiplied by time, we get

$$E = Pt = I^2Rt = (0.5\text{ A})^2(100\ \Omega)(10\text{ s}) = 250\text{ J}$$

(e) Using the equation $R = \rho L/A$, with $A = \pi r^2$, we find

$$R = \frac{\rho L}{\pi r^2} \quad \Rightarrow \quad r = \sqrt{\frac{\rho L}{\pi R}} = \sqrt{\frac{(0.45\ \Omega \cdot \text{m})(0.04\ \text{m})}{\pi(100\ \Omega)}} = 0.0076\ \text{m} = 7.6\ \text{mm}$$

2. (a) The initial current, I_0, is ε/r.

(b) Apply the equation $Q(t) = Q_f(1 - e^{-t/rC})$. We want $1 - e^{-t/rC}$ to equal 1/2, so

$$1 - e^{-t/rC} = \tfrac{1}{2} \quad \Rightarrow \quad e^{-t/rC} = \tfrac{1}{2} \quad \Rightarrow \quad -\frac{t}{rC} = \ln\tfrac{1}{2} = -\ln 2 \quad \Rightarrow \quad t = (\ln 2)rC$$

(c) Because it is connected to the positive terminal of the battery, the top plate will become positively charged.

(d) When the current through r is zero, the capacitor is fully charged, with the voltage across its plates matching the emf of the battery. Therefore,

$$U_E = \tfrac{1}{2}CV^2 = \tfrac{1}{2}C\mathcal{E}^2$$

(e) The current established by the discharging capacitor decreases exponentially according to the equation $I(t) = I_0 e^{-t/\tau} = (\varepsilon/R)e^{-t/RC}$.

(f) The power dissipated is given by the joule heating law, $P = I^2R$:

$$P(t) = [I(t)]^2 R = \left(\frac{\mathcal{E}}{R}e^{-t/RC}\right)^2 R = \frac{\mathcal{E}^2}{R}e^{-2t/RC}$$

(g) We give two solutions. First, the total energy dissipated by the resistor will equal the integral of $P(t)$ from $t = 0$ to $t = \infty$:

$$\begin{aligned} E &= \int_0^\infty P(t)\,dt \\ &= \int_0^\infty \frac{\mathcal{E}^2}{R}e^{-2t/RC}\,dt \\ &= \frac{\mathcal{E}^2}{R}\left[-\frac{RC}{2}e^{-2t/RC}\right]_0^\infty \\ &= \frac{C\mathcal{E}^2}{2} \end{aligned}$$

Alternatively, simply notice that all the energy stored in the capacitor will be dissipated as heat by the resistor R. But from part (d), we know that the initial energy stored in the capacitor (before discharging) was $\tfrac{1}{2}C\mathcal{E}^2$.

CHAPTER 12 REVIEW QUESTIONS

Section I: Multiple Choice

1. **D** Statement I is false: The magnetic field lines due to a current-carrying wire encircle the wire in closed loops. Statement II is also false: Since the magnetic force is always perpendicular to the charged particle's velocity vector, it cannot do work on the charged particle; therefore, it cannot change the particle's kinetic energy. Statement III, however, is true: If the charged particle's velocity is parallel (or antiparallel) to the magnetic field lines, then the particle will feel no magnetic force.

2. **C** The magnitude of the magnetic force is $F_B = qvB$, so the acceleration of the particle has magnitude

$$a = \frac{F_B}{m} = \frac{qvB}{m} = \frac{(4.0\times10^{-9}\text{ C})(3\times10^{4}\text{ m/s})(0.1\text{ T})}{2\times10^{-4}\text{ kg}} = 0.06\text{ m/s}^2$$

3. **D** By the right-hand rule, the direction of $\mathbf{v} \times \mathbf{B}$ is into the plane of the page. Since the particle carries a negative charge, the magnetic force it feels will be out of the page.

4. **D** Since $\mathbf{F}_B$ is always perpendicular to $\mathbf{v}$, $\mathbf{v}$ cannot be upward or downward in the plane of the page; this eliminates choices B and C. The velocity vector also cannot be to the right (choice A), since then $\mathbf{v}$ would be antiparallel to $\mathbf{B}$, and $\mathbf{F}_B$ would be zero. Because the charge is positive, the direction of $\mathbf{F}_B$ will be the same as the direction of $\mathbf{v} \times \mathbf{B}$. In order for $\mathbf{v} \times \mathbf{B}$ to be downward in the plane of the page, the right-hand rule implies that $\mathbf{v}$ must be out of the plane of the page.

5. **A** The magnetic force provides the centripetal force on the charged particle. Therefore,

$$qvB = \frac{mv^2}{r} \quad\Rightarrow\quad qB = \frac{mv}{r} \quad\Rightarrow\quad mv = qBr \quad\Rightarrow\quad p = qBr$$

6. **D** The strength of the magnetic field at a distance r from a long, straight wire carrying a current I is given by the equation $B = (\mu_0/2\pi)(I/r)$. Therefore,

$$\frac{\mu_0}{2\pi}\frac{I}{r} = \frac{(4\pi\times10^{-7}\text{ T}\cdot\text{m/A})}{2\pi}\frac{10\text{ A}}{0.02\text{ m}} = 1\times10^{-4}\text{ T}$$

7. **D** By Newton's Third Law, neither choice A nor choice B can be correct. Also, as we learned in Example 12.9, if two parallel wires carry current in the same direction, the magnetic force between them is attractive; this eliminates choices C and E. Therefore, the answer must be D. The strength of the magnetic field at a distance r from a long, straight wire carrying a current I_1 is given by the equation $B_1 = (\mu_0/2\pi)(I_1/r)$. The magnetic force on a wire of length ℓ carrying a current I through a magnetic field $\mathbf{B}$ is $I(\boldsymbol{\ell} \times \mathbf{B})$, so the force on Wire #2 (F_{B2}) due to the magnetic field of Wire #1 (B_1) is

$$F_{B2} = I_2\ell B_1 = I_2\ell\frac{\mu_0}{2\pi}\frac{I_1}{r}$$

which implies

$$\frac{F_{B2}}{\ell} = \frac{\mu_0}{2\pi}\frac{I_1 I_2}{r} = \frac{(4\pi\times10^{-7}\ \text{N/A}^2)}{2\pi}\frac{(5\ \text{A})(10\ \text{A})}{0.01\ \text{m}} = 0.001\ \text{N/m}$$

8. **E** The strength of the magnetic field at a distance r from a long, straight wire carrying a current I is given by the equation $B = (\mu_0/2\pi)(I/r)$. Therefore, the strength of the magnetic field at Point P due to either wire is $B = (\mu_0/2\pi)(I/\frac{1}{2}d)$. By the right-hand rule, the direction of the magnetic field at P due to the top wire is into the plane of the page and the direction of the magnetic field at P due to the bottom wire is out of the plane of the page. Since the two magnetic field vectors at P have the same magnitude and opposite directions, the net magnetic field at Point P is zero.

9. **B** The strength of the magnetic field within a hollow, ideal solenoid is given by the equation $B = \mu_0 nI$, where n denotes the number of turns per unit length and I is the current in the solenoid. Therefore,

$$B = \mu_0 nI = \mu_0 \frac{N}{\ell} I \quad \Rightarrow \quad N = \frac{\ell B}{\mu_0 I}$$

This gives

$$N = \frac{\ell B}{\mu_0 I} = \frac{(0.80\ \text{m})(0.2\ \text{T})}{(4\pi\times10^{-7}\ \text{T}\cdot\text{m/A})(20\ \text{A})} = 6400$$

10. **E** By Ampere's Law, $\oint_{\text{loop}} \mathbf{B}\cdot d\mathbf{s} = \mu_0 I_{\text{through loop}}$. Therefore,

$$I_{\text{through loop}} = \frac{\oint_{\text{loop}} \mathbf{B}\cdot d\mathbf{s}}{\mu_0} = \frac{6.28\times10^{-6}\ \text{T}\cdot\text{m}}{4\pi\times10^{-7}\ \text{T}\cdot\text{m/A}} = 5\ \text{A}$$

SECTION II: FREE RESPONSE

1. (a) The acceleration of an ion of charge q is equal to F_E/m. The electric force is equal to qE, where $E = V/d$. Therefore, $a = qV/(dm)$.

(b) Using $a = qV/(dm)$ and the equation $v^2 = v_0^2 + 2ad = 2ad$, we get

$$v^2 = 2\frac{qV}{dm}d \quad \Rightarrow \quad v = \sqrt{\frac{2qV}{m}}$$

As an alternate solution, notice that the change in the electrical potential energy of the ion from the source S to the entrance to the magnetic-field region is equal to qV; this is equal to the gain in the particle's kinetic energy.

Therefore,

$$qV = \tfrac{1}{2}mv^2 \quad \Rightarrow \quad v = \sqrt{\frac{2qV}{m}}$$

(c) (i) and (ii) Use the right-hand rule. Since **v** points to the right and **B** is into the plane of the page, the direction of **v** × **B** is upward. Therefore, the magnetic force on a positively-charged particle (cation) will be upward, and the magnetic force on a negatively-charged particle (anion) will be downward. The magnetic force provides the centripetal force that causes the ion to travel in a circular path. Therefore, a cation would follow Path 1 and an anion would follow Path 2.

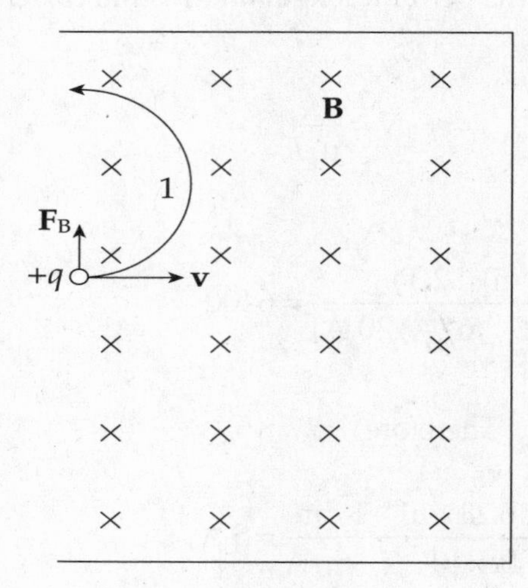

(d) Since the magnetic force on the ion provides the centripetal force,

$$qvB = \frac{mv^2}{r} \Rightarrow qvB = \frac{mv^2}{\frac{1}{2}y} \Rightarrow m = \frac{qBy}{2v}$$

Now, by the result of part (b),

$$m = \frac{qBy}{2\sqrt{\frac{2qV}{m}}} \Rightarrow m^2 = \frac{q^2B^2y^2}{\frac{8qV}{m}} \Rightarrow m^2 = \frac{mq^2B^2y^2}{8qV} \Rightarrow m = \frac{qB^2y^2}{8V}$$

(e) Since the magnetic force cannot change the speed of a charged particle, the time required for the ion to hit the photographic plate is equal to the distance traveled (the length of the semicircle) divided by the speed computed in part (b):

$$t = \frac{s}{v} = \frac{\pi \cdot \frac{1}{2}y}{\sqrt{\frac{2qV}{m}}} = \frac{1}{2}\pi y\sqrt{\frac{m}{2qV}}$$

(f) Since the magnetic force $\mathbf{F}_B$ is always perpendicular to a charged particle's velocity vector $\mathbf{v}$, it can do no work on the particle. Thus, the answer is zero.

2. (a) The current in the rectangular loop is equal to V divided by the resistance of the rectangular loop. Using the equation $R = \rho\ell/A$, we have

$$R = \frac{\rho\ell}{A} = \frac{\rho \cdot 2(a+b)}{\pi(\frac{1}{2}d)^2} = \frac{8\rho(a+b)}{\pi d^2}$$

Therefore, the current in the rectangular loop is

$$I_{\text{loop}} = \frac{V}{R} = \frac{\pi d^2 V}{8\rho(a+b)}$$

(b) We use the equation $\mathbf{F}_B = I(\boldsymbol{\ell} \times \mathbf{B})$ to find the magnetic force on each side of the rectangular loop. By symmetry, the magnetic forces on the sides of length a have the same magnitude but opposite direction, so they cancel. The total magnetic force on the loop is equal to the sum of the magnetic forces on the sides of length b. Current in the rectangle is directed clockwise, so the current in the bottom wire of length b (the one closer to the wire) is directed to the left and the current in the top wire of length b is directed to the right. Currents that are parallel feel an attractive force, while currents that are antiparallel feel a repulsive force. Since the strength of the magnetic field at a distance r from a long, straight wire carrying a current I is given by the equation $B = (\mu_0/2\pi)(I/r)$, we find that

magnetic force on bottom wire:

$$\mathbf{F}_{B1} = I_{\text{loop}}b \cdot \frac{\mu_0 I}{2\pi c}, \text{ upward (away from the long wire)}$$

magnetic force on top wire:

$$\mathbf{F}_{B2} = I_{\text{loop}}b \cdot \frac{\mu_0 I}{2\pi(c+a)}, \text{ downward (toward the long wire)}$$

Since $\mathbf{F}_{B1}$ has a greater magnitude than $\mathbf{F}_{B2}$, and the forces point in opposite directions (with the direction of the net force equaling the direction of $\mathbf{F}_{B1}$), the magnitude of the total magnetic force on the loop is equal to the difference between the magnitudes of $\mathbf{F}_{B1}$ and $\mathbf{F}_{B2}$. Therefore,

$$\begin{aligned}
\mathbf{F}_B = \mathbf{F}_{B1} + \mathbf{F}_{B2} &= I_{\text{loop}}b \cdot \frac{\mu_0 I}{2\pi c} - I_{\text{loop}}b \cdot \frac{\mu_0 I}{2\pi(c+a)}, \text{ upward} \\
&= \frac{\mu_0 I \cdot I_{\text{loop}}b}{2\pi}\left(\frac{1}{c} - \frac{1}{c+a}\right), \text{ upward} \\
&= \frac{\mu_0 I \cdot I_{\text{loop}}b}{2\pi}\frac{a}{c(c+a)}, \text{ upward}
\end{aligned}$$

Substituting the result of part (a) for I_{loop} gives

$$\mathbf{F}_B = \frac{\mu_0 I \cdot \dfrac{\pi d^2 V}{8\rho(a+b)} b}{\pi} \frac{a}{c(c+a)}, \text{ upward}$$

$$= \frac{\mu_0 I \cdot d^2 Vab}{8\rho(a+b)(a+c)c}, \text{ upward (away from long wire)}$$

(c) The circumference of the circle will be equal to the perimeter of the rectangle; thus,

$$2\pi r = 2(a+b) \quad \Rightarrow \quad r = \frac{a+b}{\pi}$$

(d) If the long, straight wire passes through the center of the circular loop, then the magnetic field line of the straight wire would coincide with the current in the loop. Whether the current in the loop is parallel or antiparallel to the direction of the magnetic field line makes no difference; the magnetic force on the current-carrying loop would be zero (because $\boldsymbol{\ell} \times \mathbf{B}$ would be $\mathbf{0}$).

(e) The current in the sliding wire is directed to the right, and the magnetic field is into the plane of the page, so the right-hand rule tells us that the direction of $\boldsymbol{\ell} \times \mathbf{B}$ [and, therefore, of $\mathbf{F}_B = I(\boldsymbol{\ell} \times \mathbf{B})$, the magnetic force on the sliding wire] will be upward. If this upward force is to balance the downward gravitational force, then, because $\ell = x$,

$$IxB = mg \quad \Rightarrow \quad I = \frac{mg}{xB}$$

3. (a) Neither of the straight portions of the wire contribute to the magnetic field at Point C; this is because $\hat{\mathbf{r}}$ is parallel (or antiparallel) to $\boldsymbol{\ell}$, so the cross product $\boldsymbol{\ell} \times \mathbf{r}$, which appears in the Biot–Savart Law, would be zero. Therefore, only the curved portion of the wire generates a magnetic field at C. Refer to the diagram below:

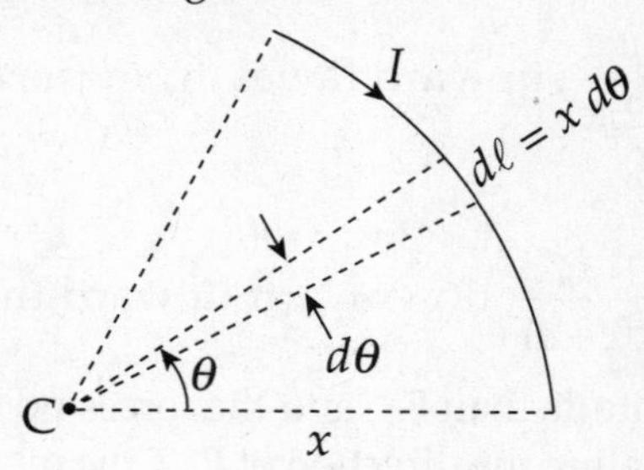

By the Biot–Savart Law, the magnetic field at C due to a length $d\ell$ of the arc has magnitude

$$dB = \frac{\mu_0}{4\pi} \frac{I \cdot d\ell}{r^2} = \frac{\mu_0}{4\pi} \frac{I \cdot (x\,d\theta)}{x^2} = \frac{\mu_0 I}{4\pi x} d\theta$$

Integrating this from $\theta = 0$ to $\theta = \phi$ gives

$$B = \int dB = \frac{\mu_0 I}{4\pi x}\int_0^{\phi} d\theta = \frac{\mu_0 I\phi}{4\pi x}$$

By the right-hand rule, the direction of **B** at C is into the plane of the page. (Either curl the fingers of your right hand in the direction of the current in the arc and notice that your thumb points into the page, or apply the right hand rule to $d\ell \times \hat{\mathbf{r}}$.)

(b) If the charged particle is placed at rest at point C, then it would feel no magnetic force. Only particles that move through magnetic fields feel a magnetic force.

(c) Since the current in the straight wire is antiparallel to the magnetic field at C, it will experience zero magnetic force ($\boldsymbol{\ell} \times \mathbf{B} = \mathbf{0}$).

4. (a) The current in the rod is equal to the current density times the cross-sectional area of the rod: $I = JA = J(\pi R^2)$.

(b) (i) We apply Ampere's Law to a circular loop of radius $r < R$. The magnetic field lines will be circles centered on the central axis of the rod (thus coinciding with the position of the Amperian loop drawn below):

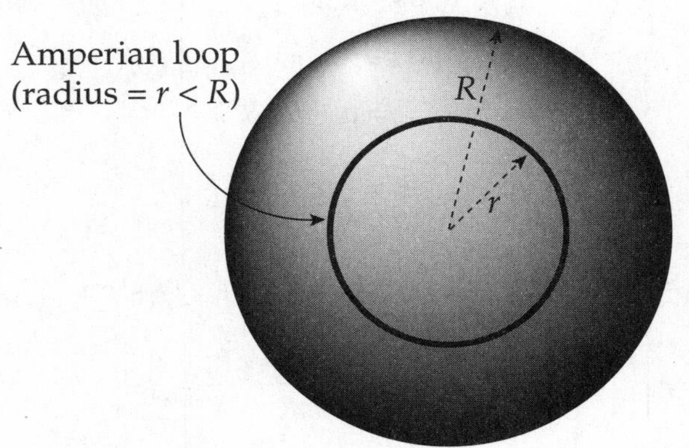

$$\oint_{\text{loop}} \mathbf{B}\cdot d\mathbf{s} = \mu_0 I_{\text{through loop}}$$

$$B(2\pi r) = \mu_0 \cdot JA_{\text{enclosed by loop}}$$

$$= \mu_0 \cdot J(\pi r^2)$$

$$B = \tfrac{1}{2}\mu_0 \cdot Jr$$

To write this result in terms of I, we simply note from part (a) that $J = I/(\pi R^2)$, so

$$B = \tfrac{1}{2}\mu_0 \cdot Jr = \tfrac{1}{2}\mu_0 \cdot \frac{I}{\pi R^2} \cdot r = \frac{\mu_0}{2\pi}\frac{I}{R^2}r$$

(b) (ii) Applying Ampere's Law to an Amperian loop of radius $r > R$,

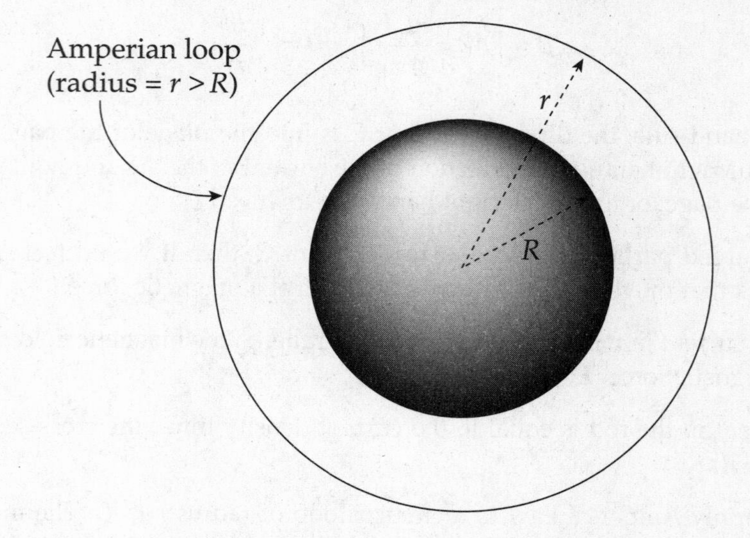

we get

$$\oint_{\text{loop}} \mathbf{B} \cdot d\mathbf{s} = \mu_0 I_{\text{through loop}}$$
$$B(2\pi r) = \mu_0 \cdot JA_{\text{enclosed by loop}}$$
$$= \mu_0 \cdot J(\pi R^2)$$
$$B = \frac{1}{2}\mu_0 J \frac{R^2}{r}$$

To write this result in terms of I, we once again use $J = I/(\pi R^2)$, giving

$$B = \tfrac{1}{2}\mu_0 J \frac{R^2}{r} = \tfrac{1}{2}\mu_0 \cdot \frac{I}{\pi R^2} \cdot \frac{R^2}{r} = \frac{\mu_0}{2\pi} \frac{I}{r}$$

which is certainly a familiar result!

(c) Since current density is current per unit area, the units of J are A/m^2. Therefore,

$$J = \sigma r \quad \Rightarrow \quad [J] = [\sigma][r] \quad \Rightarrow \quad [\sigma] = \frac{[J]}{[r]} = \frac{A/m^2}{m} = \frac{A}{m^3}$$

(d) Since the current density varies with the radial distance from the rod's center, construct a thin ring of width dr at radius r:

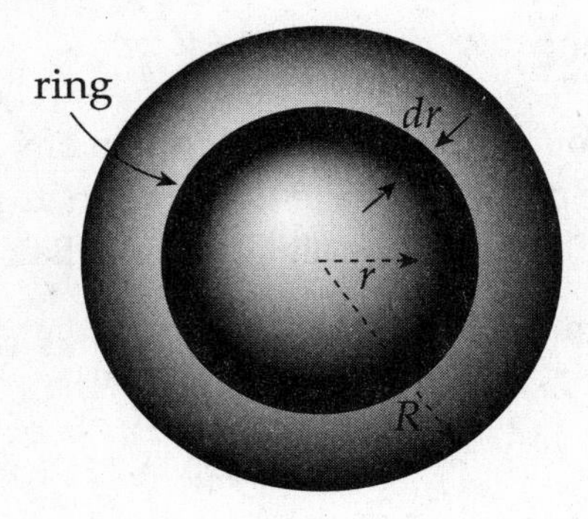

The area of this ring is $dA = (2\pi r)\,dr$, so the current through the ring is

$$dI = J\,dA = (\sigma r)(2\pi r\,dr) = 2\pi\sigma\,r^2\,dr$$

Therefore, the total current through the rod is

$$I = \int dI = \int_{r=0}^{r=R} (2\pi s \cdot r^2 dr) = 2\pi\sigma \int_0^R r^2 dr = \tfrac{2}{3}\pi\sigma R^3 \qquad (1)$$

(e) (i) As in part (b), construct an Amperian loop of radius $r < R$. The current through such a loop is given by the result of Equation (1) above with r replacing R. Therefore,

$$\oint_{\text{loop}} \mathbf{B} \cdot d\mathbf{s} = \mu_0 I_{\text{through loop}}$$

$$B(2\pi r) = \mu_0 \cdot (\tfrac{2}{3}\pi\sigma r^3)$$

$$B = \tfrac{1}{3}\mu_0 \sigma r^2$$

To write this result in terms of I, we simply note from part (d) that $\sigma = I/\tfrac{2}{3}\pi R^3$, so

$$B = \tfrac{1}{3}\mu_0 \sigma r^2 = \tfrac{1}{3}\mu_0 \cdot \frac{I}{\tfrac{2}{3}\pi R^3} \cdot r^2 = \frac{\mu_0}{2\pi}\frac{I}{R^3} r^2$$

(e) (ii) We can either follow the same procedure as in (b) (ii)—and use the results of (d) and (e) (i)—or we can simply notice that outside the wire, the full current I would pass through our circular Amperian loop of radius $r > R$, so the magnetic field at such points must be given once again by the familiar formula

$$B = \frac{\mu_0}{2\pi}\frac{I}{r}$$

CHAPTER 13 REVIEW QUESTIONS

SECTION I: MULTIPLE CHOICE

1. **E** Since **v** is upward and **B** is out of the page, the direction of **v** × **B** is to the right. Therefore, free electrons in the wire will be pushed to the left, leaving an excess of positive charge at the right. Therefore, the potential at Point b will be higher than at Point a, by $\mathcal{E} = vBL$ (motional emf).

2. **C** Consider a small radial segment of length dr as shown:

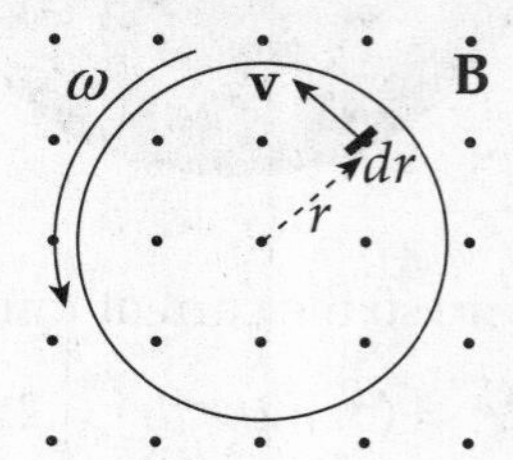

Its velocity is $v = \omega r$, so the motional emf in this little piece is $d\varepsilon = (\omega r)B\,dr$. Integrating from $r = 0$ to $r = a$ gives the induced emf between the center and the rim:

$$\mathcal{E} = \int d\mathcal{E} = \int_{r=0}^{r=a} \omega rB\,dr = \omega B\int_0^a r\,dr = \tfrac{1}{2}\omega Ba^2$$

3. **A** As shown in Example 13.3, the magnitude of the emf induced between the ends of the rod is $\mathcal{E} = BLv = (0.5\text{ T})(0.2\text{ m})(3\text{ m/s}) = 0.3\text{ V}$. Since the resistance is 10 Ω, the current induced will be $I = V/R = (0.3\text{ V})/(10\ \Omega) = 0.03\text{ A}$. To determine the direction of the current, we can note that since positive charges in the rod are moving to the left and the magnetic field points into the plane of the page, the right-hand rule tells us that the magnetic force, $q\mathbf{v} \times \mathbf{B}$, points downward. Since the resulting force on the positive charges in the rod is downward, so is the direction of the induced current.

4. **A** The magnetic field through the loop is $B = \mu_0 nI$. Since its area is $A = \pi r^2$, the magnetic flux through the loop is $\Phi_B = BA = (\mu_0 nI)(\pi r^2)$. If the current changes (with $\Delta I/\Delta t = -a$), then the magnetic flux through the loop changes, which, by Faraday's Law, implies that an emf (and a current) will be induced. We get

$$\mathcal{E} = -\frac{\Delta\Phi_B}{\Delta t} = -\frac{\Delta(\mu_0 nI \cdot \pi r^2)}{\Delta t} = -(\mu_0 n\pi r^2)\frac{\Delta I}{\Delta t} = -\mu_0 \pi n r^2(-a) = \mu_0 \pi n r^2 a$$

Since the magnetic flux into the page is decreasing, the direction of the induced current will be clockwise (opposing a *decreasing into-the-page flux* means that the induced current will create more into-the-page flux).

5. **C** By definition, magnetic field lines emerge from the north pole and enter at the south pole. Therefore, as the north pole is moved upward through the loop, the upward magnetic flux increases. To oppose an increasing upward flux, the direction of the induced current will be clockwise (as seen from above) to generate some downward magnetic flux. Now, as the south pole moves away from the center of the loop, there is a decreasing upward magnetic flux, so the direction of the induced current will be counterclockwise.

6. **E** Since the current in the straight wire is steady, there is no change in the magnetic field, no change in magnetic flux, and, therefore, no induced emf or current.

7. **A** Use the equation $I(t) = I_{max}(1 - e^{-t/\tau_L})$. In order for $I(t)$ to equal $\frac{3}{4}I_{max}$, we must have $e^{-t/\tau_L} = \frac{1}{4}$. Therefore,

$$-t/\tau_L = \ln\tfrac{1}{4} = -\ln 4 \quad\Rightarrow\quad t = (\ln 4)\tau_L = (\ln 4)\frac{L}{R}$$

8. **D** The value of I_{max} is $\mathcal{E}/R$. Since the magnetic energy stored in an inductor is given by $U_B = \frac{1}{2}LI^2$, we have

$$U_B = \tfrac{1}{2}L\left(\frac{\mathcal{E}}{R}\right)^2 = \frac{L\mathcal{E}^2}{2R^2}$$

9. **B** By definition of self-inductance, $L = N\,\Phi_B/I$. The total magnetic flux through all the windings of the solenoid is $N\Phi_B$, which is equal to LI. Since $I = \mathcal{E}/R$, we have

$$\Phi_{B,total} = L(\mathcal{E}/R)$$

10. **D** Faraday's Law shows how a changing B-field generates an E-field. The Ampere-Maxwell Law shows the reverse: how a changing E-field generates a B-field.

SECTION II: FREE RESPONSE

1. (a) Consider a small section of length dx of the rod at a distance x from the nonrotating end:

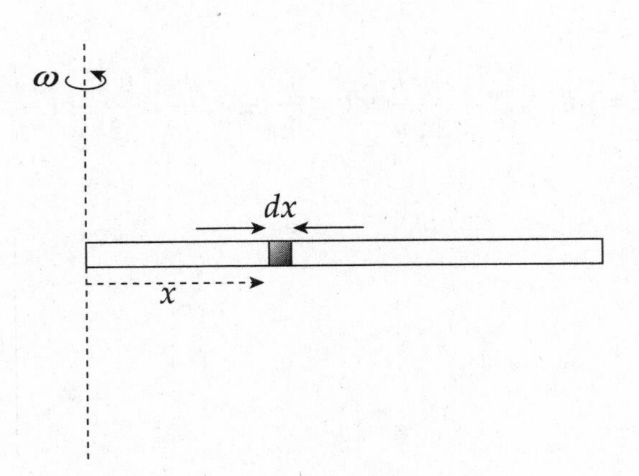

The velocity of this small piece is $v = \omega x$, so the motional emf induced between its ends is $d\varepsilon = vB\,dx = \omega xB\,dx$. Integrating this from $x = 0$ to $x = \ell$ gives

$$\mathcal{E} = \int d\mathcal{E} = \int_{x=0}^{x=\ell} \omega xB\,dx = \omega B\int_0^\ell x\,dx = \tfrac{1}{2}\omega B\ell^2$$

(b) Refer to the view from above of the rod:

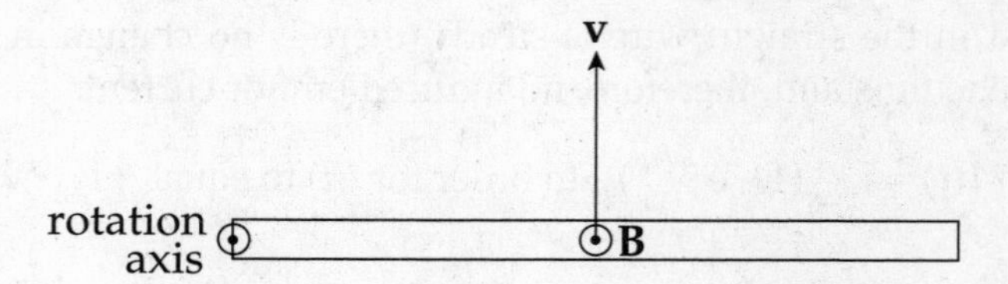

By the right-hand rule, the direction of **v** × **B** is to the right, so the magnetic force on free electrons will be to the left, leaving excess *positive* charge at the right (rotating) end.

(c) Consider a small section of length dy at a distance y from the straight wire:

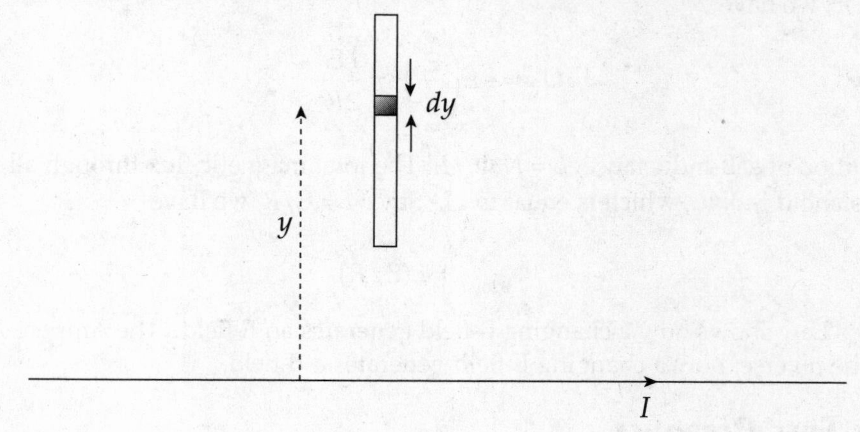

Since the magnetic field at the position of this section is $(\mu_0/2\pi)(I/y)$, the motional emf induced between its ends is $d\mathcal{E} = vB\,dy = v(\mu_0/2\pi)(I/y)\,dy$. Integrating this from $y = a$ to $y = a + \ell$ gives

$$\mathcal{E} = \int d\mathcal{E} = \int_{y=a}^{y=a+\ell} \frac{\mu_0}{2\pi}\frac{I}{y}v\,dy = \frac{\mu_0 Iv}{2\pi}\int_a^{a+\ell}\frac{dy}{y} = \frac{\mu_0 Iv}{2\pi}\ln\frac{a+\ell}{a}$$

(d) The magnetic field above the wire (where the rod is sliding) is out of the plane of the page, so the direction of $\mathbf{v} \times \mathbf{B}$ is downward.

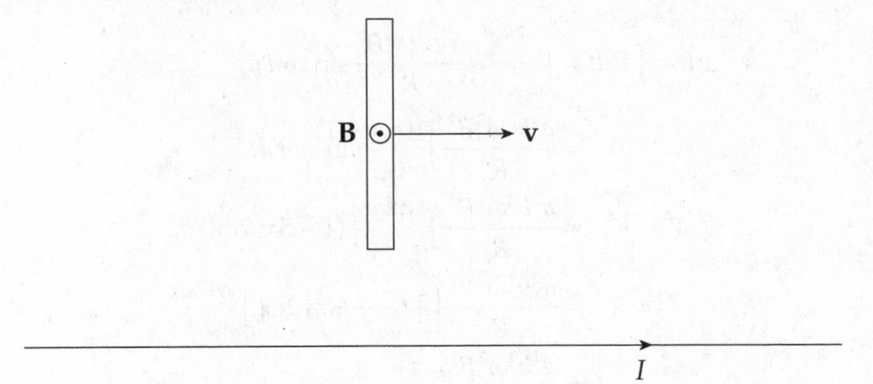

Therefore, free electrons in the rod will be pushed upward to the far end of the rod, causing it to become negatively charged.

2. (a) As the loop rotates, the angle that the normal to the loop's enclosed area makes with **B** changes according to the equation $\theta = \omega t$. Therefore, the magnetic flux through the loop is

$$\Phi_B(t) = BA\cos\theta = BA\cos\omega t = abB\cos\omega t$$

(b) According to Faraday's Law, the induced emf is equal to the rate of change of magnetic flux through the loop (with a minus sign included to conform to Lenz's Law). Therefore,

$$\mathcal{E} = -\frac{d\Phi_B}{dt} = -\frac{d}{dt}(abB\cos\omega t) = -[abB\omega(-\sin\omega t)] = ab\omega B\sin\omega t$$

(c) Using the result of part (b) with the equation $I = \mathcal{E}/R$, we find that the induced current is a function of time given by

$$I(t) = \frac{\mathcal{E}}{R} = \frac{ab\omega B\sin\omega t}{R}$$

(d) As the loop rotates from $\theta = 0$ to $\theta = \pi/2$, the magnetic flux (to the right, the $+x$ direction) through the loop decreases from $BA = abB$ to zero. To oppose this decreasing flux to the right, the current will be induced in a direction to generate a magnetic field with a component in the $+x$ direction. The current will be flowing from point 3 to 4 to 1 to 2 to 3.

(e) The rate at which energy is dissipated as heat in the loop is given by the equation $P = I^2R = (\mathcal{E}/R)^2R = \mathcal{E}^2/R$, so

$$P = \frac{\mathcal{E}^2}{R} = \frac{(ab\omega B\sin\omega t)^2}{R} = \frac{a^2b^2\omega^2B^2}{R}\sin^2\omega t$$

In one revolution, ωt increases from 0 to 2π, so the energy dissipated is

$$
\begin{aligned}
E = \int P\,dt &= \int_{t=0}^{t=2\pi/w} \frac{a^2b^2\omega^2B^2}{R}\sin^2\omega t\,dt \\
&= \frac{a^2b^2\omega^2B^2}{R}\int_{t=0}^{t=2\pi/\omega}\sin^2\omega t\,dt \\
&= \frac{a^2b^2\omega^2B^2}{R}\int_{t=0}^{t=2\pi/\omega}\tfrac{1}{2}(1-\cos 2\omega t)\,dt \\
&= \frac{a^2b^2\omega^2B^2}{R}\left[\tfrac{1}{2}t-\tfrac{1}{4\omega}\sin 2\omega t\right]_{t=0}^{t=2\pi/\omega} \\
&= \frac{a^2b^2\omega^2B^2}{R}\cdot\frac{\pi}{\omega} \\
&= \frac{\pi a^2b^2\omega B^2}{R}
\end{aligned}
$$

(f) The work done by the externally-provided torque is transformed into the heat lost in the loop. Equivalently, the rate at which the torque does work is equal to the rate at which heat energy is dissipated in the wires of the loop. Since the power associated with the external torque is $P = \tau\omega$, the result of part (e) gives

$$\tau\omega = \frac{a^2b^2\omega^2B^2}{R}\sin^2\omega t \quad\Rightarrow\quad \tau = \frac{a^2b^2\omega B^2}{R}\sin^2\omega t$$

3. (a) Construct an Amperian loop of radius R whose center coincides with the center of the toroid. Then only the current in the inner set of N windings pierces through the area bounded by the loop, so Ampere's Law becomes

$$
\begin{aligned}
\oint_{\text{loop}} \mathbf{B}\cdot d\mathbf{s} &= \mu_0 I_{\text{through loop}} \\
B(2\pi R) &= \mu_0(NI) \\
B &= \frac{\mu_0}{2\pi}\frac{NI}{R}
\end{aligned}
$$

(b) Because there is no magnetic field outside the toroid, the magnetic flux through the loop is equal to B times the cross-sectional area of the toroid (which is πa^2). Therefore, by Faraday's Law,

$$\begin{aligned}
\mathcal{E} &= -\frac{d\Phi_B}{dt} = -\frac{d}{dt}(B \cdot \pi a^2) \\
&= -\frac{d}{dt}\left(\frac{\mu_0 NI}{2\pi R} \cdot \pi a^2\right) \\
&= -\frac{\mu_0 Na^2}{2R}\frac{dI}{dt} \\
&= -\frac{\mu_0 Na^2}{2R} \cdot \frac{d}{dt}(I_0 \sin \omega t) \\
&= -\frac{\mu_0 I_0 Na^2 \omega}{2R} \cos \omega t
\end{aligned}$$

(c) Writing Faraday's Law in the form

$$\oint_{\text{loop}} \mathbf{E} \cdot d\mathbf{s} = -\frac{d\Phi_B}{dt}$$

and applying it to the circular loop of radius $2a$, we find that

$$\begin{aligned}
\oint_{\text{loop}} \mathbf{E} \cdot d\mathbf{s} &= -\frac{d\Phi_B}{dt} \\
E(2\pi \cdot r_{\text{loop}}) &= -\frac{d\Phi_B}{dt} \\
E(2\pi \cdot 2a) &= -\frac{\mu_0 I_0 Na^2 \omega}{2R} \cos \omega t \\
E &= -\frac{\mu_0 I_0 Na\omega}{8\pi R} \cos \omega t
\end{aligned}$$

(d) By definition, the self-inductance, L, is equal to $N\Phi_B/I$. Since

$$\Phi_B = B \cdot \pi a^2 = \frac{\mu_0 NI}{2\pi R} \cdot \pi a^2 = \frac{\mu_0 NIa^2}{2R}$$

as we calculated in the solution to part (b), we find that

$$L = \frac{N\Phi_B}{I} = \frac{N}{I} \cdot \frac{\mu_0 NIa^2}{2R} = \frac{\mu_0 N^2 a^2}{2R}$$

4. (a) When the switch is first moved to position A the circuit is a basic series circuit. Because there is no inductor in that part of the circuit, the current can increase to its steady state value basically instantaneously. Use Ohm's Law to determine the current

$V = IR$

$10 = I\,(10 + 5)$

$I = 0.67$ A

(b) When the switch is moved to position B there is 0.67 A going through the 5 Ω resistor. The inductor prohibits that current from changing instantaneously, so the current must remain 0.67 A for the instant after the switch is moved. Therefore the answer is 0.67 A.

(c) Take a loop around the right-hand part of the circuit and apply Kirchhoff's Loop Rule.

$$\Delta V_{loop} = 0$$
$$\varepsilon_L - IR = 0$$
$$\varepsilon_L = (.67)(5) = 3.33\ \text{V}$$

15

The Princeton Review AP Physics C Practice Exam 1

TEST TAKING ADVICE

MULTIPLE-CHOICE SECTION

Answering 35 multiple-choice questions in 45 minutes can be challenging. Make sure to pace yourself accordingly and remember that you do not need to answer every question correctly to do well. A quarter of a point is taken off for every incorrect answer so you can skip problems that you have no idea how to solve. Exploit the multiple-choice structure of this section. There are four wrong answers and only one correct one, so even if you don't know exactly which one is the right answer, you can eliminate some that you know for sure are wrong. Then you can make an educated guess from among the answers that are left and greatly increase your odds of getting that question correct.

Problems with graphs and diagrams are usually the fastest to solve and problems with an explanation for each answer usually take the longest to work through. Do not spend too much time on any one problem or you may not get to easier problems further into the test.

These practice exams are written to give you an idea of the format of the test, the difficulty of the questions, and to allow you to practice how you should pace yourself. Take them in the same circumstances as you will encounter during the real exam: 45 minutes for each of the two multiple choice sections, and no calculator or equation sheet for this section.

FREE-RESPONSE SECTION

The time constraints in answering three multiple-part free-response questions in 45 minutes can be a challenge. Again, pace yourself so you get to part of each problem. Make sure to skim all the questions quickly to determine which will be the easiest to solve and start with that one. Often several parts within a question are easier, so even if the rest of the question is very difficult, make sure to answer these easier parts. Pace yourself so that you get to answer the easiest parts of all three questions, and leave the parts that stump you for last. Often students spend too long to get all the parts of one question correct and get almost nothing correct on the other two problems because they are rushed.

In the Mechanics free-response section, there is almost always one general mechanics question that involves a variety of principles: energy, momentum, Newton's Law, etc. This problem synthesizes a lot of concepts, but each part is usually straightforward. Another of the questions is almost always a rotational motion problem: rolling motion, fixed-axis rotation, and/or angular momentum. Though you might encounter a second general mechanics question, one of the three problems usually comes from among the following topics: resistive forces, simple harmonic motion, potential energy functions, circular motion and orbits, and gravitational forces are the most common topics. One of these three questions usually involves an experimental component. This is a problem in which an experiment is explained, data or a graph is given, and you must use the information given to solve the problem. The other type of experimental problem asks you to design an experiment that would determine some value you have previously solved for. The experimental problems may make up just one part of one question, or an entire question may be centered on the experiment.

On the Electricity and Magnetism free-response section, you will almost always see one electrostatics question, one circuits question, and one magnetism question. The electrostatics question is often on Gauss's Law or several point charges distributed in a plane. The circuit question almost always involves an RC circuit and occasionally includes an inductor. The circuit will feature a switch to add different electrical components at different times. The magnetism question almost always involves induced emf, Faraday's Law, and occasionally includes Ampere's Law to determine the magnetic field of a wire, solenoid, or toroid. This portion of the test can also include an experimental component. Infrequently there are two magnetism questions and no circuits questions on the free response section of the test.

PHYSICS C

You may take the entire C Exam or Mechanics only or Electricity and Magnetism only as follows:

	Entire C Exam Both Mech. and Elect. & Mag.	Mechanics only	Electricity & Magnetism
First 45 min.	Sec. I, Mech. 35 questions	Sec. I, Mech. 35 questions	Sec. I, Elect. & Mag. 35 questions
Second 45 min.	Sec. I, Elect. & Mag. 35 questions	Sec. II, Mech. 3 questions	Sec. II, Elect. & Mag. 3 questions
Third 45 min.	Sec. II, Mech. 3 questions		
Fourth 45 min.	Sec. II, Elect. & Mag. 3 questions		

Separate grades are reported for Mechanics and for Electricity and Magnetism. Each section of each examination is 50 percent of the total grade; each question in a section has equal weight. Calculators are NOT permitted on Section I of the exam but are allowed on Section II. However, calculators cannot be shared with other students and calculators with typewriter-style (QWERTY) keyboards will not be permitted. A table of information that may be helpful is found on the following page.

The Physics C Examination contains a total of 70 multiple-choice questions. If you are taking
— *Mechanics only*, please be careful to answer numbers 1–35,
— *Electricity and magnetism only*, please be careful to answer numbers 36–70,
— *the entire examination* (Mechanics *and* Electricity and Magnetism), answer numbers 1–70 on your answer sheet.

Advanced Placement Examination
PHYSICS C
SECTION I

TABLE OF INFORMATION FOR 2008

CONSTANTS AND CONVERSION FACTORS

1 unified atomic mass unit,	$1\text{ u} = 1.66 \times 10^{-27}$ kg
	$= 931$ MeV/c^2
Proton mass,	$m_p = 1.67 \times 10^{-27}$ kg
Neutron mass,	$m_n = 1.67 \times 10^{-27}$ kg
Electron mass,	$m_e = 9.11 \times 10^{-31}$ kg
Magnitude of the electron charge,	$e = 1.60 \times 10^{-19}$ C
Avogadro's number,	$N_0 = 6.02 \times 10^{23}\text{ mol}^{-1}$
Universal gas constant,	$R = 8.31$ J/(mol · K)
Boltzmann's constant,	$k_B = 1.38 \times 10^{-23}$ J/K
Speed of light,	$c = 3.00 \times 10^{8}$ m/s
Planck's constant,	$h = 6.63 \times 10^{-34}$ J·s
	$= 4.14 \times 10^{-15}$ eV·s
	$hc = 1.99 \times 10^{-25}$ J·m
	$= 1.24 \times 10^{3}$ eV·nm
Vacuum permittivity,	$\epsilon_0 = 8.85 \times 10^{-12}\ \text{C}^2/\text{N}\cdot\text{m}^2$
Coulomb's law constant,	$k = 1/4\pi\epsilon_0 = 9.0 \times 10^{9}\ \text{N}\cdot\text{m}^2/\text{C}^2$
Vacuum permeability,	$\mu_0 = 4\pi \times 10^{-7}$ (T·m)/A
Magnetic constant,	$k' = \mu_0/4\pi = 10^{-7}$ (T·m)/A
Universal gravitational constant,	$G = 6.67 \times 10^{-11}\ \text{m}^3/\text{kg}\cdot\text{s}^2$
Acceleration due to gravity at the earth's surface,	$g = 9.8\ \text{m/s}^2$
1 atmosphere pressure,	$1\text{ atm} = 1.0 \times 10^{5}\ \text{N/m}^2$
	$= 1.0 \times 10^{5}$ Pa
1 electron volt,	$1\text{ eV} = 1.60 \times 10^{-19}$ J

UNITS

Name	Symbol
meter	m
kilogram	kg
second	s
ampere	A
kelvin	K
mole	mol
hertz	Hz
newton	N
pascal	Pa
joule	J
watt	W
coulomb	C
volt	V
ohm	Ω
henry	H
farad	F
tesla	T
degree Celsius	°C
electron-volt	eV

PREFIXES

Factor	Prefix	Symbol
10^9	giga	G
10^6	mega	M
10^3	kilo	k
10^{-2}	centi	c
10^{-3}	milli	m
10^{-6}	micro	μ
10^{-9}	nano	n
10^{-12}	pico	p

VALUES OF TRIGONOMETRIC FUNCTIONS FOR COMMON ANGLES

θ	$\sin\theta$	$\cos\theta$	$\tan\theta$
0°	0	1	0
30°	1/2	$\sqrt{3}/2$	$\sqrt{3}/3$
37°	3/5	4/5	3/4
45°	$\sqrt{2}/2$	$\sqrt{2}/2$	1
53°	4/5	3/5	4/3
60°	$\sqrt{3}/2$	1/2	$\sqrt{3}$
90°	1	0	∞

The following conventions are used in this examination.

I. Unless otherwise stated, the frame of reference of any problem is assumed to be inertial.
II. The direction of any electric current is the direction of flow of positive charge (conventional current).
III. For any isolated electric charge, the electric potential is defined as zero at an infinite distance from the charge.

PHYSICS C
SECTION I, MECHANICS
Time—45 minutes
35 Questions

Directions: Each of the questions or incomplete statements below is followed by five suggested answers or completions. Select the one that is best in each case and then mark it on your answer sheet.

1. A rock is dropped off a cliff and falls the first half of the distance to the ground in t_1 seconds. If it falls the second half of the distance in t_2 seconds, what is the value of t_2/t_1? (Ignore air resistance.)

   (A) $1/(2\sqrt{2})$
   (B) $1/\sqrt{2}$
   (C) $1/2$
   (D) $1-(1/\sqrt{2})$
   (E) $\sqrt{2}-1$

2. A box of mass m slides on a horizontal surface with initial speed v_0. It feels no forces other than gravity and the force from the surface. If the coefficient of kinetic friction between the box and the surface is μ, how far does the box slide before coming to rest?

   (A) $v_0^2/(2\mu g)$
   (B) $v_0^2/(\mu g)$
   (C) $2v_0^2/(\mu g)$
   (D) $mv_0^2/(\mu g)$
   (E) $2mv_0^2/(\mu g)$

3. An object initially at rest experiences a time-varying acceleration given by $a = (2\text{ m/s}^3)t$ for $t \geq 0$. How far does the object travel in the first 3 seconds?

   (A) 9 m
   (B) 12 m
   (C) 18 m
   (D) 24 m
   (E) 27 m

4. Which of the following conditions will ensure that angular momentum is conserved?

   I. Conservation of linear momentum
   II. Zero net external force
   III. Zero net external torque

   (A) I and II only
   (B) I and III only
   (C) II and III only
   (D) II only
   (E) III only

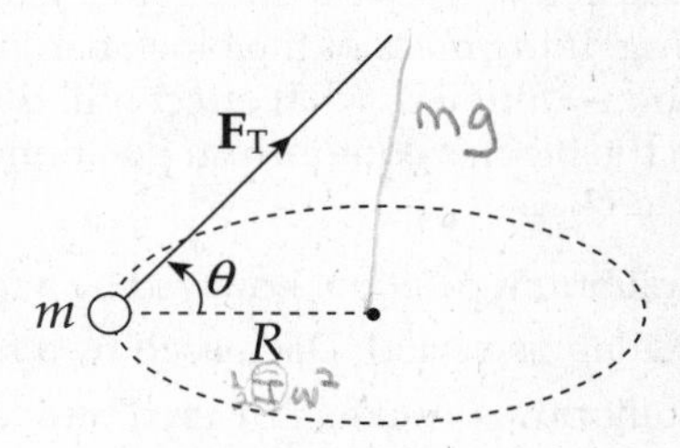

5. In the figure shown, a tension force F_T causes a particle of mass m to move with constant angular speed ω in a circular path (perpendicular to the page) of radius R. Which of the following expressions gives the magnitude of F_T?

   (A) $m\omega^2R$
   (B) $m\sqrt{\omega^4R^2 - g^2}$
   (C) $m\sqrt{\omega^4R^2 + g^2}$
   (D) $m(\omega^2R - g)$
   (E) $m(\omega^2R + g)$

GO ON TO THE NEXT PAGE

6. An object (mass = m) above the surface of the Moon (mass = M) is dropped from an altitude h equal to the Moon's radius (R). With what speed will the object strike the lunar surface?

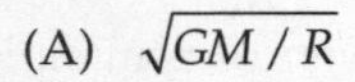

(A) $\sqrt{GM/R}$
(B) $\sqrt{GM/(2R)}$
(C) $\sqrt{2GM/R}$
(D) $\sqrt{2GMm/R}$
(E) $\sqrt{GMm/(2R)}$

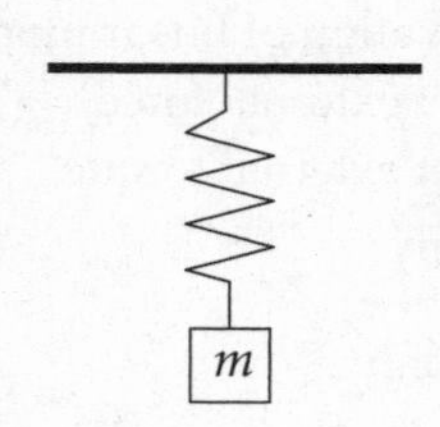

7. The figure above shows a linear spring anchored to the ceiling. If the mass of the block hanging from its lower end is doubled, what effect will this change have on the block's equilibrium position and oscillation period?

(A) Equilibrium position: Lowered by a factor of $\sqrt{2}$
Oscillation period: Decreased by a factor of $\sqrt{2}$
(B) Equilibrium position: Lowered by a factor of $\sqrt{2}$
Oscillation period: Increased by a factor of $\sqrt{2}$
(C) Equilibrium position: Lowered by a factor of $\sqrt{2}$
Oscillation period: Increased by a factor of 2
(D) Equilibrium position: Lowered by a factor of 2
Oscillation period: Decreased by a factor of $\sqrt{2}$
(E) Equilibrium position: Lowered by a factor of 2
Oscillation period: Increased by a factor of $\sqrt{2}$

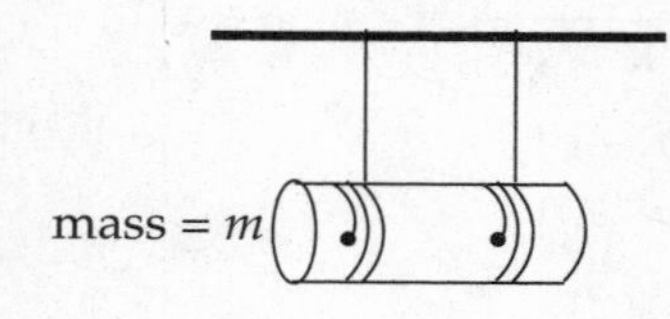

8. A uniform cylinder of mass m and radius r unrolls without slipping from two strings tied to a vertical support. If the rotational inertia of the cylinder is $\frac{1}{2}mr^2$, find the acceleration of its center of mass.

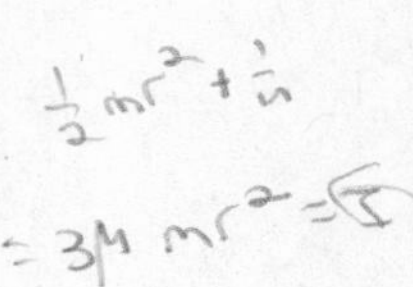
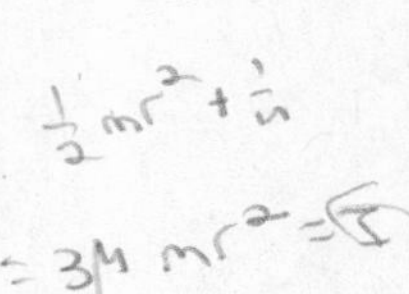

(A) $\frac{1}{4}g$
(B) $\frac{1}{2}g$
(C) $\frac{1}{3}g$
(D) $\frac{2}{3}g$
(E) $\frac{3}{4}g$

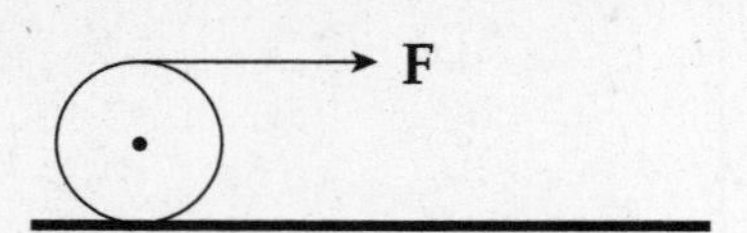

9. A uniform cylinder, initially at rest on a frictionless, horizontal surface, is pulled by a constant force **F** from time $t = 0$ to time $t = T$. From time $t = T$ on, this force is removed. Which of the following graphs best illustrates the speed, v, of the cylinder's center of mass from $t = 0$ to $t = 2T$?

(A)

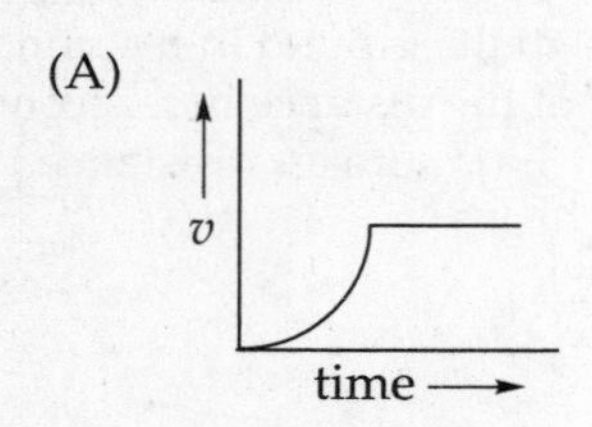

(B)

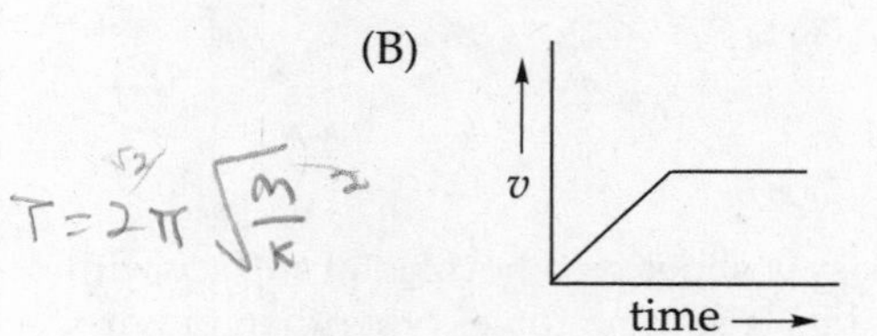

(C)

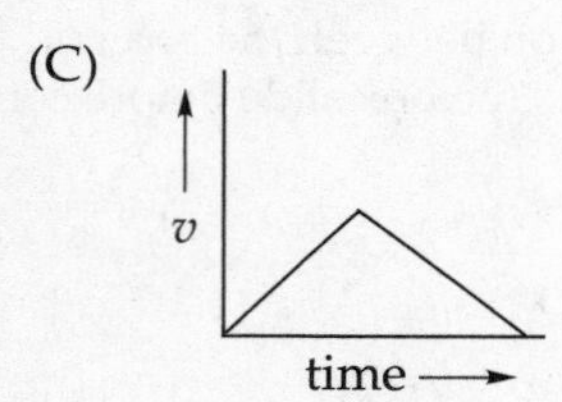

(D)

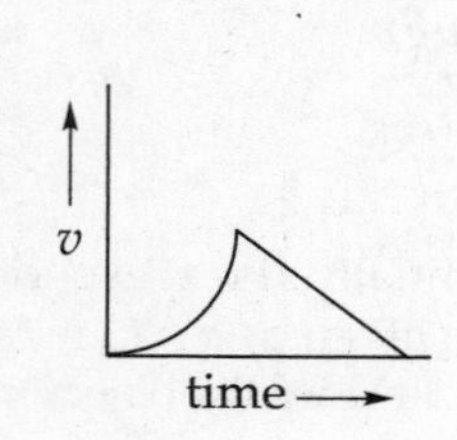

(E)

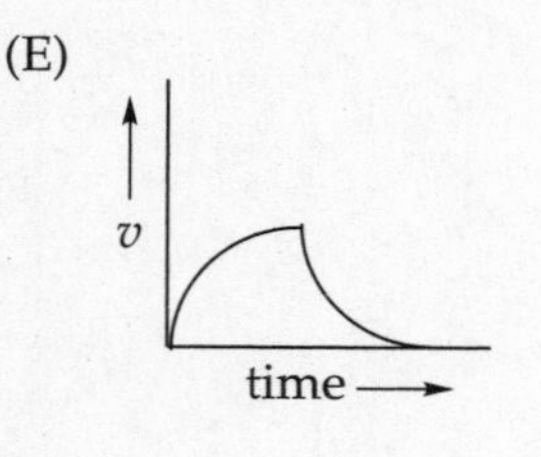

GO ON TO THE NEXT PAGE

10. An engine provides 10 kW of power to lift a heavy load at constant velocity a distance of 20 m in 5 s. What is the mass of the object being lifted?

(A) 100 kg
(B) 150 kg
(C) 200 kg
(D) 250 kg
(E) 500 kg

11. A satellite is in circular orbit around the earth. If the work required to lift the satellite to its orbit height is equal to the satellite's kinetic energy while in this orbit, how high above the surface of the earth (radius = R) is the satellite?

(A) $\frac{1}{2}R$
(B) $\frac{2}{3}R$
(C) R
(D) $\frac{3}{2}R$
(E) $2R$

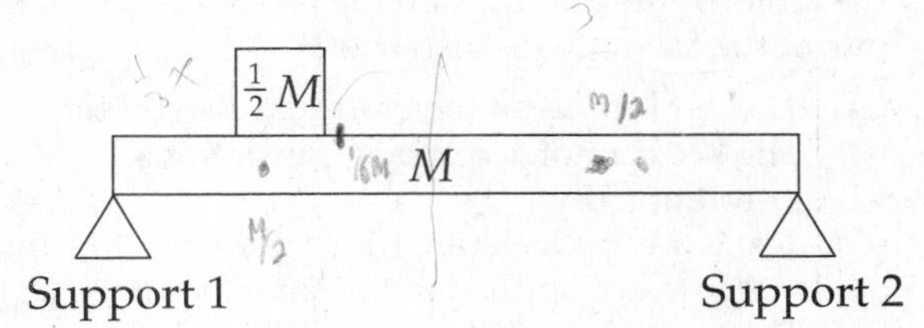

12. The figure above shows a uniform bar of mass M resting on two supports. A block of mass $\frac{1}{2}M$ is placed on the bar twice as far from Support 2 as from Support 1. If $\mathbf{F}_1$ and $\mathbf{F}_2$ denote the downward forces on Support 1 and Support 2, respectively, what is the value of F_2/F_1?

(A) 1/2
(B) 2/3
(C) 3/4
(D) 4/5
(E) 5/6

13. A rubber ball (mass = 0.08 kg) is dropped from a height of 3.2 m, and after bouncing off the floor, rises almost to its original height. If the impact time with the floor is measured to be 0.04 s, what average force did the floor exert on the ball?

(A) 0.16 N
(B) 16 N
(C) 32 N
(D) 36 N
(E) 64 N

14. A disk of radius 0.1 m initially at rest undergoes an angular acceleration of 2.0 rad/s^2. If the disk only rotates, find the total distance traveled by a point on the rim of the disk in 4.0 s.

(A) 0.4 m
(B) 0.8 m
(C) 1.2 m
(D) 1.6 m
(E) 2.0 m

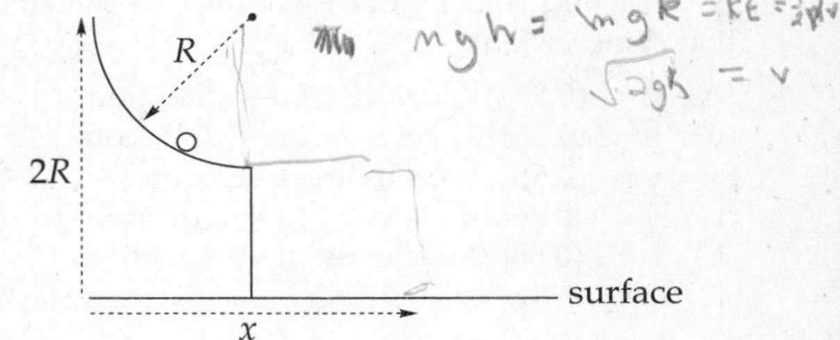

15. In the figure above, a small ball slides down a frictionless quarter-circular slide of radius R. If the ball starts from rest at a height equal to $2R$ above a horizontal surface, find its horizontal displacement, x, at the moment it strikes the surface.

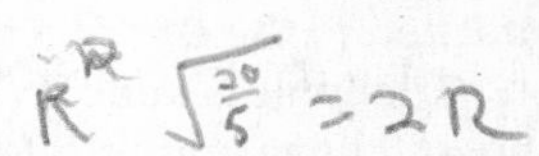

(A) $2R$
(B) $\frac{5}{2}R$
(C) $3R$
(D) $\frac{7}{2}R$
(E) $4R$

GO ON TO THE NEXT PAGE

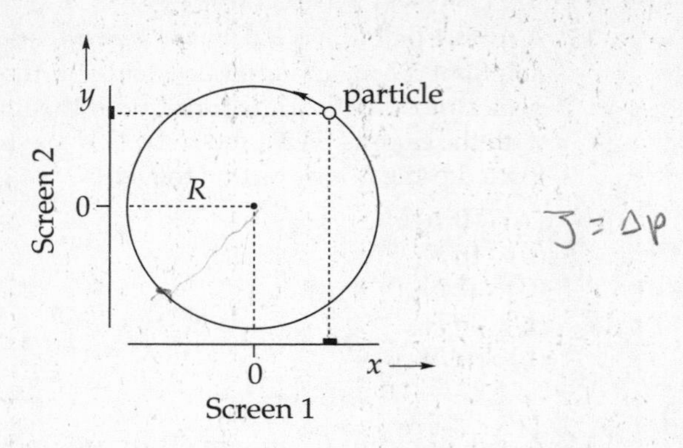

16. The figure above shows a particle executing uniform circular motion in a circle of radius R. Light sources (not shown) cause shadows of the particle to be projected onto two mutually perpendicular screens. The positive directions for x and y along the screens are denoted by the arrows. When the shadow on Screen 1 is at position $x = -(0.5)R$ and moving in the $+x$ direction, what is true about the position and velocity of the shadow on Screen 2 at that same instant?

(A) $y = -(0.866)R$; velocity in $-y$ direction
(B) $y = -(0.866)R$; velocity in $+y$ direction
(C) $y = -(0.5)R$; velocity in $-y$ direction
(D) $y = +(0.866)R$; velocity in $-y$ direction
(E) $y = +(0.866)R$; velocity in $+y$ direction

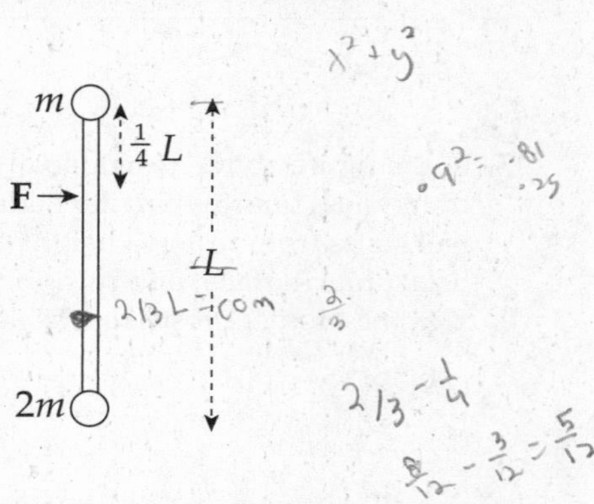

17. The figure shows a view from above of two objects attached to the end of a rigid massless rod at rest on a frictionless table. When a force **F** is applied as shown, the resulting rotational acceleration of the rod about its center of mass is $kF/(mL)$. What is k?

(A) $\frac{3}{8}$
(B) $\frac{1}{2}$
(C) $\frac{5}{8}$
(D) $\frac{3}{4}$
(E) $\frac{5}{6}$

18. A lightweight toy car crashes head-on into a heavier toy truck. Which of the following statements is true as a result of the collision?

I. The car will experience a greater impulse than the truck.
II. The car will experience a greater change in momentum than the truck.
III. The magnitude of the acceleration experienced by the car will be greater than that experienced by the truck.

(A) I and II only
(B) II only
(C) III only
(D) II and III only
(E) I, II, and III

19. A homogeneous bar is lying on a flat table. Besides the gravitational and normal forces (which cancel), the bar is acted upon by exactly two other external forces, $\mathbf{F}_1$ and $\mathbf{F}_2$, which are parallel to the surface of the table. If the net force on the rod is zero, which one of the following is also true?

(A) The net torque on the bar must also be zero.
(B) The bar cannot accelerate translationally or rotationally.
(C) The bar can accelerate translationally if $\mathbf{F}_1$ and $\mathbf{F}_2$ are not applied at the same point.
(D) The net torque will be zero if $\mathbf{F}_1$ and $\mathbf{F}_2$ are applied at the same point.
(E) None of the above

20. An astronaut lands on a planet whose mass and radius are each twice that of Earth. If the astronaut weighs 800 N on Earth, how much will he weigh on this planet?

(A) 200 N
(B) 400 N
(C) 800 N
(D) 1600 N
(E) 3200 N

GO ON TO THE NEXT PAGE

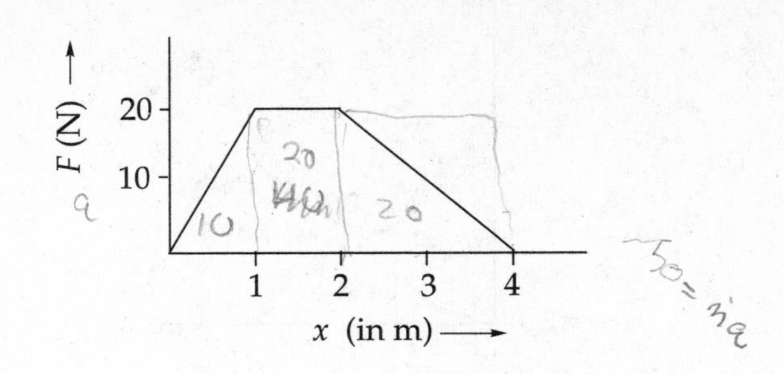

21. A particle of mass $m = 1.0$ kg is acted upon by a variable force, $F(x)$, whose strength is given by the graph given above. If the particle's speed was zero at $x = 0$, what is its speed at $x = 4$ m?

(A) 5.0 m/s
(B) 8.7 m/s
(C) 10 m/s
(D) 14 m/s
(E) 20 m/s

22. The radius of a collapsing spinning star (assumed to be a uniform sphere) decreases to $\frac{1}{16}$ its initial value. What is the ratio of the final rotational kinetic energy to the initial rotational kinetic energy?

(A) 4
(B) 16
(C) 16^2
(D) 16^3
(E) 16^4

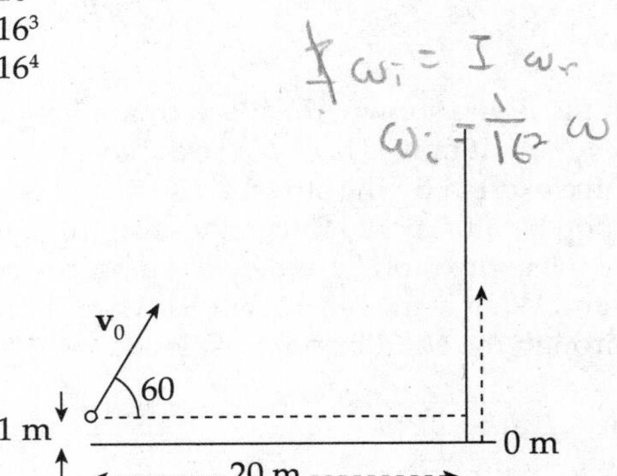

23. A ball is projected with an initial velocity of magnitude $v_0 = 40$ m/s toward a vertical wall as shown in the figure above. How long does the ball take to reach the wall?

(A) 0.25 s
(B) 0.6 s
(C) 1.0 s
(D) 2.0 s
(E) 3.0 s

24. If L, M, and T denote the dimensions of length, mass, and time, respectively, what are the dimensions of impulse?

(A) LM/T^3
(B) LM/T^2
(C) LM/T
(D) L^2M/T^2
(E) M^2L/T

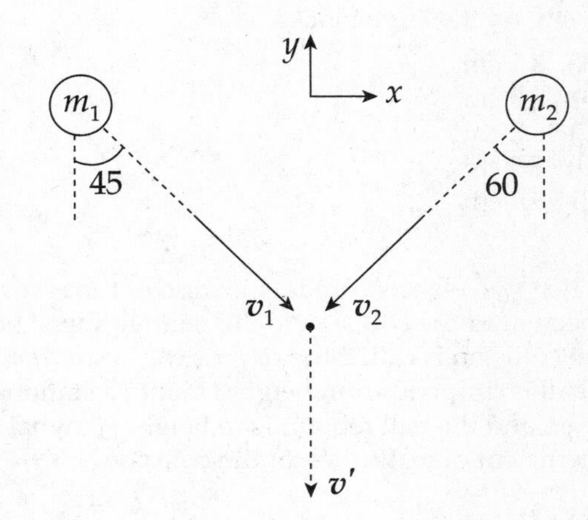

25. The figure shown is a view from above of two clay balls moving toward each other on a frictionless surface. They collide perfectly inelastically at the indicated point and are observed to then move in the direction indicated by the post-collision velocity vector, v'. If $m_1 = 2m_2$ what is v_2?

(A) $v_1(\sin 45°)/(2 \sin 60°)$
(B) $v_1(\cos 45°)/(2 \cos 60°)$
(C) $v_1(2 \cos 45°)/(\cos 60°)$
(D) $v_1(2 \sin 45°)/(\sin 60°)$
(E) $v_1(\cos 45°)/(2 \sin 60°)$

GO ON TO THE NEXT PAGE

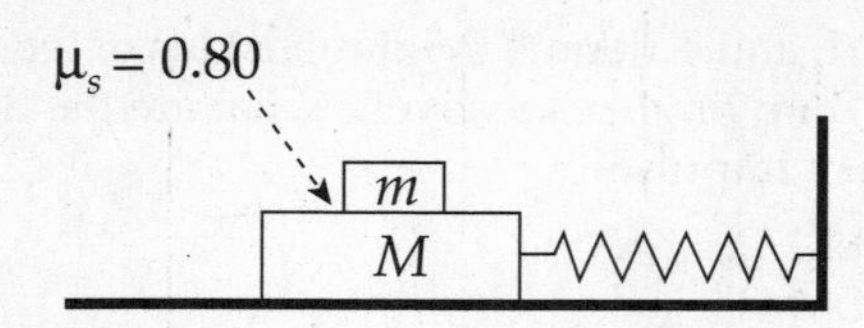

26. In the figure above, the coefficient of static friction between the two blocks is 0.80. If the blocks oscillate with a frequency of 2.0 Hz, what is the maximum amplitude of the oscillations if the small block is not to slip on the large block?

(A) 3.1 cm
(B) 5.0 cm
(C) 6.2 cm
(D) 7.5 cm
(E) 9.4 cm

27. When two objects collide, the ratio of the relative speed after the collision to the relative speed before the collision is called the *coefficient of restitution, e*. If a ball is dropped from height H_1 onto a stationary floor, and the ball rebounds to height H_2, what is the coefficient of restitution of the collision?

(A) H_2/H_1
(B) H_2/H_1
(C) $\sqrt{H_1/H_2}$
(D) $\sqrt{H_2/H_1}$
(E) $(H_1/H_2)^2$

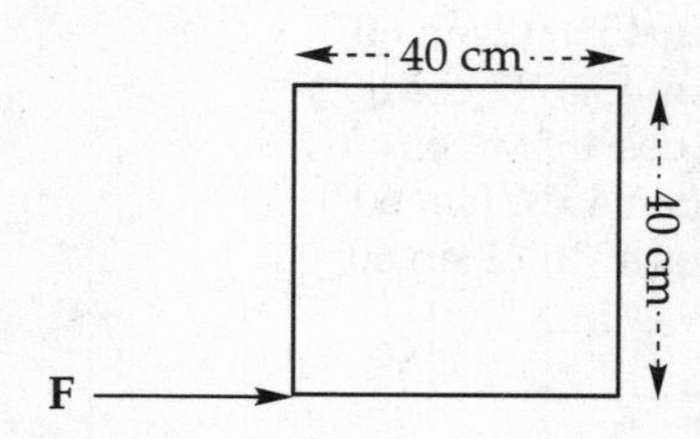

28. The figure above shows a square metal plate of side length 40 cm and uniform density, lying flat on a table. A force **F** of magnitude 10 N is applied at one of the corners, as shown. Determine the torque produced by **F** relative to the center of rotation.

(A) 0 N·m
(B) 1.0 N·m
(C) 1.4 N·m
(D) 2.0 N·m
(E) 4.0 N·m

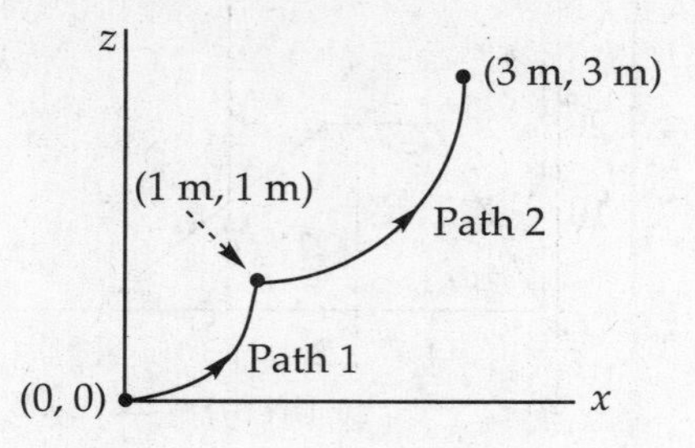

29. A small block of mass $m = 2.0$ kg is pushed from the initial point $(x_i, z_i) = (0 \text{ m}, 0 \text{ m})$ upward to the final point $(x_f, z_f) = (3 \text{ m}, 3 \text{ m})$ along the path indicated. Path 1 is a portion of the parabola $z = x^2$, and Path 2 is a quarter circle whose equation is $(x - 2)^2 + (z - 2)^2 = 2$. How much work is done by gravity during this displacement?

(A) 60 J
(B) 80 J
(C) 90 J
(D) 100 J
(E) 120 J

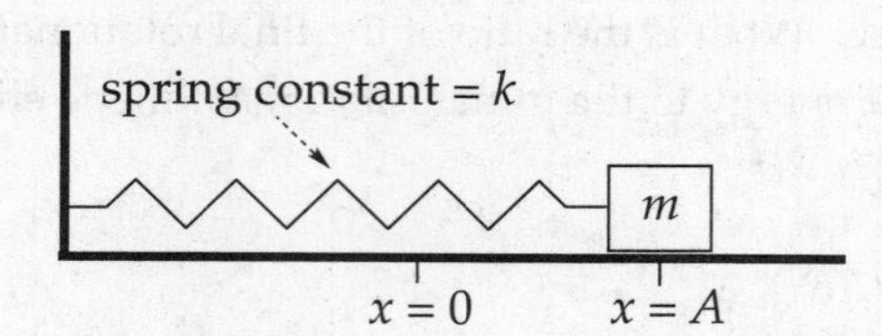

30. In the figure shown, the block (mass = m) is at rest at $x = A$. As it moves back toward the wall due to the force exerted by the stretched spring, it is also acted upon by a frictional force whose strength is given by the expression bx, where b is a positive constant. What is the block's speed when it first passes through the equilibrium position ($x = 0$)?

(A) $A\sqrt{(k + b)/m}$
(B) $A\sqrt{(k - b)/m}$
(C) $A\sqrt{(\frac{1}{2}k + b)/m}$
(D) $A\sqrt{(\frac{1}{2}k - b)/m}$
(E) $A\sqrt{\frac{1}{2}(k - b)/m}$

GO ON TO THE NEXT PAGE

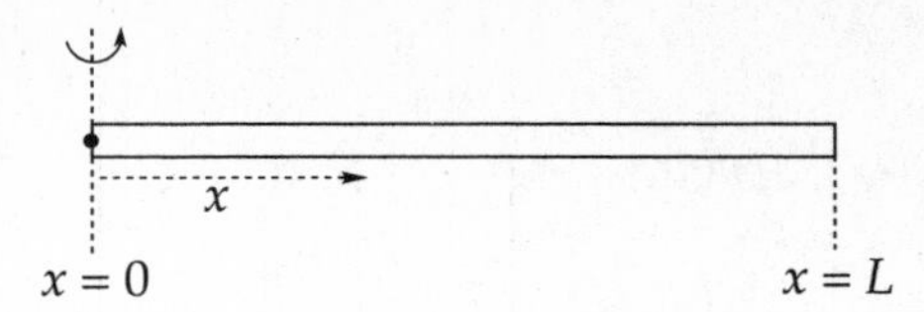

31. The rod shown above can pivot about the point $x = 0$ and rotates in a plane perpendicular to the page. Its linear density, λ, increases with x such that $\lambda(x) = kx$, where k is a positive constant. Determine the rod's moment of inertia in terms of its length, L, and its total mass, M.

(A) $\frac{1}{6}ML^2$

(B) $\frac{1}{4}ML^2$

(C) $\frac{1}{3}ML^2$

(D) $\frac{1}{2}ML^2$

(E) $2ML^2$

32. A particle is subjected to a conservative force whose potential energy function is

$$U(x) = (x - 2)^3 - 12x$$

where U is given in joules when x is measured in meters. Which of the following represents a position of stable equilibrium?

(A) $x = -4$
(B) $x = -2$
(C) $x = 0$
(D) $x = 2$
(E) $x = 4$

33. At what angle to the horizontal should an ideal projectile be launched so that its horizontal displacement (the range) is equal to its maximum vertical displacement?

(A) $\sin^{-1}(1/g)$
(B) $\cos^{-1}(1/g)$
(C) 45°
(D) $\tan^{-1} 2$
(E) $\tan^{-1} 4$

34. A particle's kinetic energy is changing at a rate of –6.0 J/s when its speed is 3.0 m/s. What is the magnitude of the force on the particle at this moment?

(A) 0.5 N
(B) 2.0 N
(C) 4.5 N
(D) 9.0 N
(E) 18 N

35. An object of mass 2 kg is acted upon by three external forces, each of magnitude 4 N. Which of the following could NOT be the resulting acceleration of the object?

(A) 0 m/s²
(B) 2 m/s²
(C) 4 m/s²
(D) 6 m/s²
(E) 8 m/s²

STOP

END OF SECTION I, MECHANICS

IF YOU FINISH BEFORE TIME IS CALLED, YOU MAY CHECK YOUR WORK ON SECTION I, MECHANICS, ONLY.

DO NOT TURN TO ANY OTHER TEST MATERIALS.

PHYSICS C

SECTION I, ELECTRICITY AND MAGNETISM

Time—45 minutes

35 Questions

Directions: Each of the questions or incomplete statements below is followed by five suggested answers or completions. Select the one that is best in each case and mark it on your answer sheet.

36. A nonconducting sphere is given a nonzero net electric charge, $+Q$, and then brought close to a neutral conducting sphere of the same radius. Which of the following will be true?

    (A) An electric field will be induced within the conducting sphere.
    (B) The conducting sphere will develop a net electric charge of $-Q$.
    (C) The spheres will experience an electrostatic attraction.
    (D) The spheres will experience an electrostatic repulsion.
    (E) The spheres will experience no electrostatic interaction.

37. Which of the following would increase the capacitance of a parallel-plate capacitor?

    (A) Using smaller plates
    (B) Replacing the dielectric material between the plates with one that has a smaller dielectric constant
    (C) Decreasing the voltage between the plates
    (D) Increasing the voltage between the plates
    (E) Moving the plates closer together

38. Each of the following particles is projected with the same speed into a uniform magnetic field **B** such that the particle's initial velocity is perpendicular to **B**. Which one would move in a circular path with the largest radius?

    (A) Proton
    (B) Beta particle
    (C) Alpha particle
    (D) Electron
    (E) Positron

GO ON TO THE NEXT PAGE

39. An ellipsoid-shaped conductor is negatively charged. Which one of the following diagrams best illustrates the charge distribution and electric field lines?

(A)

(B)

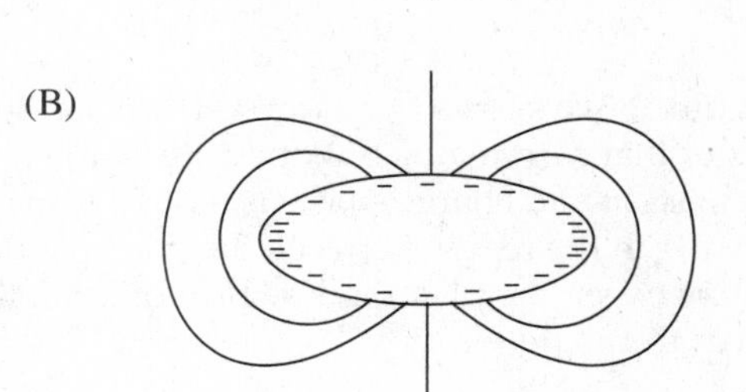

(C)

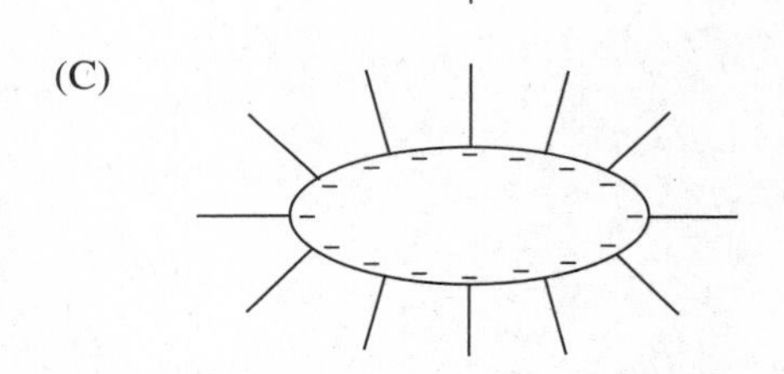

(D)

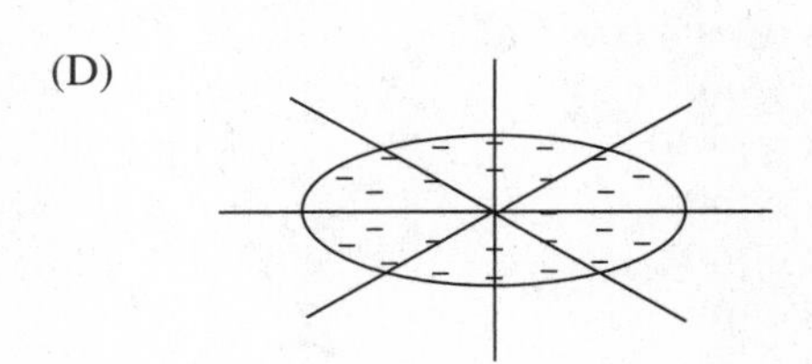

(E)

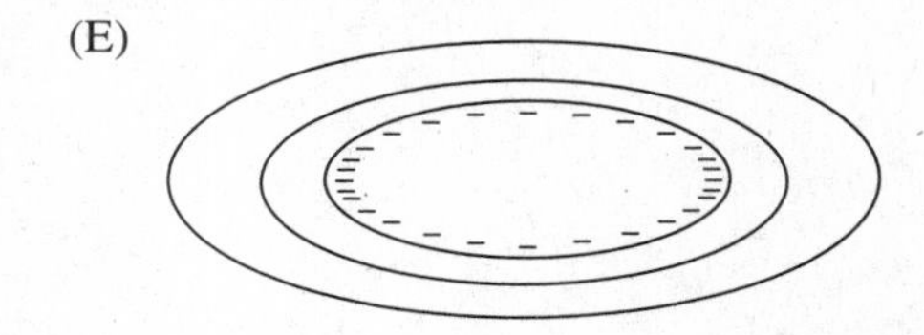

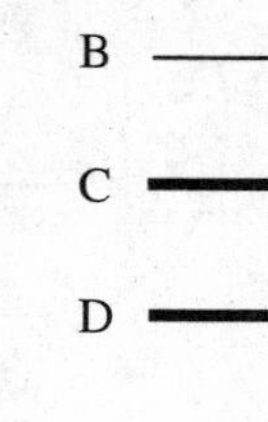

40. The four wires shown above are each made of aluminum. Which wire will have the greatest resistance?

(A) Wire A
(B) Wire B
(C) Wire C
(D) Wire D
(E) All the wires have the same resistance, because they're all composed of the same material.

41. Which of the following is NOT equal to one tesla?

(A) 1 J/(A·m^2)
(B) 1 kg/(C·s)
(C) 1 N/(A·m)
(D) 1 V·s/m^2
(E) 1 A·N/V

GO ON TO THE NEXT PAGE

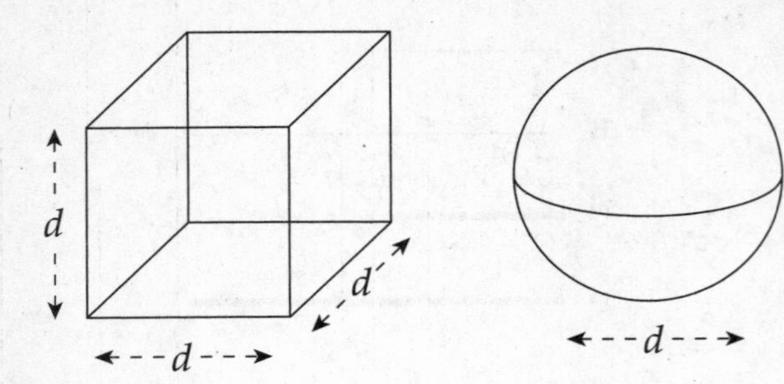

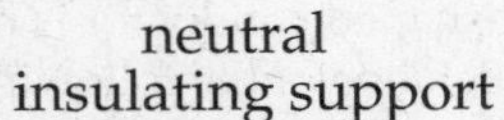

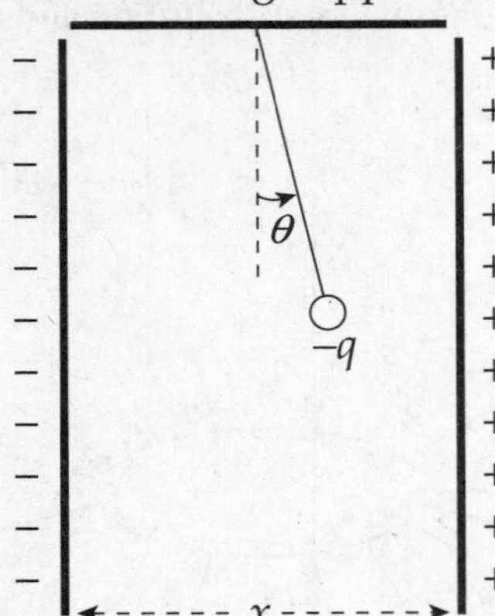

42. The figure above shows two Gaussian surfaces: a cube with side length d and a sphere with diameter d. The net electric charge enclosed within each surface is the same, $+Q$. If Φ_C denotes the total electric flux through the cubical surface, and Φ_S denotes the total electric flux through the spherical surface, then which of the following is true?

(A) $\Phi_C = (\pi/6)\Phi_S$
(B) $\Phi_C = (\pi/3)\Phi_S$
(C) $\Phi_C = \Phi_S$
(D) $\Phi_C = (3/\pi)\Phi_S$
(E) $\Phi_C = (6/\pi)\Phi_S$

43. The figure above shows two large vertical conducting plates that carry equal but opposite charges. A ball of mass m and charge $-q$ is suspended from a light string in the region between the plates. If the voltage between the plates is V, which of the following gives the angle θ?

(A) $\cos^{-1}(qV/mgx)$
(B) $\sin^{-1}(qV/mgx)$
(C) $\tan^{-1}(qV/mgx)$
(D) $\cos^{-1}(qV/x)$
(E) $\sin^{-1}(qV/x)$

44. An object carries a charge of −1 C. How many excess electrons does it contain?

(A) 6.25×10^{18}
(B) 8.00×10^{18}
(C) 1.60×10^{19}
(D) 3.20×10^{19}
(E) 6.25×10^{19}

GO ON TO THE NEXT PAGE

Questions 45–46

Each of the resistors shown in the circuit below has a resistance of 200 Ω. The emf of the ideal battery is 24 V.

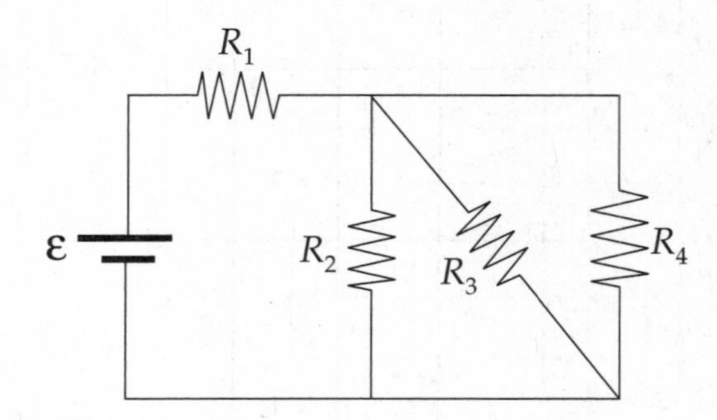

45. How much current is provided by the source?

(A) 30 mA
(B) 48 mA
(C) 64 mA
(D) 72 mA
(E) 90 mA

46. What is the ratio of the power dissipated by R_1 to the power dissipated by R_4?

(A) 1/9
(B) 1/4
(C) 1
(D) 4
(E) 9

47. What is the value of the following product?

$$20\ \mu\text{F} \times 500\ \Omega$$

(A) 0.01 henry
(B) 0.01 ampere per coulomb
(C) 0.01 weber
(D) 0.01 second
(E) 0.01 volt per ampere

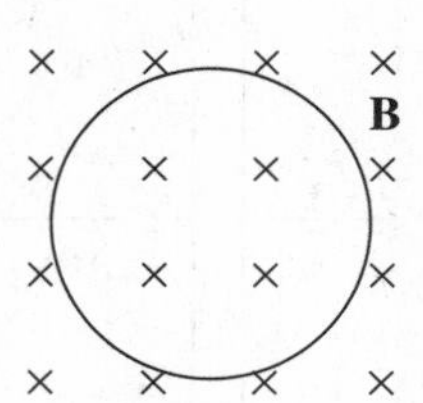

48. A copper wire in the shape of a circle of radius 1 m, lying in the plane of the page, is immersed in a magnetic field, **B**, that points into the plane of the page. The strength of **B** varies with time, t, according to the equation

$$B(t) = 2t(1 - t)$$

where B is given in teslas when t is measured in seconds. What is the magnitude of the induced electric field in the wire at time $t = 1$ s?

(A) $(1/\pi)$ N/C
(B) 1 N/C
(C) 2 N/C
(D) π N/C
(E) 2π N/C

GO ON TO THE NEXT PAGE

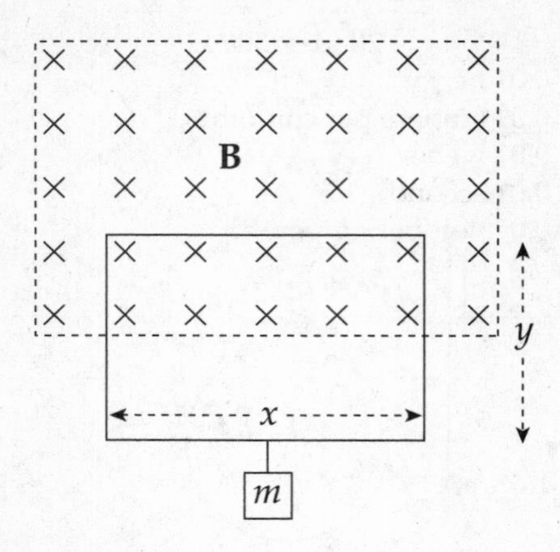

49. In the figure above, the top half of a rectangular loop of wire, x meters by y meters, hangs vertically in a uniform magnetic field, **B**. Describe the magnitude and direction of the current in the loop necessary for the magnetic force to balance the weight of the mass m supported by the loop.

(A) $I = mg/xB$, clockwise
(B) $I = mg/xB$, counterclockwise
(C) $I = mg / \left(x + \frac{1}{2}y\right)B$, clockwise
(D) $I = mg / \left(x + \frac{1}{2}y\right)B$, counterclockwise
(E) $I = mg/(x+y)B$, clockwise

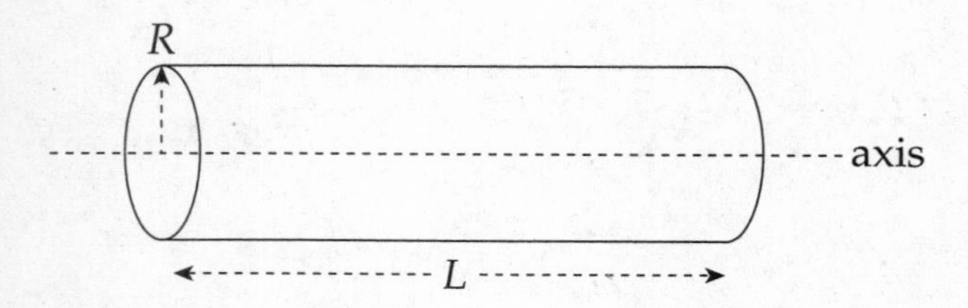

50. A solid nonconducting cylinder of radius R and length L contains a volume charge density given by the equation $\rho(r) = (+3\ \text{C/m}^4)r$, where r is the radial distance from the cylinder's central axis. This means that the total charge contained within a concentric cylinder of radius $r < R$ and length $\ell < L$ is equal to $2\pi\ell r^3$. Find an expression for the strength of the electric field inside this cylinder.

(A) $1/\varepsilon_0 r^2$
(B) $r/2\varepsilon_0$
(C) $2r/\varepsilon_0$
(D) r^2/ε_0
(E) $r^2/2\varepsilon_0$

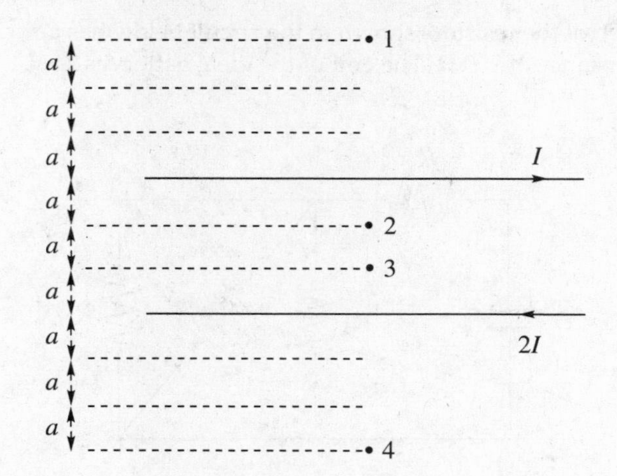

51. The figure above shows a pair of long, straight current-carrying wires and four marked points. At which of these points is the net magnetic field zero?

(A) Point 1 only
(B) Points 1 and 2 only
(C) Point 2 only
(D) Points 3 and 4 only
(E) Point 3 only

GO ON TO THE NEXT PAGE

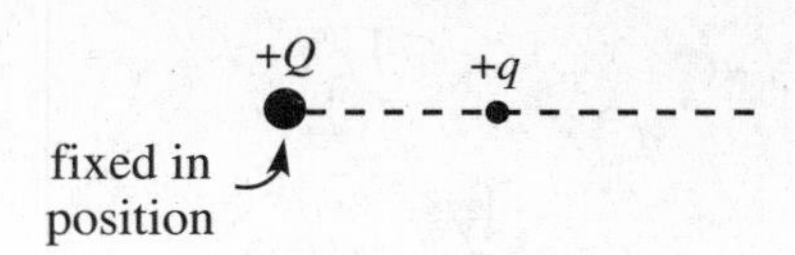

52. The figure above shows two positively-charged particles. The $+Q$ charge is fixed in position, and the $+q$ charge is brought close to $+Q$ and released from rest. Which of the following graphs best depicts the acceleration of the $+q$ charge as a function of its distance r from $+Q$?

(A)
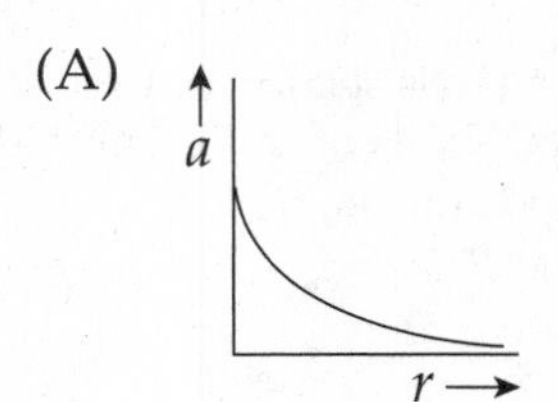

(B)
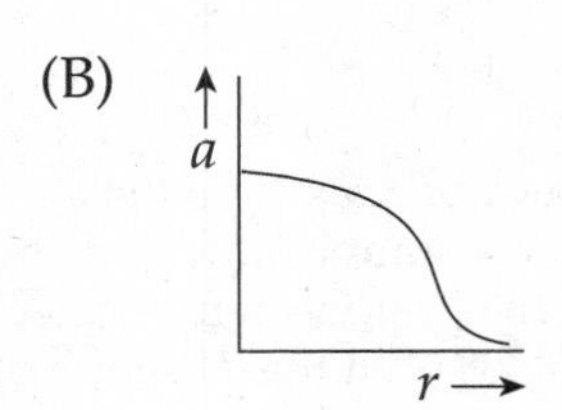

(C)
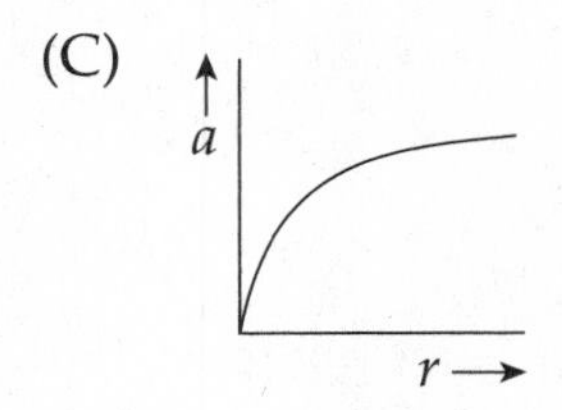

(D)
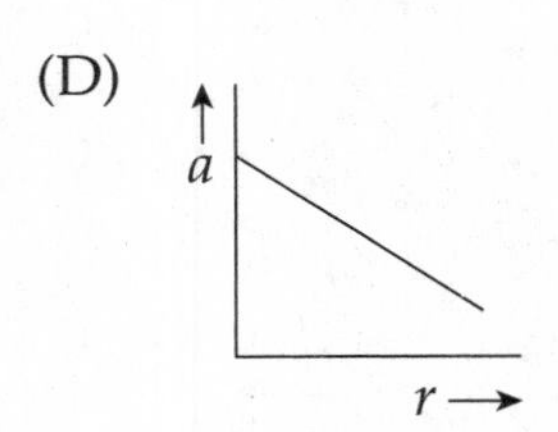

(E)
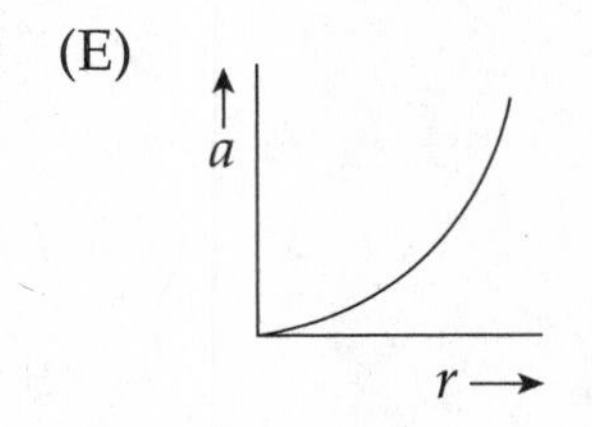

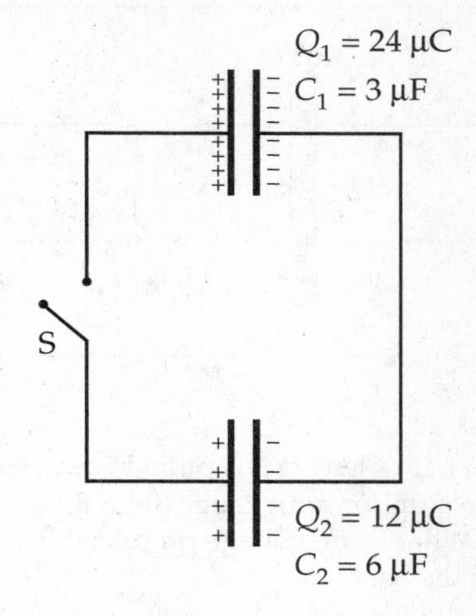

53. Once the switch S in the figure above is closed and electrostatic equilibrium is regained, how much charge will be stored on the positive plate of the 6 μF capacitor?

(A) 9 μC
(B) 18 μC
(C) 24 μC
(D) 27 μC
(E) 36 μC

GO ON TO THE NEXT PAGE

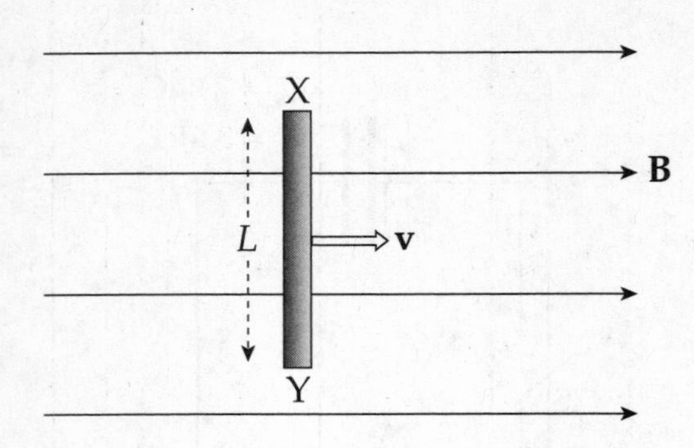

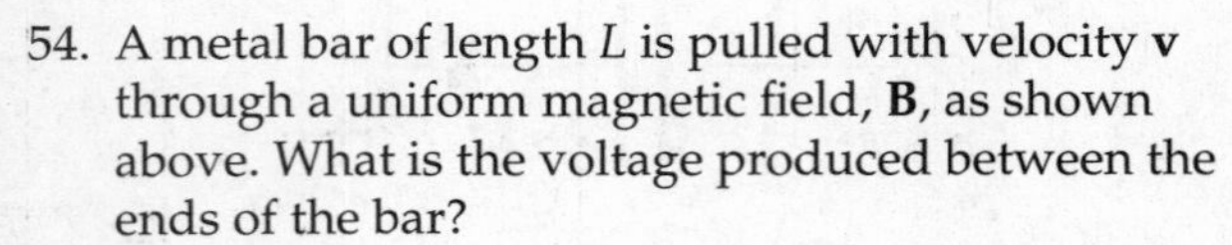

54. A metal bar of length L is pulled with velocity $\mathbf{v}$ through a uniform magnetic field, $\mathbf{B}$, as shown above. What is the voltage produced between the ends of the bar?

(A) vB, with Point X at a higher potential than Point Y
(B) vB, with Point Y at a higher potential than Point X
(C) vBL, with Point X at a higher potential than Point Y
(D) vBL, with Point Y at a higher potential than Point X
(E) None of the above

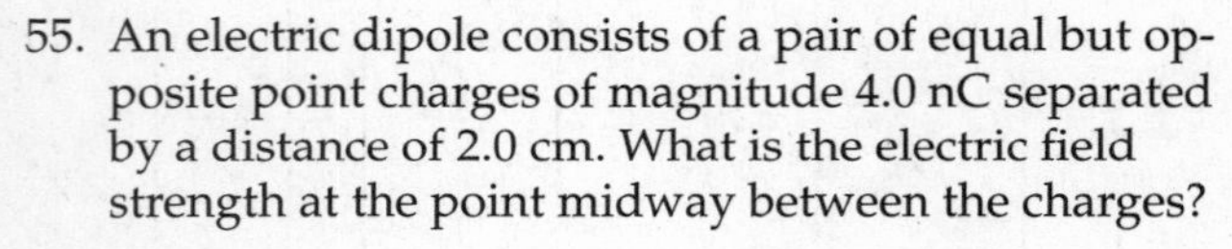

55. An electric dipole consists of a pair of equal but opposite point charges of magnitude 4.0 nC separated by a distance of 2.0 cm. What is the electric field strength at the point midway between the charges?

(A) 0
(B) 9.0×10^4 V/m
(C) 1.8×10^5 V/m
(D) 3.6×10^5 V/m
(E) 7.2×10^5 V/m

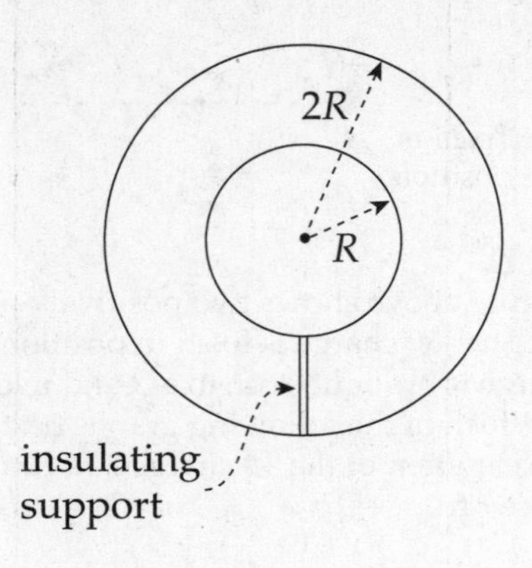

56. The figure above shows a cross section of two concentric spherical metal shells of radii R and $2R$, respectively. Find the capacitance.

(A) $1/(8\pi\varepsilon_0 R)$
(B) $1/(4\pi\varepsilon_0 R)$
(C) $2\pi\varepsilon_0 R$
(D) $4\pi\varepsilon_0 R$
(E) $8\pi\varepsilon_0 R$

57. Traveling at an initial speed of 1.5×10^6 m/s, a proton enters a region of constant magnetic field, $\mathbf{B}$, of magnitude 1.0 T. If the proton's initial velocity vector makes an angle of 30° with the direction of $\mathbf{B}$, compute the proton's speed 4 s after entering the magnetic field.

(A) 5.0×10^5 m/s
(B) 7.5×10^5 m/s
(C) 1.5×10^6 m/s
(D) 3.0×10^6 m/s
(E) 6.0×10^6 m/s

GO ON TO THE NEXT PAGE

Questions 58–60

There is initially no current through any circuit element in the following diagram.

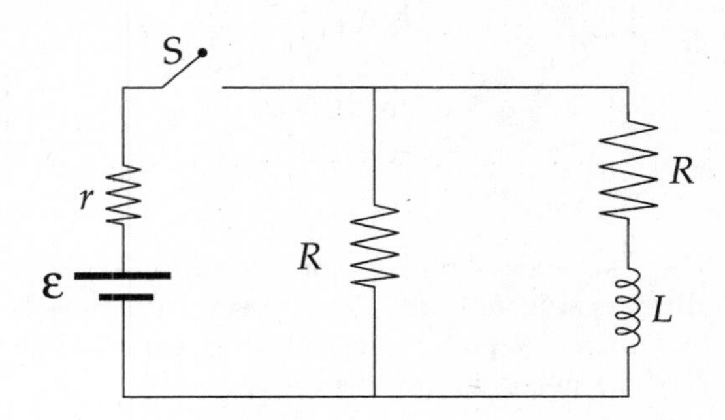

58. What is the current through r immediately after the switch S is closed?

(A) 0

(B) $\dfrac{\mathcal{E}}{r+R}$

(C) $\dfrac{\mathcal{E}}{r+2R}$

(D) $\dfrac{\mathcal{E}(r+R)}{rR}$

(E) $\dfrac{\mathcal{E}(2R)}{2Rr+2R}$

59. After the switch has been kept closed for a long time, how much energy is stored in the inductor?

(A) $\dfrac{L\mathcal{E}^2}{2(r+R)^2}$

(B) $\dfrac{L\mathcal{E}^2}{2(r+2R)^2}$

(C) $\dfrac{L\mathcal{E}^2}{4(2r+R)^2}$

(D) $\dfrac{L(\mathcal{E}R)^2}{8(2r+R)^2}$

(E) $\dfrac{L\mathcal{E}^2}{8(2r+R)^2}$

60. After having been closed for a long time, the switch is suddenly opened. What is the current through r immediately after S is opened?

(A) 0

(B) $\dfrac{\mathcal{E}}{r+R}$

(C) $\dfrac{\mathcal{E}}{r+2R}$

(D) $\dfrac{\mathcal{E}(r+R)}{rR}$

(E) $\dfrac{\mathcal{E}(2R)}{r(2R)+2R}$

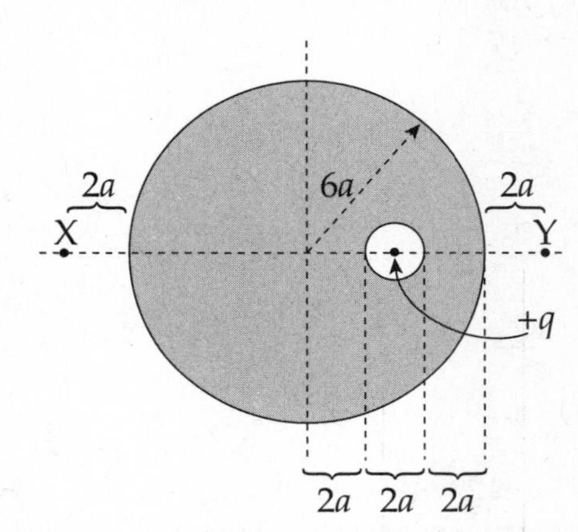

61. A solid, neutral metal sphere of radius $6a$ contains a small cavity, a spherical hole of radius a as shown above. Within this cavity is a charge, $+q$. If E_X and E_Y denote the strength of the electric field at points X and Y respectively, which of the following is true?

(A) $E_Y = 4E_X$
(B) $E_Y = 16E_X$
(C) $E_Y = E_X$
(D) $E_Y = (11/5)E_X$
(E) $E_Y = (11/5)^2E_X$

GO ON TO THE NEXT PAGE

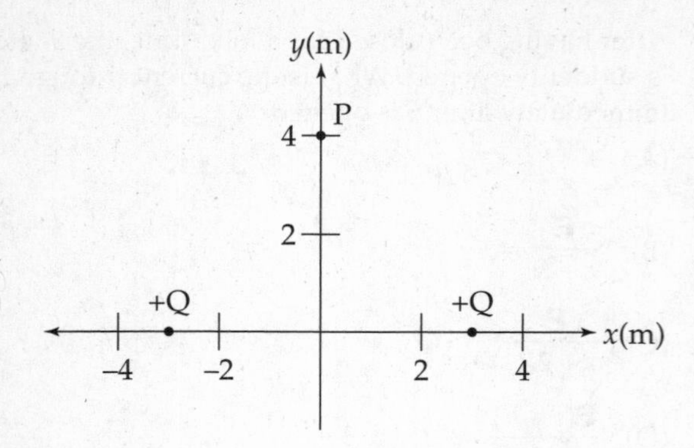

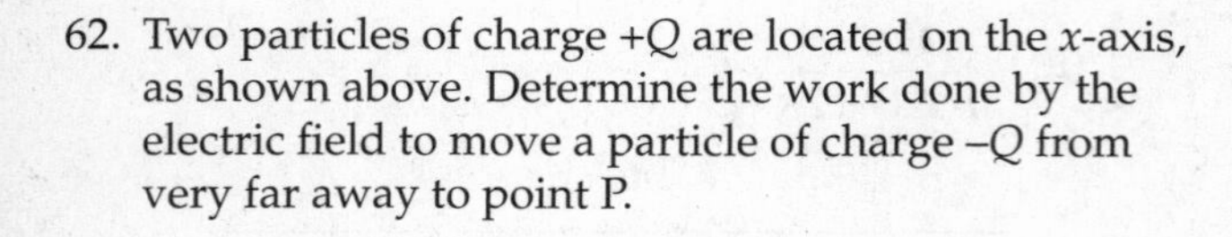

62. Two particles of charge $+Q$ are located on the x-axis, as shown above. Determine the work done by the electric field to move a particle of charge $-Q$ from very far away to point P.

(A) $\frac{2kQ}{5}$

(B) $\frac{2kQ^2}{5}$

(C) $-\frac{2kQ^2}{5}$

(D) $\frac{kQ^2}{5}$

(E) $-\frac{3kQ^2}{5}$

63. A battery is connected in series with a switch, a resistor of resistance R, and an inductor of inductance L. Initially, there is no current in the circuit. Once the switch is closed and the circuit is completed, how long will it take for the current to reach 99% of its maximum value?

(A) $(\ln\frac{99}{100})RL$

(B) $L\mathcal{E}^2$

(C) $(\ln\frac{1}{100})L / R$

(D) $L / R(\ln\frac{100}{99})$

(E) $(\ln 100)\frac{L}{R}$

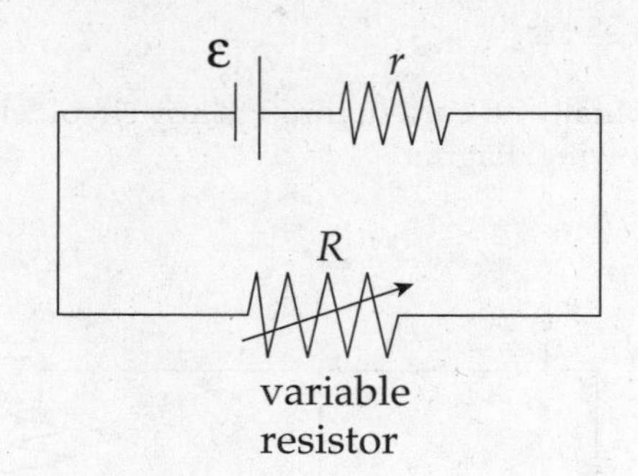

64. The resistance of the variable resistor R in the circuit above is adjusted until the power it dissipates is maximized. Which of the following expressions gives its maximum power dissipation?

(A) $\mathcal{E}^2/8r$

(B) $\mathcal{E}^2/4r$

(C) $\mathcal{E}^2/2r$

(D) $\mathcal{E}^2/r$

(E) None of the above

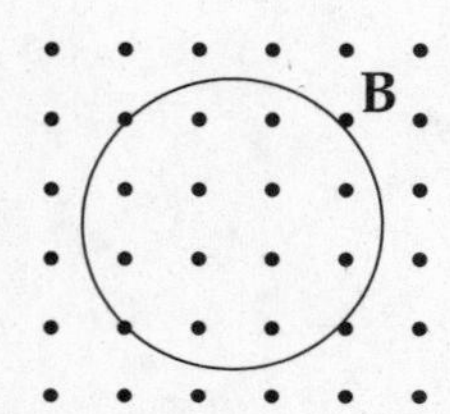

65. The metal loop of wire shown above is situated in a magnetic field **B** pointing out of the plane of the page. If **B** decreases uniformly in strength, the induced electric current within the loop is

(A) clockwise and decreasing
(B) clockwise and increasing
(C) counterclockwise and decreasing
(D) counterclockwise and constant
(E) counterclockwise and increasing

GO ON TO THE NEXT PAGE

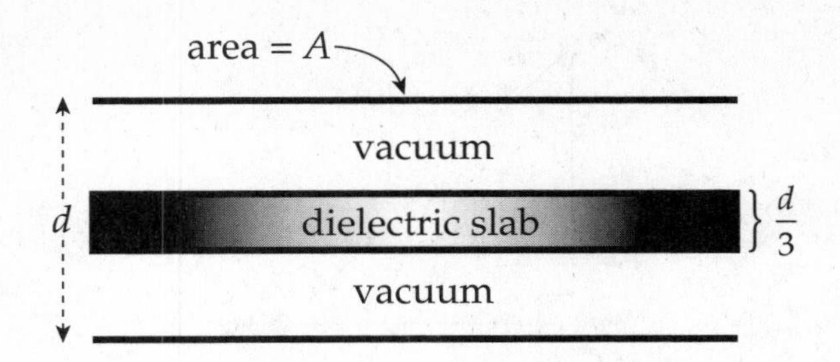

66. A dielectric of thickness $\frac{d}{3}$ is placed between the plates of a parallel-plate capacitor, as shown above. If K is the dielectric constant of the slab, what is the capacitance?

(A) $\dfrac{\varepsilon_0 A(2+3K)}{d}$

(B) $\dfrac{d}{\varepsilon_0 A(2+3K)}$

(C) $\dfrac{3\varepsilon_0 A}{d(2K+1)}$

(D) $\dfrac{3K\varepsilon_0 A}{d(2K+1)}$

(E) $\dfrac{3K\varepsilon_0 A}{d}$

$-Q$

67. Consider the two source charges shown above. At how many points in the plane of the page, in a region around these charges are both the electric field and the electric potential equal to zero?

(A) 0
(B) 1
(C) 2
(D) 3
(E) 4

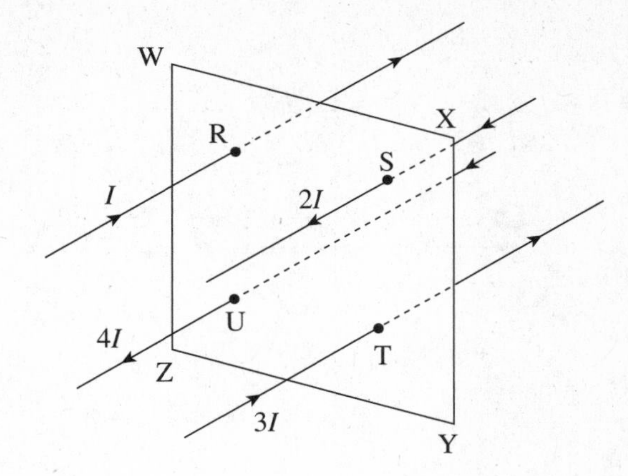

68. The figure above shows four current-carrying wires passing perpendicularly through the interior of a square whose vertices are W, X, Y, and Z. The points where the wires pierce the plane of the square (namely, R, S, T, and U) themselves form the vertices of a square each side of which has half the length of each side of WXYZ. If the currents are as labeled in the figure, what is the absolute value of

$$\oint \mathbf{B} \cdot d\boldsymbol{\ell}$$

where the integral is taken around WXYZ?

(A) $\frac{1}{2}\mu_0 I$
(B) $\mu_0 I$
(C) $\sqrt{2}\,\mu_0 I$
(D) $2\mu_0 I$
(E) $5\mu_0 I$

GO ON TO THE NEXT PAGE

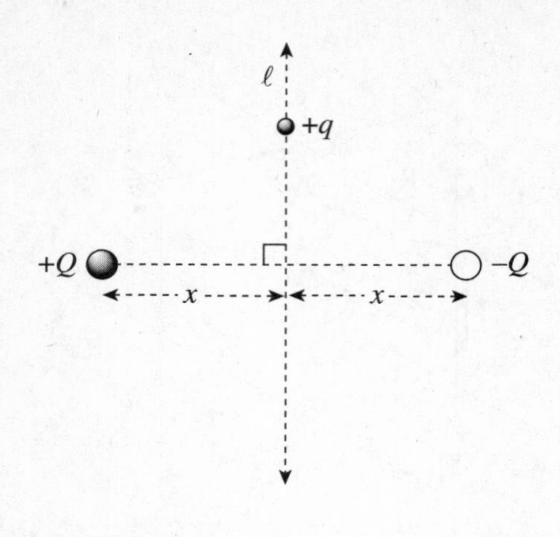

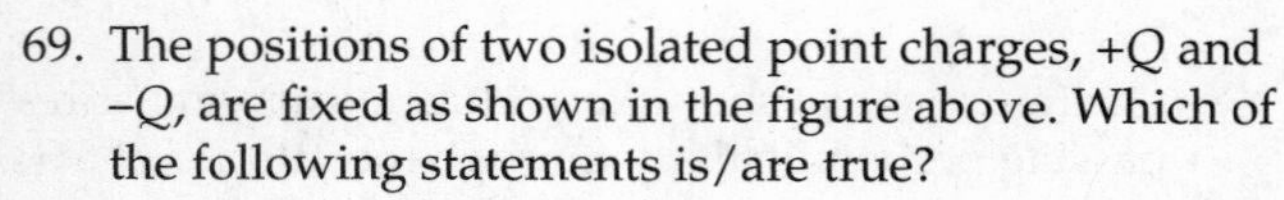

69. The positions of two isolated point charges, $+Q$ and $-Q$, are fixed as shown in the figure above. Which of the following statements is/are true?

    I. A charge $+q$ moving along line ℓ would experience no change in electrical potential energy.

    II. There is exactly one point on line ℓ where the total electrostatic force on $+q$ is equal to zero.

    III. At any point P on line ℓ, the intersection of the equipotential surface containing P with the plane of the figure is an ellipse with the charges $+Q$ and $-Q$ at the foci.

    (A) I only
    (B) I and II only
    (C) II only
    (D) I and III only
    (E) II and III only

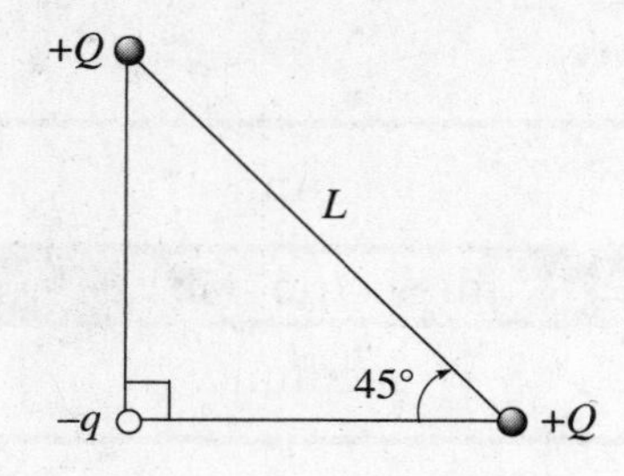

70. Two point charges, each $+Q$, are fixed a distance L apart. A particle of charge $-q$ and mass m is placed as shown in the figure above. What is this particle's initial acceleration when released from rest?

    (A) $\dfrac{\sqrt{2}Qq}{2\pi\varepsilon_0 L^2 m}$

    (B) $\dfrac{\sqrt{2}Qq}{\pi\varepsilon_0 L^2 m}$

    (C) $\dfrac{2Qq}{\pi\varepsilon_0 L^2 m}$

    (D) $\dfrac{2\sqrt{2}Qq}{\pi\varepsilon_0 L^2 m}$

    (E) $\dfrac{4Qq}{\pi\varepsilon_0 L^2 m}$

STOP

END OF SECTION I, ELECTRICITY AND MAGNETISM

IF YOU FINISH BEFORE TIME IS CALLED, YOU MAY
CHECK YOUR WORK ON SECTION I, ELECTRICITY AND MAGNETISM, ONLY.

DO NOT GO ON TO SECTION II UNTIL YOU ARE TOLD TO DO SO.

PHYSICS C

SECTION II

Free-Response Questions

Mechanics	45 minutes	3 required questions of equal weight
Electricity and Magnetism	45 minutes	3 required questions of equal weight

Section II is 50 percent of the total grade for each of the two examinations.
Mark one of the boxes below to indicate which questions you are answering.

☐ Mechanics only

☐ Electricity and Magnetism only

☐ Both Mechanics and Electricity and Magnetism

GO ON TO THE NEXT PAGE

ADVANCED PLACEMENT PHYSICS C EQUATIONS FOR 2008

MECHANICS

$v = v_0 + at$

$x = x_0 + v_0 t + \frac{1}{2}at^2$

$v^2 = v_0^2 + 2a(x - x_0)$

$\Sigma \mathbf{F} = \mathbf{F}_{net} = m\mathbf{a}$

$\mathbf{F} = \frac{d\mathbf{p}}{dt}$

$\mathbf{J} = \int \mathbf{F}\, dt = \Delta \mathbf{p}$

$\mathbf{p} = m\mathbf{v}$

$F_{fric} \le \mu N$

$W = \int \mathbf{F} \cdot d\mathbf{r}$

$K = \frac{1}{2}mv^2$

$P = \frac{dW}{dt}$

$P = \mathbf{F} \cdot \mathbf{v}$

$\Delta U_g = mgh$

$a_c = \frac{v^2}{r} = \omega^2 r$

$\boldsymbol{\tau} = \mathbf{r} \times \mathbf{F}$

$\Sigma \boldsymbol{\tau} = \boldsymbol{\tau}_{net} = I\boldsymbol{\alpha}$

$I = \int r^2 dm = \Sigma mr^2$

$\mathbf{r}_{cm} = \Sigma m\mathbf{r} / \Sigma m$

$v = r\omega$

$\mathbf{L} = \mathbf{r} \times \mathbf{p} = I\boldsymbol{\omega}$

$K = \frac{1}{2}I\omega^2$

$\omega = \omega_0 + \alpha t$

$\theta = \theta_0 + \omega_0 t + \frac{1}{2}\alpha t^2$

a = acceleration
F = force
f = frequency
h = height
I = rotational inertia
J = impulse
K = kinetic energy
k = spring constant
ℓ = length
L = angular momentum
m = mass
N = normal force
P = power
p = momentum
r = radius or distance
$\mathbf{r}$ = position vector
T = period
t = time
U = potential energy
v = velocity or speed
W = work done on a system
x = position
μ = coefficient of friction
θ = angle
τ = torque
ω = angular speed
α = angular acceleration

$\mathbf{F}_s = -k\mathbf{x}$

$U_s = \frac{1}{2}kx^2$

$T = \frac{2\pi}{\omega} = \frac{1}{f}$

$T_s = 2\pi\sqrt{\frac{m}{k}}$

$T_p = 2\pi\sqrt{\frac{\ell}{g}}$

$\mathbf{F}_G = -\frac{Gm_1m_2}{r^2}\hat{\mathbf{r}}$

$U_G = -\frac{Gm_1m_2}{r}$

ELECTRICITY AND MAGNETISM

$F = \frac{1}{4\pi\epsilon_0}\frac{q_1q_2}{r^2}$

$\mathbf{E} = \frac{\mathbf{F}}{q}$

$\oint \mathbf{E} \cdot d\mathbf{A} = \frac{Q}{\epsilon_0}$

$E = -\frac{dV}{dr}$

$V = \frac{1}{4\pi\epsilon_0}\sum_i \frac{q_i}{r_i}$

$U_E = qV = \frac{1}{4\pi\epsilon_0}\frac{q_1q_2}{r}$

$C = \frac{Q}{V}$

$C = \frac{\kappa\epsilon_0 A}{d}$

$C_p = \sum_i C_i$

$\frac{1}{C_s} = \sum_i \frac{1}{C_i}$

$I = \frac{dQ}{dt}$

$U_c = \frac{1}{2}QV = \frac{1}{2}CV^2$

$R = \frac{\rho\ell}{A}$

$\mathbf{E} = \rho\mathbf{J}$

$I = Nev_d A$

$V = IR$

$R_s = \sum_i R_i$

$\frac{1}{R_p} = \sum_i \frac{1}{R_i}$

$P = IV$

$\mathbf{F}_M = q\mathbf{v} \times \mathbf{B}$

A = area
B = magnetic field
C = capacitance
d = distance
E = electric field
$\mathcal{E}$ = emf
F = force
I = current
J = current density
L = inductance
ℓ = length
n = number of loops of wire per unit length
N = number of charge carriers per unit volume
P = power
Q = charge
q = point charge
R = resistance
r = distance
t = time
U = potential or stored energy
V = electric potential
v = velocity or speed
ρ = resistivity
ϕ_m = magnetic flux
κ = dielectric constant

$\oint \mathbf{B} \cdot d\boldsymbol{\ell} = \mu_0 I$

$d\mathbf{B} = \frac{\mu_0}{4\pi}\frac{I\, d\boldsymbol{\ell} \times \mathbf{r}}{r^3}$

$\mathbf{F} = \int I\, d\boldsymbol{\ell} \times \mathbf{B}$

$B_s = \mu_0 nI$

$\phi_m = \int \mathbf{B} \cdot d\mathbf{A}$

$\mathcal{E} = -\frac{d\phi_m}{dt}$

$\mathcal{E} = -L\frac{dI}{dt}$

$U_L = \frac{1}{2}LI^2$

ADVANCED PLACEMENT PHYSICS C EQUATIONS FOR 2008

GEOMETRY AND TRIGONOMETRY

Rectangle

$A = bh$

Triangle

$A = \frac{1}{2}bh$

Circle

$A = \pi r^2$

$C = 2\pi r$

Parallelepiped

$V = \ell wh$

Cylinder

$V = \pi r^2 \ell$

$S = 2\pi r\ell + 2\pi r^2$

Sphere

$V = \frac{4}{3}\pi r^3$

$S = 4\pi r^2$

Right Triangle

$a^2 + b^2 = c^2$

$\sin\theta = \frac{a}{c}$

$\cos\theta = \frac{b}{c}$

$\tan\theta = \frac{a}{b}$

A = area
C = circumference
V = volume
S = surface area
b = base
h = height
ℓ = length
w = width
r = radius

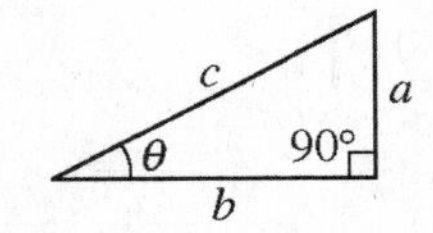

CALCULUS

$$\frac{df}{dx} = \frac{df}{du}\frac{du}{dx}$$

$$\frac{d}{dx}\left(x^n\right) = nx^{n-1}$$

$$\frac{d}{dx}\left(e^x\right) = e^x$$

$$\frac{d}{dx}(\ln x) = \frac{1}{x}$$

$$\frac{d}{dx}(\sin x) = \cos x$$

$$\frac{d}{dx}(\cos x) = -\sin x$$

$$\int x^n\,dx = \frac{1}{n+1}x^{n+1},\ n \neq -1$$

$$\int e^x\,dx = e^x$$

$$\int \frac{dx}{x} = \ln|x|$$

$$\int \cos x\,dx = \sin x$$

$$\int \sin x\,dx = -\cos x$$

PHYSICS C

SECTION II, MECHANICS

Time—45 minutes

3 Questions

Directions: Answer all three questions. The suggested time is about 15 minutes per question for answering each of the questions, which are worth 15 points each. The parts within a question may not have equal weight.

Mech 1. A massless, frictionless pulley is suspended from a rigid rod attached to the roof of an elevator car. Two masses, m and M (with $M > m$), are suspended on either side of the pulley by a light, inextendable cord. The distance from the top of the elevator car to the center of mass of the pulley, y_p, is fixed.

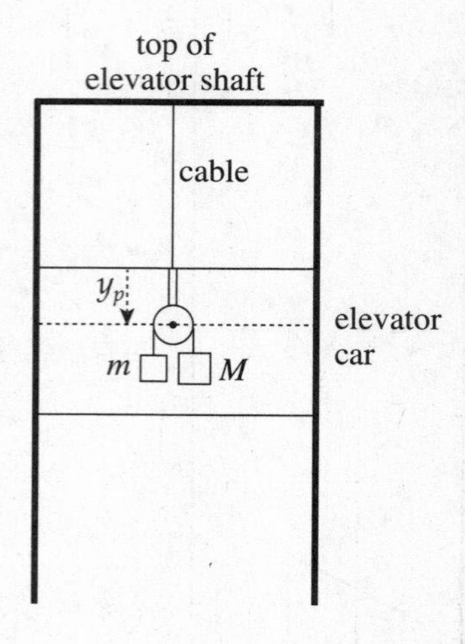

(a) Assume that the elevator car is descending at constant velocity. For each of the following, write your answer in terms of m, M, and g.

(i) Determine the accelerations of the masses.

(ii) Find the tension in the cord.

(b) Now assume that the elevator car is descending with a nonzero (but constant) acceleration, b.

(i) State why the coordinate system that uses the center of the pulley as its origin is not an inertial reference system.

For parts (ii) and (iii), use the top of the elevator shaft as a reference. Write your answers in terms of m, M, g, and b.

(ii) Find the acceleration of the masses.

(iii) Determine the tension in the cord.

(iv) For what value of b would the answer to part (iii) be zero?

GO ON TO THE NEXT PAGE

Mech 2. A narrow tunnel is drilled through the earth (mass = M, radius = R), connecting points P and Q, as shown in the diagram on the left below. The perpendicular distance from the earth's center, C, to the tunnel is x. A package (mass = m) is dropped from Point P into the tunnel; its distance from P is denoted y and its distance from C is denoted r. See the diagram on the right.

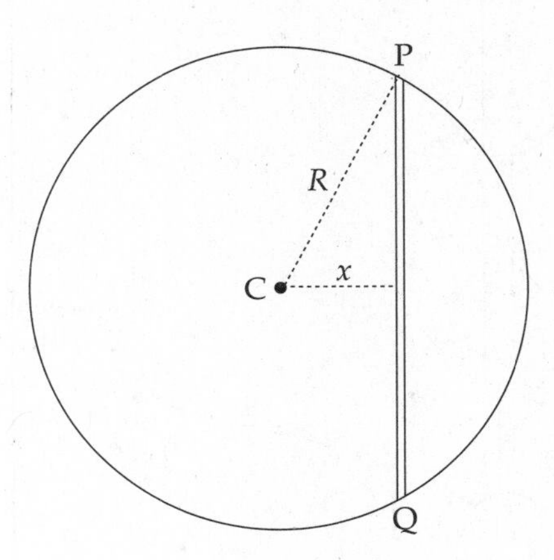

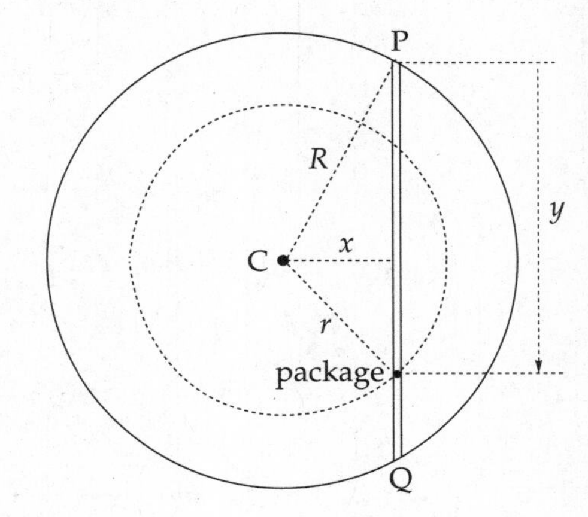

(a) Assuming that the earth is a homogeneous sphere, use the fact that the gravitational force F on the package is due to m and the mass contained within the sphere of radius $r < R$ to show that

$$F = -\frac{GMm}{R^3}r$$

(b) Use the equation $F(r) = -dU/dr$ to find an expression for the change in gravitational potential energy of the package as it moves from Point P to a point where its distance from the earth's center is r. Write your answer in terms of G, M, m, R, and r.

(c) Apply Conservation of Energy to determine the speed of the package in terms of G, M, R, x, and y. (Ignore friction.)

(d) (i) At what point in the tunnel—that is, for what value of y—will the speed of the package be maximized?

(ii) What is this maximum speed? (Write your answer in terms of G, M, R, and x.)

GO ON TO THE NEXT PAGE

Mech 3. The diagram below is a view from above of three sticky hockey pucks on a frictionless horizontal surface. The pucks with masses m and $2m$ are connected by a massless rigid rod of length L and are initially at rest. The puck of mass $3m$ is moving with velocity v directly toward puck m. When puck $3m$ strikes puck m, the collision is perfectly inelastic.

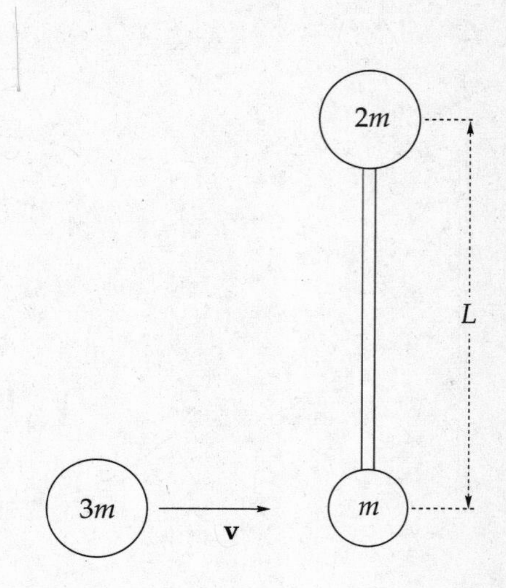

(a) Immediately after the collision,

(i) where is the center of mass of the system?

(ii) what is the speed of the center of mass? (Write your answer in terms of v.)

(iii) what is the angular speed of the system? (Write your answer in terms of v and L.)

(b) What fraction of the system's initial kinetic energy is lost as a result of the collision?

STOP

END OF SECTION II, MECHANICS

IF YOU FINISH BEFORE TIME IS CALLED, YOU MAY CHECK YOUR WORK ON SECTION II, MECHANICS, ONLY. DO NOT TURN TO ANY OTHER TEST MATERIALS.

PHYSICS C

SECTION II, ELECTRICITY AND MAGNETISM

Time—45 minutes

3 Questions

Directions: Answer all three questions. The suggested time is about 15 minutes per question for answering each of the questions, which are worth 15 points each. The parts within a question may not have equal weight.

E & M 1. A uniformly-charged, nonconducting, circular ring of radius R carries a charge $+Q$. Its central axis is labeled the z axis, and the center of the ring is $z = 0$. Points above the ring on the z axis have positive z coordinates; those below have negative z coordinates.

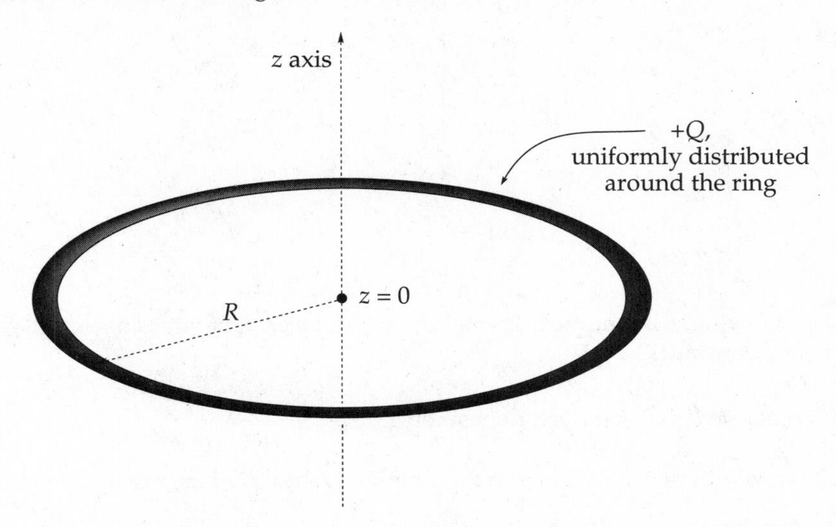

(a) Calculate the electric potential at an arbitrary point on the z axis. Write your answer in terms of Q, z, R, and fundamental constants.

(b) (i) At what point(s) on the z axis will the potential have its greatest value?

(ii) What is this maximum potential value?

(c) Find an expression for the electric field (magnitude and direction) at an arbitrary point on the z axis. Write your answer in terms of Q, z, R, and fundamental constants.

(d) (i) At what point(s) on the z axis will the electric field strength have its greatest value?

(ii) What is this maximum electric field strength?

A particle of charge $-q$ and mass m is released from rest at $z = +A$ (where $A << R$).

(e) What is its speed as it passes through $z = 0$? Give an answer that has been derived by applying the approximation $(1 + x)^{-n} \approx 1 - nx$ if $x < 1$. Your answer should be written in terms of Q, q, m, A, R, and fundamental constants.

(f) The particle will execute (approximately) simple harmonic motion. Derive [from your answer to part (e) or otherwise] an expression for its frequency of oscillation.

GO ON TO THE NEXT PAGE

E & M 2. In the circuit shown below, the capacitor is uncharged and there is no current in any circuit element.

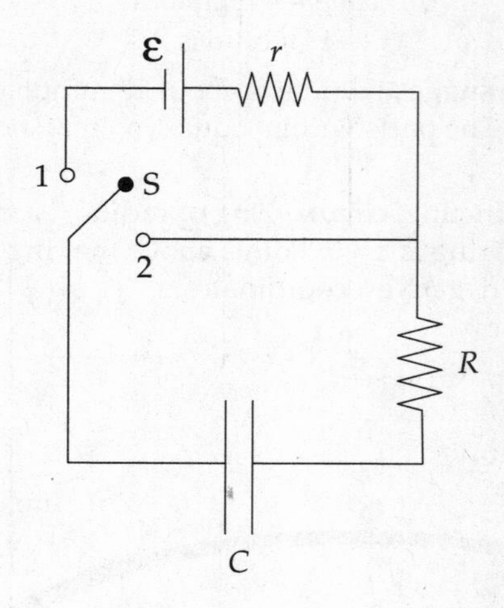

In each of the following, k is a number greater than 1; write each of your answers in terms of ℰ, r, R, C, k, and fundamental constants.

(a) At $t = 0$, the switch S is moved to position 1.

(i) At what time t is the current through R equal to $\frac{1}{k}$ of its initial value?

(ii) At what time t is the charge on the capacitor equal to $\frac{1}{k}$ of its maximum value?

(iii) At what time t is the energy stored in the capacitor equal to $\frac{1}{k}$ of its maximum value?

(b) After the switch has been at position 1 for a very long time, it is then moved to position 2. Let this redefine $t = 0$ for purposes of the following questions.

(i) How long will it take for the current through R to equal $\frac{1}{k}$ of its initial value?

(ii) At what time t is the charge on the capacitor equal to $\frac{1}{k}$ of its initial value?

GO ON TO THE NEXT PAGE

E & M 3. The diagram below shows two parallel conducting rails connected by a third rail of length L, raised to an angle of θ with the horizontal (supported by a pair of insulating columns). A metal strip of length L, mass m, and resistance R is free to slide without friction down the rails. The apparatus is immersed in a vertical, uniform magnetic field, **B**. The resistance of the stationary rails may be neglected.

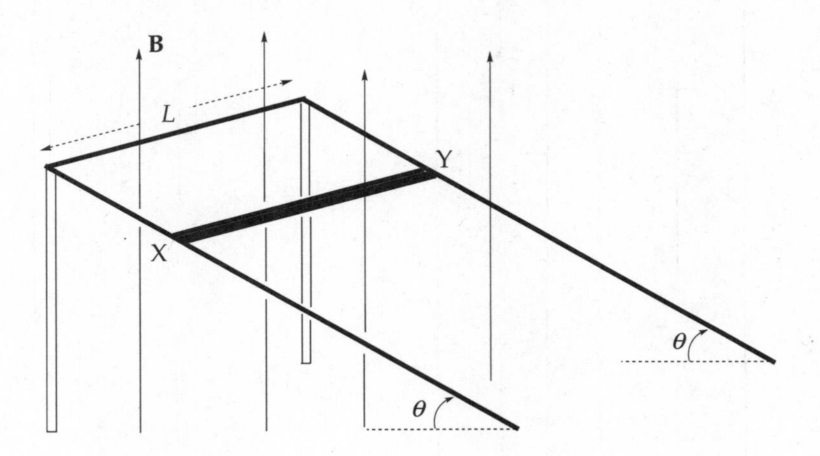

(a) As the strip slides down the rails with instantaneous speed v, determine

(i) the induced emf. (Write your answer in terms of L, B, v, and θ.)

(ii) the direction of the induced current in the strip (X to Y, or Y to X?).

(iii) the value of the induced current. (Write your answer in terms of L, B, R, v, and θ.)

(b) The strip XY will eventually slide down the rails at a constant velocity. Derive an expression for this velocity in terms of L, B, R, m, g, and θ.

(c) When the strip is sliding with the constant velocity determined in part (b), show that the power dissipation in the strip is equal to the rate at which gravity is doing work on the strip.

END OF EXAMINATION

STOP

END OF SECTION II, ELECTRICITY AND MAGNETISM

IF YOU FINISH BEFORE TIME IS CALLED, YOU MAY CHECK YOUR WORK ON SECTION II, ELECTRICITY AND MAGNETISM, ONLY. DO NOT TURN TO ANY OTHER TEST MATERIALS.

16

AP Physics C Practice Exam 1: Answers and Explanations

SECTION I, MECHANICS

1. **E** Let y denote the total distance that the rock falls and let T denote the total time of the fall . Then $\frac{1}{2}y = \frac{1}{2}gt_1^2$ and $y = \frac{1}{2}gT^2$, so

$$\frac{t_2}{t_1} = \frac{T - t_1}{t_1} = \frac{T}{t_1} - 1 = \frac{\sqrt{2y/g}}{\sqrt{y/g}} - 1 = \sqrt{2} - 1$$

2. **A** The work-energy theorem says that the work done by friction is equal to the change in kinetic energy of the box . Therefore,

$$-\mu mgd = -\frac{1}{2}mv_0^2 \Rightarrow d = \frac{v_0^2}{2\mu g}$$

3. **A** Integrating $a(t)$ gives $v(t)$:

$$v(t) = \int a(t)dt = \int 2t\ dt = t^2 + v_0$$

Since $v_0 = 0$, we have $v(t) = t^2$. Now integrating $v(t)$ from $t = 0$ to $t = 3$ s gives the object's displacement during this time interval:

$$s = \int_0^3 v(t)\ dt = \int_0^3 t^2 dt = \frac{1}{3}t^3 \Big|_0^3 = 9m$$

4. **E** Angular momentum is conserved when no net external torque is applied (Item III) . However, a body can experience a net nonzero torque even when the net external force is zero (which is the condition that guarantees conservation of linear momentum), so neither Item I nor Item II necessarily ensure conservation of angular momentum.

5. **C** The figure below shows that $F_T \sin\theta = mg$ and $F_T \cos\theta = mv^2/R = m\omega^2 R$:

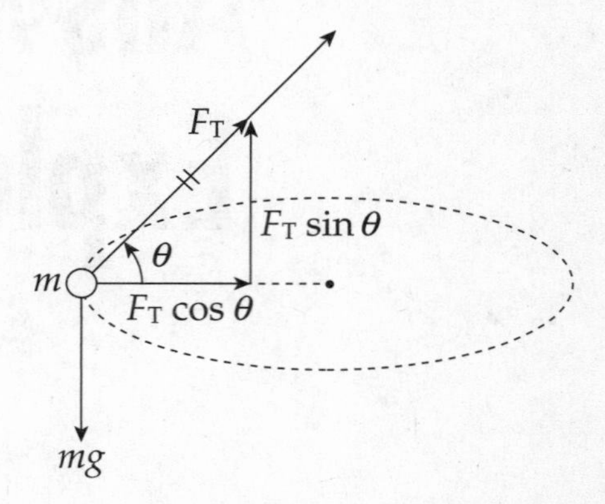

We can eliminate θ from these equations by remembering that $\sin^2\theta + \cos^2\theta$ is always equal to 1:

$$\left(\frac{mg}{F_T}\right)^2 + \left(\frac{m\omega^2 R}{F_T}\right)^2 = 1 \Rightarrow \frac{m^2g^2 + m^2\omega^4R^2}{F_T^2} = 1 \Rightarrow F_T = m\sqrt{\omega^4R^2 + g^2}$$

6. **A** Because the initial height of the object is comparable to the radius of the Moon, we cannot simply use mgh for its initial potential energy. Instead, we must use the more general expression $U = -GMm/r$, where r is the distance from the center of the Moon. Notice that since $h = R$, the object's initial distance from the Moon's center is $h + R = R + R = 2R$. Conservation of Energy then gives

$$K_i + U_i = K_f + U_f$$
$$0 - \frac{GMm}{2R} = \frac{1}{2}mv^2 - \frac{GMm}{R}$$
$$\frac{GMm}{2R} = \frac{1}{2}mv^2$$
$$v = \sqrt{\frac{GM}{R}}$$

7. **E** At equilibrium, the net force on the block is zero (by definition), so $ky = mg$, which gives $y = mg/k$. If m doubles, then so does y. This eliminates choices A, B, and C. Since $T = 2\pi\sqrt{m/k}$, T is proportional to $\sqrt{m}$. Therefore, if m increases by a factor of 2, then T increases by a factor of $\sqrt{2}$.

8. **D** Let F_T be the tension in each string. Then $F_{net} = ma$ becomes $mg - 2F_T = ma$. Also, the total torque exerted by the tension forces on the cylinder is $2rF_T$, so $\tau_{net} = I\alpha$ becomes $2rF_T = (\frac{1}{2}mr^2)\alpha$. Because the cylinder doesn't slip, $\alpha = a/r$, so $2F_T = \frac{1}{2}ma$. These equations can be combined to give $mg - \frac{1}{2}ma = ma$, which implies $mg = \frac{3}{2}ma$. Therefore, $a = \frac{2}{3}g$.

9. **B** The cylinder slides across the surface with acceleration $a = F/m$ until time $t = T$, when a drops to zero (because F becomes zero). Therefore, from time $t = 0$ to $t = T$, the velocity is steadily increasing (because the acceleration is a positive constant), but, at $t = T$, the velocity remains constant. This is illustrated in graph (B).

10. **D** The speed of the object being lifted is $v = (20\text{ m})/(5\text{ s}) = 4\text{ m/s}$. Since $P = Fv$, we find that $F = P/v = 10\text{ kW}/(4\text{ m/s}) = 2.5\text{ kN} = 2500\text{ N}$. Since the object is being lifted with constant velocity, the net force on it must be zero, so F must be equal to the object's weight, mg. This gives $m = F/g = 2500\text{ N}/(10\text{ N/kg}) = 250\text{ kg}$.

11. **A** Let m and M denote the mass of the satellite and the earth, respectively. The work required to lift the satellite to height h above the surface of the earth is

$$W = \Delta U = -\frac{GMm}{R+h} - \left(-\frac{GMm}{R}\right) = \frac{GMmh}{R(R+h)}$$

The earth's gravitational pull provides the necessary centripetal force on the satellite, so

$$\frac{GMm}{(R+h)^2} = \frac{mv^2}{R+h} \Rightarrow mv^2 = \frac{GMm}{R+h} \Rightarrow K = \frac{GMm}{2(R+h)}$$

Because we're told that $W = K$, we find that

$$\frac{GMmh}{R(R+h)} = \frac{GMm}{2(R+h)} \Rightarrow \frac{h}{R} = \frac{1}{2} \Rightarrow h = \frac{R}{2}$$

12. **D** First note that if L is the total length of the bar, then the distance of the block from Support 1 is $\frac{1}{3}L$ and its distance from Support 2 is $\frac{2}{3}L$. Let P_1 denote the point at which Support 1 touches the bar. With respect to P_1, the upward force exerted by Support 1, $\mathbf{F}_1$, produces no torque, but the upward force exerted by Support 2, $\mathbf{F}_2$, does. Since the net torque must be zero if the system is in static equilibrium, the counterclockwise torque of $\mathbf{F}_2$ must balance the total clockwise torque produced by the weight of the block and of the bar (which acts at the bar's midpoint):

$$L \cdot F_2 = \frac{1}{3}L \cdot \frac{1}{2}Mg + \frac{1}{2}L \cdot Mg \Rightarrow F_2 = \frac{2}{3}Mg$$

Now, with respect to P_2, the point at which Support 2 touches the bar,

$$L \cdot F_1 = \frac{2}{3}L \cdot \frac{1}{2}Mg + \frac{1}{2}L \cdot Mg \Rightarrow F_1 = \frac{5}{6}Mg$$

These results give $F_2 / F_1 = \frac{2}{3}Mg / \frac{5}{6}Mg = \frac{4}{5}$.

13. **C** Because the ball rebounded to the same height from which it was dropped, its takeoff speed from the floor must be the same as the impact speed. Calling *up* the positive direction, the change in linear momentum of the ball is $p_f - p_i = mv - (-mv) = 2mv$, where $v = \sqrt{2gh}$ (this last equation comes from the equation $\frac{1}{2}mv^2 = mgh$). The impulse–momentum theorem, $J = \Delta p$, now gives

$$F_{avg}\Delta t = 2m\sqrt{2gh} \Rightarrow F_{avg} = \frac{2m\sqrt{2gh}}{\Delta t} = \frac{2(0.08 \text{ kg})\sqrt{2(10 \text{ m/s}^2)(3.2 \text{ m})}}{0.04 \text{ s}} = 32 \text{ N}$$

14. **D** Using the Big Five equation $\Delta\theta = \omega_0 t + \frac{1}{2}\alpha t^2$, we find that

$$\Delta\theta = \frac{1}{2}(2 \text{ rad/s}^2)(4.0 \text{ s})^2 = 16 \text{ rad}$$

Therefore, $\Delta s = r\Delta\theta = (01 \text{ m})(16 \text{ rad}) = 1.6 \text{ m}$.

15. **A** The ball's initial velocity from the slide is horizontal, so $v_{0y} = 0$, which implies that $\Delta y = -\frac{1}{2}gt^2$. Since $\Delta y = -R$,

$$-R = -\frac{1}{2}gt^2 \Rightarrow t = \sqrt{\frac{2R}{g}}$$

The (horizontal) speed with which the ball leaves the slide is found from the equation $mgR = \frac{1}{2}mv^2$, which gives $v = \sqrt{2gR}$. Therefore,

$$\Delta x = v_{0x}t = \sqrt{2gR} \cdot \sqrt{\frac{2R}{g}} = 2R$$

Since the ball travels a horizontal distance of $2R$ from the end of the slide, the total horizontal distance from the ball's starting point is $R + 2R = 3R$.

16. **A** From the diagram below,

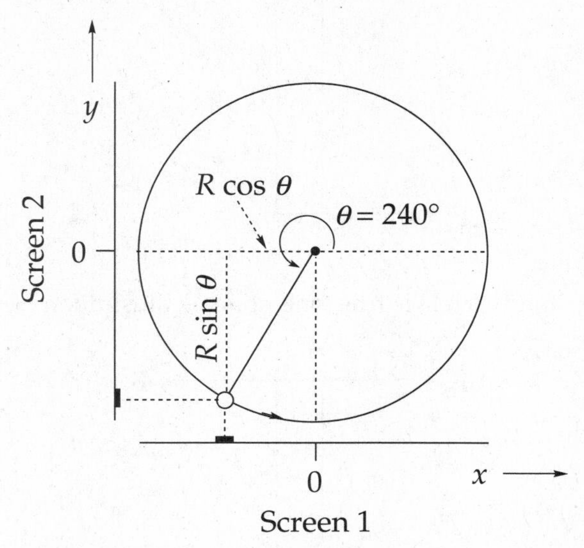

if $R \cos \theta = -(0.5)R$, then $\theta = 240°$. Therefore, $y = R \sin \theta = R \sin 240° = -(0.866)R$. Also, it is clear that the particle's subsequent motion will cause the shadow on Screen 2 to continue moving in the $-y$ direction.

17. **C** The center of mass of the system is at a distance of

$$y_{cm} = \frac{(m)(0) + (2m)(L)}{m + 2m} = \frac{2}{3}L$$

below the mass m. With respect to this point, the clockwise torque produced by the force **F** has magnitude

$$\tau = rF = \left(\frac{2}{3}L - \frac{1}{4}L\right)F = \frac{5}{12}LF$$

Since the rotational inertia of the system about its center of mass is

$$I = \Sigma m_i r_i^2 = m\left(\frac{2}{3}L\right)^2 + (2m)\left(\frac{1}{3}L\right)^2 = \frac{2}{3}mL^2$$

the equation $\tau = I\alpha$ becomes

$$\frac{5}{12}LF = \left(\frac{2}{3}mL^2\right)\alpha \Rightarrow \alpha = \frac{\frac{5}{8}F}{mL}$$

18. **C** By Newton's Third Law, both vehicles experience the same magnitude of force and, therefore, the same impulse; so Statement I is false . Invoking Newton's Second Law, in the form *impulse = change in momentum*, we see that Statement II is therefore also false. However, since the car has a smaller mass than the truck, its acceleration will be greater in magnitude than that of the truck, so Statement III is true.

19. **D** Since $\mathbf{F}_{net} = \mathbf{F}_1 + \mathbf{F}_2 = \mathbf{0}$, the bar cannot accelerate translationally, so C is false. The net torque does *not* need to be zero, as the following diagram shows (eliminating choices A and B):

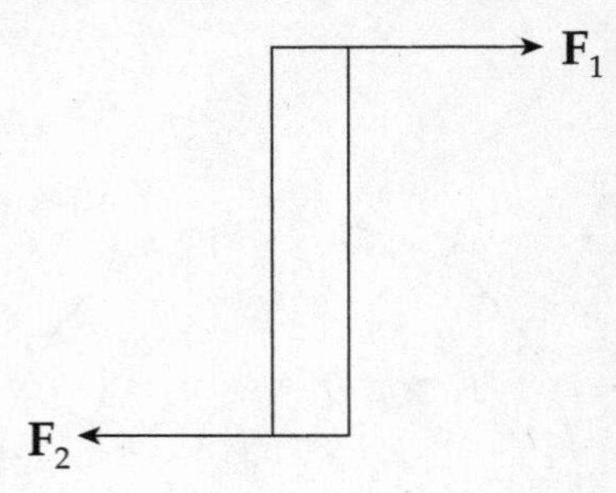

However, since $\mathbf{F}_2 = -\mathbf{F}_1$, choice D is true; one possible illustration of this is given below:

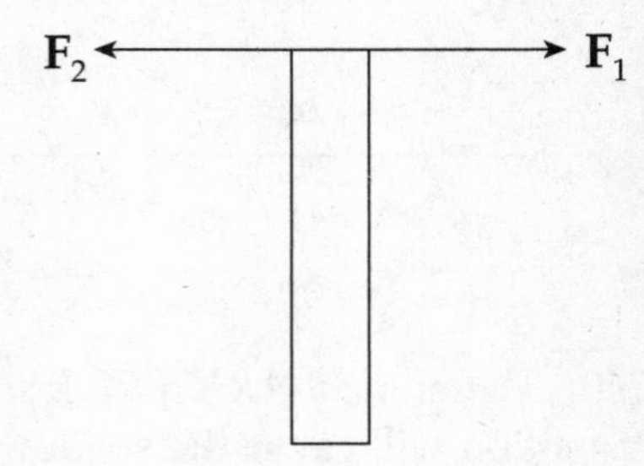

20. **B** The value of g near the surface of a planet depends on the planet's mass and radius:

$$mg = \frac{GMm}{r^2} \Rightarrow g = \frac{GM}{r^2}$$

Therefore, calling this Planet X, we find that

$$g_X = \frac{GM_X}{r_X^2} = \frac{G(2M_{\text{Earth}})}{(2r_{\text{Earth}})^2} = \frac{1}{2}\frac{GM_{\text{Earth}}}{r_{\text{Earth}}^2} = \frac{1}{2}g_{\text{Earth}}$$

Since g is half as much on the planet as it is on Earth, the astronaut's weight (mg) will be half as much as well.

21. **C** The work done by the force $F(x)$ is equal to the area under the given curve. The region under the graph can be broken into two triangles and a rectangle, and it is then easy to figure out that the total area is 50 N·m. Since work is equal to change in kinetic energy (and $v_i = 0$),

$$W = \Delta K = \frac{1}{2}m(v_f^2 - v_i^2) \Rightarrow v_f = \sqrt{\frac{2W}{m}} = \sqrt{\frac{2(50\text{ N}\cdot\text{m})}{1.0\text{ kg}}} = 10\text{ m/s}$$

22. **C** Since no external torques act on the star as it collapses (just like a skater when she pulls in her arms), angular momentum, $I\omega$, is conserved, and the star's rotational speed increases. The moment of inertia of a sphere of mass m and radius r is given by the equation $I = kmr^2$ (with $k = \frac{2}{5}$, but the actual value is irrelevant), so we have:

$$\omega_f = \frac{I_i}{I_f}\omega_i = \frac{kmr_i^2}{kmr_f^2}\omega_i = \frac{r_i^2}{r_f^2}\omega_i = \frac{r_i^2}{\left(\frac{1}{16}r_i\right)^2}\omega_i = 16^2\omega_i$$

Therefore,

$$\frac{K_f}{K_i} = \frac{\frac{1}{2}I_f\omega_f^2}{\frac{1}{2}I_i\omega_i^2} = \frac{\frac{1}{2}kmr_f^2\omega_f^2}{\frac{1}{2}kmr_i^2\omega_i^2} = \frac{r_f^2\omega_f^2}{r_i^2\omega_i^2} = \frac{\left(\frac{1}{16}r_i\right)^2\left((16^2\omega_i)\right)^2}{r_i^2\omega_i^2} = 16^2$$

23. **C** The projectile's horizontal speed is $v_0 \cos 60° = v_0 = 20$ m/s, so it reaches the wall in 1 s.

24. **C** Impulse, *J*, is defined to be force × time, so

$$[J] = [F][t] = \text{N}\cdot\text{s} = \frac{\text{kg}\cdot\text{m}}{\text{s}^2}\cdot s = \frac{\text{kg}\cdot\text{m}}{\text{s}} = \frac{\text{ML}}{\text{T}}$$

25. **D** The diagram given with the question shows that after the clay balls collide, they move in the $-y$ direction, which means that the horizontal components of their linear momenta canceled. In other words, $\mathbf{p}_{1x} + \mathbf{p}_{2x} = \mathbf{0}$:

$$m_1 v_{1x} + m_2 v_{2x} = \mathbf{0}$$

$$(m_1 v_1 \sin 45°)\hat{\mathbf{i}} + (m_2 v_2 \sin 60°)(-\hat{\mathbf{i}}) = \mathbf{0}$$

$$v_2 = \frac{m_1 v_1 \sin 45°}{m_2 \sin 60°}$$

$$= \frac{2 \sin 45°}{\sin 60°} v_1$$

26. **B** The horizontal position of the blocks can be given by the equation

$$x = A \sin(\omega t + \phi_0)$$

where A is the amplitude, ω is the angular frequency, and ϕ_0 is the initial phase. Differentiating this twice gives the acceleration:

$$a = \frac{d^2x}{dt^2} = A\omega^2 \sin(\omega t + \phi_0) \Rightarrow a_{\text{max}} = A\omega^2$$

This means that the maximum force on block m is $F_{\text{max}} = ma_{\text{max}} = mA\omega^2$. The static friction force must be able to provide this much force; otherwise, the block will slip. Therefore,

$$F_{\text{static, max}} \geq mA\omega^2$$

$$\mu mg \geq mA\omega^2$$

$$A \leq \frac{\mu mg}{m\omega^2} = \frac{\mu g}{(2\pi f)^2} = \frac{(0.80)(10 \text{ m/s}^2)}{4\pi^2(2.0 \text{ Hz})^2} \approx 0.05 \text{ m}$$

27. **D** The ball strikes the floor with speed $v_1 = \sqrt{2gH_1}$, and it leaves the floor with speed $v_2 = \sqrt{2gH_2}$. Therefore,

$$e = \frac{v_2}{v_1} = \frac{\sqrt{2gH_2}}{\sqrt{2gH_1}} = \sqrt{\frac{H_2}{H_1}}$$

28. **D** The center of rotation is the center of mass of the plate, which is at the geometric center of the square because the plate is homogeneous. Since the line of action of the force coincides with one of the sides of the square, the lever arm of the force, ℓ, is simply equal to $\frac{1}{2}$ s. Therefore,

$$\tau = \ell F = \frac{1}{2}sF = (0.20 \text{ m})(10 \text{ N}) = 2.0 \text{ N·m}$$

29. **A** Since gravity is a conservative force, the actual path taken is irrelevant. If the block rises a vertical distance of $z = 3$ m, then the work done by gravity is

$$W = -mgz = -(2.0 \text{ kg})(10 \text{ N/kg})(3 \text{ m}) = -60 \text{ J}$$

30. **B** The work done by the friction force as the block slides from $x = A$ to $x = 0$ is

$$W_{\text{friction}} = \int_A^0 bx\,dx = \frac{1}{2}bx^2\Big|_A^0 = -\frac{1}{2}bA^2$$

Now use Conservation of Energy, using $U(x) = \frac{1}{2}kx^2$:

$$K_i + U_i + W_{\text{friction}} = K_f + U_f$$

$$0 + \frac{1}{2}kA^2 - \frac{1}{2}bA^2 = \frac{1}{2}mv^2 + 0$$

$$v = A\sqrt{\frac{k-b}{m}}$$

31. **D** Consider an infinitesimal slice of width dx at position x; its mass is $dm = \lambda\, dx = kx\, dx$. Then, by definition of rotational inertia,

$$I = \int r^2 dm = \int_0^L x^2(kx\,dx) = \frac{1}{4}kx^4\Big|_0^L = \frac{1}{4}kL^4$$

Because the total mass of the rod is

$$M = \int dm = \int_0^L kx\,dx = \frac{1}{2}kx^2\Big|_0^L = \frac{1}{2}kL^2$$

we see that $I = \frac{1}{2}ML^2$.

32. **E** Points of equilibrium occur when dU/dx is equal to zero:

$$\frac{dU}{dx}=0 \Rightarrow 3(x-2)^2-12=0 \Rightarrow (x-2)^2=4 \Rightarrow x=0,4$$

Points of *stable* equilibrium occur when $dU/dx = 0$ and d^2U/dx^2 is positive (because then the equilibrium point is a relative minimum). Since

$$\frac{d^2U}{dx^2}=6(x-2)$$

the point $x = 4$ is a point of stable equilibrium ($x = 0$ is unstable).

33. **E** The range of a projectile is given by the equation

$$R=\frac{2v_0^2 \sin\theta_0 \cos\theta_0}{g}$$

and the maximum height attained by the projectile is

$$H=\frac{v_0^2 \sin^2\theta_0}{2g}$$

In order for R to equal H, it must be true that

$$\frac{2v_0^2 \sin\theta_0 \cos\theta_0}{g}=\frac{v_0^2 \sin^2\theta_0}{2g} \Rightarrow 2\cos\theta_0=\frac{\sin\theta_0}{2} \Rightarrow \tan\theta_0=4 \Rightarrow \theta_0=\tan^{-1}4$$

34. **B** Differentiate the equation $K = \frac{1}{2}mv^2$ with respect to time:

$$\frac{dK}{dt}=\frac{1}{2}m\cdot 2v\frac{dv}{dt}=mv\frac{dv}{dt}=v\cdot m\frac{dv}{dt}=v\cdot ma=vF$$

This tells us that

$$F=\frac{dK/dt}{v}=\frac{-6.0\ \text{J/s}}{3.0\ \text{m/s}}=-2\text{N}$$

35. **E** The maximum net force on the object occurs when all three forces act in the same direction, giving $F_{net} = 3F = 3(4\ \text{N}) = 12\ \text{N}$, and a resulting acceleration of $a = F_{net}/m = (12\ \text{N})/(2\ \text{kg}) = 6\ \text{m/s}^2$. These three forces could not give the object an acceleration greater than this.

SECTION I, ELECTRICITY AND MAGNETISM

36. **C** The proximity of the charged sphere will induce negative charge to move to the side of the uncharged sphere closer to the charged sphere. Since the induced negative charge is closer than the induced positive charge to the charged sphere, there will be a net electrostatic attraction between the spheres.

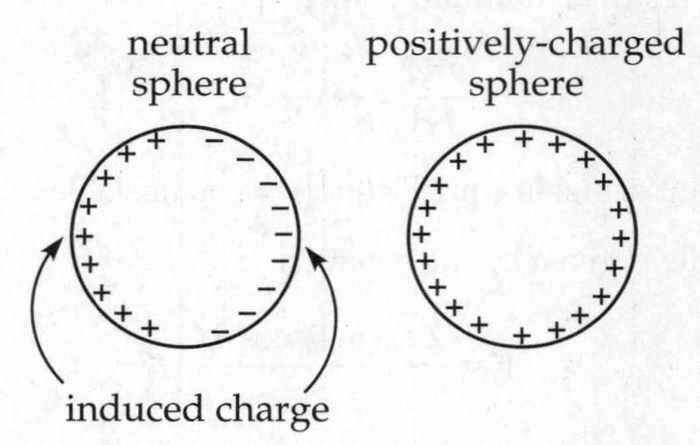

37. **E** The capacitance of a parallel-plate capacitor is $C = K\varepsilon_0 A/d$, where K is the dielectric constant, A is the area of each plate, and d is their separation distance. Decreasing d will cause C to increase.

38. **C** When the particle enters the magnetic field, the magnetic force provides the centripetal force to cause the particle to execute uniform circular motion:

$$|q|vB = \frac{mv^2}{r} \Rightarrow r = \frac{mv}{|q|B}$$

Since v and B are the same for all the particles, the largest r is found by maximizing the ratio $m/|q|$. The value of $m/|q|$ for an alpha particle is about twice that for a proton and thousands of times greater than that of an electron or positron.

39. **A** Electric field lines are always perpendicular to the surface of a conductor, eliminating choices B and E. Excess charge on a conductor always resides on the surface (eliminating choice D), and there is a greater density of charge at points where the radius of curvature is smaller (eliminating choice C).

40. **B** The resistance of a wire made of a material with resistivity ρ and with length L and cross-sectional area A is given by the equation $R = \rho L/A$. Since Wire B has the greatest length and smallest cross-sectional area, it has the greatest resistance.

41. **E** From the equation $F = qvB$, we see that 1 tesla is equal to 1 N·s/(C·m). Since

$$\frac{1\text{N}\cdot\text{s}}{\text{C}\cdot\text{m}} = \frac{1\text{N}}{\text{A}\cdot\text{m}} = \frac{1\text{J}}{\text{A}\cdot\text{m}^2}$$

choices A and C are eliminated.

Furthermore,

$$\frac{1\ \text{N}\cdot\text{s}}{\text{C}\cdot\text{m}} = \frac{1\ \text{kg}\cdot\text{m/s}}{\text{C}\cdot\text{m}} = \frac{1\ \text{kg}}{\text{C}\cdot\text{s}}$$

eliminating choice B. Finally, since

$$\frac{1\ \text{J}}{\text{A}\cdot\text{m}^2} = \frac{1\ \text{C}\cdot\text{V}}{(\text{C}/\text{s})\cdot\text{m}^2} = \frac{1\ \text{V}\cdot\text{s}}{\text{m}^2}$$

choice D is also eliminated.

42. **C** Gauss's Law states that the total electric flux through a closed surface is equal to $(1/\varepsilon_0)$ times the net charge enclosed by the surface. Both the cube and the sphere contain the same net charge, so Φ_C must be equal to Φ_S.

43. **C** The electric field between the plates is uniform and equal to V/x, so the magnitude of the electric force on the charge is $F_E = qE = qV/x$. From the following free-body diagram

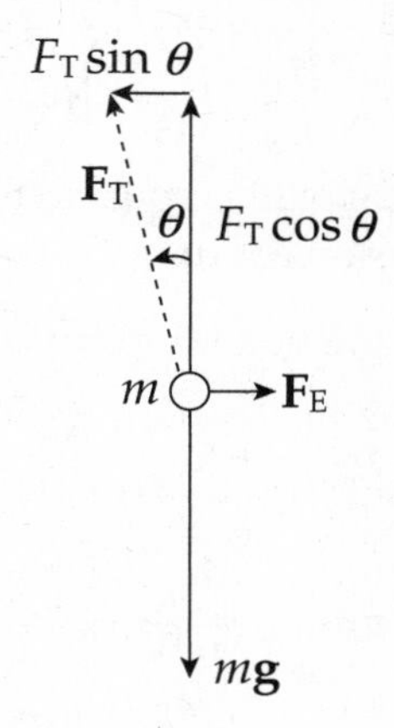

we see that

$$F_T \sin\theta = F_E = qE = \frac{qV}{x} \quad \text{and} \quad F_T \cos\theta = mg$$

Dividing the first equation by the second one, we find that

$$\frac{F_T \sin\theta}{F_T \cos\theta} = \frac{qV/x}{mg} \Rightarrow \tan\theta = \frac{qV}{mgx} \Rightarrow \theta = \tan^{-1}\frac{qV}{mgx}$$

44. **A** Because charge is quantized, a negatively-charged object must have a charge q, equal to whole number, n, times the elementary charge, $-e$. Therefore,

$$q = n(-e) \Rightarrow n = \frac{q}{-e} = \frac{-1\text{C}}{-1.60\times10^{-19}} = \frac{1}{-1.60}\times10^{19} = 6.25\times10^{18}$$

45. **E** Resistors R_2, R_3, and R_4 are in parallel, so their equivalent resistance, $R_{2\text{-}3\text{-}4}$, satisfies

$$\frac{1}{R_{2-3-4}} = \frac{1}{R_2} + \frac{1}{R_3} + \frac{1}{R_4} = \frac{3}{200\ \Omega} \Rightarrow R_{2-3-4} = \frac{200}{3}\Omega$$

Since this is in series with R_1, the total resistance in the circuit is

$$R = R_1 + R_{2-3-4} = \frac{800}{3}\Omega$$

The current provided by the source must therefore be

$$I = \frac{\varepsilon}{R} = \frac{24\ V}{\frac{800}{3}\ \Omega} = \frac{9}{100} A = \frac{90}{100} A = 90\text{mA}$$

46. **E** The current through R_4 is $\frac{1}{3}$ the current through R_1, and $R_4 = R_1$, so

$$\frac{P_1}{P_4} = \frac{I_1^2 R_1}{I_4^2 R_4} = \frac{I_1^2 R_1}{\left(\frac{1}{3} I_1\right) R_1} = 9$$

47. **D** Because the time constant for an RC circuit is equal to the product of resistance and capacitance, $\tau = RC$, this product has the dimensions of time.

48. **B** Apply Faraday's Law of Electromagnetic Induction:

$$\oint E \cdot d\ell = -\frac{d\Phi_B}{dt}$$

$$E(2\pi r) = -\frac{d}{dt}(BA)$$

$$= -A\frac{dB}{dt}$$

$$= \pi r^2 \cdot \frac{d}{dt}(2t - 2t^2)$$

$$= \pi r^2 (2 - 4t)$$

$$E = -r(1 - 2t)$$

Since $r = 1$ m, the value of E at $t = 1$ s is $E = 1$ N/C.

49. **A** The magnetic force, $\mathbf{F}_B$, on the top horizontal wire of the loop must be directed upward and have magnitude mg. (The magnetic forces on the vertical portions of the wire will cancel.) By the right-hand rule, the current in the top horizontal wire must be directed to the right—because **B** is directed into the plane of the page—in order for $\mathbf{F}_B$ to be directed upward; therefore, the current in the loop must be clockwise (eliminating choices B and D). Since $F_B = IxB$, we must have

$$IxB = mg \Rightarrow I = \frac{mg}{xB}$$

50. **D** Construct a cylindrical Gaussian surface of radius $r < R$ and length $\ell < L$ within the given cylinder. By symmetry, the electric field must be radial, so the electric flux through the lids of the Gaussian cylinder is zero. The lateral surface area of the Gaussian cylinder is $2\pi r\ell$, so by Gauss's Law,

$$\Phi_E = \frac{Q_{\text{enclosed}}}{\varepsilon_0} \Rightarrow E(2\pi r\ell) = \frac{2\pi\ell r^3}{\varepsilon_0} \Rightarrow E = \frac{r^2}{\varepsilon_0}$$

51. **A** Call the top wire (the one carrying a current I to the right) Wire 1, and call the bottom wire (carrying a current $2I$ to the left) Wire 2. Then in the region between the wires, the individual magnetic field vectors due to the wires are both directed into the plane of the page, so they could not cancel in this region. Therefore, the total magnetic field could not be zero at either Point 2 or Point 3. This eliminates choices B, C, D, and E, so the answer must be A. Since the magnetic field created by a current-carrying wire is proportional to the current and inversely proportional to the distance from the wire, the fact that Point 1 is in a region where the individual magnetic field vectors created by the wires point in opposite directions and that Point 1 is twice as far from Wire 2 as from Wire 1 implies that the total magnetic field there will be zero.

52. **A** If the mass of the $+q$ charge is m, then its acceleration is

$$a = \frac{F_E}{m} = \frac{1}{4\pi\varepsilon_0}\frac{Qq}{mr^2}$$

The graph in A best depicts an inverse-square relationship between a and r.

53. **C** Once the switch is closed, the capacitors are in parallel, which means the voltage across C_1 must equal the voltage across C_2. Since $V = Q/C$,

$$\frac{Q_2'}{C_2} = \frac{Q_1'}{C_1} \Rightarrow Q_2' = \frac{C_2}{C_1}Q_1' = \frac{6\mu F}{3\mu F}Q_1' = 2Q_1'$$

The total charge, 24 μC + 12 μC = 36 μC, must be redistributed so that $Q_2' = 2Q_1'$. Therefore, we see that $Q_2' = 24$ μC (and $Q_1' = 12$ μC).

54. **E** Because **v** is parallel to **B**, the charges in the bar feel no magnetic force, so there will be no movement of charge in the bar and no motional emf.

55. **E** Along the line joining the two charges of an electric dipole, the individual electric field vectors point in the same direction, so they add constructively. At the point equidistant from both charges, the total electric field vector has magnitude

$$E_+ + E_- = 2E = 2\frac{kQ}{\left(\frac{1}{2}d\right)^2} = \frac{8kQ}{d^2}$$

$$= \frac{8(9\times10^9\,\text{N}\cdot\text{m}^2/\text{C}^2)(4.0\times10^{-9}\,\text{C})}{(2.0\times10^{-2}\,\text{m}^2)^2}$$

$$= 72\times10^4\ \text{N/C}$$

$$= 7.2\times10^5\ \text{N/C (or V/m)}$$

56. **E** Imagine placing equal but opposite charges on the spheres, say, $+Q$ on the inner sphere and $-Q$ on the outer sphere. Then the electric field between the spheres is radial and depends only on $+Q$, and its strength is

$$E = \frac{1}{4\pi\varepsilon_0}\frac{Q}{r^2}$$

Therefore, the potential difference between the spheres is

$$V = \int_R^{2R} \mathbf{E} \cdot dr = \frac{Q}{4\pi\varepsilon_0}\int_R^{2R}\frac{1}{r^2}dr = \frac{Q}{4\pi\varepsilon_0}\left[-\frac{1}{r}\right]_R^{2R} = \frac{Q}{4\pi\varepsilon_0}\left(\frac{1}{R}-\frac{1}{2R}\right) = \frac{Q}{8\pi\varepsilon_0 R}$$

Now, by definition, $C = Q/V$, so

$$C = \frac{Q}{V} = Q \cdot \frac{1}{V} = Q \cdot \frac{8\pi\varepsilon_0 R}{Q} = 8\pi\varepsilon_0 R$$

57. **C** Since the magnetic force is always perpendicular to the object's velocity, it does zero work on any charged particle. Zero work means zero change in kinetic energy, so the speed remains the same. Remember: The magnetic force can only change the direction of a charged particle's velocity, not its speed.

58. **B** The presence of the inductor in the rightmost branch effectively removes that branch from the circuit at time $t = 0$ (the inductor produces a large back emf when the switch is closed and the current jumps abruptly). Initially, then, current only flows in the loop containing the resistors r and R. Since their total resistance is $r + R$, the initial current is $\mathcal{E}/(r + R)$.

59. **E** After a long time, the current through the branch containing the inductor increases to its maximum value. With all three resistors in play, the total resistance is $R_{eq} = r + \frac{R}{2}$, since $\frac{R}{2}$ is the resistance of the parallel branch.

The current provided by the source, $\frac{\mathcal{E}}{R_{eq}}$, splits at the junction leading to the parallel combination. The amount which flows through the inductors is half the total.

$$I = \frac{1}{2} \bullet \frac{\mathcal{E}}{\left(r + \frac{R}{2}\right)} = \frac{\mathcal{E}}{2r + R}$$

so the energy stored in the inductor is

$$U_L = \frac{LI^2}{2} = \frac{L}{2}\left[\frac{1}{2} \bullet \frac{\mathcal{E}}{(2r+R)}\right]^2 = \frac{L\mathcal{E}^2}{8(2r+R)^2}$$

60. **A** Once the switch is opened, the resistor r is cut off from the circuit, so no current passes through it. (Current *will* flow around the loop containing the resistors labeled R, gradually decreasing until the energy stored in the inductor is exhausted.)

61. **C** If a conducting sphere contains a charge of $+q$ within an inner cavity, a charge of $-q$ will move to the wall of the cavity to "guard" the interior of the sphere from an electrostatic field, regardless of the size, shape, or location of the cavity. As a result, a charge of $+q$ is left on the exterior of the sphere (and it will be uniform). So, at points outside the sphere, the sphere behaves as if this charge $+q$ were concentrated at its center, so the electric field outside is simply kQ/r^2. Since points X and Y are at the same distance from the center of the sphere, the electric field strength at Y will be the same as at X.

62. **B** First determine the electric potential at Point P. Realize that Point P is 5 meters away from each charge because it is a 3-4-5 right triangle.

$$V = \sum \frac{kQ}{r}$$

$$V_P = \frac{kQ}{5} + \frac{kQ}{5} = \frac{2kQ}{5}$$

The work done by the electric field will be positive because a $-Q$ charge would be attracted to point P by the two positive Q charges. Also the work done by the force is the negative of the change in potential energy.

$$\Delta V = \frac{\Delta U}{q}$$

$$V_P - V_\infty = \frac{\Delta U}{-Q}$$

$$\frac{2kQ}{5} - 0 = \frac{\Delta U}{-Q}$$

$$\Delta U = \frac{-2kQ^2}{5}$$

$$W = -\Delta U = \frac{2kQ^2}{5}$$

63. **E** The current in an L–R series circuit—in which initially no current flows—increases with time t according to the equation

$$I(t) = I_{max}(1 - e^{-t/\tau})$$

where $I_{max} = \mathcal{E}/R$ and τ is the inductive time constant, L/R. The time t at which $I = (99\%)I_{max}$ is found as follows:

$$1 - e^{-t/\tau} = \frac{99}{100}$$

$$e^{-t/\tau} = \frac{1}{100}$$

$$-t/\tau = \ln\frac{1}{100}$$

$$t = (-\ln\frac{1}{100})\tau = (\ln 100)\tau = (\ln 100)L/R$$

64. **B** Since the total resistance in this series circuit is $r + R$, the current through R is $I = \mathcal{E}/(r + R)$. Therefore, the power dissipated by R is

$$P = \frac{\mathcal{E}^2 R}{(r+R)^2}$$

To find the value of R at which P is maximized, we set dP/dR equal to zero and solve:

$$\frac{dP}{dR} = \frac{(r+R)^2\mathcal{E}^2 - \mathcal{E}^2 R \cdot 2(r+R)}{(r+R)^4} = 0$$

$$(r+R)^2\mathcal{E}^2 - \mathcal{E}^2 R \cdot 2(r+R) = 0$$

$$\mathcal{E}^2(r+R)[(r+R) - 2R] = 0$$

$$R = r$$

Therefore, maximum power dissipation occurs when R equals r. For this value of R, the power dissipated by R is

$$P_{\max} = \frac{\mathcal{E}^2 r}{(r+r)^2} = \frac{\mathcal{E}^2}{4r}$$

65. **D** Since the magnetic flux out of the page is decreasing, the induced current will oppose this change (as always), attempting to create more magnetic flux out of the page. In order to do this, the current must circulate in the counterclockwise direction (remember the right-hand rule). As B decreases *uniformly* (that is, while dB/dt is negative and *constant*), the induced emf,

$$\mathcal{E} = \frac{d\Phi_B}{dt} = -A\frac{dB}{dt}$$

is nonzero and constant, which implies that the induced current, $I = \mathcal{E}/R$, is also nonzero and constant.

66. **D** Treat the configuration as three capacitors in parallel, each with a distance between the plates of $\frac{d}{3}$. The capacitance of the two vacuum capacitors will be $\frac{\varepsilon_o A}{\left(\frac{d}{3}\right)} = \frac{3\varepsilon_o A}{d}$, and the capacitance of the dielectric capacitor will be that value times K.

Now solve for the total capacitance for the three capacitors in parallel.

$$\frac{1}{C_{eq}} = \frac{1}{C_1} + \frac{1}{C_2} + \frac{1}{C_3}$$

$$\frac{1}{C_{eq}} = \frac{1}{\left(\frac{3\varepsilon_o A}{d}\right)} + \frac{1}{\left(\frac{3K\varepsilon_o A}{d}\right)} + \frac{1}{\left(\frac{3\varepsilon_o A}{d}\right)}$$

$$\frac{1}{C_{eq}} = \frac{d}{3\varepsilon_o A} + \frac{d}{3K\varepsilon_o A} + \frac{d}{3\varepsilon_o A}$$

$$\frac{1}{C_{eq}} = \frac{2Kd + d}{3K\varepsilon_o A}$$

$$C_{eq} = \frac{3K\varepsilon_o A}{d(2K+1)}$$

67. **A** The potential is zero at the point midway between the charges, but nowhere is the electric field equal to zero (except at infinity).

68. **D** Ampere's Law states that

$$\oint \mathbf{B} \cdot d\ell = \mu_0 I_{\text{net through loop}}$$

Because a total current of $2I + 4I = 6I$ passes through the interior of the loop in one direction, and a total current of $I + 3I = 4I$ passes through in the opposite direction, the net current passing through the loop is $6I - 4I = 2I$. Therefore, the absolute value of the integral of **B** around the loop WXYZ is equal to $\mu_0(2I)$.

69. **A** Statement I is true: Since the line ℓ is the perpendicular bisector of the line segment joining the two charges, every point on ℓ is equidistant from the two charges. Therefore, the electric potential at *any* point on this line is

$$k(+Q)/r + k(-Q)/r = 0$$

so ℓ is an equipotential. By definition, then, a charge moving along ℓ would experience no change in electrical potential energy. Statement II is false: The individual field vectors created by each source charge do not point in opposite directions at any point on ℓ, so at no point on ℓ would the total electric field be zero, immediately implying that at no point on ℓ would the electrostatic force on a charge be zero. Statement III is also false: The equipotentials in the plane of the page are not ellipses with the source charges as foci. In fact, as we showed in the discussion of Statement I, the line ℓ itself is an equipotential (and it's not an ellipse!).

70. **A** First, note that the distance between $-q$ and each charge $+Q$ is $\frac{1}{\sqrt{2}}L$. Now, refer to the following diagram, where we've invoked Coulomb's Law to determine the two electrostatic forces:

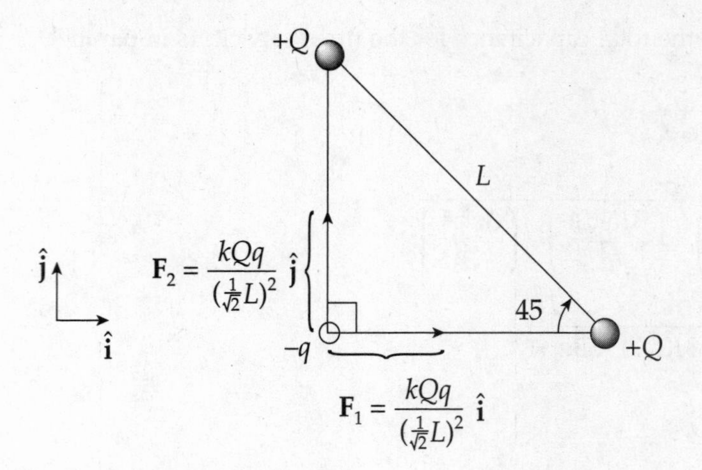

We find that the magnitude of **F** is

$$F = \frac{kQq}{\left(\frac{1}{\sqrt{2}}L\right)^2}\sqrt{1^2+1^2} = k\frac{2\sqrt{2}\cdot Qq}{L^2} = \frac{1}{4\pi\varepsilon_0}\frac{2\sqrt{2}\cdot Qq}{L^2} = \frac{\sqrt{2}\cdot Qq}{2\pi\varepsilon_0 L^2}$$

so dividing this by m gives the initial acceleration of the charge $-q$.

SECTION II, MECHANICS

1. (a) (i) Since the pulley is not accelerating, it can serve as the origin of an inertial reference frame. Applying Newton's Second Law to the masses using the following free-body diagrams,

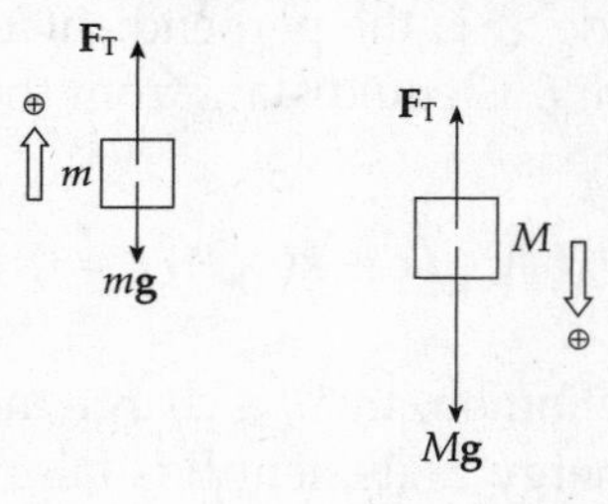

gives

$$F_T - mg = ma \quad \text{and} \quad Mg - F_T = Ma$$

Adding these equations eliminates F_T, and we find that

$$Mg - mg = (m+M)a \quad \Rightarrow \quad a = \frac{M-m}{M+m}g$$

The acceleration of block m is $(M - m)g/(M + m)$ upward, and the acceleration of block M is this value, $(M - m)g/(M + m)$, downward.

(a) (ii) Substitute the value for a into either one of the equations containing F_T:

$$F_T = mg + ma = mg + m \cdot \frac{M-m}{M+m} g = \frac{2Mmg}{M+m}$$

(b) (i) Because the center of the pulley is accelerating, it cannot serve as an inertial reference frame. Inertial frames are non-accelerating.

(b) (ii) Refer to the following diagram:

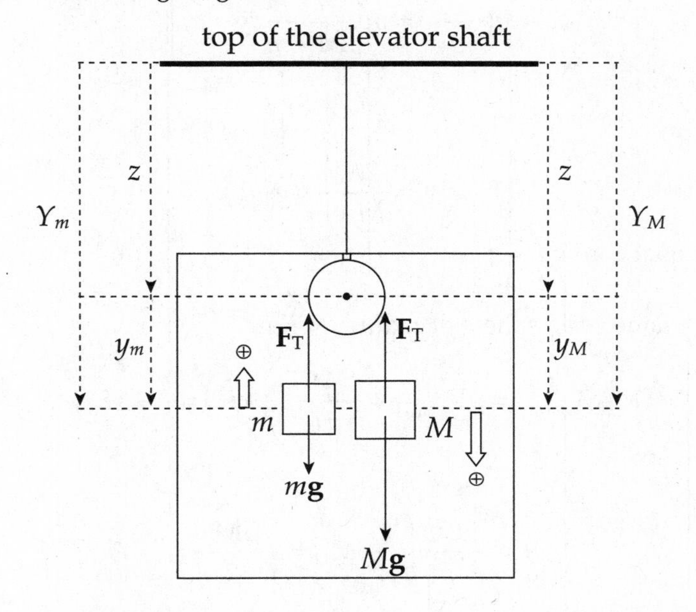

Since the second time derivative of displacement is acceleration,

$$F_T - mg = m\ddot{Y}_m$$
$$Mg - F_T = M\ddot{Y}_M$$

Now since $Y_m = z + y_m$ and $Y_M = z + y_M$, it is also true that

$$\ddot{Y}_m = \ddot{z} + \ddot{y}_m \quad \text{and} \quad \ddot{Y}_M = \ddot{z} + \ddot{y}_M$$

Because our positive direction is *up* for m and *down* for M, and $\ddot{z}$ is equal to b downward,

$$\ddot{Y}_m = -b + \ddot{y}_m \quad \text{and} \quad \ddot{Y}_M = b + \ddot{y}_M$$

Therefore,

$$F_T - mg = m(-b + \ddot{y}_m) \Rightarrow F_T - m(g-b) = m\ddot{y}_m$$
$$Mg - F_T = M(b + \ddot{y}_M) \Rightarrow M(g-b) - F_T = M\ddot{y}_M$$

Adding these equations (and remembering that $\ddot{y}_m = \ddot{y}_M$), we find that

$$M(g-b) - m(g-b) = (M+m)\ddot{y}$$

$$\ddot{y} = a = \frac{M-m}{M+m}(g-b)$$

(b) (iii) Substitute the value for a into either one of the equations containing F_T:

$$F_T - m(g-b) = m\left[\frac{M-m}{M+m}(g-b)\right]$$

$$F_T = m(g-b)\left[1 + \frac{M-m}{M+m}\right]$$

$$= \frac{2Mm}{M+m}(g-b)$$

(b) (iv) F_T will equal zero if $b = g$.

2. (a) The mass contained in a sphere of radius $r < R$ is

$$M_{\text{within } r} = \rho V_{\text{within } r} = \frac{M}{\frac{4}{3}\pi R^3}\left(\tfrac{4}{3}\pi r^3\right) = \frac{r^3}{R^3}M$$

so

$$F = -\frac{Gm}{r^2}\left(\frac{r^3}{R^3}M\right) = -\frac{GMm}{R^3}r$$

(b) Since $dU = -F(r)\,dr$,

$$\Delta U\Big|_R^r = \int_R^r dU = \int_R^r -F(r)\,dr = \int_R^r \frac{GMm}{R^3}r\,dr = \frac{GMm}{R^3}\left[\tfrac{1}{2}r^2\right]_R^r = \frac{GMm}{2R^3}(r^2 - R^2)$$

(c) Conservation of energy, $K_i + U_i = K_f + U_f$, can be written in the form $\Delta K = -\Delta U$, which gives

$$\tfrac{1}{2}mv^2 = -\frac{GMm}{2R^3}(r^2 - R^2)$$

$$v = \sqrt{\frac{GM}{R^3}(R^2 - r^2)}$$

By applying the Pythagorean theorem to the two right triangles in the figure,

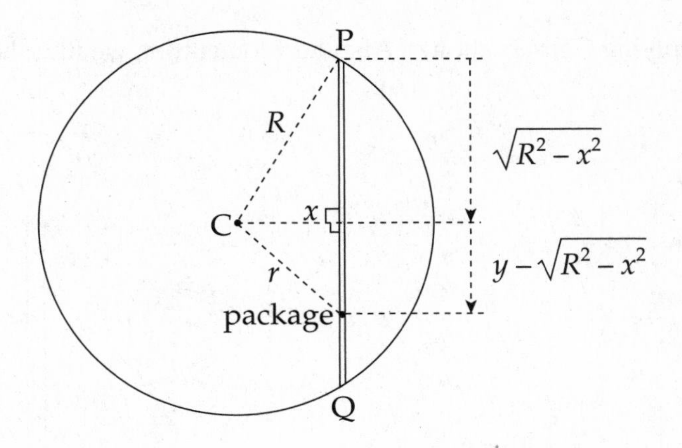

we see that

$$r^2 = x^2 + \left(y - \sqrt{R^2 - x^2}\right)^2$$

so the expression given above for v can be rewritten in the form

$$v = \sqrt{\frac{GM}{R^3}\left[R^2 - x^2 - \left(y - \sqrt{R^2 - x^2}\right)^2\right]} \quad (*)$$

$$= \sqrt{\frac{GM}{R^3}\left[2y\sqrt{R^2 - x^2} - y^2\right]}$$

(d) (i) From expression (*) given in part (c) above, we can see that the value of y which will maximize v is

$$y = \sqrt{R^2 - x^2}$$

This is the midpoint of the tunnel.

(d) (ii) The maximum value for v is

$$v_{\text{max}} = \sqrt{\frac{GM}{R^3}\left[R^2 - x^2\right]}$$

3. (a) (i) Once the $3m$ puck sticks to the m puck, the center of mass of the system is

$$y_{\text{cm}} = \frac{(m + 3m)(0) + (2m)(L)}{(m + 3m) + 2m} = \tfrac{1}{3}L$$

above the bottom pucks on the rod.

(a) (ii) Applying Conservation of Linear Momentum, we find that

$$(3m)v = (3m + m + 2m)v'_{cm} \quad \Rightarrow \quad v'_{cm} = \tfrac{1}{2}v$$

(a) (iii) Now, applying Conservation of Angular Momentum, we find that

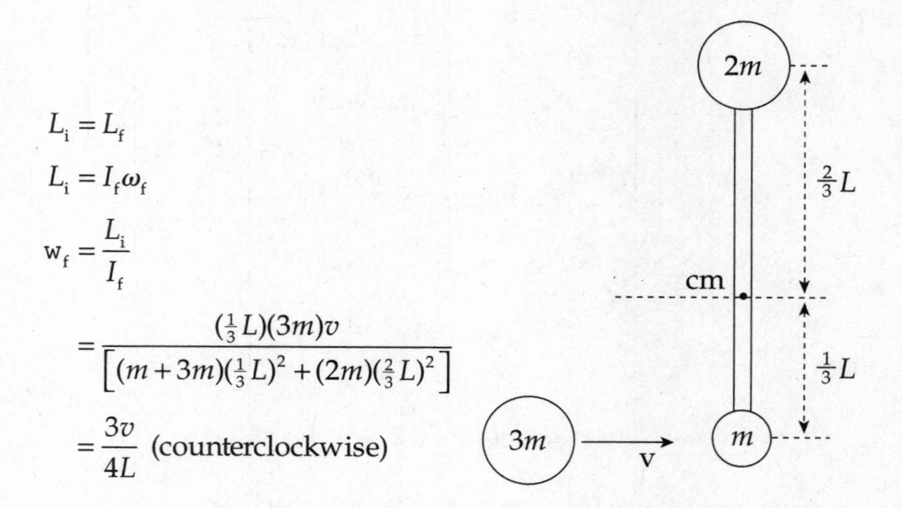

$$L_i = L_f$$

$$L_i = I_f \omega_f$$

$$w_f = \frac{L_i}{I_f}$$

$$= \frac{(\frac{1}{3}L)(3m)v}{\left[(m+3m)(\frac{1}{3}L)^2 + (2m)(\frac{2}{3}L)^2\right]}$$

$$= \frac{3v}{4L} \text{ (counterclockwise)}$$

(b) The kinetic energy before the collision was the initial translational kinetic energy of the $3m$ puck:

$$K_i = \tfrac{1}{2}(3m)v^2$$

After the collision, the total kinetic energy is the sum of the system's translational kinetic energy and its rotational kinetic energy:

$$K' = K'_{\text{translational}} + K'_{\text{rotational}}$$

$$= \tfrac{1}{2}Mv_{cm}^2 + \tfrac{1}{2}I\omega^2$$

$$= \tfrac{1}{2}(3m+m+2m)(\tfrac{1}{2}v)^2 + \tfrac{1}{2}\left[(3m+m)(\tfrac{1}{3}L)^2 + (2m)(\tfrac{2}{3}L)^2\right]\left(\tfrac{3v}{4L}\right)^2$$

$$= \tfrac{9}{8}mv^2$$

Therefore, the fraction of the initial kinetic energy that is lost due to the collision is

$$\frac{K - K'}{K} = \frac{\frac{3}{2}mv^2 - \frac{9}{8}mv^2}{\frac{3}{2}mv^2} = \frac{1}{4}$$

SECTION II, ELECTRICITY AND MAGNETISM

1. (a) Consider a small arc on the ring, subtended by an angle $d\theta$. The length of the arc is $ds = R\,d\theta$, so the charge it carries is $dQ = (Q/2\pi R)(R\,d\theta) = (Q/2\pi)\,d\theta$, since $\lambda = Q/(2\pi R)$ is the linear charge density.

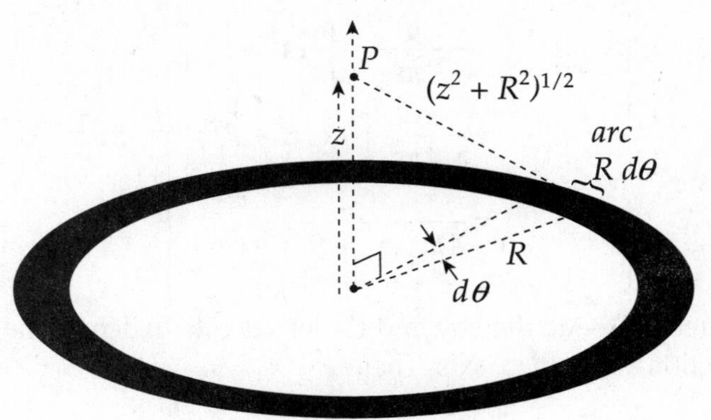

The distance from this arc to an arbitrary Point P on the z axis is $(z^2 + R^2)^{1/2}$, so the contribution to the electric potential at P due to this arc is

$$dV = \frac{1}{4\pi\varepsilon_0}\frac{dQ}{r} = \frac{Q\,d\theta}{8\pi^2\varepsilon_0(z^2+R^2)^{1/2}}$$

Integrating this from $\theta = 0$ to $\theta = 2\pi$ (i.e., all the way around the ring), we find that

$$V = \int_0^{2\pi}\frac{Q\,d\theta}{8\pi^2\varepsilon_0(z^2+R^2)^{1/2}} = \frac{Q}{8\pi^2\varepsilon_0(z^2+R^2)^{1/2}}\cdot 2\pi = \frac{Q}{4\pi\varepsilon_0(z^2+R^2)^{1/2}}$$

(b) (i) From the expression derived in part (a), we see that V will be maximized when $z = 0$. That is, the point on the z axis where the potential due to the ring is greatest is at the center of the ring.

(b) (ii) Putting $z = 0$ into the expression from part (a), we find that

$$V_{\text{max}} = \frac{Q}{4\pi\varepsilon_0(0^2+R^2)^{1/2}} = \frac{Q}{4\pi\varepsilon_0 R}$$

(c) The easiest way to obtain an expression for the electric field is to use the relationship $E(z) = -dV/dz$:

$$E = -\frac{dV}{dz} = -\frac{d}{dz}\left[\frac{Q}{4\pi\varepsilon_0(z^2+R^2)^{1/2}}\right]$$
$$= -\frac{Q}{4\pi\varepsilon_0} \cdot \frac{d}{dz}(z^2+R^2)^{-1/2}$$
$$= -\frac{Q}{4\pi\varepsilon_0}\left[-\tfrac{1}{2}(z^2+R^2)^{-3/2} \cdot 2z\right]$$
$$= \frac{Qz}{4\pi\varepsilon_0(z^2+R^2)^{3/2}}$$

To include the direction—and thereby find **E**—let's use $\hat{\mathbf{k}}$ to denote the unit vector that defines the positive direction along the z axis. Then

$$\mathbf{E} = \frac{Qz}{4\pi\varepsilon_0(z^2+R^2)^{3/2}}\hat{\mathbf{k}}$$

(d) (i) To maximize E, we set dE/dz equal to zero and solve for z:

$$\frac{dE}{dz} = \frac{Q}{4\pi\varepsilon_0} \cdot \frac{d}{dz}\left[\frac{z}{(z^2+R^2)^{3/2}}\right] = \frac{Q}{4\pi\varepsilon_0} \cdot \frac{(z^2+R^2)^{3/2} - z \cdot \frac{3}{2}(z^2+R^2)^{1/2} \cdot 2z}{(z^2+R^2)^3} \stackrel{\text{set}}{=} 0$$

This gives

$$(z^2+R^2)^{3/2} - z \cdot \tfrac{3}{2}(z^2+R^2)^{1/2} \cdot 2z = 0$$
$$(z^2+R^2)^{1/2}[(z^2+R^2) - 3z^2] = 0$$
$$(z^2+R^2) - 3z^2 = 0$$
$$z = \pm\tfrac{1}{\sqrt{2}}R$$

[Note: To verify that these values of z do indeed maximize E, we could check that the value of d^2E/dz^2 is negative at these points, or simply notice that on physical grounds, any extremum must be a maximum because $E = 0$ at $z = 0$ and $z = \pm\infty$, so $E(z)$ must rise then fall back to zero.]

(d) (ii) At either of the two points found above, the value of E is

$$E_{\max} = \frac{Q \cdot \frac{1}{\sqrt{2}}R}{4\pi\varepsilon_0\left[(\frac{1}{\sqrt{2}}R)^2 + R^2\right]^{3/2}} = \frac{Q}{4\pi\varepsilon_0} \cdot \frac{\frac{1}{\sqrt{2}}R}{(\frac{3}{2})^{3/2}R^3} = \frac{Q}{6\sqrt{3}\pi\varepsilon_0 R^2}$$

(e) The change in kinetic energy of the particle is equal to the work done by the electric field, which is equal to – (–q)ΔV. Using the expression for V derived in part (a), we find that

$$
\begin{aligned}
\tfrac{1}{2}mv^2 &= -(-q)\Delta V \\
&= -(-q)\left[\frac{Q}{4\pi\varepsilon_0(0^2+R^2)^{1/2}} - \frac{Q}{4\pi\varepsilon_0(A^2+R^2)^{1/2}}\right] \\
&= \frac{qQ}{4\pi\varepsilon_0}\left[\frac{1}{R} - \frac{1}{(A^2+R^2)^{1/2}}\right] \\
&= \frac{qQ}{4\pi\varepsilon_0}\left[\frac{1}{R} - \frac{1}{\left[R^2\left(1+\frac{A^2}{R^2}\right)\right]^{1/2}}\right] \\
&= \frac{qQ}{4\pi\varepsilon_0 R}\left[1-\left(1+\tfrac{A^2}{R^2}\right)^{-1/2}\right] \\
&\approx \frac{qQ}{4\pi\varepsilon_0 R}\left[1-\left(1-\tfrac{A^2}{2R^2}\right)\right] \\
&= \frac{qQA^2}{8\pi\varepsilon_0 R^3} \\
v &= A\sqrt{\frac{qQ}{4\pi\varepsilon_0 R^3 m}}
\end{aligned}
$$

(f) In simple harmonic motion, the position of an oscillator with amplitude A is given by the equation $z = A\sin(\omega t + \phi_0)$, where ω is the angular frequency. Differentiating $z(t)$ with respect to time, we get the velocity: $v = A\omega\cos(\omega t + \phi_0)$, which implies that the maximum speed is $A\omega$. The speed computed in part (e) is the maximum speed of the oscillator as it passes through the equilibrium position ($z = 0$, where $F = 0$), so

$$
\begin{aligned}
A\omega &= v_{max} \\
&= A\sqrt{\frac{qQ}{4\pi\varepsilon_0 R^3 m}} \\
\omega &= \sqrt{\frac{qQ}{4\pi\varepsilon_0 R^3 m}} \\
f &= \frac{1}{2\pi}\sqrt{\frac{qQ}{4\pi\varepsilon_0 R^3 m}}
\end{aligned}
$$

2. (a) (i) The current decays exponentially according to the equation

$$I(t) = \frac{\mathcal{E}}{r+R} e^{-t/(r+R)C}$$

where the initial current, $I(0)$, is equal to $\mathcal{E}/(r+R)$. To find the time t at which $I(t) = I(0)/k$, we solve the following equation:

$$\begin{aligned}
\frac{\mathcal{E}}{r+R} e^{-t/(r+R)C} &= \tfrac{1}{k} \cdot \frac{\mathcal{E}}{r+R} \\
e^{-t/(r+R)C} &= \tfrac{1}{k} \\
-\frac{t}{(r+R)C} &= \ln \tfrac{1}{k} \\
-\frac{t}{(r+R)C} &= -\ln k \\
t &= (\ln k)(r+R)C
\end{aligned}$$

(a) (ii) The charge on the capacitor increases according to the equation

$$Q(t) = C\mathcal{E}\left[1 - e^{-t/(r+R)C}\right]$$

The maximum value is $C\mathcal{E}$ (when the capacitor is fully charged). To find the time t at which $Q(t) = C\mathcal{E}/k$, we solve the following equation:

$$\begin{aligned}
C\mathcal{E}\left[1 - e^{-t/(r+R)C}\right] &= \tfrac{1}{k} C\mathcal{E} \\
1 - e^{-t/(r+R)C} &= \tfrac{1}{k} \\
e^{-t/(r+R)C} &= 1 - \tfrac{1}{k} \\
-\frac{t}{(r+R)C} &= \ln \tfrac{k-1}{k} \\
\frac{t}{(r+R)C} &= \ln \tfrac{k}{k-1} \\
t &= (\ln \tfrac{k}{k-1})(r+R)C
\end{aligned}$$

(a) (iii) The field energy stored in the capacitor can be written in the form $U_E(t) = [Q(t)]^2/(2C)$, where $Q(t)$ is the function of time given in part (ii) above. We are asked to find the time t at which

$$\begin{aligned}
U(t) &= \tfrac{1}{k} U_{\text{max}} \\
\frac{[Q(t)]^2}{2C} &= \tfrac{1}{k} \cdot \frac{(C\mathcal{E})^2}{2C} \\
\left\{C\mathcal{E}\left[1 - e^{-t/(r+R)C}\right]\right\}^2 &= \tfrac{1}{k}(C\mathcal{E})^2 \\
1 - e^{-t/(r+R)C} &= \tfrac{1}{\sqrt{k}} \\
e^{-t/(r+R)C} &= 1 - \tfrac{1}{\sqrt{k}} \\
-\frac{t}{(r+R)C} &= \ln \tfrac{\sqrt{k}-1}{\sqrt{k}} \\
\frac{t}{(r+R)C} &= \ln \tfrac{\sqrt{k}}{\sqrt{k}-1} \\
t &= \left(\ln \tfrac{\sqrt{k}}{\sqrt{k}-1}\right)(r+R)C
\end{aligned}$$

(b) (i) Once the switch is turned to position 2, the capacitor discharges through resistor R (only). Since the capacitor is fully charged (to $V = \mathcal{E}$) when the switch is moved to position 2, the current will decrease exponentially according to the equation

$$I(t) = I_0 e^{-t/RC}$$

where $I_0 = I_{max} = \varepsilon/R$. To find the time at which I is equal to I_0/k, we solve the following equation:

$$\begin{aligned} I_0 e^{-t/RC} &= \tfrac{1}{k} I_0 \\ e^{-t/RC} &= \tfrac{1}{k} \\ -\frac{t}{RC} &= \ln \tfrac{1}{k} \\ \frac{t}{RC} &= \ln k \\ t &= (\ln k)RC \end{aligned}$$

(b) (ii) Since the charge on the capacitor obeys the same equation as the current—as given in (b) (i) above, simply replacing I by Q—the time t at which the charge drops to $1/k$ its initial value is the same as the time t at which the current drops to $1/k$ times its initial value: $t = (\ln k)RC$.

3. (a) (i) As the strip XY slides down the rails, the magnetic flux through the rectangular region bounded by the strip increases. This change in magnetic flux induces an emf. If ℓ denotes the distance the strip has slid down the rails, then the magnetic flux through the rectangular region is

$$\Phi_B = BA \cos\theta = BL\ell \cos\theta$$

The rate of change of Φ_B is

$$\frac{d\Phi_B}{dt} = \frac{d}{dt}(BL\ell \cos\theta) = BL\cos\theta \frac{d\ell}{dt} = BL\cos\theta \cdot v$$

where $v = d\ell/dt$ is the speed with which the strip slides. By Faraday's Law, this is the (magnitude of the) induced emf.

(a) (ii) As the strip slides down, the area increases, so the magnetic flux upward increases. To oppose this change (in accordance with Lenz's Law), the induced current must flow in such a way as to create some magnetic flux downward. By the right-hand rule, then, current must flow from Y to X.

(a) (iii) The induced current depends on the speed of the sliding strip and is equal to

$$I = \frac{\mathcal{E}}{R} = \frac{BL\cos\theta \cdot v}{R}$$

(b) In order for the strip to slide with constant velocity, the net force on it must be zero. The force it feels parallel to the rails and downward is $F_w \sin \theta = mg \sin \theta$. The following diagram

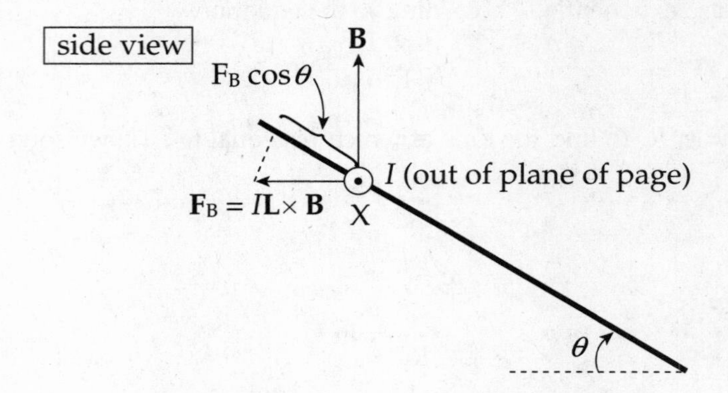

which is a side view of the apparatus looking at the X end of the strip, shows that the force the strip feels parallel to the rails and upward is

$$F_B \cos \theta = ILB \cos \theta = \frac{BL \cos \theta \cdot v}{R} LB \cos \theta = \frac{B^2 L^2 v \cos^2 \theta}{R}$$

The net force on the strip is zero when these two opposing forces balance; the speed at which this occurs will be denoted v_f:

$$\frac{B^2 L^2 v_f \cos^2 \theta}{R} = mg \sin \theta$$

$$v_f = \frac{mgR \sin \theta}{B^2 L^2 \cos^2 \theta}$$

(c) The rate at which thermal energy is produced in the strip is I^2R. When $v = v_f$, this is equal to

$$\begin{aligned} P_{\text{dissipated by strip}} &= I^2 R \\ &= \left(\frac{BL \cos \theta}{R} \cdot v_f \right)^2 R \\ &= \left(\frac{BL \cos \theta}{R} \cdot \frac{mgR \sin \theta}{B^2 L^2 \cos^2 \theta} \right)^2 R \\ &= \frac{m^2 g^2 R \sin^2 \theta}{B^2 L^2 \cos^2 \theta} \\ &= \frac{mgR \sin \theta}{B^2 L^2 \cos^2 \theta} \cdot mg \sin \theta \\ &= v_f \cdot mg \sin \theta \\ &= v_f \cdot (F_w \sin \theta) \\ &= \text{rate at which gravity does work on strip} \end{aligned}$$

where the last equality follows from the formula $P = Fv$ for the power produced by a constant force **F** acting on an object whose velocity, **v**, is parallel to **F**.

17

The Princeton Review AP Physics C Practice Exam 2

PHYSICS C

You may take the entire C Exam or Mechanics only or Electricity and Magnetism only as follows:

	Entire C Exam Both Mech. and Elect. & Mag.	Mechanics only	Electricity & Magnetism
First 45 min.	Sec. I, Mech. 35 questions	Sec. I, Mech. 35 questions	Sec. I, Elect. & Mag. 35 questions
Second 45 min.	Sec. I, Elect. & Mag. 35 questions	Sec. II, Mech. 3 questions	Sec. II, Elect. & Mag. 3 questions
Third 45 min.	Sec. II, Mech. 3 questions		
Fourth 45 min.	Sec. II, Elect. & Mag. 3 questions		

Separate grades are reported for Mechanics and for Electricity and Magnetism. Each section of each examination is 50 percent of the total grade; each question in a section has equal weight. Calculators are NOT permitted on Section I of the exam but are allowed on Section II. However, calculators cannot be shared with other students and calculators with typewriter-style (QWERTY) keyboards will not be permitted. A table of information that may be helpful is found on the following page.

The Physics C Examination contains a total of 70 multiple-choice questions. If you are taking
— *Mechanics only*, please be careful to answer numbers 1–35,
— *Electricity and Magnetism only*, please be careful to answer numbers 36–70,
— *the entire examination* (Mechanics *and* Electricity and Magnetism), answer numbers 1–70 on your answer sheet.

Advanced Placement Examination

PHYSICS C

SECTION I

TABLE OF INFORMATION FOR 2008

CONSTANTS AND CONVERSION FACTORS

1 unified atomic mass unit,	$1\ u = 1.66 \times 10^{-27}$ kg
	$= 931\ MeV/c^2$
Proton mass,	$m_p = 1.67 \times 10^{-27}$ kg
Neutron mass,	$m_n = 1.67 \times 10^{-27}$ kg
Electron mass,	$m_e = 9.11 \times 10^{-31}$ kg
Magnitude of the electron charge,	$e = 1.60 \times 10^{-19}$ C
Avogadro's number,	$N_0 = 6.02 \times 10^{23}\ mol^{-1}$
Universal gas constant,	$R = 8.31\ J/(mol \cdot K)$
Boltzmann's constant,	$k_B = 1.38 \times 10^{-23}\ J/K$
Speed of light,	$c = 3.00 \times 10^8$ m/s
Planck's constant,	$h = 6.63 \times 10^{-34}\ J \cdot s$
	$= 4.14 \times 10^{-15}\ eV \cdot s$
	$hc = 1.99 \times 10^{-25}\ J \cdot m$
	$= 1.24 \times 10^3\ eV \cdot nm$
Vacuum permittivity,	$\epsilon_0 = 8.85 \times 10^{-12}\ C^2/N \cdot m^2$
Coulomb's law constant,	$k = 1/4\pi\epsilon_0 = 9.0 \times 10^9\ N \cdot m^2/C^2$
Vacuum permeability,	$\mu_0 = 4\pi \times 10^{-7} (T \cdot m)/A$
Magnetic constant,	$k' = \mu_0/4\pi = 10^{-7} (T \cdot m)/A$
Universal gravitational constant,	$G = 6.67 \times 10^{-11}\ m^3/kg \cdot s^2$
Acceleration due to gravity at the earth's surface,	$g = 9.8\ m/s^2$
1 atmosphere pressure,	$1\ atm = 1.0 \times 10^5\ N/m^2$
	$= 1.0 \times 10^5$ Pa
1 electron volt,	$1\ eV = 1.60 \times 10^{-19}$ J

UNITS

Name	Symbol
meter	m
kilogram	kg
second	s
ampere	A
kelvin	K
mole	mol
hertz	Hz
newton	N
pascal	Pa
joule	J
watt	W
coulomb	C
volt	V
ohm	Ω
henry	H
farad	F
tesla	T
degree Celsius	°C
electron-volt	eV

PREFIXES

Factor	Prefix	Symbol
10^9	giga	G
10^6	mega	M
10^3	kilo	k
10^{-2}	centi	c
10^{-3}	milli	m
10^{-6}	micro	μ
10^{-9}	nano	n
10^{-12}	pico	p

VALUES OF TRIGONOMETRIC FUNCTIONS FOR COMMON ANGLES

θ	sin θ	cos θ	tan θ
0°	0	1	0
30°	1/2	$\sqrt{3}/2$	$\sqrt{3}/3$
37°	3/5	4/5	3/4
45°	$\sqrt{2}/2$	$\sqrt{2}/2$	1
53°	4/5	3/5	4/3
60°	$\sqrt{3}/2$	1/2	$\sqrt{3}$
90°	1	0	∞

The following conventions are used in this examination.

I. Unless otherwise stated, the frame of reference of any problem is assumed to be inertial.

II. The direction of any electric current is the direction of flow of positive charge (conventional current).

III. For any isolated electric charge, the electric potential is defined as zero at an infinite distance from the charge.

PHYSICS C
SECTION I, MECHANICS
Time – 45 minutes
35 Questions

Directions: Each of the following questions or incomplete statements below is followed by five suggested answers or completions. Select the one that is best in each case and then mark it on your answer sheet.

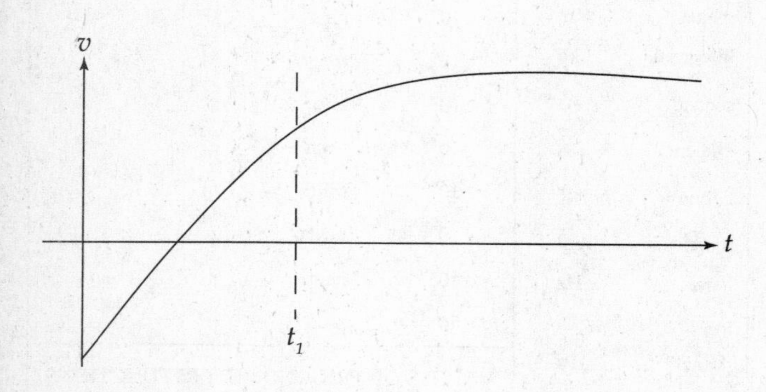

1. The graph above shows the velocity vs. time graph for a 3 kg object moving in one dimension. Which of the following is a possible graph of position versus time for this object?

(A)

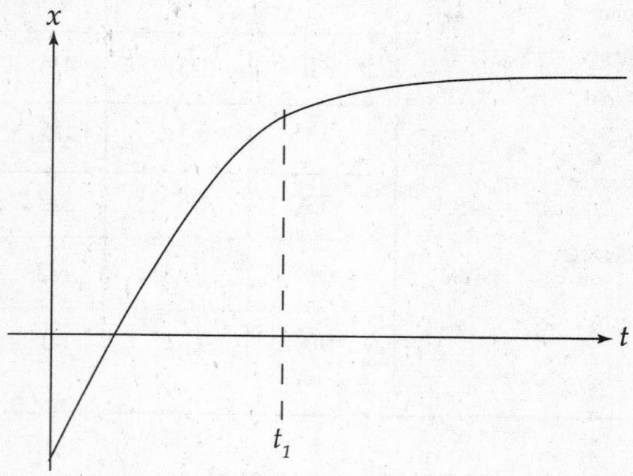

(B)

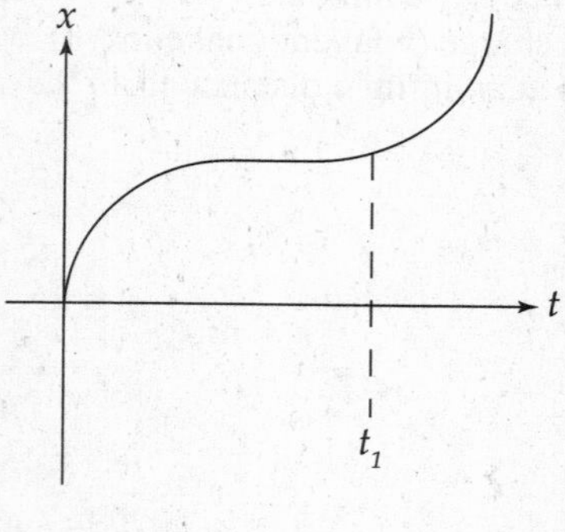

(C)

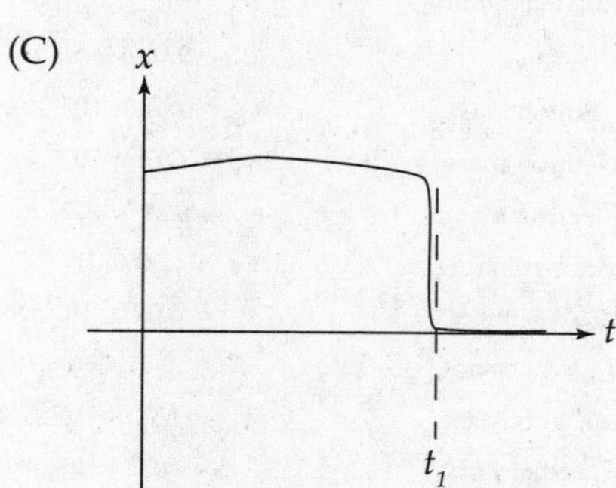

(D)

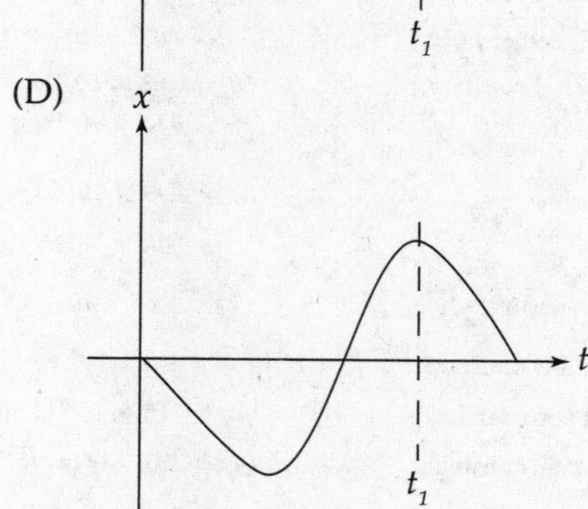

(E)

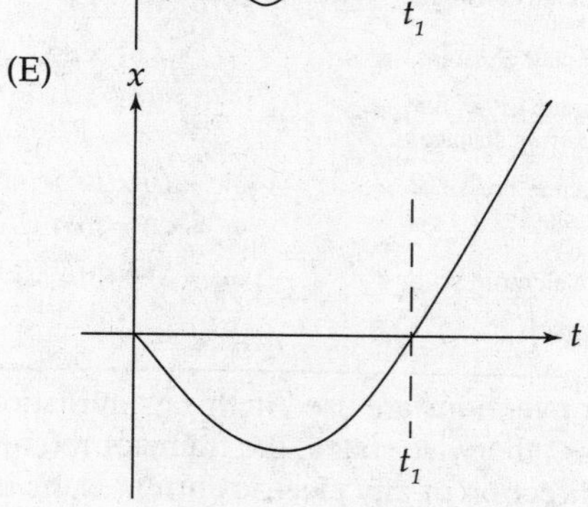

2. A ball is dropped from an 80 m tall building. How long does the ball take to reach the ground?

(A) 8 seconds
(B) 16 seconds
(C) 8.9 seconds
(D) 4 seconds
(E) 2.8 seconds

GO ON TO THE NEXT PAGE

Velocity before the collision

Velocity after the collision

3. The velocity of an object before a collision is directed straight north and the velocity after the collision is directed straight west, as shown above. Which of the following vectors represents the change in momentum of the object?

(A) 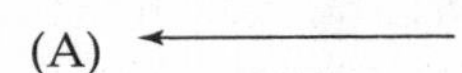

(B)

(C) 

(D)

(E) 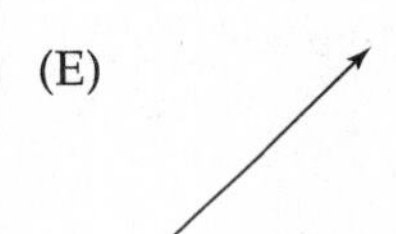

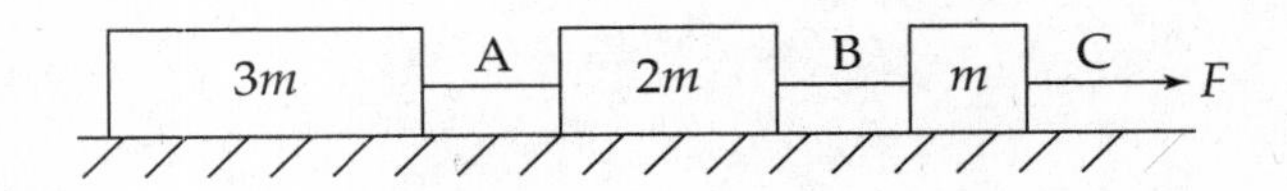

4. Three blocks of masses $3m$, $2m$, and m are connected to strings A, B, and C as shown above. The blocks are pulled along a frictionless, horizontal floor with a force, F. Determine the acceleration of the $2m$ block.

(A) $\frac{F}{2m}$

(B) $\frac{F}{6m}$

(C) $2Fm$

(D) $6Fm$

(E) $\frac{F}{m}$

Questions 5–6

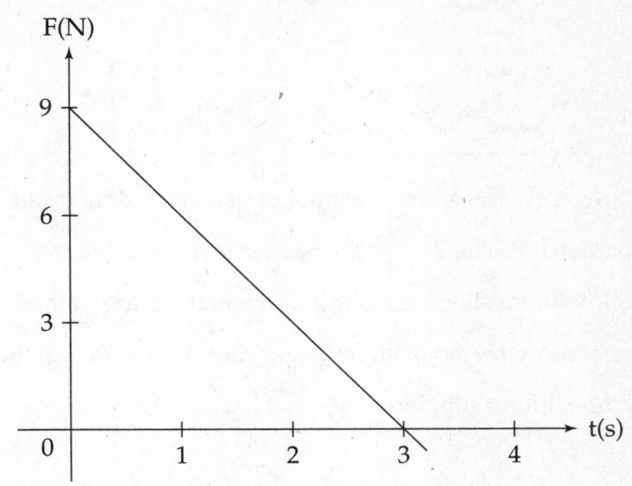

A block of mass 2 kg, initially at rest, is pulled along a frictionless, horizontal surface with a force shown as a function of time by the graph above.

5. The acceleration of the block at t = 2 s is

(A) 0 m/s²
(B) 1.5 m/s²
(C) 2.0 m/s²
(D) 2.5 m/s²
(E) 3.0 m/s²

6. The speed of the block at t = 3 s is

(A) 0 m/s
(B) 4.5 m/s
(C) 6.75 m/s
(D) 13.5 m/s
(E) 54 m/s

GO ON TO THE NEXT PAGE

Questions 7–8

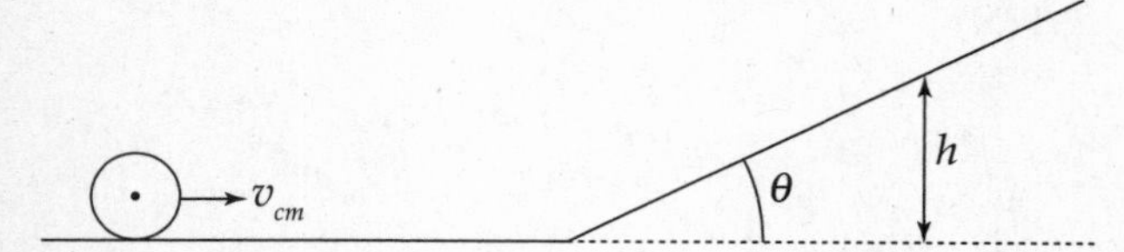

The center of mass of a cylinder of mass m, radius r, and rotational inertia $I = \frac{1}{2}mr^2$ has a velocity of v_{cm} as it rolls without slipping along a horizontal surface. It then encounters a ramp of angle θ, and continues to roll up the ramp without slipping.

7. What is the maximum height the cylinder reaches?

(A) $\frac{v^2}{4g}$

(B) $\frac{4v^2}{3g}$

(C) $\frac{v^2}{3g}$

(D) $\frac{3v^2}{4g}$

(E) $\frac{4g}{3v^2}$

8. Now the cylinder is replaced with a hoop that has the same mass and radius. The hoop's rotational inertia is mr^2. The center of mass of the hoop has the same velocity as the cylinder when it is rolling along the horizontal surface and the hoop also rolls up the ramp without slipping. How would the maximum height of the hoop compare to the maximum height of the cylinder?

(A) The hoop would reach a greater maximum height than the cylinder.
(B) The hoop and cylinder would reach the same maximum height.
(C) The cylinder would reach a greater maximum height than the hoop.
(D) The cylinder would reach less than half the height of the hoop.
(E) None of the above

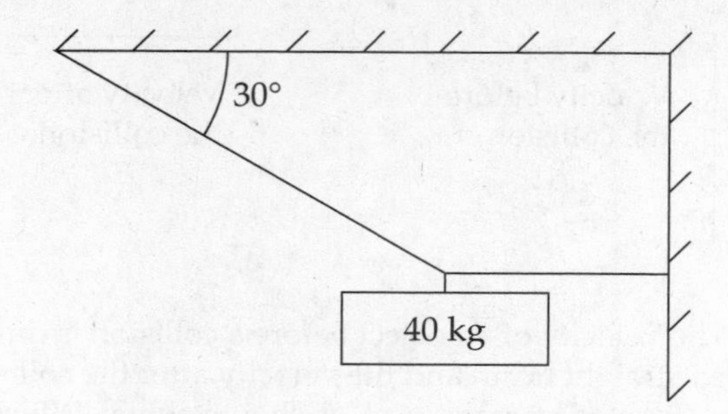

9. An object of mass 40 kg is suspended by means of two cords, as shown above. The tension in the angled cord is

(A) 400 N
(B) 80 N
(C) 800 N
(D) 690 N
(E) 230 N

GO ON TO THE NEXT PAGE

Questions 10–12

A box is on an incline of angle θ above the horizontal. The box may be subject to the following forces: frictional (f), gravitational (F_g), tension from a string connected to it (F_T) and normal (N). In the following free-body diagrams for the box, the lengths of the vectors are proportional to the magnitudes of the forces.

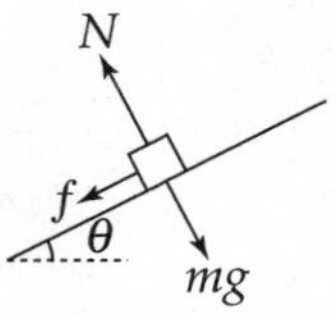

N
f
θ
mg
Figure A

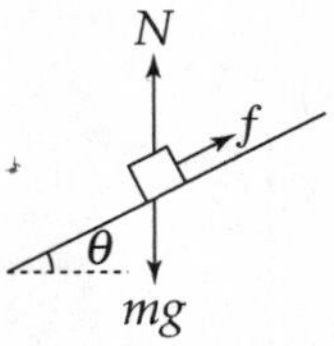

N
f
θ
mg
Figure B

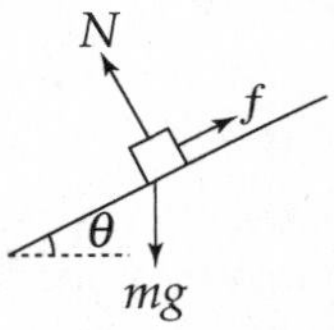

N
f
θ
mg
Figure C

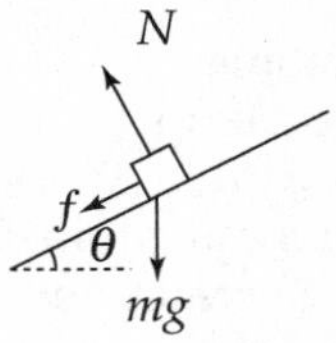

N
f
θ
mg
Figure D

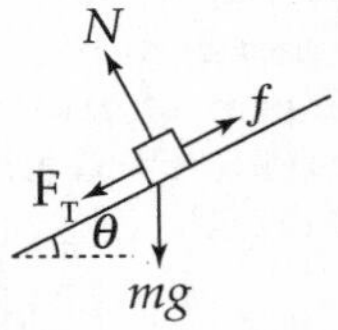

N
f
F_T
θ
mg
Figure E

10. Which of the following best represents the free-body diagram for the box if it is decelerating as it goes up the incline?

(A) Figure A
(B) Figure B
(C) Figure C
(D) Figure D
(E) Figure E

11. Which of the following best represents the free-body diagram for the box if it is moving at a constant velocity down the ramp?

A) Figure A
(B) Figure B
(C) Figure C
(D) Figure D
(E) Figure E

12. Which of the following best represents the free-body diagram for the box if its speed is increasing as it moves down the incline?

(A) Figure A
(B) Figure B
(C) Figure C
(D) Figure D
(E) Figure E

---

13. The force on an object as a function of time t is given by the expression $F = Ct^3$, where C is a constant. Determine the change in momentum for the time interval 0 to t_1.

(A) $\frac{Ct_1^2}{2}$

(B) $\frac{Ct_1^3}{3}$

(C) $\frac{Ct_1^4}{3}$

(D) $\frac{Ct_1^3}{4}$

(E) $\frac{Ct_1^4}{4}$

GO ON TO THE NEXT PAGE

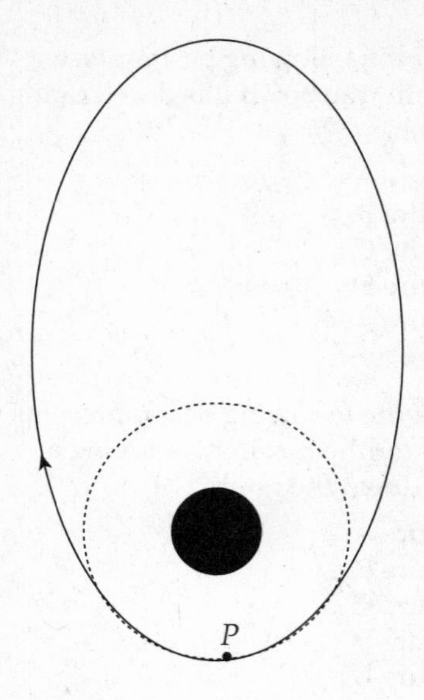

14. A spaceship orbits Earth in a clockwise, elliptical orbit as shown above. The spaceship needs to change to a circular orbit. When the spaceship passes point P, a short burst of the ship's engine will change its orbit. What direction should the engine burst be directed?

(A) →

(B) ←

(C)

(D)

(E)

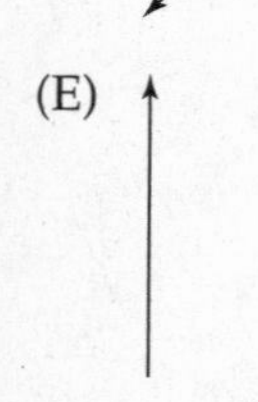

15. A motorcycle of mass 200 kg completes a vertical, circular loop of radius 5 m, with a constant speed of 10 m/s. How much work is done on the motorcycle by the normal force of the track?

(A) 0 J
(B) 1×10^5 J
(C) 1×10^6 J
(D) 4 J
(E) 10π J

16. A ball with a radius of 0.2 m rolls without slipping on a level surface. The center of mass of the ball moves at a constant velocity, moving a distance of 30 meters in 10 seconds. The angular speed of the ball about its point of contact on the surface is

(A) 3 m/s
(B) 0.6 m/s
(C) 60 m/s
(D) 15 m/s
(E) 8 m/s

17. A bullet is moving with a velocity v_0 when it collides with and becomes embedded in a wooden bar that is hinged at one end, as shown above. Consider the bullet and the wooden bar to be the system. For this scenario, which of the following is true?

(A) The linear momentum of the system is conserved because the net force on the system is zero.
(B) The angular momentum of the system is conserved because the net torque on the system is zero.
(C) The kinetic energy of the system is conserved because it is an inelastic collision.
(D) The kinetic energy of the system is conserved because it is an elastic collision.
(E) Linear momentum and angular momentum are both conserved.

GO ON TO THE NEXT PAGE

Questions 18–19

A spring mass system is vibrating along a frictionless, horizontal floor. The spring constant is 8 N/m, the amplitude is 5 cm and the period is 4 seconds.

18. The mass of the system is

(A) $32\pi^2$

(B) $\frac{32}{\pi^2}$

(C) $\frac{16}{\pi}$

(D) $\frac{0.2}{\pi^2}$

(E) $\frac{20}{\pi^2}$

19. Which of the following equations could represent the position of the mass from equilibrium x as a function of time t, where x is in meters and t is in seconds.

(A) $x = 0.05 \cos \pi t$

(B) $x = 0.05 \cos 2\pi t$

(C) $x = 0.05 \cos \frac{\pi}{2} t$

(D) $x = 8 \cos \frac{\pi}{2} t$

(E) $x = 0.05 \cos \frac{\pi}{4} t$

20. Two blocks of masses M and $3M$ are connected by a light string. The string passes over a frictionless pulley of negligible mass so that the blocks hang vertically. The blocks are then released from rest. What is the acceleration of the mass M?

(A) $\frac{g}{4}$

(B) $\frac{g}{3}$

(C) g

(D) $\frac{2g}{3}$

(E) $\frac{g}{2}$

21. For a particular nonlinear spring, the relationship between the magnitude of the applied force F and the stretch of the spring x is given by the equation $F = kx^{1.5}$. How much energy is stored in the spring when is it stretched a distance x_1?

(A) $\frac{2kx_1^{2.5}}{5}$

(B) $\frac{kx_1^{1.5}}{5}$

(C) $kx_1^{2.5}$

(D) $\frac{1}{2}kx_1^2$

(E) $1.5kx_1^{.5}$

GO ON TO THE NEXT PAGE

Questions 22–23

Two ice skaters are moving on frictionless ice and are about to collide. The 50-kg skater is moving directly west at 4 m/s. The 75-kg skater is moving directly north at 2 m/s. After the collision they stick together.

22. What is the magnitude of the momentum of the two-skater system after the collision?
    (A) 150 kg m/s
    (B) 350 kg m/s
    (C) 50 kg m/s
    (D) 200 kg m/s
    (E) 250 kg m/s

23. For this scenario, which of the following is true?
    (A) The linear momentum of the system is conserved because the net force on the system is nonzero during the collision.
    (B) Only the kinetic energy of the system is conserved because it is an inelastic collision.
    (C) Only the kinetic energy of the system is conserved because it is an elastic collision.
    (D) The linear momentum of the system is conserved because the net force on the system is zero.
    (E) Both the linear momentum and the kinetic energy of the system are conserved.

---

24. The position of an object is given by the equations $x = 2.0t^3 + 4.0t + 6.25$, where x is in meters and t is in seconds. What is the acceleration of the object at $t = 1.50$ s?
    (A) 12 m/s²
    (B) 24 m/s²
    (C) 18 m/s²
    (D) 6 m/s²
    (E) 32 m/s²

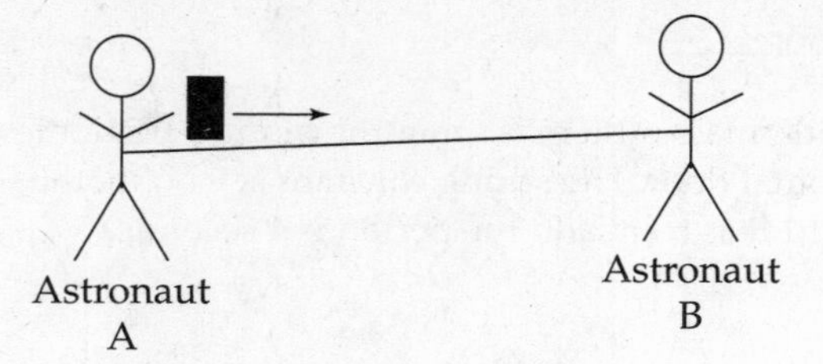

25. Two astronauts are at rest out in space and connected to a rope that is taut, as shown above. Astronaut A throws a heavy oxygen container towards Astronaut B, who then catches the container. Which of the following describes the motion of Astronaut B?

	Immediately After the Throw	After the Catch
(A)	Moves to the right	Does not move
(B)	Moves to the left	Moves to the right
(C)	Moves to the left	Does not move
(D)	Does not move	Moves to the right
(E)	Moves to the left	Moves to the left

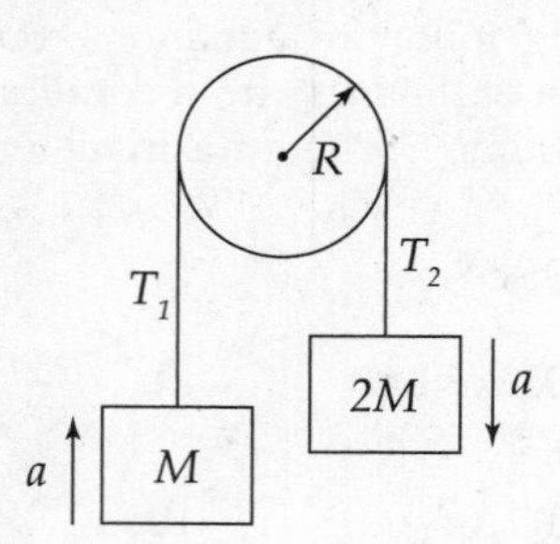

26. Two blocks of masses M and $2M$ are connected by a light string. The string passes over a pulley, as shown above. The pulley has a radius R and moment of inertia I about its center. T_1 and T_2 are the tensions in the string on each side of the pulley and a is the acceleration of the masses. Which of the following equations best describes the pulley's rotational motion during the time the blocks accelerate?

    (A) $(T_2 + T_1)R = I\alpha$
    (B) $(T_2 - T_1)R = I\alpha$
    (C) $(T_2 - T_1)R = I\frac{a}{R}$
    (D) $MgR = I\frac{a}{R}$
    (E) $3MgR = I\frac{a}{R}$

GO ON TO THE NEXT PAGE

27. A solid sphere of uniform density with mass M and radius R is located far out in space. A test mass, m, is placed at various locations both within the sphere and outside the sphere. Which graph correctly shows the force of gravity on the test mass vs. the distance from the center of the sphere?

(A)

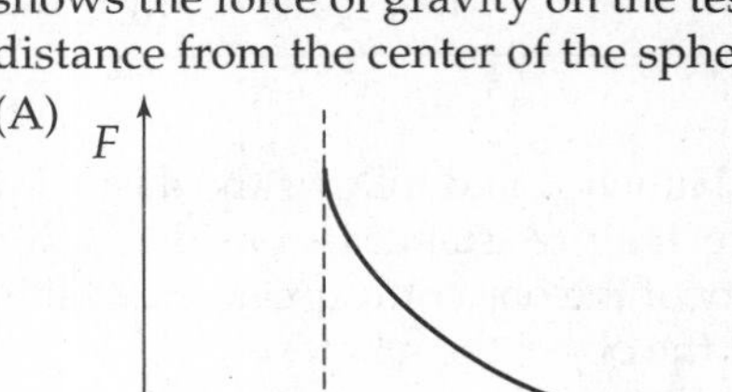

(B)

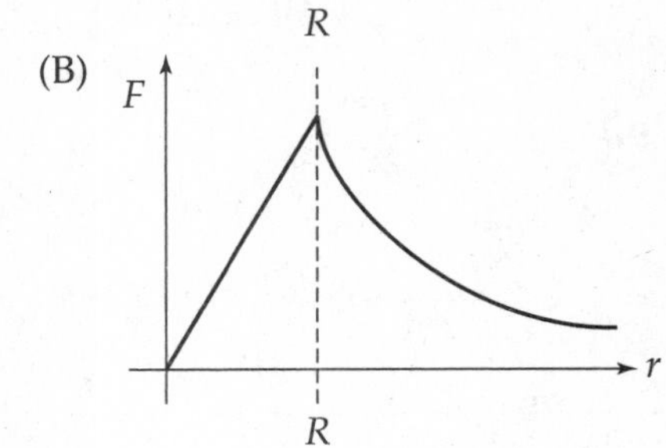

(C)

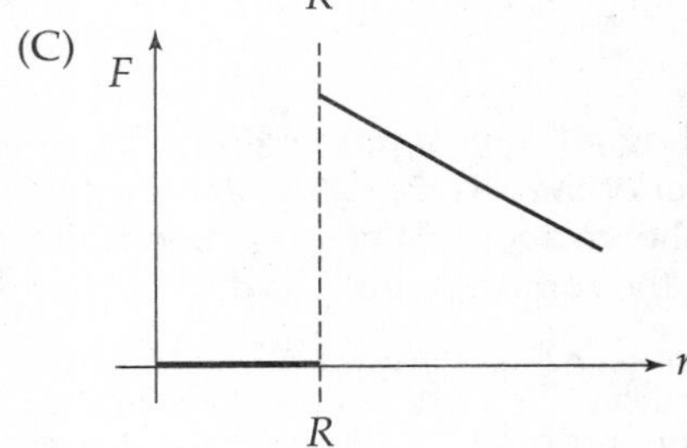

(D)

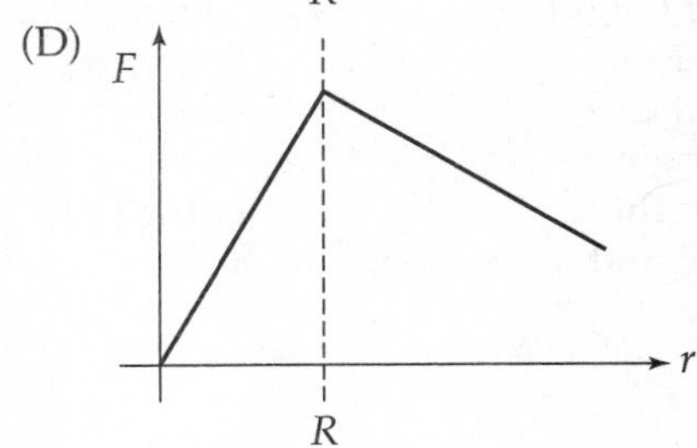

(E)

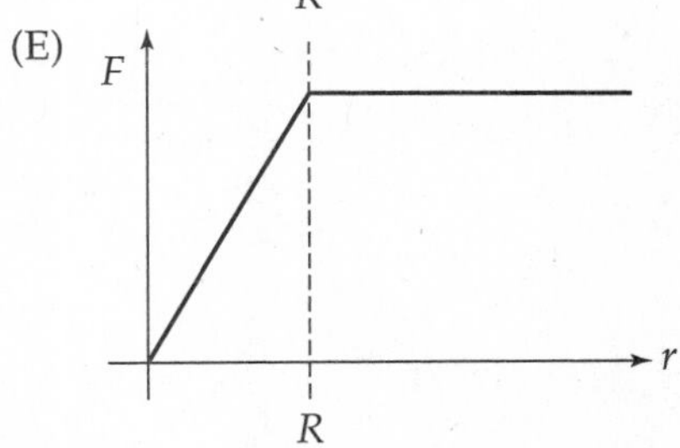

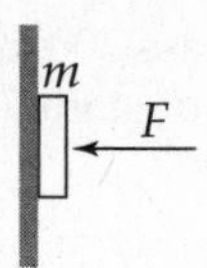

28. A horizontal force F pushes a book of mass m against a vertical wall, as shown above. How large does the coefficient of friction μ need to be between the wall and the book to prevent the book from slipping down the wall?

(A) mgF

(B) mg

(C) $\frac{mg}{F}$

(D) $\frac{mg}{2F}$

(E) $\frac{F}{mg}$

29. A simple pendulum of length ℓ and mass M is oscillating with a period T with very small amplitude. Now the amplitude is halved. The new period is most nearly

(A) T

(B) $2T$

(C) $\frac{T}{\sqrt{2}}$

(D) $\sqrt{2T}$

(E) $\frac{T}{2}$

GO ON TO THE NEXT PAGE

30. An electric car of mass 300 kg delivers 400 W as it moves the car at a constant 20 m/s. The force delivered by the motor is

(A) 20 N
(B) $\frac{4}{3}$ N
(C) 6000 N
(D) 600 N
(E) 8000 N

31. A 2.0 kg mass is attached to the end of a vertical ideal spring with a spring constant of 800 N/m. The mass is pulled down 10 cm from the equilibrium position and then released, so that it oscillates. The kinetic energy of the 2.0 kg mass at the equilibrium position is

(A) 40 J
(B) $\frac{2}{3}$ J
(C) 2 J
(D) 4 J
(E) 12 J

32. Physics students are checking the constant acceleration equations of kinematics by measuring the velocity of a tennis ball that is dropped and falls 6 meters and then passes through a photogate. The predicted velocity is 20% above the measured velocity of the photogate. Which of the following best describes the cause of the large percent difference?

(A) The ball changes its shape while falling.
(B) The acceleration of gravity varies as the ball is falling.
(C) Air resistance increases the acceleration of the ball.
(D) The acceleration of the balls varies with the velocity.
(E) The acceleration of gravity changes due to air resistance.

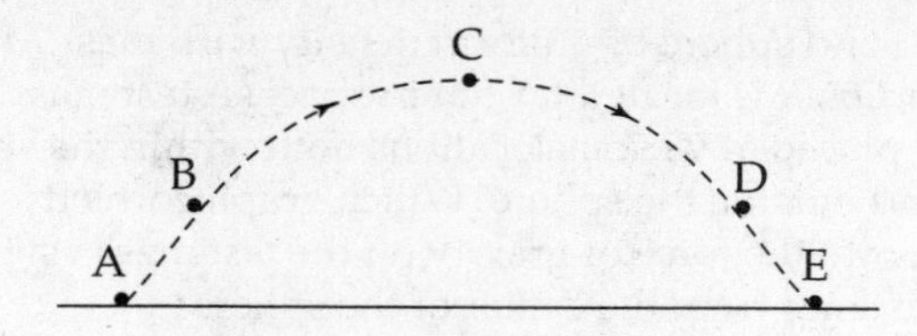

33. An object is launched and follows the dashed path shown above. If air resistance is considered, when is the velocity of the object the greatest and the acceleration of the object the greatest?

	Greatest Velocity	Greatest Acceleration
(A)	A	All equal to g
(B)	C	All equal to g
(C)	A	A
(D)	E	E
(E)	A	E

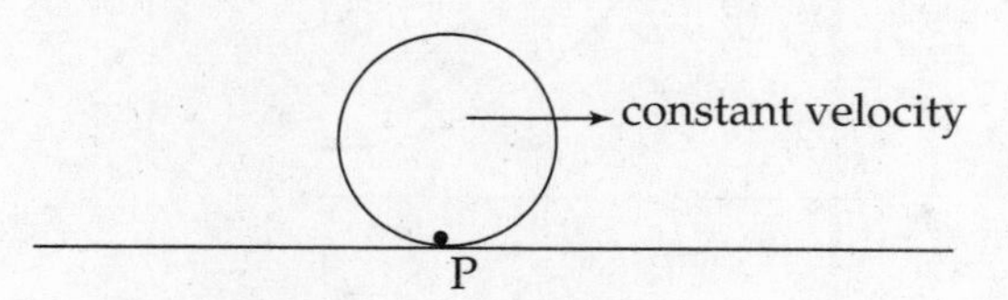

34. A disk is rolling without slipping along the ground and the center of mass is traveling at a constant velocity, as shown above. What direction is the acceleration of the contact point P, and the center of mass?

	Acceleration of Contact Point P	Acceleration of Center of Mass
(A)	Upward	To the right
(B)	Upward	Zero
(C)	To the right	Zero
(D)	To the right	To the right
(E)	Upward and to the right	Zero

GO ON TO THE NEXT PAGE

35. The escape velocity for a rocket launched from the surface of a planet is v_0. Determine the escape velocity for another planet that has twice the mass and twice the radius of this planet.

(A) $2v_0$

(B) $\frac{1}{2}v_0$

(C) $\frac{1}{\sqrt{2}}v_0$

(D) $\sqrt{2}v_0$

(E) v_0

STOP

END OF SECTION I, MECHANICS

IF YOU FINISH BEFORE TIME IS CALLED, YOU MAY CHECK YOUR WORK ON SECTION I, MECHANICS, ONLY.

DO NOT TURN TO ANY OTHER TEST MATERIALS.

PHYSICS C
SECTION I, ELECTRICITY AND MAGNETISM
Time – 45 minutes
35 Questions

Directions: Each of the following questions or incomplete statements below is followed by five suggested answers or completions. Select the one that is best in each case and then mark it on your answer sheet.

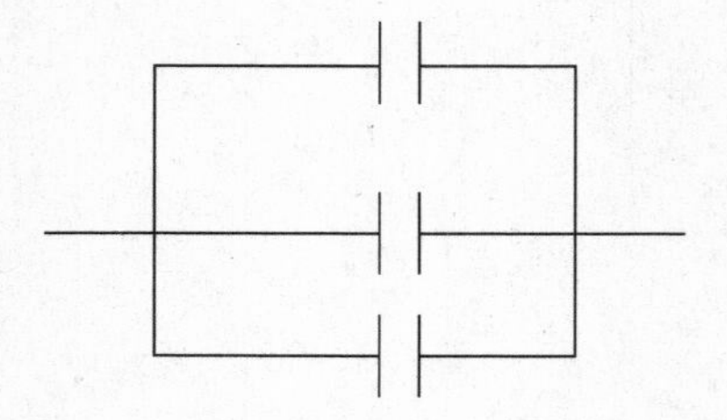

36. Three 3 μF capacitors are connected in parallel as shown above. Determine the equivalent capacitance of the arrangement.

(A) 3 μF

(B) 6 μF

(C) 9 μF

(D) $\frac{1}{3}$ μF

(E) 1 μF

37. A microwave is connected to an outlet, 120 V, and draws a current of 2 amps. At what rate is energy being used by the microwave?

(A) 240 W
(B) 60 W
(C) 480 W
(D) 30 W
(E) 10 W

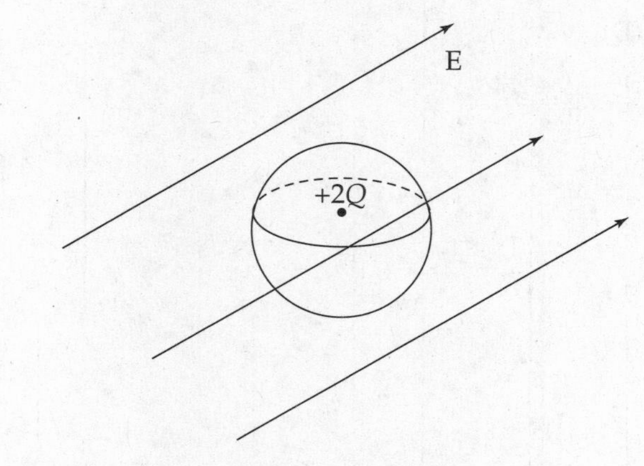

38. A uniform electric field exists in a region, and then a neutral, conducting, spherical shell with a stationary charge +2*Q* at its center is placed in the region, as shown above. The radius of the sphere is *R*. The flux through the sphere depends on the value of

(A) E, Q and R
(B) Only *R*
(C) E and *Q*
(D) *R* and *Q*
(E) Only *Q*

39. In a certain region, the electric field varies with the radius away from origin by the equation $E_r = -6r^2 + 4r + 3$, where *r* is given in meters and *E* in N/C.
The potential difference between the origin and the point (3, 4) is

(A) 315 V
(B) 185 V
(C) 64 V
(D) –165 V
(E) –120 V

GO ON TO THE NEXT PAGE

Questions 40–41

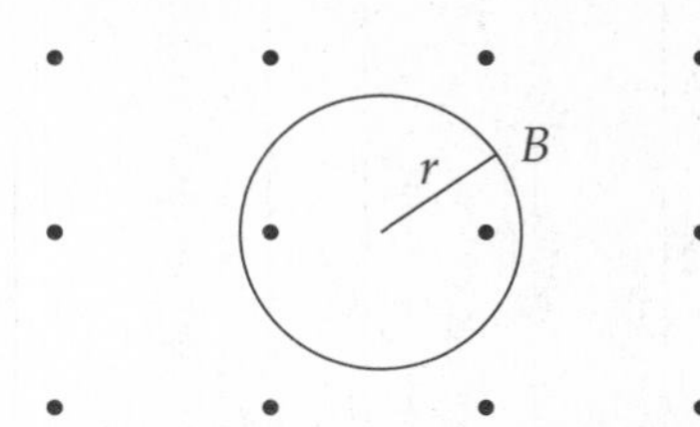

A particle of charge $-q$ and mass m moves with speed v perpendicular to a uniform magnetic field B directed out of the page. The path of the particle is a circle of radius r, as shown above.

40. Which of the following correctly gives the direction of motion and the magnitude of the acceleration of the charge?

	Direction	Acceleration of Charge
(A)	Clockwise	qBv
(B)	Clockwise	$\frac{qBv}{m}$
(C)	Counterclockwise	$\frac{qBv}{m}$
(D)	Counterclockwise	qBv
(E)	Counterclockwise	$\frac{qBv}{r}$

41. The frequency with which the particle completes the circular path is

(A) $\frac{v}{2r}$

(B) $\frac{mv}{2\pi r}$

(C) $\frac{2\pi r}{v}$

(D) $\frac{v}{2\pi r}$

(E) $\frac{v}{2\pi}$

42. A 30 μF capacitor has 6 millicoulombs of charge on each plate. The energy stored in the capacitor is most nearly

(A) 5.4×10^{-10} J
(B) 9.0×10^{-8} J
(C) 0.6 J
(D) 12.5 J
(E) 100 J

43. Two large, parallel conducting plates have a potential difference of V maintained across them. A proton starts at rest on the surface of one plate and accelerates toward the other plate. Its acceleration in the region between the plates is proportional to

(A) $\frac{1}{V}$

(B) $\frac{1}{\sqrt{V}}$

(C) $\sqrt{V}$

(D) V

(E) V^2

GO ON TO THE NEXT PAGE

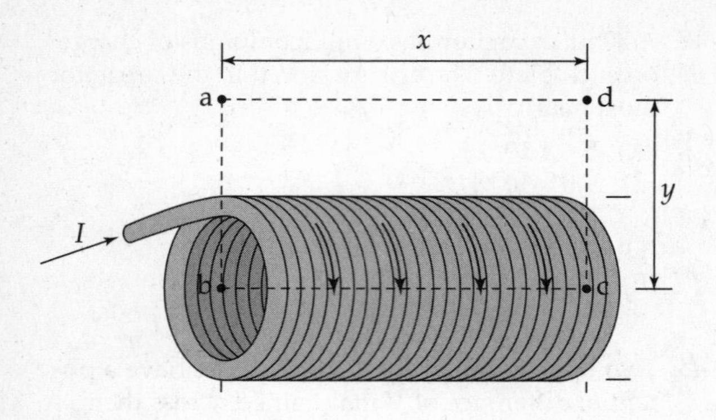

44. An ideal solenoid with N total turns has a current I passing through the helical wires that make up the solenoid. Ampere's Law is used with a rectangular path *abcd* as shown above, to calculate the magnitude of the magnetic field B within the solenoid. The horizontal distances of the path are length x and the vertical distances of the path are length y. Which of the following equations results from the correct application of Ampere's law in this situation?

(A) $B(2x + 2y) = \mu_o NI$

(B) $B(2x) = \mu_o NI$

(C) $B(x + 2y) = \mu_o NI$

(D) $B(2y) = \mu_o NI$

(E) $B(x) = \mu_o NI$

Questions 45–46

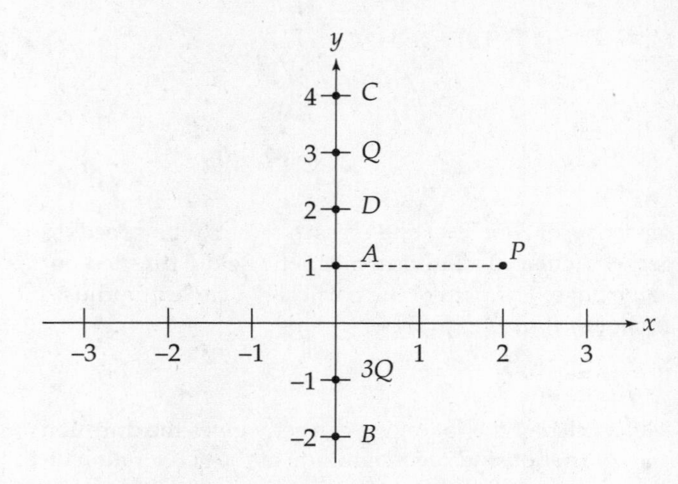

Particles of charge $+3Q$ and $+Q$ are located on the y-axis as shown above. Assume the particles are isolated from other particles and are stationary. A, B, C, D, and P are points in the plane as indicated in the diagram.

45. Which of the following describes the direction of the electric field at point P?

(A) $+x$ direction
(B) $-y$ direction
(C) components in both the $+x$ and $-y$ direction
(D) components in both $-x$ and $+y$ direction
(E) components in both $+x$ and $+y$ direction

46. At which of the labeled points is the electric potential zero?

(A) A
(B) B
(C) C
(D) D
(E) None of the points

GO ON TO THE NEXT PAGE

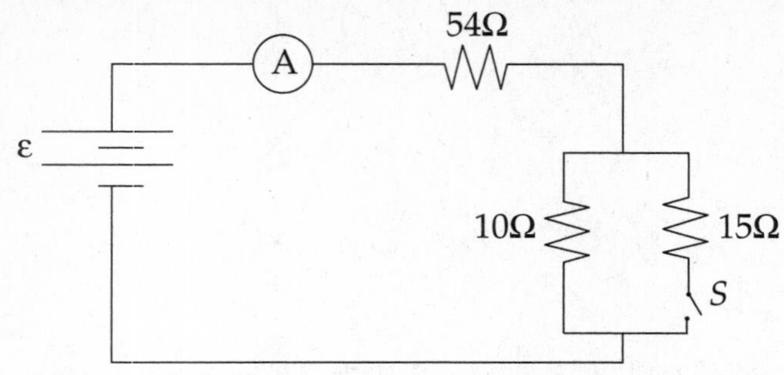

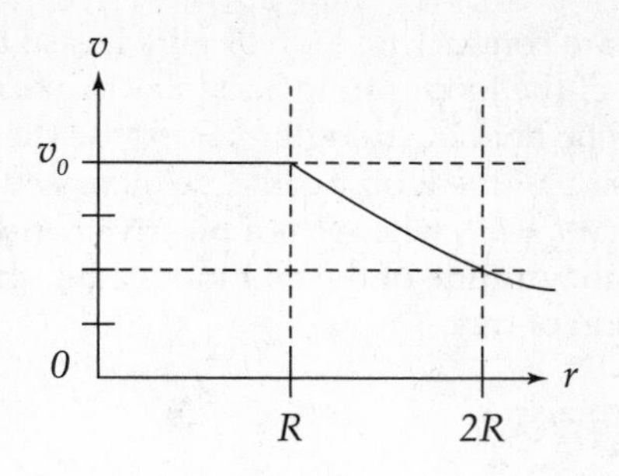

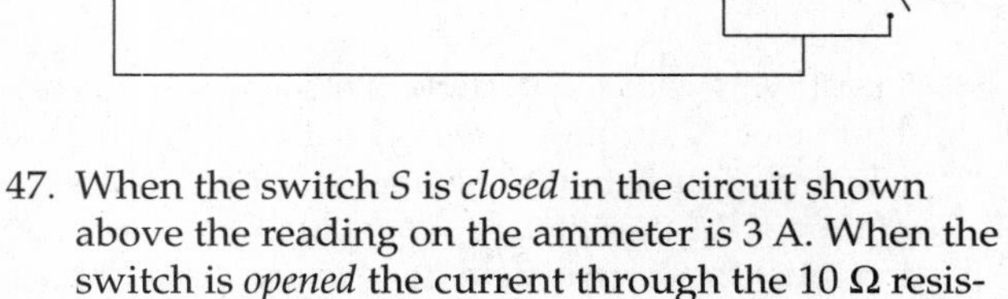

47. When the switch S is *closed* in the circuit shown above the reading on the ammeter is 3 A. When the switch is *opened* the current through the 10 Ω resistor will

(A) double
(B) increase but not double
(C) remain the same
(D) decrease but not be halved
(E) be halved

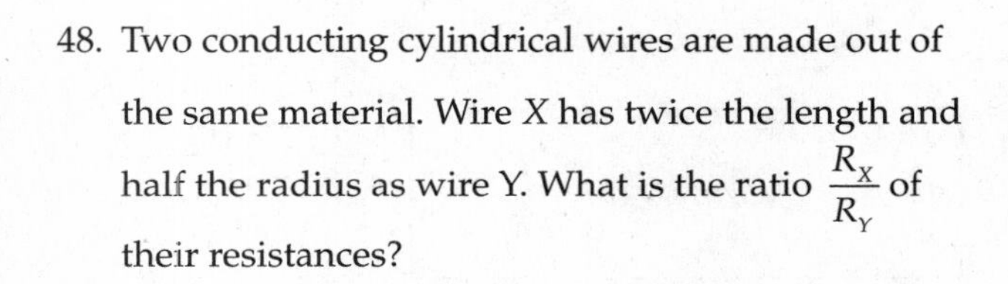

48. Two conducting cylindrical wires are made out of the same material. Wire X has twice the length and half the radius as wire Y. What is the ratio $\frac{R_X}{R_Y}$ of their resistances?

(A) 8

(B) 4

(C) 1

(D) $\frac{1}{2}$

(E) $\frac{1}{4}$

49. A graph of electric potential V as a function of the radius from the origin r is shown above. What can be concluded about the electric field in the region $0 < r < R$?

(A) It increases linearly as r increases.
(B) It decreases linearly as r increases.
(C) It is zero.
(D) It increases non-linearly as r increases.
(E) It decreases non-linearly as r increases.

50. Two parallel wires, each carrying a current I, attract each other with a force F. If both currents are halved the attractive force is

(A) $4F$

(B) $\frac{1}{\sqrt{2}}F$

(C) $\frac{1}{2}F$

(D) $\sqrt{2}F$

(E) $\frac{1}{4}F$

GO ON TO THE NEXT PAGE

51. A square conducting loop of wire lies so that the plane of the loop is perpendicular to a constant magnetic field of strength B. Suppose the length of each side of the loop ℓ could be increased with time t so that $\ell = kt^2$, where k is a positive constant. What is the magnitude of the emf induced in the loop as a function of time?

(A) $4Bk^2t^3$

(B) $2Bk^2t$

(C) $4Bkt^3$

(D) $2Bkt$

(E) $\frac{Bk^2t^5}{5}$

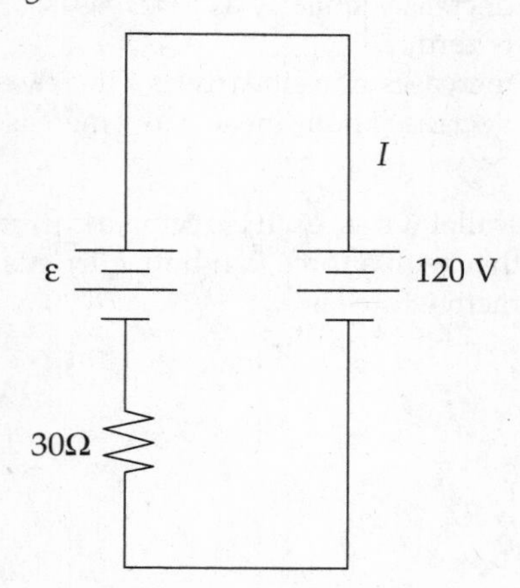

52. A battery with emf ε and internal resistance of 30 Ω is being recharged by connecting it to an outlet with a potential difference of 120 V as shown above. While it is being recharged, 3 A flows through the battery. Determine the emf of the battery.

(A) 210 V
(B) 150 V
(C) 90 V
(D) 30 V
(E) 9 V

× × × B into

× × ×

⊕ ⟶ V

× × ×

53. A positively charged particle is moving with a constant velocity through a region with both a magnetic field and electric field. The magnetic field and the motion of the particle are shown above. What direction must the electric field be to cause the particle to travel at a constant velocity?

(A) upward
(B) downward
(C) left
(D) right
(E) out of the page

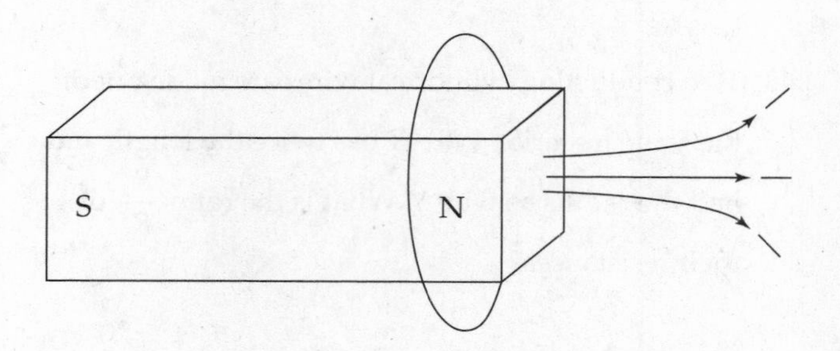

54. A conducting loop of wire is initially around a magnet as shown above. The magnet is moved to the left. What is the direction of the force on the loop and the direction of the magnetic field at the center of the loop due to the induced current?

	Direction of Force on Loop	Direction of Magnetic Field at Center of Loop Due to Induced Current
(A)	To the right	To the right
(B)	To the right	To the left
(C)	To the left	To the right
(D)	To the left	To the left
(E)	No direction; the force is zero	To the left

GO ON TO THE NEXT PAGE

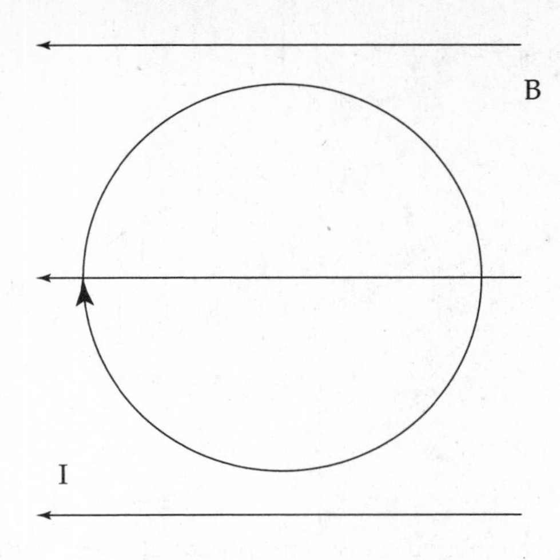

55. A loop of wire carrying a current I is initially in the plane of the page and is located in a uniform magnetic field B which points toward the left side of the page, as shown above. Which of the following shows the correct initial rotation of the loop due to the force exerted by the magnetic field?

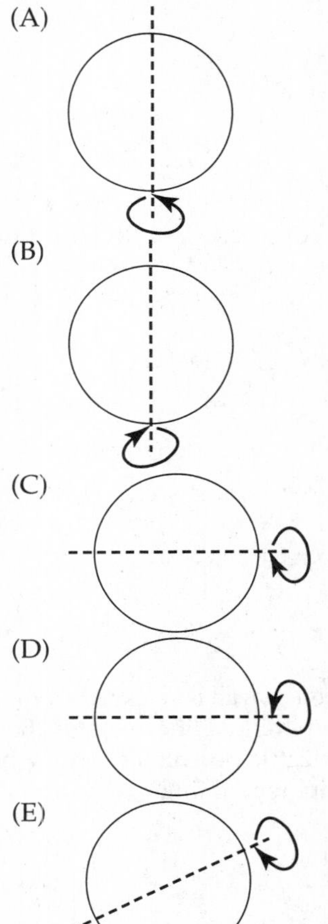

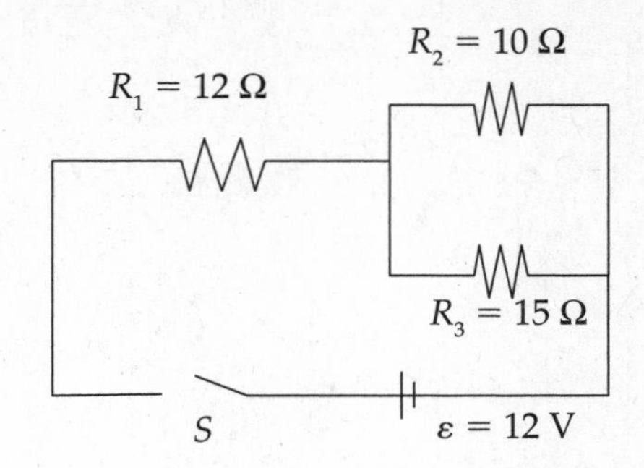

56. What is the equivalent resistance of the three resistors shown in the circuit above?

(A) 37 Ω
(B) 8.1 Ω
(C) 6 Ω
(D) 18 Ω
(E) 22 Ω

GO ON TO THE NEXT PAGE

Questions 57–58

+Q +Q

ℓ

+Q +Q

Four particles, each with a charge $+Q$, are held fixed at the corners of a square, as shown above. The distance from each charge to the center of the square is ℓ.

57. What is the magnitude of the electric field at the center of the square?

(A) 0

(B) $\frac{4kQ}{\ell^2}$

(C) $\frac{2kQ}{\ell^2}$

(D) $\frac{4kQ}{\sqrt{2}\ell^2}$

(E) $\frac{kQ}{\sqrt{2}\ell^2}$

58. What is the magnitude of the work required to move a charge of +3Q from the center of the square to very far away?

(A) $\frac{12kQ^2}{\ell}$

(B) $\frac{12kQ^2}{\ell^2}$

(C) $\frac{4kQ^2}{\ell^2}$

(D) $\frac{3kQ^2}{\ell}$

(E) $\frac{4kQ}{\ell}$

Questions 59–61

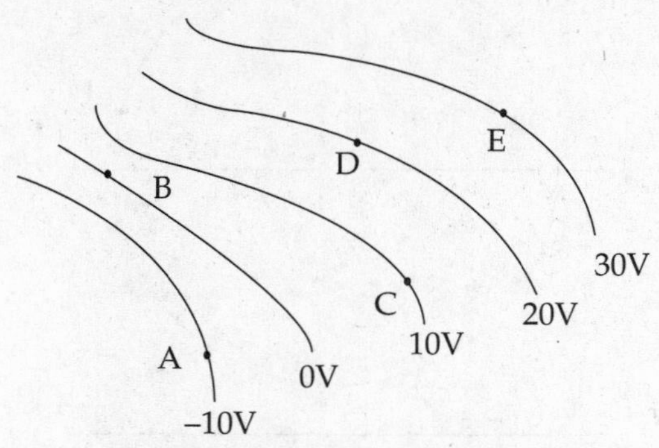

The diagram above shows equipotential lines produced by a charge distribution. *A*, *B*, *C*, *D*, and *E* are points in the plane.

59. At which point is the magnitude of the electric field the greatest?

(A) A
(B) B
(C) C
(D) D
(E) E

60. Which vector below bests describes the direction of the electric field at point *D* ?

(A) 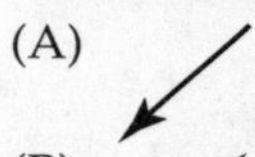

(B) 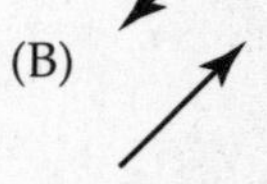

(C) 

(D)

(E) 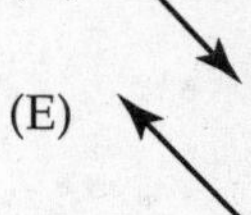

61. A particle with a –3 μC charge is released from rest on the –10 V equipotential line. What is the particle's change in electric potential energy when it reaches the 20 V equipotential line?

(A) 90 μJ
(B) 60 μJ
(C) 30 μJ
(D) –60 μJ
(E) –90 μJ

GO ON TO THE NEXT PAGE

62. Which of Maxwell's equations allows for the calculation of a magnetic field due to a changing electric field?

(A) $\oint E \bullet dA = \frac{q}{\varepsilon_0}$

(B) $\oint E \bullet d\ell = \frac{d\phi_B}{dt}$

(C) $\oint B \bullet dA = 0$

(D) $\oint B \bullet d\ell = \mu_0 i + \mu_0 \varepsilon_0 \frac{d\phi_E}{dt}$

(E) None of the above

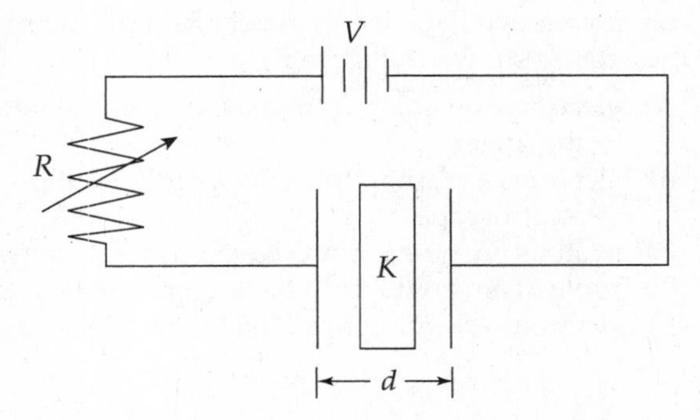

63. A parallel plate capacitor has a dielectric material between the plates with a constant K. The capacitor is connected to a variable resistor R and a power supply of potential difference V. Each plate of the capacitor has a cross-sectional area A and the plates are separated by a distance d. Which of the following changes could increase the capacitance and decrease the amount of charge stored on the capacitor?

(A) Increase R and increase A
(B) Decrease V and decrease d
(C) Decrease R and increase d
(D) Increase K and increase V
(E) Increase K and increase R

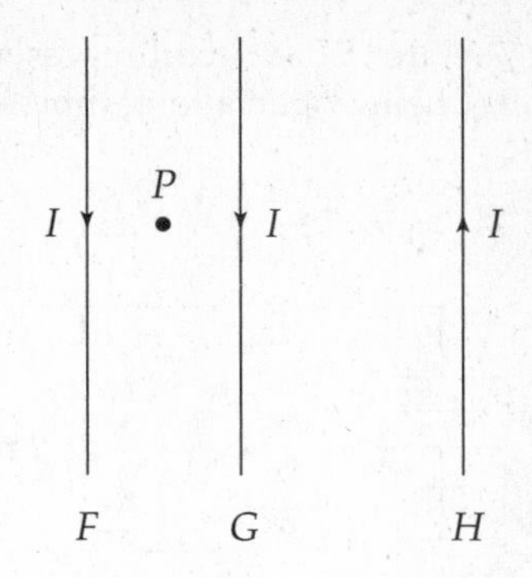

64. Three parallel wires, F, G, and H, all carry equal current I, in the directions shown above. Wire G is closer to wire F than to wire H. The magnetic field at point P is directed

(A) into the page
(B) out of the page
(C) to the left
(D) to the right
(E) toward the top of the page

65. A solid, metal object is isolated from other charges and has charge distributed on its surface. The charge distribution is not uniform. It may be correctly concluded that the

(A) electric field outside the object is zero
(B) the electric field outside the object is equal to the electric field inside the object
(C) the electric field outside the object is directly proportional to the distance away from the center of mass of the object
(D) the electric field outside the object, but very close to the surface, is equal to the surface charge density at any location divided by the permittivity of free space
(E) the electric potential on the surface of the object is not constant

GO ON TO THE NEXT PAGE

Questions 66–67 relate to the circuit represented below. The switch S, after being open a long time, is then closed.

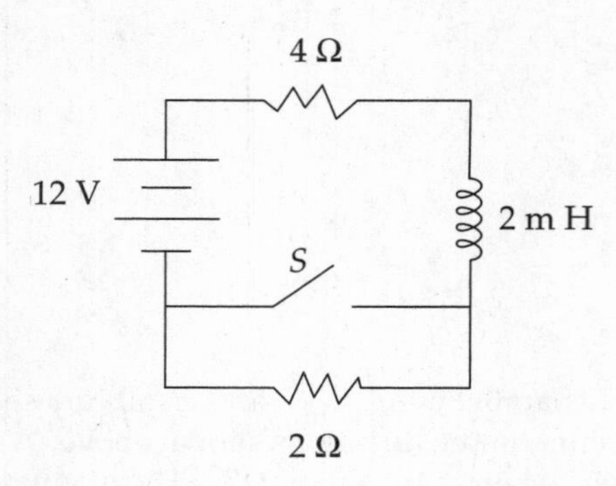

66. What is the potential difference across the inductor immediately after the switch is closed?

(A) 4 V
(B) 12 V
(C) 0 V
(D) 2 V
(E) 8 V

67. What is the current through the 4 Ω resistor after the switch has been closed a long time?

(A) 2 A
(B) 1.2 A
(C) 6 A
(D) 3 A
(E) 1.5 A

---

68. A spherical charge distribution varies with the radius r by the equation

$\rho = ar$, where ρ is the volume charge density and a is a positive constant. The distribution goes out to a radius R.

Which of the following is true of the electric field strength due to this charge distribution at a distance r from the center?

(A) It increases as r approaches infinity.
(B) It decreases linearly for $r > R$.
(C) It increases linearly for $r > R$.
(D) It increases linearly for $r < R$.
(E) It increases non-linearly for $r < R$.

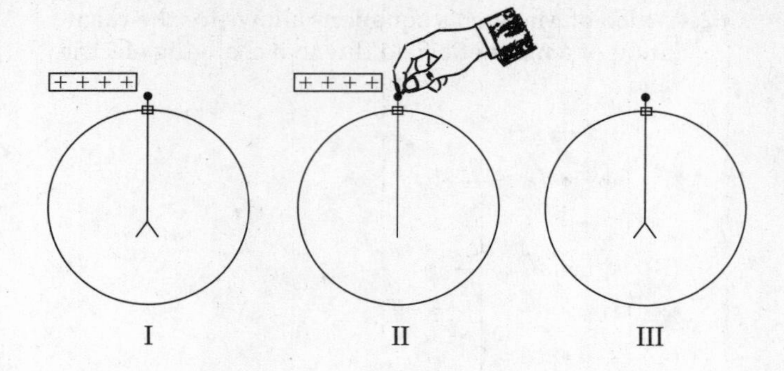

69. When a positively charged rod is brought near, but does not touch, the initially neutral electroscope shown above, the leaves repel (I). When the electroscope is then touched with a finger, the leaves hang vertically (II). Next when the finger and finally the rod are removed, the leaves repel again (III). During the process shown in Figure II

(A) electrons are going from the electroscope into the finger
(B) electrons are going from the finger into the electroscope
(C) protons are going from the rod into the finger
(D) protons are going from the finger into the rod
(E) electrons are going from the finger into the rod

GO ON TO THE NEXT PAGE

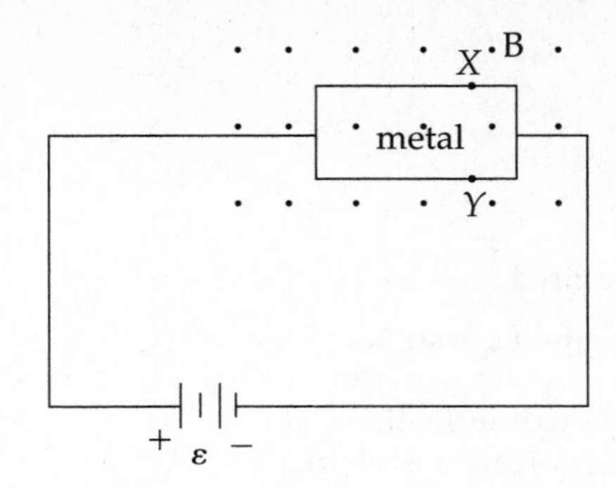

70. A piece of metal in the plane of the page is connected in a circuit as shown above, causing electrons to move through the metal to the left. The piece of metal is in a magnetic field B directed out of the page. X and Y are points on the edge of the metal. Which of the following statements is true?

(A) The current will decrease to zero due to the magnetic field.
(B) The potentials at X and Y are equal.
(C) X is at a higher potential than Y.
(D) Y is at a higher potential than X.
(E) The current will increase exponentially due to the magnetic field.

STOP

END OF SECTION I, ELECTRICITY AND MAGNETISM

IF YOU FINISH BEFORE TIME IS CALLED, YOU MAY CHECK YOUR WORK ON SECTION I, ELECTRICITY AND MAGNETISM, ONLY.

DO NOT GO ON TO SECTION II UNTIL YOU ARE TOLD TO DO SO.

PHYSICS C

SECTION II

Free-Response Questions

Mechanics	45 minutes	3 required questions of equal weight
Electricity and Magnetism	45 minutes	3 required questions of equal weight

Section II is 50 percent of the total grade for each of the two examinations.
Mark one of the boxes below to indicate which questions you are answering.

☐ Mechanics only

☐ Electricity and Magnetism only

☐ Both Mechanics and Electricity and Magnetism

ADVANCED PLACEMENT PHYSICS C EQUATIONS FOR 2008

MECHANICS

$v = v_0 + at$

$x = x_0 + v_0 t + \frac{1}{2}at^2$

$v^2 = v_0^2 + 2a(x - x_0)$

$\Sigma \mathbf{F} = \mathbf{F}_{net} = m\mathbf{a}$

$\mathbf{F} = \frac{d\mathbf{p}}{dt}$

$\mathbf{J} = \int \mathbf{F}\,dt = \Delta\mathbf{p}$

$\mathbf{p} = m\mathbf{v}$

$F_{fric} \le \mu N$

$W = \int \mathbf{F} \cdot d\mathbf{r}$

$K = \frac{1}{2}mv^2$

$P = \frac{dW}{dt}$

$P = \mathbf{F} \cdot \mathbf{v}$

$\Delta U_g = mgh$

$a_c = \frac{v^2}{r} = \omega^2 r$

$\boldsymbol{\tau} = \mathbf{r} \times \mathbf{F}$

$\Sigma \boldsymbol{\tau} = \boldsymbol{\tau}_{net} = I\boldsymbol{\alpha}$

$I = \int r^2 dm = \Sigma mr^2$

$\mathbf{r}_{cm} = \Sigma m\mathbf{r}/\Sigma m$

$v = r\omega$

$\mathbf{L} = \mathbf{r} \times \mathbf{p} = I\boldsymbol{\omega}$

$K = \frac{1}{2}I\omega^2$

$\omega = \omega_0 + \alpha t$

$\theta = \theta_0 + \omega_0 t + \frac{1}{2}\alpha t^2$

$\mathbf{F}_s = -k\mathbf{x}$

$U_s = \frac{1}{2}kx^2$

$T = \frac{2\pi}{\omega} = \frac{1}{f}$

$T_s = 2\pi\sqrt{\frac{m}{k}}$

$T_p = 2\pi\sqrt{\frac{\ell}{g}}$

$\mathbf{F}_G = -\frac{Gm_1m_2}{r^2}\hat{\mathbf{r}}$

$U_G = -\frac{Gm_1m_2}{r}$

a = acceleration
F = force
f = frequency
h = height
I = rotational inertia
J = impulse
K = kinetic energy
k = spring constant
ℓ = length
L = angular momentum
m = mass
N = normal force
P = power
p = momentum
r = radius or distance
$\mathbf{r}$ = position vector
T = period
t = time
U = potential energy
v = velocity or speed
W = work done on a system
x = position
μ = coefficient of friction
θ = angle
τ = torque
ω = angular speed
α = angular acceleration

ELECTRICITY AND MAGNETISM

$F = \frac{1}{4\pi\epsilon_0}\frac{q_1q_2}{r^2}$

$\mathbf{E} = \frac{\mathbf{F}}{q}$

$\oint \mathbf{E} \cdot d\mathbf{A} = \frac{Q}{\epsilon_0}$

$E = -\frac{dV}{dr}$

$V = \frac{1}{4\pi\epsilon_0}\sum_i \frac{q_i}{r_i}$

$U_E = qV = \frac{1}{4\pi\epsilon_0}\frac{q_1q_2}{r}$

$C = \frac{Q}{V}$

$C = \frac{\kappa\epsilon_0 A}{d}$

$C_p = \sum_i C_i$

$\frac{1}{C_s} = \sum_i \frac{1}{C_i}$

$I = \frac{dQ}{dt}$

$U_c = \frac{1}{2}QV = \frac{1}{2}CV^2$

$R = \frac{\rho\ell}{A}$

$\mathbf{E} = \rho\mathbf{J}$

$I = Nev_d A$

$V = IR$

$R_s = \sum_i R_i$

$\frac{1}{R_p} = \sum_i \frac{1}{R_i}$

$P = IV$

$\mathbf{F}_M = q\mathbf{v} \times \mathbf{B}$

A = area
B = magnetic field
C = capacitance
d = distance
E = electric field
$\mathcal{E}$ = emf
F = force
I = current
J = current density
L = inductance
ℓ = length
n = number of loops of wire per unit length
N = number of charge carriers per unit volume
P = power
Q = charge
q = point charge
R = resistance
r = distance
t = time
U = potential or stored energy
V = electric potential
v = velocity or speed
ρ = resistivity
ϕ_m = magnetic flux
κ = dielectric constant

$\oint \mathbf{B} \cdot d\boldsymbol{\ell} = \mu_0 I$

$d\mathbf{B} = \frac{\mu_0}{4\pi}\frac{I\,d\boldsymbol{\ell} \times \mathbf{r}}{r^3}$

$\mathbf{F} = \int I\,d\boldsymbol{\ell} \times \mathbf{B}$

$B_s = \mu_0 nI$

$\phi_m = \int \mathbf{B} \cdot d\mathbf{A}$

$\mathcal{E} = -\frac{d\phi_m}{dt}$

$\mathcal{E} = -L\frac{dI}{dt}$

$U_L = \frac{1}{2}LI^2$

ADVANCED PLACEMENT PHYSICS C EQUATIONS FOR 2008

GEOMETRY AND TRIGONOMETRY

Rectangle

$A = bh$

Triangle

$A = \frac{1}{2}bh$

Circle

$A = \pi r^2$

$C = 2\pi r$

Parallelepiped

$V = \ell wh$

Cylinder

$V = \pi r^2 \ell$

$S = 2\pi r\ell + 2\pi r^2$

Sphere

$V = \frac{4}{3}\pi r^3$

$S = 4\pi r^2$

Right Triangle

$a^2 + b^2 = c^2$

$\sin\theta = \frac{a}{c}$

$\cos\theta = \frac{b}{c}$

$\tan\theta = \frac{a}{b}$

A = area
C = circumference
V = volume
S = surface area
b = base
h = height
ℓ = length
w = width
r = radius

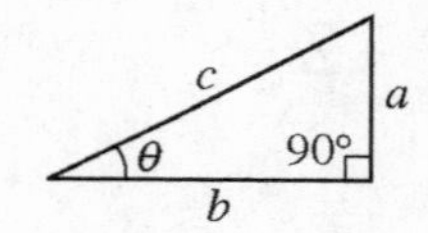

CALCULUS

$$\frac{df}{dx} = \frac{df}{du}\frac{du}{dx}$$

$$\frac{d}{dx}(x^n) = nx^{n-1}$$

$$\frac{d}{dx}(e^x) = e^x$$

$$\frac{d}{dx}(\ln x) = \frac{1}{x}$$

$$\frac{d}{dx}(\sin x) = \cos x$$

$$\frac{d}{dx}(\cos x) = -\sin x$$

$$\int x^n\,dx = \frac{1}{n+1}x^{n+1},\ n \neq -1$$

$$\int e^x\,dx = e^x$$

$$\int \frac{dx}{x} = \ln|x|$$

$$\int \cos x\,dx = \sin x$$

$$\int \sin x\,dx = -\cos x$$

PHYSICS C
SECTION II, MECHANICS
Time – 45 minutes
3 Questions

Directions: Answer all three questions. The suggested time is about 15 minutes per question for answering each of the questions, which are worth 15 points each. The parts within a question may not have equal weight.

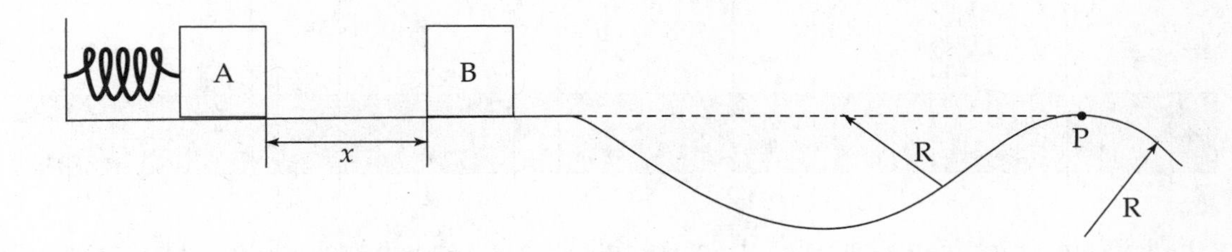

1. A massless spring with force constant k is attached at its left end to a wall, as shown above. Initially, block A and block B, each of mass M, are at rest on a frictionless, level surface, with block A in contact with the spring (but not compressing it) and block B a distance x from block A. Block A is then moved to the left, compressing the spring a distance of d, and held in place while block B remains at rest. First block A is released, then as it passes the equilibrium position loses contact with the spring. After block A is released it moves forward and has a perfectly inelastic collision with block B and then follows the frictionless, curved path shown above. The radius of the valley and the hill in the diagram are both R. Answer the following in terms of M, k, d, x, g and R.

   (a) Determine the speed of block A just before it collides with block B.
   (b) Determine the speed of block B just after the collision occurs.
   (c) Determine the change in kinetic energy for the collision.
   (d) Determine the normal force on the boxes when they are at position P, the top of the hill.

GO ON TO THE NEXT PAGE

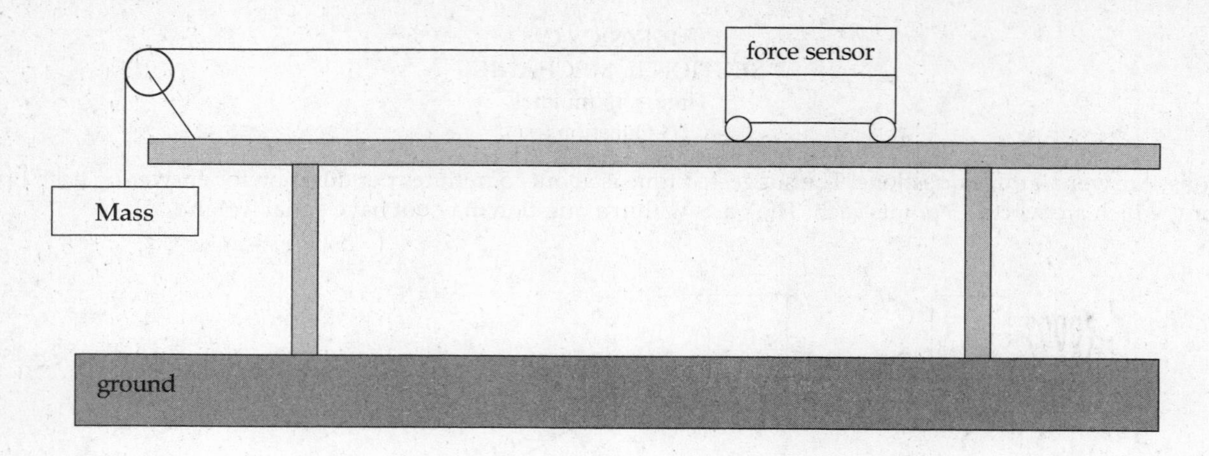

2. Physics students are performing a lab using the mass-pulley system shown above. The cart has a force sensor attached to it, and the total mass of the force sensor and cart is 1.2 kg. The force sensor is attached to a 300-gram hanging mass by a string that is placed over a pulley. When the system is released, a motion detector recording the motion of the cart produces the following velocity vs. time graph.

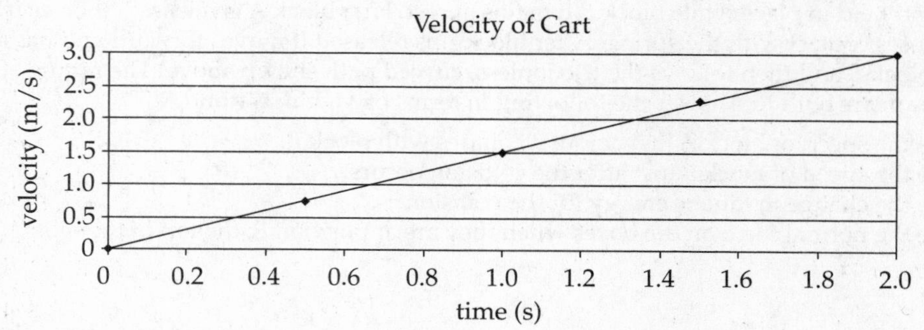

(a) Determine the acceleration of the cart.

(b) Assume tension is the only horizontal force on the force sensor/cart combination (ignore friction). Determine the tension in the string for this scenario.

(c) The measured tension is actually 20% greater than the tension predicted in (b). Explain why this might be the case. The force sensor and the motion sensor are working properly, so do not use faulty data to explain the result.

(d) Describe a process you could use to determine the rotational inertia of the pulley with this system, a meterstick, the force sensor, and motion sensor.

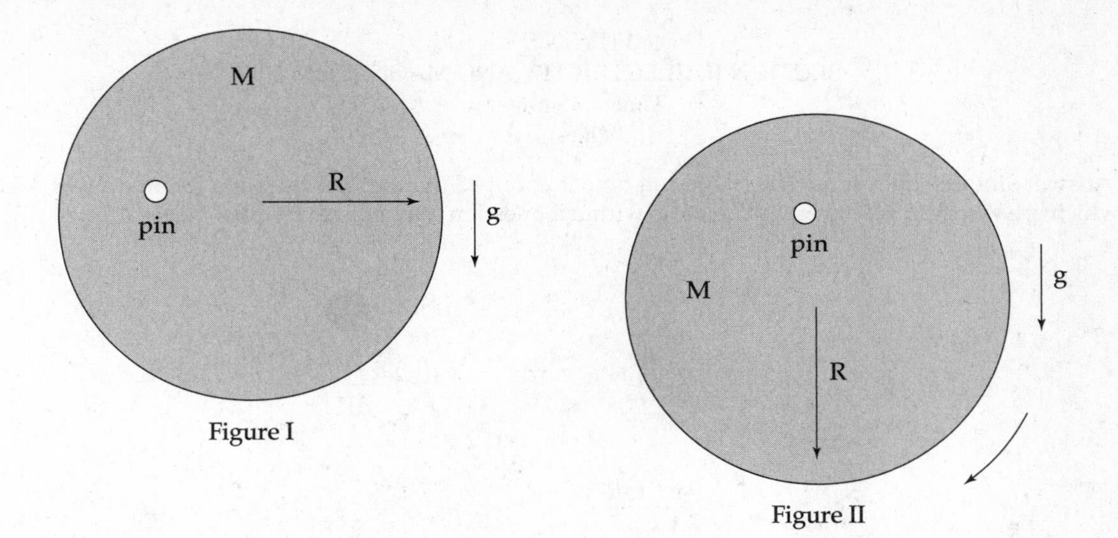

3. A disk of mass M and radius R is pinned half of the way along its radius, and held in a horizontal position, as shown in Figure I. The rotational inertia of the disk about its center is $\frac{1}{2}MR^2$. The disk is released at $t = 0$ s, and falls to the vertical position shown in Figure II, and continues to rotate about the pin. Answer the following in terms of M, R, and g.

(a) Determine the rotational inertia of the disk about the pin.
(b) Determine the angular acceleration of the disk at $t = 0$ s.
(c) Determine angular velocity of the disk when it is in the vertical position shown in Figure II.

Now the disk is stopped and brought to rest in the vertical position shown in Figure II. It is given a slight disturbance to an angle θ_0.

(d) Determine the angular frequency of the oscillation.

STOP

END OF SECTION II, MECHANICS

IF YOU FINISH BEFORE TIME IS CALLED, YOU MAY CHECK YOUR WORK ON SECTION II, MECHANICS, ONLY. DO NOT TURN TO ANY OTHER TEST MATERIALS.

PHYSICS C
SECTION II, ELECTRICITY AND MAGNETISM
Time – 45 minutes
3 Questions

Directions: Answer all three questions. The suggested time is about 15 minutes per question for answering each of the questions, which are worth 15 points each. The parts within a question may not have equal weight.

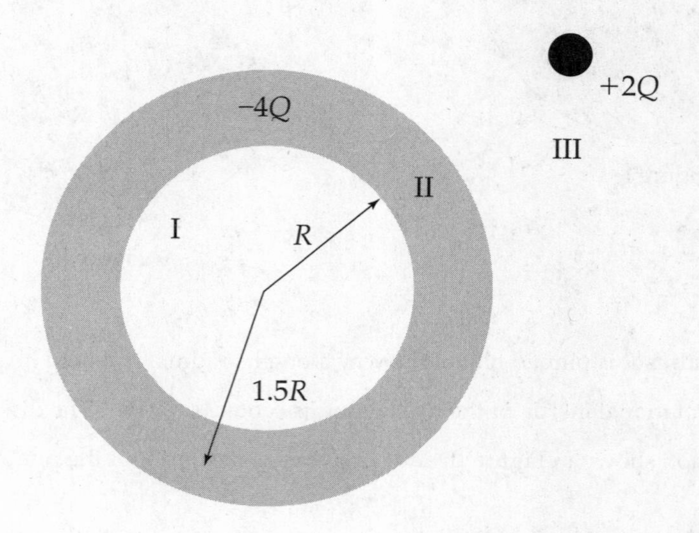

1. A spherical, metal shell of inner radius R and outer radius $1.5R$ has a charge of $-4Q$. A point charge of $+2Q$ is initially located outside the shell as shown above. Express all answers in terms of fundamental constants and given values.

   (a) (i) Determine the charge on each surface of the spherical shell.
   (ii) Sketch the electric field in regions I, II and III.

GO ON TO THE NEXT PAGE

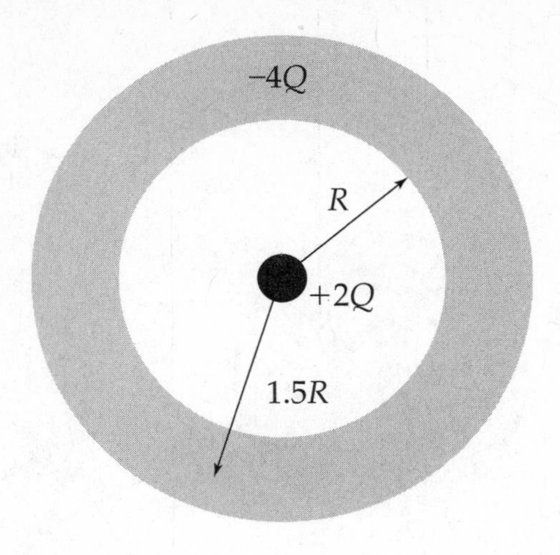

Now the +2Q point charge is moved to the center of the spherical shell as shown above.

(b) Determine the electric field strength for the following radii.
- (i) $r < R$
- (ii) $R < r < 1.5R$
- (iii) $r > 1.5R$

(c) Determine the potential difference between infinity and the outside surface of the spherical shell.

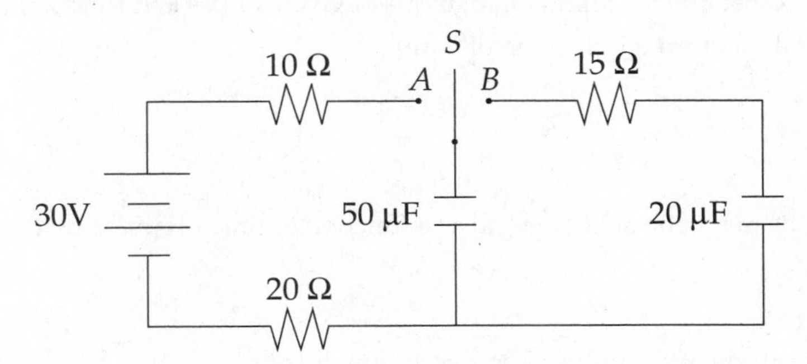

2. In the circuit shown above, the switch S is initially in the open position and both capacitors are initially uncharged. Then the switch is moved to position A.

(a) Determine the current through the 20 Ω resistor immediately after the switch is moved to position A.

(b) Sketch a graph of voltage vs. time for the voltage across the 10 Ω resistor.

After a long time the switch is moved to position B.

(c) Determine the current through the 15 Ω resistor immediately after the switch is moved to position B.

(d) Determine the amount of charge stored on the upper plate of the 20 μF capacitor after a long time.

GO ON TO THE NEXT PAGE

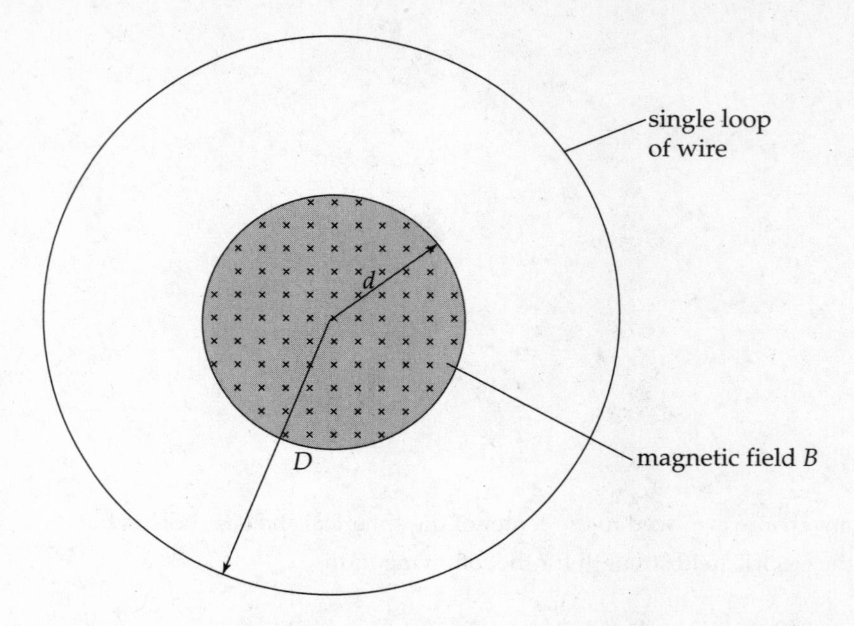

3. A uniform magnetic field B is directed into the page, and exists in a circular region of radius d. A single loop of wire of radius D is placed concentrically around the magnetic field region in the plane of the page. The initial magnetic field strength is B_o. Calculate the following in terms of given values and fundamental constants.

(a) Determine the initial flux through the loop of wire.

Starting at time $t = 0$ s, the magnetic field strength as a function of time t is given by the equation $B(t) = B_0 t^2$.

(b) Determine the magnitude of the induced emf in the single loop.
(c) Determine the direction of the induced current in the loop.

The loop of wire has a resistance R.

(d) Determine the energy dissipated in the loop up until a given time t_1.

END OF EXAMINATION

STOP

END OF SECTION II, ELECTRICITY AND MAGNETISM

IF YOU FINISH BEFORE TIME IS CALLED YOU MAY CHECK YOUR WORK ON SECTION II, ELECTRICITY AND MAGNETISM, ONLY.

DO NOT TURN TO ANY OTHER TEST MATERIALS.

18

AP Physics C Practice Exam 2: Answers and Explanations

SECTION I, MECHANICS

1. **E** The initial velocity is negative, so the initial displacement must start out negative. The object also continues at a constant speed after time t_1, so that position vs. time should be a straight line with a positive slope at this time. (E) is the only answer choice that meets these two conditions.

2. **D** When an object is dropped its initial vertical velocity is zero. We can solve for the time to land using the following constant acceleration equation and plugging in appropriate values.

$$y = y_o + v_0 t + \frac{1}{2} g t^2$$

$$0 = 80 + 0 + \frac{1}{2}(-10)t^2$$

$$t = 4\,\text{s}$$

3. **D** The direction of the change in momentum of an object will be the direction of Δv. The resultant of $v_2 - v_1$ is shown below.

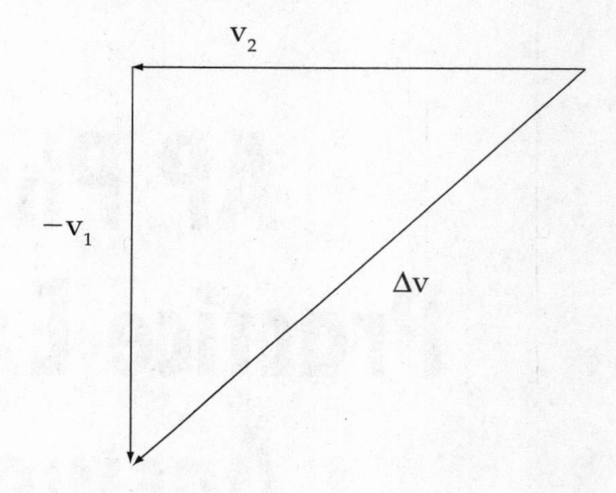

4. **B** The acceleration of the $2m$ mass is the same as the acceleration of the m and $3m$ mass because they are connected to each other. Use Newton's Second Law for the entire system to solve for the acceleration.

$$\sum F = ma$$

$$F = (3m + 2m + m)a$$

$$a = \frac{F}{6m}$$

5. **B** Read the force directly off the graph. At $t = 2$ s the force is 3 N. Use Newton's Second Law to solve for the acceleration.

$$\sum F = ma$$
$$3 = 2a$$
$$a = 1.5 m/s^2$$

6. **C** We will use the impulse-momentum theorem here. We cannot use the constant acceleration equations with the acceleration from question 5 because the acceleration is not constant. The area between the graph and the *x*-axis is the impulse.

$$\int F dt = \Delta p$$
$$Area = m\Delta v$$
$$\frac{1}{2}(3)(9) = 2(v_2 - v_1)$$
$$13.5 = 2v_2$$
$$v_2 = 6.75 m/s$$

7. **D** We will use the law of conservation of energy for this problem. The cylinder has rotational and translational kinetic energy as it rolls along the horizontal surface. All of that energy will be converted into gravitational potential energy when it reaches its maximum height up the ramp.

$$K_R + K_T = U_g$$
$$\frac{1}{2}I\omega^2 + \frac{1}{2}mv^2 = mgh$$
$$\frac{1}{2}\left(\frac{1}{2}mr^2\right)\left(\frac{v}{r}\right)^2 + \frac{1}{2}mv^2 = mgh$$
$$\frac{1}{4}mv^2 + \frac{1}{2}mv^2 = mgh$$
$$\frac{3}{4}v^2 = gh$$
$$h = \frac{3v^2}{4g}$$

8. **A** Again we will use the law of conservation of energy. The easiest way to determine which object will reach a greater height is to determine which object has more kinetic energy as it rolls along the horizontal surface. If both the cylinder and the hoop have the same velocity then the hoop will have more total kinetic energy because it has a greater rotational inertia. The calculation below shows the calculation for the height which also shows that the hoop reaches a greater height.

$$K_R + K_T = U_g$$

$$\frac{1}{2}I\omega^2 + \frac{1}{2}mv^2 = mgh$$

$$\frac{1}{2}\left(mr^2\right)\left(\frac{v}{r}\right)^2 + \frac{1}{2}mv^2 = mgh$$

$$\frac{1}{2}mv^2 + \frac{1}{2}mv^2 = mgh$$

$$v^2 = gh$$

$$h = \frac{v^2}{g}$$

While both (A) and (D) indicate that the hoop reaches a greater maximum height than the cylinder, (D) is incorrect as the cylinder reaches 3/4 the height of the hoop.

9. **C** The mass is in static equilibrium so the net force in all directions is zero. Using the sum of the forces in the vertical direction, one can conclude that the vertical component of the slanted cord must equal the force of gravity of the mass. That produces the following triangle for the tension in the slanted cord.

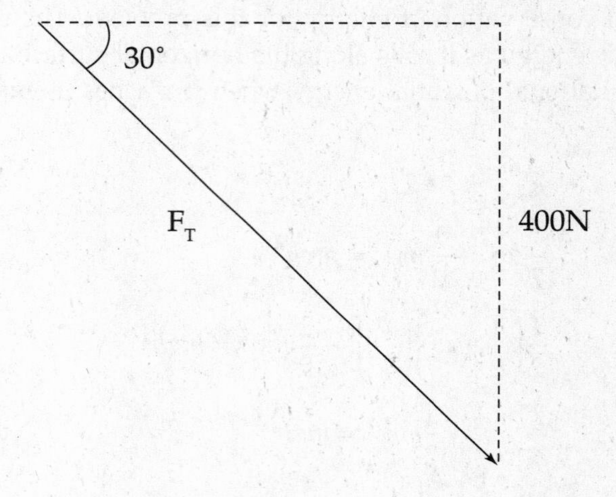

Using trigonometry we can solve for F_T.

$$\sin 30^o = \frac{400}{F_T}$$

$$F_T = 800\,N$$

10. **D** For all three of these problems, (A) cannot be the correct choice because the gravitational force mg must be directed straight down. (B) cannot be the correct choice because the normal force N must be perpendicular to the surface. If the box is going up the incline then the frictional force must be directed down the incline, which is only true for (D).

11. **C** For all three of these problems, (A) cannot be the correct choice because the gravitational force mg must be directed straight down. (B) cannot be the correct choice because the normal force N must be perpendicular to the surface. If the box is moving at a constant velocity the net force in the x-direction, along the incline, must be zero. This is true only for (C).

12. **E** For all three of these problems, (A) cannot be the correct choice because the gravitational force mg must be directed straight down. (B) cannot be the correct choice because the normal force N must be perpendicular to the surface. If the box is speeding up while moving down the incline then the net force must be directed down the incline, and the frictional force must be directed up the incline to oppose the motion. This is true only for (E).

13. **E** According to the impulse-momentum theorem, the change in momentum of an object is equal to the impulse on an object. The impulse on an object is equal to the integral of $F \bullet dt$.

$$\int Fdt = \textit{change in momentum}$$

$$\int_0^{t_1} Ct^3 dt = \textit{change in momentum}$$

$$\left.\frac{Ct^4}{4}\right|_0^{t_1} = \textit{change in momentum}$$

$$\frac{Ct_1^{\,4}}{4} = \textit{change in momentum}$$

14. **B** To reduce this elliptical orbit to circular the spaceship must slow down. To accomplish this, the engine burst must be in the same direction of the velocity which will cause the spaceship to accelerate in the opposite direction. This will cause the spaceship to slow down. The velocity at point P is directed to the left, so the engine burst must be to the left.

15. **A** The normal force of the track would point toward the center of the circle for the entire time the motorcycle was in the circular loop and the velocity would always be tangent to the path. The normal force and the displacement of the motorcycle are always perpendicular to each other. Therefore the work done by the normal force would be zero since work is the dot product of the force and the displacement.

16. **D** The speed of the center of mass of the ball can be calculated using the equation, $x = vt$. Inserting the appropriate numbers gives

$$30 = v(10)$$
$$v = 3 \text{ m/s}$$

The angular speed of the ball about the contact point can be calculated using the equation, $v = \omega r$. The value for r is the distance from the contact point to the center of mass, which is the radius of the ball. Inserting the appropriate numbers gives

$$3 = \omega(.2)$$
$$\omega = 15 \text{ rad/s}^2$$

17. **B** When objects stick together, a perfectly inelastic collision has occurred. For this situation kinetic energy is not conserved. This eliminates (C) and (D). Linear momentum is conserved when the net force on the system is zero. In this situation the hinge in the bar is exerting a force on it

prohibiting it from translating to the right. Therefore linear momentum is not conserved. This eliminates (A) and (E). Angular momentum is conserved when the net torque on a system is zero. The force exerted by the hinge provides no torque because the lever arm is zero. Therefore (B) is correct.

18. **B** The period for a spring-mass system is given by the equation $T = 2\pi\sqrt{\frac{m}{k}}$. Use that equation to solve for the mass, as shown below.

$$4 = 2\pi\sqrt{\frac{m}{8}}$$

$$\frac{2}{\pi} = \sqrt{\frac{m}{8}}$$

$$\frac{4}{\pi^2} = \frac{m}{8}$$

$$\frac{32}{\pi^2} = m$$

19. **C** The equation for an object undergoing simple harmonic motion will be of the form $x = A\cos(\omega t + \phi)$, where A is the amplitude, ω is the angular frequency and ϕ is the phase constant. None of the solutions involve the phase constant, so we will ignore that part of the equation. The amplitude of the oscillation is 5 cm, which is 0.05 m. This eliminates (D). The angular frequency can be calculated by using the fact that the period is 2π divided by the angular frequency.

$$T = \frac{2\pi}{\omega}$$

$$4 = \frac{2\pi}{\omega}$$

$$\omega = \frac{\pi}{2}$$

20. **E** Apply Newton's Second Law to the entire system to determine the acceleration of the system, and therefore each mass. The net force on the system is the gravitational force on mass $3M$ minus the gravitational force on mass M because they are trying to cause the system to move opposite directions even though they both point downward. F_g for the $3M$ mass is trying to move that mass down, which would cause mass M to move upwards, and F_g for mass M is trying to move that mass down, which would cause mass $3M$ to move upwards. Therefore these forces must be subtracted from each other to determine the net force. The solution is shown below.

$$\sum F = ma$$

$$3Mg - Mg = (3M + M)a$$

$$\frac{2Mg}{4M} = a$$

$$\frac{g}{2} = a$$

21. **A** For a spring that is not linear (i.e. does not obey Hooke's Law) the energy stored is not $\frac{1}{2}kx^2$. The magnitude of the energy stored will be equal to the magnitude of the work done to stretch the spring to x_1. The steps to calculate the work are shown below.

$$W = \int F dx$$

$$W = \int_0^{x_1} (kx^{1.5})dx$$

$$W = \left. \frac{kx^{2.5}}{2.5} \right|_0^{x_1}$$

$$W = \frac{2kx_1^{\ 2.5}}{5}$$

22. **E** Applying the law of conservation of linear momentum tells us that the total momentum of the system after the collision is equal to the total momentum of the system before the collision. Momentum can be calculated by multiplying the mass times the velocity for each skater and adding the vectors as shown below. You can use the Pythagorean theorem to solve for the total momentum, or realize this is a 3-4-5 right triangle, so the total momentum is equal to 250 kg•m/s.

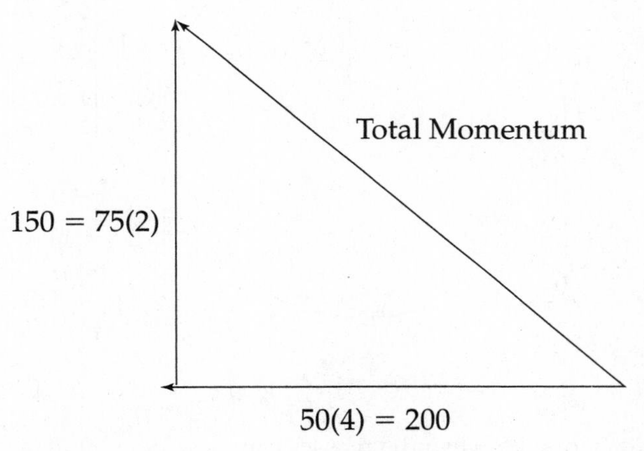

23. **D** When objects stick together, a perfectly inelastic collision has occurred. For this situation kinetic energy is not conserved. This eliminates (B), (C), and (E). Linear momentum is conserved when the net force on the system is zero, so only (D) combines all this information correctly.

24. **C** Acceleration is the second derivative of the position. Take the derivative of the position function twice and then plug in t = 1.5 s to get the correct answer, as shown below.

$$\frac{dx}{dt} = 6t^2 + 4$$

$$\frac{d^2x}{dt^2} = 12t$$

$$\frac{d^2x}{dt^2} = 12(1.5) = 18$$

25. **C** There is no net force on the system of the two astronauts, the rope, and oxygen tank, so the total linear momentum of the system will be conserved. Immediately after the throw, Astronaut A will recoil to the left and make Astronaut B also come to the left since they are attached by a rope. After Astronaut B catches the tank the total linear momentum of the system must return to zero since it was initially zero before the tank was thrown. If the linear momentum is zero when the tank is caught all objects must come to rest because Astronauts A and B move together. Combining these two pieces of information means (C) is correct.

26. **C** The pulley will rotate because there is a net torque on the pulley due to tension 1 and tension 2. Apply Newton's Second Law for rotation to the diagram below and then substitute $\alpha = \frac{a}{r}$. Take clockwise and downward to be positive, and remember that torque is the cross product of force and distance.

$$\sum \tau = I\alpha$$

$$(T_2 - T_1)R = I\frac{a}{R}$$

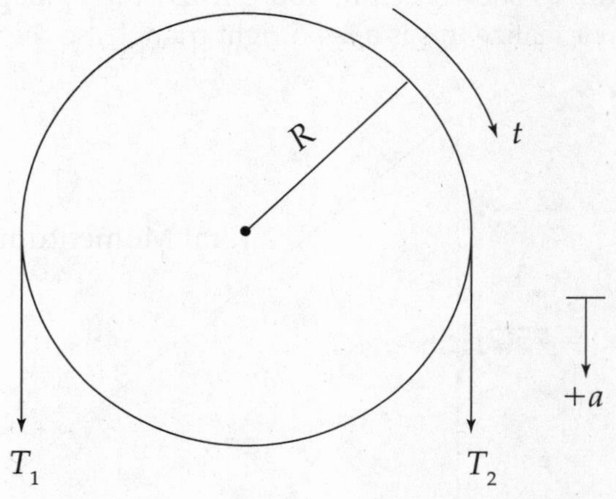

R

t

+a

T_1 T_2

27. **B** The force of gravity acting between the masses can be calculated using Newton's Law of Gravity: $F = \frac{Gm_1m_2}{r^2}$. When the test mass is outside of the sphere it is an inverse square relationship, so (C), (D), and (E) can be eliminated. When the test mass is inside the sphere, at a radius $r < R$, the sphere still attracts the test mass with amount of mass, M', that is at a radius less than the test mass. M' is proportional to the total mass by the volume contained compared to the total volume as shown below.

$$\frac{M'}{M} = \frac{\frac{4}{3}\pi r^3}{\frac{4}{3}\pi R^3}$$

$$M' = M\frac{r^3}{R^3}$$

This produces a gravitational force that is linear for the region $r < R$, as shown below.

$$F = \frac{GmM'}{r^2}$$

$$F = \frac{Gm\left(M\frac{r^3}{R^3}\right)}{r^2}$$

$$F = \frac{GmMr}{R^3}$$

28. **C** The net force of the book is zero, because the book is at rest. A free-diagram is shown below. The normal force is equal to F because those are the only two forces in the horizontal direction. Friction needs to be equal to gravity for the book to not slide down the wall because those are the only forces in the vertical direction. Therefore,

$$F_{fr} = mg$$
$$F_N = F$$
$$\mu = \frac{mg}{F}$$

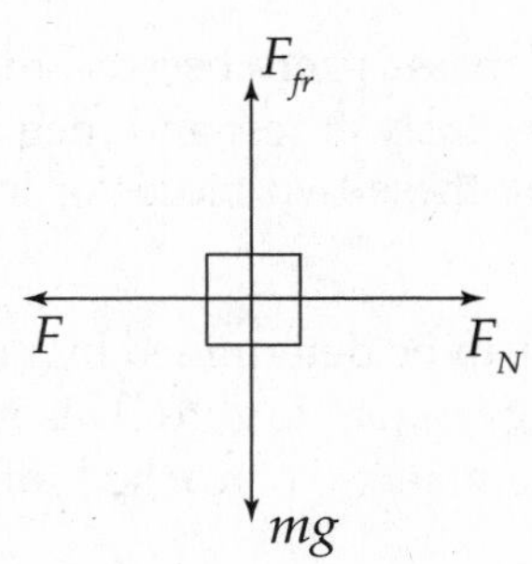

29. **A** For small amplitude oscillations, the period of a pendulum is given by the equation, $T = 2\pi\sqrt{\frac{\ell}{g}}$.

This indicates the period is independent of the amplitude, so the new period will equal the previous period.

30. **A** Power is equal to the rate work is done and also the dot product of force and velocity.

$$P = Fv$$
$$400 = F(20)$$
$$F = 20 \text{ N}$$

31. **C** The energy of the oscillating spring-mass system will remain constant. When it is pulled down 10 cm the energy will be stored in the spring, and when it passes the equilibrium position, all of the energy will be kinetic energy.

$$E_1 = E_2$$
$$\frac{1}{2}kx^2 = k_2$$
$$\frac{1}{2}(800)(0.1)^2 = k_2$$
$$4\text{J} = k_2$$

32. **D** The constant kinematics equations ignore the air resistance that decreases the total mechanical energy of the ball as it falls. The force due to air resistance also increases the faster the ball is going, so the force is increasing with time. This would make the acceleration of the ball begin at 9.8 m/s^2 and then decrease as the ball falls. This eliminates (C). The ball is rigid so it will not change shape when falling, which eliminates (A). The acceleration of gravity will be constant over the 6 meters that the ball falls, so this eliminates (B) and (E). Therefore (D) is left and is correct because the fall is speeding up at a decreasing rate of acceleration.

33. **C** The acceleration of the object will be due to gravity and air resistance. Air resistance is proportional to the velocity of the object. Since air resistance decreases the total mechanical energy of the object, the greatest velocity will occur at the point closest to the initial launch, point A. The greatest acceleration will occur when the air resistance and gravity are in the same direction and when the object is traveling the fastest. This is also at point A. Therefore (C) is correct.

34. **B** The acceleration of the center of mass is zero because the disk is rolling at a constant velocity. The contact point, P, is instantaneously at rest and then moves upward, which implies the acceleration is upward. Combining these two pieces of information indicates (B) is the correct answer.

35. **E** The escape velocity for a planet can be determined by making the total energy (kinetic energy plus gravitational potential energy) equal to zero. This implies that the spaceship has escaped from the gravitational pull of the planet (i.e. reached infinity) with no more velocity. This calculation is shown below.

$$\frac{1}{2}m_1 v_{esc}^{\ 2} - \frac{Gm_1 m_2}{r} = 0$$
$$v_{esc}^{\ 2} = \frac{2Gm_2}{r}$$
$$v_{esc} = \sqrt{\frac{2Gm_2}{r}}$$

Examining the last equation, one can see that the effects of doubling the mass and doubling the radius will cancel each other out. Therefore the escape velocity will remain v_0.

SECTION I, ELECTRICITY AND MAGNETISM

36. **C** The equivalent capacitance of capacitors connected in parallel is the sum of each capacitor. Therefore $C_{eq} = 3 + 3 + 3 = 9\mu\text{F}$

37. **A** The rate energy is being used is equal to the power. For an electrical device $P = IV$, so $P = (120)(2) = 240$ W.

38. **E** The flux is proportional to the number of field lines that pass through the surface of the sphere. Since the background electric field E is uniform, every background field line that enters the sphere also exits the sphere. Therefore, it has no contribution to the flux. The field lines from the $+2Q$ charge will radiate outward and every field line that exists due to the $+2Q$ will pass through the sphere, regardless of R. Therefore, the flux only depends on the value of Q.

39. **B** The potential difference can be calculated using the equation that relates the potential difference to the electric field, $\Delta V = -\int E \bullet dr$. The limits of the integration will be from 0 to 5, because the radius is 5 for the point (3, 4).

$$\Delta V = -\int_0^5 (-6r^2 + 4r + 3)dr$$

$$\Delta V = -(-2r^3 + 2r^2 + 3r)\Big|_0^5$$

$$\Delta V = -(-250 + 50 + 15) = 185$$

40. **C** Determine the direction of the force on the charge using the left-hand rule because the charge is negative. Consider the leftmost point of the circle. The force must be toward the center of the circle, to the right, and the magnetic field is out of the page, so the velocity must be down. This indicates the particle is moving counterclockwise. The force on the charge is $F = qvB$, and use Newton's Second Law to solve for the acceleration as shown below.

$$\sum F = ma$$

$$qvB = ma$$

$$a = \frac{qvB}{m}$$

(C) has these two pieces of information correct.

41. **D** For a particle traveling in uniform circular motion, $2\pi r = vT$ because the particle travels one revolution in one period. The period is the inverse of the frequency. Applying this information and solving for f is shown below.

$$2\pi r = \frac{v}{f}$$

$$f = \frac{v}{2\pi r}$$

42. **C** The energy stored in a capacitor is given by the equations $U_c = \frac{1}{2}QV = \frac{1}{2}CV^2 = \frac{Q^2}{2C}$. Select the last part of the equation because the charge, in millicoulombs, and the capacitance, in microfarads, are given. The solution is shown below

$$U_c = \frac{(6 \times 10^{-3})^2}{2(30 \times 10^{-6})} = 0.6 \text{ J}$$

43. **D** The electric field is uniform in between the plates, so the force and the acceleration are constant. Relating the potential difference to the electric field gives us $\Delta V = -Ex$, where x is the separation of the plates. The definition of the electric field is the force divided by the charge. Use these two pieces of information and Newton's Second Law to solve for the acceleration in terms of V.

$$E = \frac{F}{q}$$

$$\frac{V}{x} = \frac{F}{q}$$

$$\frac{Vq}{x} = ma$$

$$\frac{Vq}{mx} = a$$

This shows the acceleration is directly proportional to the potential difference V.

44. **E** Ampere's Law is $\oint B \bullet ds = \mu_0 I$. For an ideal solenoid the magnetic field is uniform and directed along the central axis within the solenoid and zero outside. In this case only segment *bc* contributes to the integral because *ad* is outside the solenoid where the field is zero and *ab* and *cd* are perpendicular to the field, so the dot product of those segments with B is zero. The number of times the current will pass through our path is equal to the number of coils in the solenoid. This give us the equation $B(x) = \mu_o NI$.

45. **E** The electric field will radiate outward from both $+3Q$ and $+Q$, and the field due to the $+3Q$ charge will be greater because point P is the same distance for both charges. This vector diagram is shown below. The resultant will be in the $+x$ and $+y$ directions.

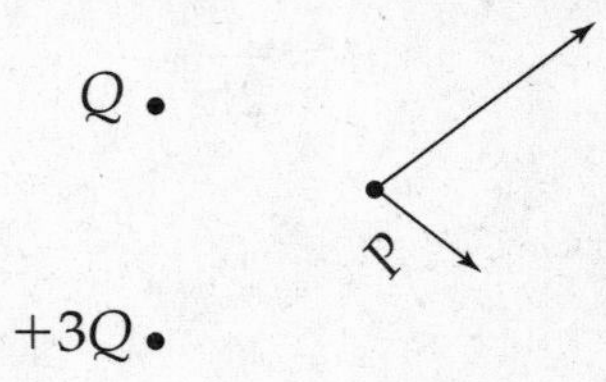

46. **E** The electric potential can be calculated using the equation $V = \sum \frac{kQ}{r}$. Because there are only positive charges, the electric potential due to each charge will be positive, so the sum cannot be zero.

47. **D** You may be tempted to do circuit analysis to solve for the potential difference of the battery and then solve for the current when the switch is open to compare it to the original. However, time can be saved by comparing the circuit with the switch closed and then open. Most of the resistance of the circuit comes from the 54 Ω resistor, so when the switch is opened the total resistance increases because the parallel branch is not operable anymore. However, the total resistance does not nearly double, so the current should decrease but will not be halved. This means (D) is correct.

48. **A** The equation for the resistance of a wire is $R = \frac{\rho L}{A}$, and $A = \pi r^2$. If the length is twice as long the wire will have twice the resistance and if the radius is half, the wire will have 4 times the resistance per unit length because the radius is squared in the equation. Conceptually this makes sense also because the charge is trying to flow through a cross-sectional area one-fourth as large. These two factors mean the total resistance is 8 times the resistance of wire Y.

49. **C** For the region $0 < r < R$ the electric potential is constant. For the potential to be constant the electric field must be zero because $E = -\frac{dV}{dr}$.

50. **E** The force between two parallel wires is given by the equation $F_B = \frac{\mu_o \ell I_1 I_2}{2\pi x}$, where x is the distance between the wires and ℓ is the length of the wires. Therefore, if the currents are halved the force will be one fourth the original value.

51. **A** The area of the square is given by $A = \ell^2 = k^2t^4$. Use Faraday's Law to determine the induced emf as shown below. Since the question asks for the magnitude of the emf, it is not necessary to include the negative sign in your answer.

$$\varepsilon = -\frac{d\Phi}{dt}$$

$$\varepsilon = -B\frac{d}{dt}(k^2t^4)$$

$$\varepsilon = -4Bk^2t^3$$

52. **D** The potential difference around the loop must be zero according to Kirchhoff's Loop Law. The current in the circuit is going counterclockwise if the battery is being charged by the 120 V outlet. Take a counterclockwise loop starting right below the 120 V outlet will produce the following, $120 - \varepsilon - (3)(30) = 0$. Therefore $\varepsilon = 30$ V.

53. **B** Use the right-hand rule to determine that the magnetic force on the charge is directed upward. Therefore the electric force must be directed down, and the electric field then must be down because the electric field indicates the direction of force on a positive charge.

54. **C** The magnetic field through the loop decreases when the magnet is pulled away. To oppose the change in the flux through the loop, the loop will induce its own current that creates a magnetic field to the right. This will cause the loop to be attracted to the magnet with a force to the left (again to oppose the change in flux). (C) combines these two pieces of information correctly.

55. **A** Use the right-hand rule to determine the force on each part of the loop. A diagram is shown below with the forces indicated. These forces would produce a torque that causes the loop to rotate about the indicated axis. This would cause the loop to rotate as shown in (A).

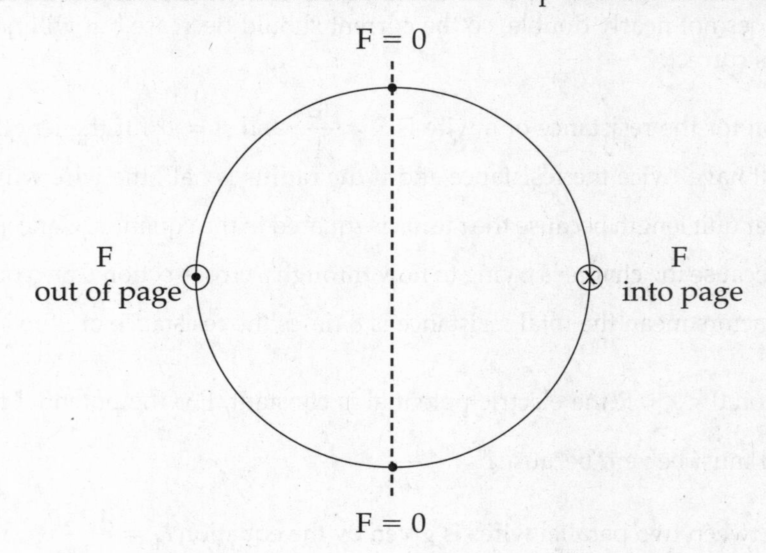

56. **D** Calculate the resistance of the parallel branch first and then add that to the 12 Ω resistor in series with the parallel branch as shown below.

$$\frac{1}{R_{\|}} = \frac{1}{R_2} + \frac{1}{R_3}$$

$$\frac{1}{R_{\|}} = \frac{1}{10} + \frac{1}{15} = \frac{3}{30} + \frac{2}{30} = \frac{5}{30}$$

$$R_{\|} = 6$$

$$R_{eq} = 12 + 6 = 18$$

57. **A** The strength of the electric field at the center of the square is the same for each charge because they are each the same distance away from the center. The electric field due to each one radiates outward. The field created by each is cancelled by the charge diagonal from it, so the total field is zero.

58. **A** The work done will be equal to the change in electric potential energy. The electric potential is the algebraic sum of the potential due to each one, $V = \sum \frac{kQ}{r}$. Therefore the total electric potential in the center is $V = \frac{4kQ}{\ell}$ because they are all positive and the same distance ℓ from the center. The electric potential far away is zero, so $\Delta V = V$ at the center. Now relate the potential difference to the potential energy and solve for ΔV.

$$\Delta V = \frac{\Delta U}{Q}$$
$$\Delta V(3Q) = \Delta U$$
$$\frac{4kQ}{\ell}(3Q) = \Delta U$$
$$\frac{12kQ^2}{\ell} = \Delta U$$

59. **B** The strength of the electric field is given by $E = -\frac{dV}{dx}$. Therefore, the closer the equipotential lines are the greater the strength of the electric field. B is the point where the equipotential lines have the greatest density, so (B) is correct.

60. **A** The equation relating the electric field to potential difference is $E = -\frac{dV}{dx}$. The negative sign indicates that the electric field points toward decreasing electric potential. The electric field is also perpendicular to equipotential lines. Therefore (A) indicates the correct direction for the field at point D.

61. **E** If a negative charge were released from the –10 V equipotential line it would want to travel to a higher potential, so when it goes to the 20 V line it must be losing electric potential energy. This means the answer should be negative: eliminate (A), (B), and (C). We can also use the definition of electric potential to solve for the value of ΔU.

$$\Delta V = \frac{\Delta U}{q}$$
$$30(-3) = \Delta U$$
$$-90 = \Delta U$$

62. **D** Maxwell's equations relates electric and magnetic fields. A changing electric field would create a changing electric flux, which in turn would induce a magnetic field. This is Ampere-Maxwell's Law, which is (D).

63. **B** For this circuit, R would change how long the capacitor took to get charged, but would not alter the capacitance or the charge on the plates. Therefore any answer that changes R will not

be correct, so (A), (C) and (E) can be eliminated. The capacitance of a parallel plate capacitor can be increased by increasing the dielectric constant K, increasing the area of the plates A, or decreasing the distance between the plates d. If we increase V, more charge will be stored on the capacitor. Therefore (D) would certainly increase the charge stored and can be eliminated. (B) is correct because decreasing d would increase the capacitance and we could decrease V such that less charge would be stored on the capacitor even though the capacitance increased.

64. **B** Use the right-hand rule to determine the direction of the magnetic field due to each wire. The magnetic fields due to F and G cancel because the one due to F is out of the page at point P and the one due to G is into the page at point P. The field due to wire H is out of the page, so (B) is the answer.

65. **D** For a solid, metal object the electric field inside is equal to zero. The electric field outside the object will not be zero because some excess charge is contained on the object and Gauss's Law can be applied to show that Q_{in} would generate an electric field. This information eliminates (A) and (B). The surface of a conductor is an equipotential, so (E) is not correct. The strength of the electric field should decrease as the distance away from the center increases, so (C) is not correct. (D) is correct, and we can apply Gauss's Law as shown below to the odd-shaped figure as indicated. The Q_{in} will be equal to the surface charge density at the location times the area of the endcap of the Gaussian cylinder. This is only true very close to the surface of the object so that the Gaussian cylinder is perpendicular to the surface.

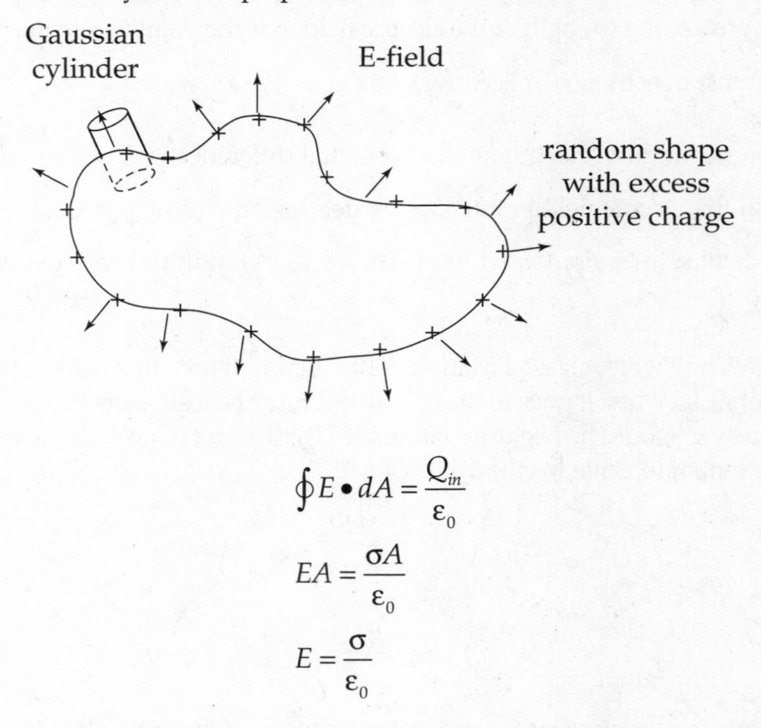

$$\oint E \bullet dA = \frac{Q_{in}}{\varepsilon_0}$$

$$EA = \frac{\sigma A}{\varepsilon_0}$$

$$E = \frac{\sigma}{\varepsilon_0}$$

66. **A** Apply Ohm's Law to solve for the current before the switch is closed. $\Delta V = IR$, $12 = I(6)$, which give I is 2 A. Immediately after the switch is closed the current through the circuit must remain 2 A because the inductor opposes changes in current that flowed before the switch was closed. The switch also shorts out the 2 Ω resistor. Using a loop around the top part of the circuit and applying Kirchhoff's Loop rule we get:

$$12 - IR - \varepsilon_L = 0$$
$$12 - (4)(2) - \varepsilon_L = 0$$
$$\varepsilon_L = -4 \text{ V}$$

The negative sign indicates the inductor is acting as an opposing battery trying to stop the current from increasing, so the potential difference across the inductor is 4 V.

67. **D** After the switch has been closed a long time the inductor acts like a wire, and the switch shorts out the 2 Ω resistor. Use Ohm's Law to solve for the current through the 4 Ω resistor, as follows.

$$\Delta V = IR$$
$$12 = I(4)$$
$$I = 3 \text{ A}$$

68. **E** The electric field will decrease as $\frac{1}{r^2}$ when $r > R$. This eliminates (A), (B), and (C). When ρ is constant for a spherical charge distribution, the electric field strength varies linearly with r for $r < R$. Therefore, if ρ increases with r, as in this problem, more charge is being contained as r increases. This will make the electric field increase less rapidly than if ρ were constant. Therefore (E) is correct. The general shapes of E vs. r for a uniform ρ and our problem are shown below. This sketch assumes the total charge is equal for both.

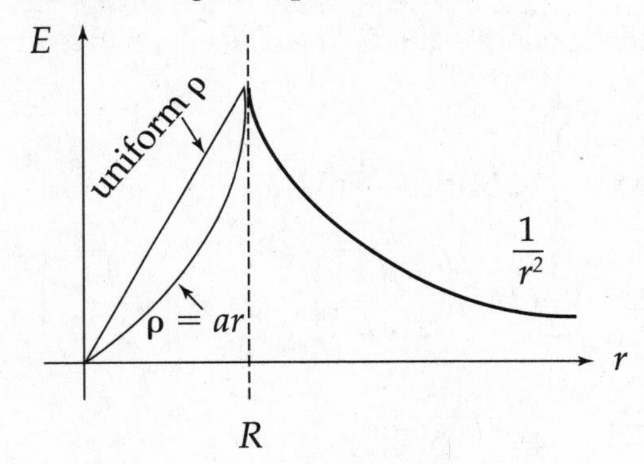

69. **B** For figure II, the electrons in the finger are attracted to the rod and will go from the finger into the electroscope. Therefore (B) is correct.

70. **C** This is essentially a Hall's voltage question. Use the left-hand rule to determine the direction of the magnetic force on the electrons traveling to the left in the diagram. The force is downward, which means the potential at X will be positive and the potential at Y will be negative. Therefore (C) is correct.

SECTION II, MECHANICS

1. (a) Use conservation of energy since all of the energy stored in the spring will be transferred into kinetic energy of block *A*. Realize the compression of the spring is given as *d*.

$$E_1 = E_2$$

$$\frac{1}{2}kd^2 = \frac{1}{2}Mv^2$$

$$\frac{kd^2}{M} = v^2$$

$$v = d\sqrt{\frac{k}{M}}$$

(b) Use conservation of linear momentum for the collision. During a perfectly inelastic collision the two blocks stick together and continue at the same velocity.

$$p_1 = p_2$$

$$Md\sqrt{\frac{k}{M}} = 2Mv_2$$

$$v_2 = \frac{d}{2}\sqrt{\frac{k}{M}}$$

(c) The change in the kinetic energy during the collision is given by $\Delta K = K_2 - K_1$.

$$\Delta K = K_2 - K_1$$

$$\Delta K = \frac{1}{2}(2M){v_2}^2 - \frac{1}{2}(M){v_1}^2$$

$$\Delta K = \frac{1}{2}(2M)\left(\frac{d}{2}\sqrt{\frac{k}{M}}\right)^2 - \frac{1}{2}(M)\left(d\sqrt{\frac{k}{M}}\right)^2$$

$$\Delta K = \frac{d^2k}{4} - \frac{d^2k}{2}$$

$$\Delta K = -\frac{d^2k}{4}$$

(d) Now the blocks are traveling along a curved path and at the top of the hill they are traveling with the same velocity as in (B) since it is at the same elevation. They are also traveling in circular motion at the top of the hill. The free-body diagram for the blocks is shown on the next page. The gravitational force is pointing toward the center of the circle, so it will be positive and the normal force is pointing away from the center of the circle, so it will be negative.

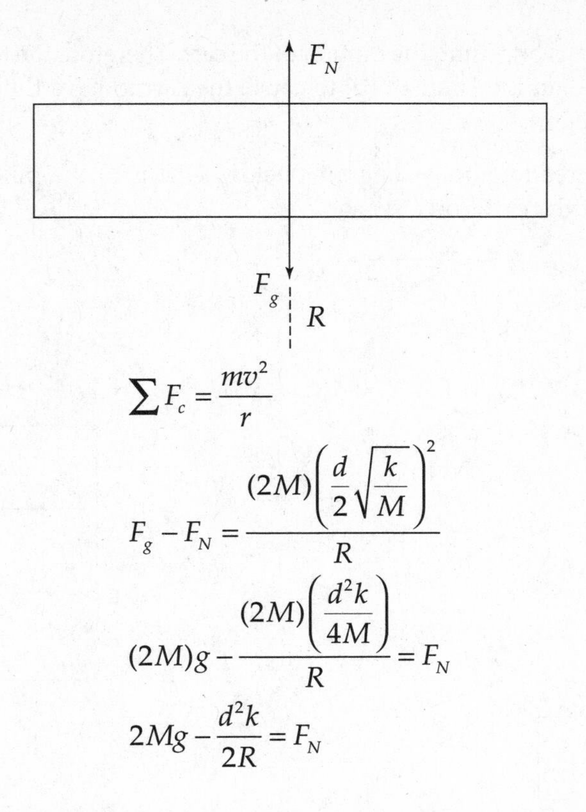

$$\sum F_c = \frac{mv^2}{r}$$

$$F_g - F_N = \frac{(2M)\left(\frac{d}{2}\sqrt{\frac{k}{M}}\right)^2}{R}$$

$$(2M)g - \frac{(2M)\left(\frac{d^2k}{4M}\right)}{R} = F_N$$

$$2Mg - \frac{d^2k}{2R} = F_N$$

2. (a) The acceleration of the system is equal to the slope of the velocity vs. time graph. One might try to use Newton's Second Law to calculate the acceleration, but for a laboratory problem, one can almost always use the data or graph given.

$$a = slope = \frac{3}{2}$$

$$a = 1.5 \text{ m/s}^2$$

(b) According to the problem, tension is the only force causing the cart/force sensor to accelerate. A free-body diagram of the cart/force sensor is shown below.

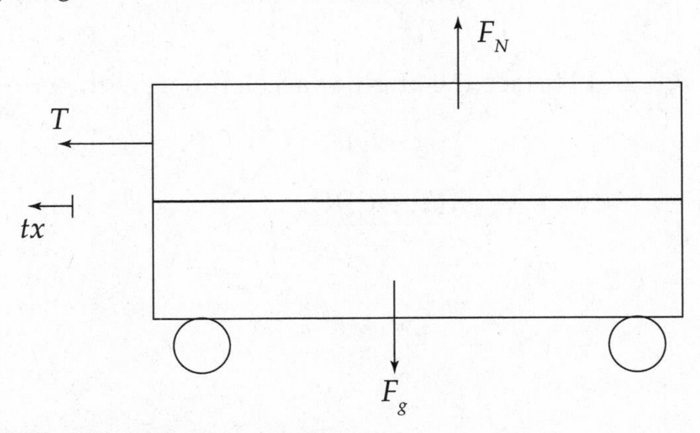

Use Newton's Second Law to solve for the tension.

$$F_x = ma_x$$

$$T = (1.2)(1.5)$$

$$T = 1.8 \text{ N}$$

(c) In reality, friction is opposing the motion of the cart. Therefore the tension in the string would have to be greater than the value in (b) to cause the cart to have the acceleration calculated in (a) to overcome friction.

(d) Consider the three free-body diagrams below. One is of the pulley, one is of the hanging mass, and one is of the cart/force sensor.

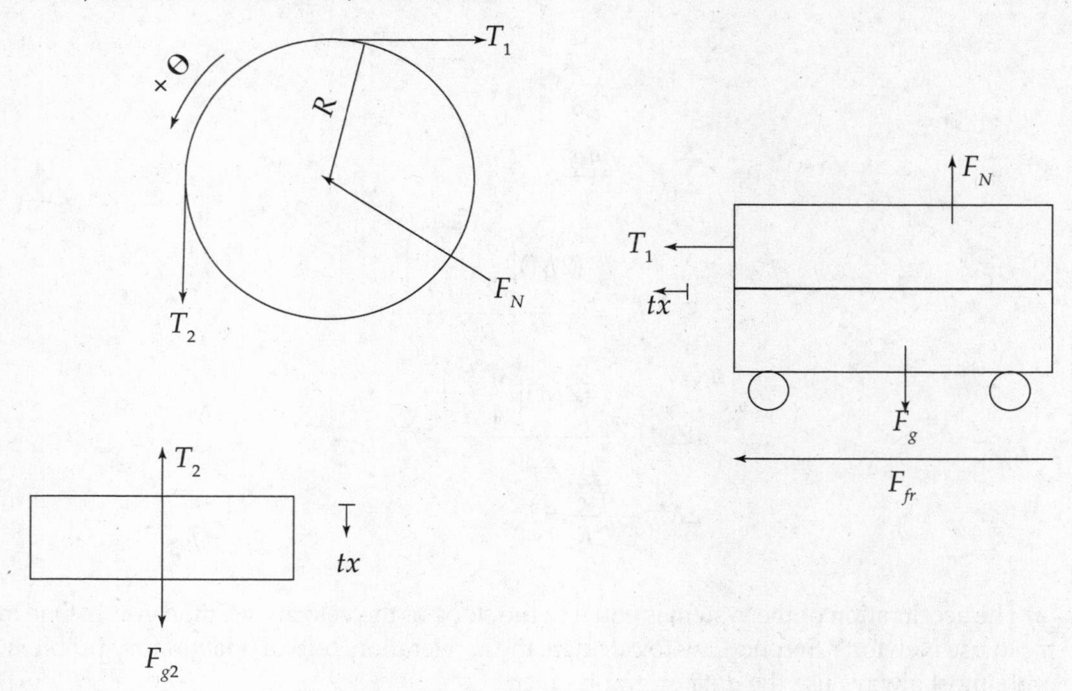

Use the motion detector to measure the velocity of the cart and find the acceleration of the cart, and also the hanging mass since they are attached, by taking the slope of the v vs. t graph. Use the force sensor to measure T_1. Measure the radius of the pulley with the meterstick. Apply Newton's Second Law to the hanging mass since you already know F_{g2} and its acceleration, to determine T_2. Apply Newton's Second Law of Rotation to the pulley, to determine the rotational inertia of the pulley as shown below.

$\tau = I$, and replace the angular acceleration with $\alpha = \dfrac{a}{R}$

$$(T_2 - T_1)R = I\frac{a}{R}$$

$$\frac{(T_2 - T_1)R^2}{a} = I$$

3a i $F_{net} = m_2g - m_1g\sin\theta$

$0 < m_2g - m_1g\sin\theta$ $\sin\theta < \frac{m_2}{m_1}$

b

5a f_s F_N F_g

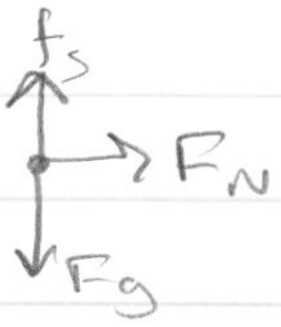

b) $f_s = F_g$ $\mu F_N = mg$

$\mu \frac{mv^2}{r} = mg$ $\frac{\mu v^2}{r} = g$

c same

6 a 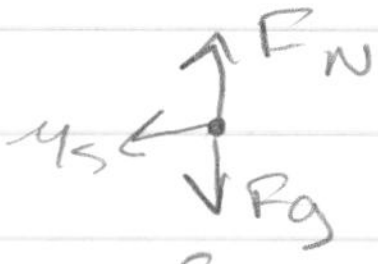 F_N μ_s F_g

$f_s = \mu_s mg$

$\mu_s mg = \frac{mv^2}{r}$ $v = \sqrt{\mu_s gr}$

c F_N θ F_g θ

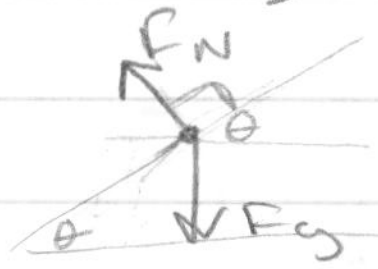

$F_c = F_{Nx}$

$F_N = F_g\cos\theta$; $90+\theta$

$F_{Nx} = mg\cos\theta\sin\theta = \frac{1}{2}mg\sin2\theta$

$\frac{1}{2}mg\sin2\theta = \frac{mv^2}{r}$ $\frac{2gv^2}{r} = \sin2\theta$

$\frac{1}{2}\sin^{-1}\frac{2gv^2}{r} = \theta$

1 B ② D ③  10 5 θ 30 $\sin\theta = \frac{1}{4}$ A

④ $20 \cdot \frac{4}{10} = 8$ 10N C ⑤ ~~mg~~ sin 30 .5mg cos 30 E

$\frac{1}{2}$ $\frac{1}{2}$

6 E ⑦ E ⑧ D ⑨ $F_N = 1000$ $f_s = 400$ D

10 E ⑪ 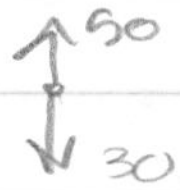 50 30 $20 = \frac{3v^2}{.6}$ = B

12 $30 = 5v^2$ C ⑬ $v = 2\pi r$ $a = 4\pi^2 r$ D

3. (a) Use the parallel axis theorem to determine the rotational inertia of the object about the pin.

$$I = I_{cm} + md^2$$

$$I = \frac{1}{2}MR^2 + M\left(\frac{R}{2}\right)^2$$

$$I = \frac{3}{4}MR^2$$

(b) When the object is first released, gravity provides a torque about the pin causing the disk to rotate about the pin. A free-body diagram for the disk is shown below.

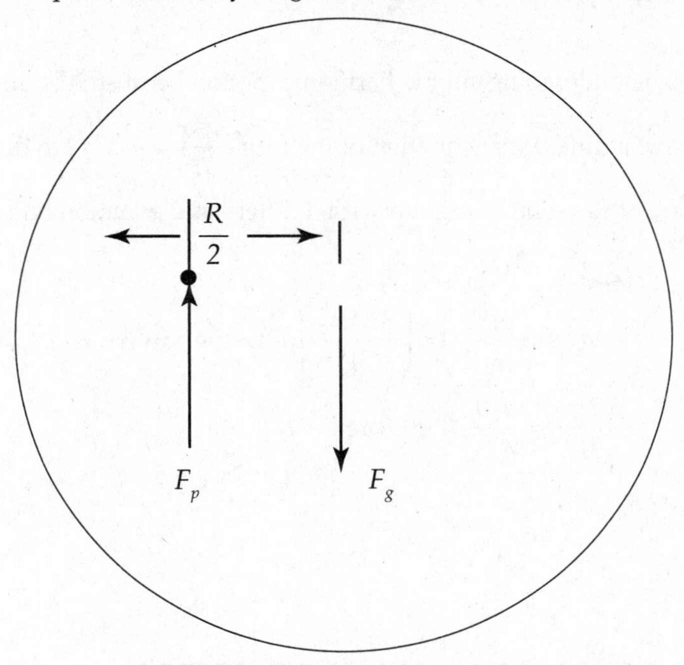

Use Newton's Second Law of Rotation about the pin to determine the angular acceleration.

$$\sum \tau_{pin} = I\alpha$$

$$Mg\left(\frac{R}{2}\right) = \left(\frac{3MR^2}{4}\right)\alpha$$

$$\frac{2g}{3R} = \alpha$$

(c) Use the Law of Conservation of Mechanical Energy to determine the angular velocity at the vertical position. All the initial gravitational potential energy is converted into rotational kinetic energy. The center of mass falls half the radius.

$$E_1 = E_2$$

$$Mg\left(\frac{R}{2}\right) = \frac{1}{2}I\omega^2$$

$$\frac{MgR}{2} = \frac{1}{2}\left(\frac{3MR^2}{4}\right)\omega^2$$

$$\frac{4g}{3R} = \omega^2$$

$$\sqrt{\frac{4g}{3R}} = \omega$$

(d) Now the disk is undergoing simple harmonic motion because it is undergoing small angle oscillations. Derive a differential equation of the form $\frac{d^2x}{dt^2} = -\omega^2 x$, to determine ω. Start with Newton's 2nd Law of Rotation to end up with a differential equation because $\alpha = \frac{d^2\theta}{dt^2}$.

$$\sum \tau_{pin} = I\alpha$$

$$-Mg\sin\theta\left(\frac{R}{2}\right) = \left(\frac{3MR^2}{4}\right)\alpha, \text{ using } \sin\theta = \theta,$$

$$-\frac{2g}{3R}\theta = \frac{d^2\theta}{dt^2}, \text{ therefore}$$

$$\omega = \sqrt{\frac{2g}{3R}}$$

SECTION II, ELECTRICITY AND MAGNETISM

1. (a) (i) Consider the Gaussian sphere shown on the next page as a dashed line. Since the electric field is zero inside a conductor, the charge enclosed must be zero. Therefore the charge on the inner surface is zero. This leaves the $-4Q$ charge on the outer surface, as shown below.

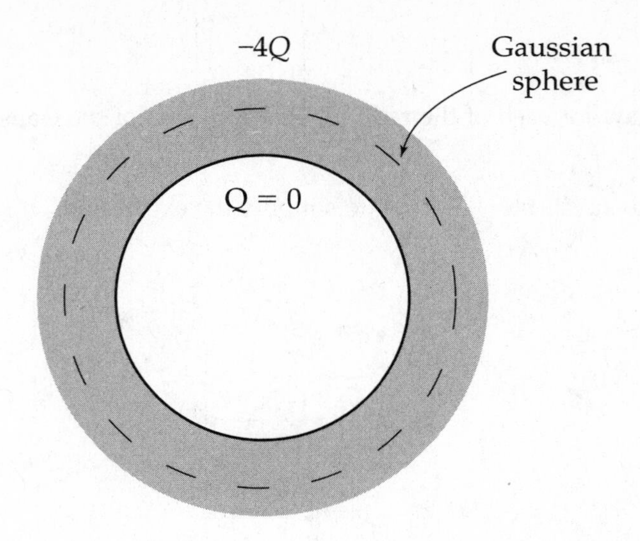

(ii) The electric field inside a conductor at equilibrium is always zero, so the electric field in region II is zero. Field lines originate on positive charges and end on negative charges. Since there is no excess charge on the inner surface of the sphere or within region I the electric field in region I is also zero. The electric field in region III is shown below. The electric field lines will originate at the +2Q charge and terminate at the outer surface of the spherical shell because the outer surface will have a charge of –4Q. Some of the -4Q charge will be closer to the +2Q charge, but that still leaves an excess of negative charge around the rest of the outer surface of the shell, so the field lines should be ending on the outer surface. All field lines need to be perpendicular to the surface of the shell.

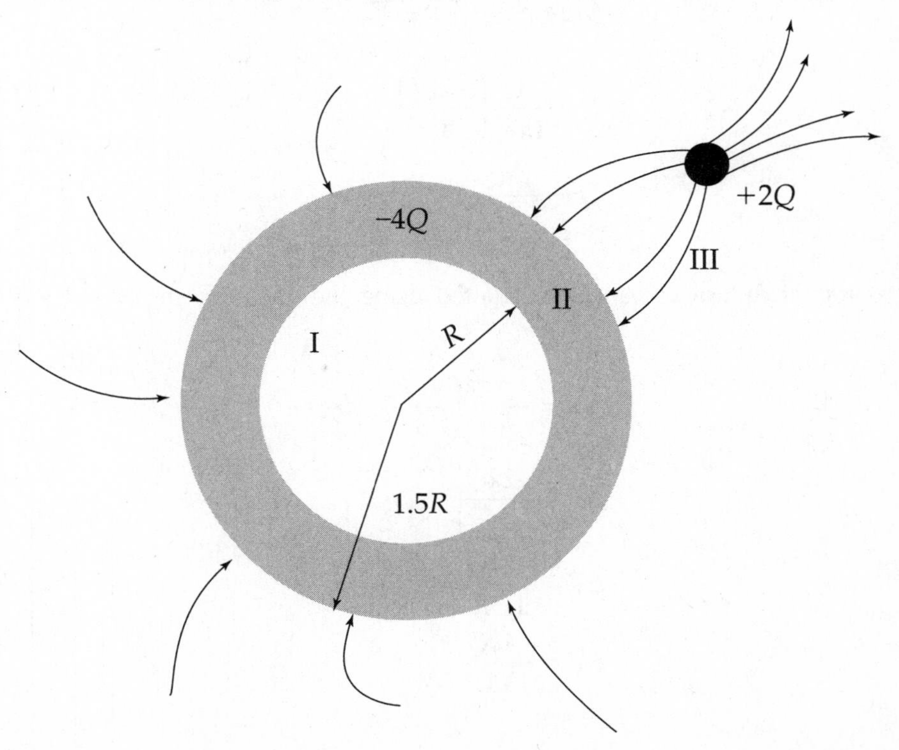

(b) Use Gauss's Law for each of the radii. The surface area of the Gaussian sphere is $4\pi r^2$, so

$\oint dA = 4\pi r^2$. Also substitute $\frac{1}{4\pi\varepsilon_o} = k$ to simplify the expression.

(i)

$$\oint E \bullet dA = \frac{q_{in}}{\varepsilon_0}$$

$$E(4\pi r^2) = \frac{(+2Q)}{\varepsilon_0}$$

$$E = \frac{1}{4\pi r^2}\left(\frac{+2Q}{\varepsilon_0}\right)$$

$$E = \frac{2kQ}{r^2}$$

(ii) $E = 0$, within a conductor.

(iii)

$$\oint E \bullet dA = \frac{q_{in}}{\varepsilon_0}$$

$$E(4\pi r^2) = \frac{(+2Q - 4Q)}{\varepsilon_0}$$

$$E = \frac{1}{4\pi r^2}\left(\frac{-2Q}{\varepsilon_0}\right)$$

$$E = \frac{-2kQ}{r^2}$$

(c) The potential difference can be calculated using the equation $\Delta V = -\int E \bullet dr$ as shown below.

$$\Delta V = -\int_{\infty}^{1.5R}\left(\frac{-2kQ}{r^2}\right)dr$$

$$\Delta V = \left.\frac{-2kQ}{r}\right|_{\infty}^{1.5R}$$

$$\Delta V = \frac{-2kQ}{1.5R} - 0$$

$$\Delta V = \frac{-4kQ}{3R}$$

2. (a) When the switch is first moved to position A the 50 μF capacitor is acting like a wire because it does not oppose the current when it is uncharged. The current through the 20 Ω resistor is the same as the entire left side of the circuit because it is a series circuit. Use Ohm's Laws to solve for the current.

$$V = IR$$
$$30 = I\,(10 + 20)$$
$$I = 1.0\text{ A}$$

(b) The initial voltage across the 10 Ω resistor can be calculated using Ohm's Law.

$$V = IR$$
$$V = 1(10)$$
$$V = 10\ V$$

For an RC circuit the voltage across the resistor will decrease exponentially because the current decreases exponentially as the capacitor is charging. The sketch of the voltage is shown below.

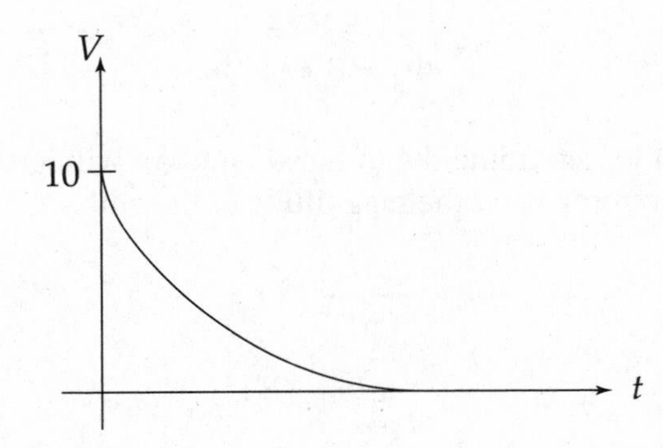

(c) When the switch is in position A, the capacitor will continue being charged until the voltage across the 50 μF becomes 30 V, which will happen after a long time. When the switch is moved to position B, the 20 μF capacitor acts like a wire because it is uncharged. Use Ohm's Law to solve for the current through the 15 Ω resistor immediately after the switch is moved to position B.

$$V = IR$$

$$30 = I\,(15)$$

$$I = 2\text{ A}$$

(d) The total charge stored on the 50 μF capacitor when the switch is at position A can be calculated using the equation

$$C = \frac{Q}{V}$$

$$50 = \frac{Q}{30}$$

$$Q = 1{,}500\mu\text{C}$$

When the switch is moved to position *B*, this charge will distribute between the two capacitors until the system reaches equilibrium. The voltage across each capacitor will be same when the system is in equilibrium. Use this information and the fact that $\Delta V = \frac{Q}{C}$ for a capacitor. Let q be the charge on the 20 μF capacitor, which means 1,500 – q is the charge on the 50 μF capacitor.

$$\frac{1,500 - q}{50} = \frac{q}{20}$$
$$30,000 - 20q = 50q$$
$$30,000 = 70q$$
$$q = 429\mu C$$

3. (a) The magnetic flux is the dot product of the magnetic field and area vector that the magnetic field goes through. For this problem the area goes out to a radius d because the field does not go out to the radius of the loop.

$$\Phi_B = B \bullet A$$
$$\Phi_B = B_o \bullet \pi d^2$$

(b) Use Faraday's Law to determine the induced emf. We will ignore the negative sign in the final answer because we only need the magnitude of the emf.

$$\varepsilon = -\frac{d\phi_B}{dt}$$
$$\varepsilon = -\pi d^2 \frac{d}{dt}(B_o t^2)$$
$$\varepsilon = -\pi d^2 (2B_o t)$$
$$\varepsilon = -2B_o \pi d^2 t$$
$$|\varepsilon| = 2B_o \pi d^2 t$$

(c) Lenz's Law states that the induced emf will produce a magnetic field to oppose the change in the external magnetic field. The external magnetic field going into the page is increasing with time, so to oppose the change in the external magnetic field the induced current in the loop will create a magnetic field out of the page. To do this the induced current must be counterclockwise.

(d) Power is the rate at which energy is dissipated. Power for a circuit is given by the equations $P = IV = \frac{V^2}{R} = I^2R$. We will use $\frac{V^2}{R}$ for power since we know the voltage and the resistance.

$$P = \frac{dE}{dt}$$

$$\frac{V^2}{R}dt = dE$$

$$\int_0^{t_1} \frac{V^2}{R}dt = E$$

$$\int_0^{t_1} \frac{(2B_o \pi d^2 t)^2}{R}dt = E$$

$$\frac{4B_o{}^2 \pi^2 d^4 t^1}{3R} = E$$

ABOUT THE AUTHORS

Paul Waechtler has taught introductory and Advanced Placement physics courses for the past decade. He is currently a physics teacher and the technology staff development coordinator at New Trier Township High School in Illinois. He received his Bachelor's and Master's degrees in Mechanical Engineering from the University of California, Santa Barbara. He joined the faculty at New Trier High School eight years ago after teaching in California for two years.When he's not in the classroom or on his computer, he can be found hiking, canoeing, playing volleyball, or playing with his two daughters and wife at the beach, park, or swimming pool.

Steve Leduc has been teaching at the university level since the age of 19, earned his Sc.B. in Theoretical Mathematics from MIT at 20 and his M.A. in Mathematics from UCSD at 22. After his graduate studies, Steve co-founded Hyperlearning, Inc., an educational services company that provided supplemental courses in undergraduate math and science for students from the University of California, where he lectured seventeen different courses in mathematics and physics. He has published four math books, *Differential Equations* in 1995, *Linear Algebra* in 1996, The Princeton Review's *Cracking the GRE Math Subject Test* in 2000, and *Cracking the Virginia SOL Algebra II* in 2001, as well as a physics book, The Princeton Review's *Cracking the SAT II Physics* in 2000. Through Hyperlearning, Steve has directed the creation and administration of the most successful preparation course for the medical school entrance exam (the MCAT) in California, where he has taught mathematics and physics to thousands of undergraduates. Hyperlearning merged with The Princeton Review in 1996. Steve currently owns two-to-the-eleventh power CDs and has seen *Monty Python and The Holy Grail, The Lord of the Rings* trilogy (extended versions), and *Blade Runner* about two-to-the-eleventh power times.

INDEX

J

K

L

M

N

O

P

R

S

T

U

V

W

The Princeton Review

Completely darken bubbles with a No. 2 pencil. If you make a mistake, be sure to erase mark completely. Erase all stray marks.

1. YOUR NAME: ______________________________
(Print) Last First M.I.

SIGNATURE: ______________________________ **DATE:** ____/____/____

HOME ADDRESS: ______________________________
(Print) Number and Street

____
City State Zip Code

PHONE NO. : ______________________________
(Print)

IMPORTANT: Please fill in these boxes exactly as shown on the back cover of your test book.

2. TEST FORM

3. TEST CODE

A B C D E F G (second column)
0 1 2 3 4 5 7 8 9 (other columns)

4. REGISTRATION NUMBER

0 1 2 3 4 5 7 8 9 (each column)

5. YOUR NAME

First 4 letters of last name | FIRST INIT | MID INIT

A B C D E F G H I J K L M N O P Q R S T U V W X Y Z (each column)

6. DATE OF BIRTH

Month	Day	Year
JAN	0 1 2 3	0 1 2 3 4 5 7 8 9
FEB	0 1 2 3 4 5 7 8 9	0 1 2 3 4 5 7 8 9
MAR		
APR		
MAY		
JUN		
JUL		
AUG		
SEP		
OCT		
NOV		
DEC		

7. SEX

MALE

FEMALE

The Princeton Review

FORM NO. 00001-PR

Section 1

Start with number 1 for each new section.
If a section has fewer questions than answer spaces, leave the extra answer spaces blank.

1. A B C D E	31. A B C D E	61. A B C D E
2. A B C D E	32. A B C D E	62. A B C D E
3. A B C D E	33. A B C D E	63. A B C D E
4. A B C D E	34. A B C D E	64. A B C D E
5. A B C D E	35. A B C D E	65. A B C D E
6. A B C D E	36. A B C D E	66. A B C D E
7. A B C D E	37. A B C D E	67. A B C D E
8. A B C D E	38. A B C D E	68. A B C D E
9. A B C D E	39. A B C D E	69. A B C D E
10. A B C D E	40. A B C D E	70. A B C D E
11. A B C D E	41. A B C D E	71. A B C D E
12. A B C D E	42. A B C D E	72. A B C D E
13. A B C D E	43. A B C D E	73. A B C D E
14. A B C D E	44. A B C D E	74. A B C D E
15. A B C D E	45. A B C D E	75. A B C D E
16. A B C D E	46. A B C D E	76. A B C D E
17. A B C D E	47. A B C D E	77. A B C D E
18. A B C D E	48. A B C D E	78. A B C D E
19. A B C D E	49. A B C D E	79. A B C D E
20. A B C D E	50. A B C D E	80. A B C D E
21. A B C D E	51. A B C D E	81. A B C D E
22. A B C D E	52. A B C D E	82. A B C D E
23. A B C D E	53. A B C D E	83. A B C D E
24. A B C D E	54. A B C D E	84. A B C D E
25. A B C D E	55. A B C D E	85. A B C D E
26. A B C D E	56. A B C D E	86. A B C D E
27. A B C D E	57. A B C D E	87. A B C D E
28. A B C D E	58. A B C D E	88. A B C D E
29. A B C D E	59. A B C D E	89. A B C D E
30. A B C D E	60. A B C D E	90. A B C D E

NOTES

NOTES

NOTES

NOTES

NOTES

NOTES

NOTES

AP Exams

Cracking the AP Biology Exam,
2008 Edition
978-0-375-76640-4 • $18.00/C$22.00

Cracking the AP Calculus AB & BC Exams,
2008 Edition
978-0-375-76641-1 • $19.00/C$23.00

Cracking the AP Chemistry Exam,
2008 Edition
978-0-375-76642-8 • $18.00/C$22.00

Cracking the AP Computer Science A & AB Exams, 2006–2007 Edition
978-0-375-76528-5 • $19.00/C$27.00

Cracking the AP Economics (Macro & Micro) Exams, 2008 Edition
978-0-375-42841-8 • $18.00/C$22.00

Cracking the AP English Language and Composition Exam, 2008 Edition
978-0-375-42842-5 • $18.00/C$22.00

Cracking the AP English Literature Exam,
2008 Edition
978-0-375-42843-2 • $18.00/C$22.00

Cracking the AP Environmental Science Exam, 2008 Edition
978-0-375-42844-9 • $18.00/C$22.00

Cracking the AP European History Exam,
2008 Edition
978-0-375-42845-6 • $18.00/C$22.00

Cracking the AP Physics B Exam,
2008 Edition
978-0-375-42846-3 • $18.00/C$22.00

Cracking the AP Physics C Exam,
2008 Edition
978-0-375-42854-8 • $18.00/C$22.00

Cracking the AP Psychology Exam,
2008 Edition
978-0-375-42847-0 • $18.00/C$22.00

Cracking the AP Spanish Exam, with Audio CD, 2008 Edition
978-0-375-42848-7 • $24.95/$29.95

Cracking the AP Statistics Exam,
2008 Edition
978-0-375-42849-4 • $19.00/C$23.00

Cracking the AP U.S. Government and Politics Exam, 2008 Edition
978-0-375-42850-0 • $18.00/C$22.00

Cracking the AP U.S. History Exam,
2008 Edition
978-0-375-42851-7 • $18.00/C$22.00

Cracking the AP World History Exam,
2008 Edition
978-0-375-42852-4 • $18.00/C$22.00

SAT Subject Tests

Cracking the SAT Biology E/M Subject Test, 2007–2008 Edition
978-0-375-76588-9 • $19.00/C$25.00

Cracking the SAT Chemistry Subject Test,
2007–2008 Edition
978-0-375-76589-6 • $18.00/C$22.00

Cracking the SAT French Subject Test,
2007–2008 Edition
978-0-375-76590-2 • $18.00/C$22.00

Cracking the SAT Literature Subject Test,
2007–2008 Edition
978-0-375-76592-6 • $18.00/C$22.00

Cracking the SAT Math 1 and 2 Subject Tests, 2007–2008 Edition
978-0-375-76593-3 • $19.00/C$25.00

Cracking the SAT Physics Subject Test,
2007–2008 Edition
978-0-375-76594-0 • $19.00/C$25.00

Cracking the SAT Spanish Subject Test,
2007–2008 Edition
978-0-375-76595-7 • $18.00/C$22.00

Cracking the SAT U.S. & World History Subject Tests, 2007–2008 Edition
978-0-375-76591-9 • $19.00/C$25.00